Course Study Guide Solutions Manual
for Genetics,
Fifth Edition

Leland M. Hartwell Janice A. Fischer
Michael L. Goldberg
Leroy Hood

Leroy Hood

http://create.mheducation.com

Copyright 2016 by McGraw-Hill Education. All rights reserved. Printed in the United States of America. Except as permitted under the United States Copyright Act of 1976, no part of this publication may be reproduced or distributed in any form or by any means, or stored in a database or retrieval system, without prior written permission of the publisher.

This McGraw-Hill Create text may include materials submitted to McGraw-Hill for publication by the instructor of this course. The instructor is solely responsible for the editorial content of such materials. Instructors retain copyright of these additional materials.

ISBN-10: 0077515072   ISBN-13: 9780077515072

Contents

1. Genetics: The Study of Biological Information   1
2. Mendel's Principles of Heredity   9
3. Extensions to Mendel's Laws   33
4. The Chromosome Theory of Inheritance   57
5. Linkage, Recombination, and the Mapping of Genes on Chromosomes   85
6. DNA Structure, Replication, and Recombination   139
7. Anatomy and Function of a Gene: Dissection Through Mutation   155
8. Gene Expression: The Flow of Information from DNA to RNA to Protein   185
9. Digital Analysis of Genomes   215
10. Analyzing Genomic Variation   243
11. The Eukaryotic Chromosome   269
12. Chromosomal Rearrangements and Changes in Chromosome Number   283
13. Bacterial Genetics   325
14. Organellar Genetics   347
15. Gene Regulation in Prokaryotes   361
16. Gene Regulation in Eukaryotes   385
17. Manipulating the Genomes of Eukaryotes   405
18. The Genetic Analysis of Development   425
19. The Genetics of Cancer   449
20. Variation and Selection in Populations   469
21. Genetics of Complex Traits   493

Credits

1. Genetics: The Study of Biological Information: *Chapter 1 from Study Guide Solutions Manual for Genetics, Fifth Edition by Hartwell, 2015* 1
2. Mendel's Principles of Heredity: *Chapter 2 from Study Guide Solutions Manual for Genetics, Fifth Edition by Hartwell, 2015* 9
3. Extensions to Mendel's Laws: *Chapter 3 from Study Guide Solutions Manual for Genetics, Fifth Edition by Hartwell, 2015* 33
4. The Chromosome Theory of Inheritance: *Chapter 4 from Study Guide Solutions Manual for Genetics, Fifth Edition by Hartwell, 2015* 57
5. Linkage, Recombination, and the Mapping of Genes on Chromosomes: *Chapter 5 from Study Guide Solutions Manual for Genetics, Fifth Edition by Hartwell, 2015* 85
6. DNA Structure, Replication, and Recombination: *Chapter 6 from Study Guide Solutions Manual for Genetics, Fifth Edition by Hartwell, 2015* 139
7. Anatomy and Function of a Gene: Dissection Through Mutation: *Chapter 7 from Study Guide Solutions Manual for Genetics, Fifth Edition by Hartwell, 2015* 155
8. Gene Expression: The Flow of Information from DNA to RNA to Protein: *Chapter 8 from Study Guide Solutions Manual for Genetics, Fifth Edition by Hartwell, 2015* 185
9. Digital Analysis of Genomes: *Chapter 9 from Study Guide Solutions Manual for Genetics, Fifth Edition by Hartwell, 2015* 215
10. Analyzing Genomic Variation: *Chapter 10 from Study Guide Solutions Manual for Genetics, Fifth Edition by Hartwell, 2015* 243
11. The Eukaryotic Chromosome: *Chapter 11 from Study Guide Solutions Manual for Genetics, Fifth Edition by Hartwell, 2015* 269
12. Chromosomal Rearrangements and Changes in Chromosome Number: *Chapter 12 from Study Guide Solutions Manual for Genetics, Fifth Edition by Hartwell, 2015* 283
13. Bacterial Genetics: *Chapter 13 from Study Guide Solutions Manual for Genetics, Fifth Edition by Hartwell, 2015* 325
14. Organellar Genetics: *Chapter 14 from Study Guide Solutions Manual for Genetics, Fifth Edition by Hartwell, 2015* 347
15. Gene Regulation in Prokaryotes: *Chapter 15 from Study Guide Solutions Manual for Genetics, Fifth Edition by Hartwell, 2015* 361
16. Gene Regulation in Eukaryotes: *Chapter 16 from Study Guide Solutions Manual for Genetics, Fifth Edition by Hartwell, 2015* 385
17. Manipulating the Genomes of Eukaryotes: *Chapter 17 from Study Guide Solutions Manual for Genetics, Fifth Edition by Hartwell, 2015* 405
18. The Genetic Analysis of Development: *Chapter 18 from Study Guide Solutions Manual for Genetics, Fifth Edition by Hartwell, 2015* 425

19. The Genetics of Cancer: *Chapter 19 from Study Guide Solutions Manual for Genetics, Fifth Edition by Hartwell, 2015* 449
20. Variation and Selection in Populations: *Chapter 20 from Study Guide Solutions Manual for Genetics, Fifth Edition by Hartwell, 2015* 469
21. Genetics of Complex Traits: *Chapter 21 from Study Guide Solutions Manual for Genetics, Fifth Edition by Hartwell, 2015* 493

chapter 1

Genetics: The Study of Biological Information

Synopsis

Chapter 1 is an introduction to the study of modern-day genetics. Genetics is the study of genes: how genes are segments of DNA molecules; how genes are inherited; and how genes direct an organism's characteristics. The most important insight from this chapter is that the basic function of most (but not all) genes is to direct the synthesis of (to **encode**) a particular type of protein.

Key terms

DNA – the macromolecular polymer that constitutes genes

nucleotides – the chemical building blocks of DNA

bases – components of nucleotides that are of four different types in DNA; abbreviated as A, G, C, and T

base pair – DNA is double-stranded; two nucleotide polymers are held together by hydrogen bonds between A-T and G-C base pairs.

genes – segments of DNA that, in most cases, encode proteins

chromosomes – large DNA molecules that can contain hundreds or thousands of genes

genome – all of the DNA, and thus all the genes, in a particular organism

metabolism – the chemical reactions by which organisms use energy and matter to construct their bodies

genetic code – the way that genes are "read" by the molecular machines that use genes to make proteins

RNA – a polymer structurally similar to DNA that serves as a chemical intermediate in the pathway from genes to proteins

proteins – linear polymers of amino acids that fold into complex three dimensional shapes. Proteins constitute the structures of cells, and also carry out the chemical reactions of metabolism.

amino acids – the chemical subunits of proteins. Twenty different common amino acids exist in proteins.

mutation – a heritable chemical change in the base sequence of DNA that enables evolution to take place

> **evolution** – the change in characteristics of populations of organisms over time due to the accumulation of mutations in genes
>
> **convergent evolution** – the evolution of similar structures independently in the lineages leading to different species
>
> **model organisms** – species used commonly for genetic analysis by scientists
>
> **gene family** – two or more genes with similar DNA sequences and similar functions that most likely arose from a single ancestral gene by a series of duplication and divergence events. A **multigene family** is a large gene family; a **gene superfamily** is a group of gene families and multigene families that share a common ancestral gene.
>
> **exons** and **introns** – the portions of genes that are used to make proteins (*exon*s) and the regions of DNA that separate them (*introns*)
>
> **prokaryotic cells** – single-cell organisms like bacteria whose genomes are not enclosed within a membrane (not inside a nucleus)
>
> **eukaryotic cells** – cells such as human cells whose genomes are within a nucleus, a membrane-enclosed organelle
>
> **Human Genome Project** – the effort to determine the DNA base sequence of every human chromosome and to analyze the genes making up the human genome

Problem Solving

The first chapter of this book provides a broad overview of genetics. Chapter 1 covers a lot of ground, but only superficially. Don't worry if at this point you don't understand all of the information given at a deep level – you will later on. However, you are likely familiar already (from introductory biology classes) with some of the fundamentals of what a gene is and how genes are used to make proteins. The problems in this chapter are meant to get you started in the habit of thinking like a geneticist – quantitatively, analytically, carefully, and logically.

Vocabulary

1.

    a. complementarity   4. G-C and A-T base pairing in DNA through hydrogen bonds

    b. nucleotide   11. subunit of the DNA macromolecule

    c. chromosomes   7. DNA/protein structures that contain genes

    d. protein   1. a linear polymer of amino acids that folds into a particular shape

    e. genome   9. the entirety of an organism's hereditary information

chapter 1

f.	gene	8. DNA information for a single function, such as a protein
g.	uracil	12. the one of the four bases in RNA that is not in DNA
h.	exon	6. part of a gene that contains protein coding information
i.	intron	2. part of a gene that does not contain protein coding information
j.	DNA	10. a double-stranded polymer of nucleotides that stores the inherited blueprint of an organism
k.	RNA	3. a polymer of nucleotides that is an intermediary in the synthesis of proteins from DNA
l.	mutation	5. alteration of DNA sequence

Section 1.1

2. The complementary strand of a DNA molecule is simply the strand with which the original DNA molecule forms base pairs. Remember two things: (1) The two strands of a double-stranded DNA molecule are oriented in the opposite direction with respect to each other (their 5' and 3' ends run in opposite directions), and (2) the base pairs are A-T and G-C. Therefore, the DNA strand complementary to the one shown is:

    5' AGCTTAATGCT 3'

3. a. If the 3 billion (3,000,000,000) base pairs of the human genome is divided into 23 chromosomes, the average size of a human chromosome is **3,000,000,000 base pairs/23 chromosomes ≈ 135,435,000 base pairs per chromosome.**

    b. The human genome contains about 25,000 genes, and assuming that they are spread evenly over the 23 chromosomes, on average there are **25,000 genes/23 chromosomes ≈ 1087 genes per chromosome.**

    c. About half the DNA of the human genome contains genes, meaning that all the genes are found within 1.5 billion (1,500,000,000) base pairs. Therefore, on average there are **1,500,000,000 base pairs / 25,000 genes ≈ 60,000 base pairs per gene.**

Section 1.2

4. a. **Both.** Each protein is composed of a "string" of amino acids, and DNA is a "string" of nucleotides.

    b. **DNA.** DNA is double-stranded through complementary base pairing of single strands in opposite orientations. A protein is a single strand of linked amino acids, and the strand folds into a particular shape.

c. **DNA.** Four different kinds of nucleotides – A, G, C, and T – are present in the DNA polymer. Twenty different common amino acids are present in almost all proteins.

d. **Protein.** Twenty distinct amino acid subunits are the building blocks of almost all proteins. DNA is made up of only four different types of nucleotides.

e. **Protein.** Proteins are polymers of amino acids; DNA is a polymer of nucleotides.

f. **DNA.** DNA is a polymer of nucleotides; proteins are polymers of amino acids.

g. **DNA.** Genes are segments of DNA; by using the genetic code, most genes encode proteins.

h. **Protein.** Some proteins (*enzymes*) perform chemical reactions.

5. a. Each base in a single strand of a DNA molecule can be either an A, G, C or T. Therefore, a specific 100-nucleotide DNA strand could start with any one of the four nucleotides, the second nucleotide could be any one of the four nucleotides, etc. The number of different possible sequences increases by a factor of 4 at each successive step in the addition of a base (see the following figure). Thus, **the number of different possible sequences of a 100-nucleotide DNA strand is $4^{100} = \sim 1.6 \times 10^{60}$**. We need not consider the second, complementary strand of DNA, as its base sequence is determined by the sequence of the first strand.

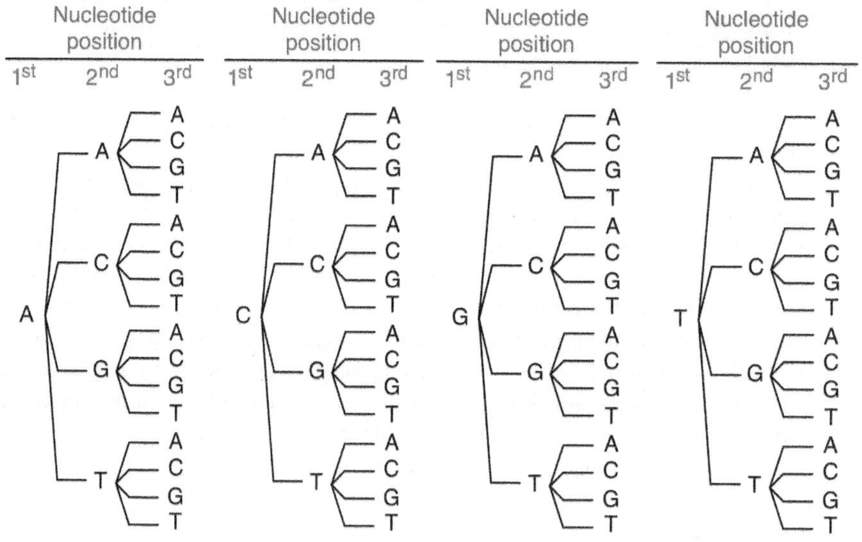

b. Because each amino acid can be 1 of 20 different amino acids, by the same logic as in part (a), **the number of different 100-amino acid proteins is $20^{100} = \sim 1.3 \times 10^{130}$**.

chapter 1

Section 1.3

6. Scientists think that all forms of life on earth have a common origin because **organisms as distant as humans and bacteria share the same genetic code, and many of their proteins are similar in amino acid sequence and biochemical function.**

7. Scientists study model organisms like yeast and fruit flies in order to understand universal biochemical pathways. Because of their common origin and because they have similar genes and proteins, all organisms share certain universal pathways. For example, many of the genes that help regulate cell division are similar in yeast and humans. Obviously, **scientists cannot perform experiments on humans, but researchers can manipulate organisms like yeast, fruit flies, and mice in the laboratory in many useful ways. Universal principles of biology may be learned from these model organisms because of the common origin of all life.**

8. To detect proteins in different organisms that have a common origin, scientists use computer analysis of the DNA sequences of genomes to look for genes that encode proteins with large stretches of amino acids that are identical or similar. To assess whether related genes in different organisms have similar functions, **scientists can generate mutations in the genes and see if the mutations have similar effects.** For example, suppose bacteria with a mutation in a particular gene are unable to grow because the cells cannot divide. If fruit flies with a mutation in a gene with related DNA sequences that encode a similar protein die as very young embryos with very few cells, you could conclude that the genes in each organism have a key function in cell division.

    In some cases, you could go one step further by **placing the normal fruit fly gene into the genome of the mutant bacterial cells (or the normal bacterial gene into the genome of the mutant fruit flies).** If the mutant organisms with the gene from the other species were able to grow properly, you could then conclude that the genes from the different organisms do in fact encode proteins that fulfill the same biochemical role in cell division. Because bacteria and fruit flies are so distantly related to each other, this type of "gene rescue" experiment is only rarely successful. But for more closely related species (like fruit flies and yeast cells, both of which are eukaryotic organisms), such experiments have often demonstrated that genes from different species that have related DNA sequences also have similar gene function.

Section 1.4

9. Scientists think that new genes arise by duplication of an original gene and divergence by mutation because **the genomes of all organisms have gene families and superfamilies.** These gene families and superfamilies contain genes that encode proteins with with similar amino acid sequences; the proteins in these families fold into similar three-dimensional structures and they perform related functions. The genomes

of more complex organisms usually contain more members of the same gene/protein families that exist in the genomes of simpler organisms. It is unlikely that all of these gene/protein families arose anew in each organism.

10. Genes have *exons* that include protein coding regions, and also regions of DNA between the exons called *introns*. **Exons from different genes could be "shuffled" by chromosome rearrangements. Modules from different proteins could thus reassort to form new proteins with new functions.**

11. **A protein is likely to perform the same type of biochemical reaction in different cell types.** For example, if a protein is a *kinase* (a kind of enzyme that adds a phosphate group to other molecules called *substrates*) it would probably be a kinase in all cells. However, the kinase might add a phosphate group to one substrate in one cell type but a different substrate in other kinds of cells. Therefore, **a protein with a particular biochemical activity could function in the same or in different pathways in various cell types.**

Section 1.5

12. a. **Untrue;** the zebrafish that lacks a functional version of the gene is viable.

    b. **True;** the zebrafish that lacks a functional version of the gene lacks stripes.

    c. **Insufficient information;** no information is given as to why the stripes are absent in the mutant zebrafish and many explanations for this observation are possible.

    d. **Insufficient information;** the gene is not required for viability because the fish lacking a functional version of it are alive. However, no information is given about possible abnormalities in the mutant zebrafish other than a failure to form horizontal stripes.

13. a. **The DNA sequence of the *WDR62* gene would have enabled scientists to predict the amino acid sequence of the protein it encodes. Conserved regions of amino acid sequence often reveal structural features indicative of the biochemical function of the protein.** In fact, *WDR62* is so named because the protein it encodes contains "WD repeats": regions with similar amino acid sequences that are found in several proteins. These WD repeats allow the proteins that contain them to bind to other proteins.

    b. **Knowing the *WDR62* mutations cause microcephaly indicates that at the level of the organism, the gene and the protein it encodes are required for brain development.**

    c. If the mutant mice had a syndrome similar to people with microcephaly, then we would know for sure that *WDR62* is the microcephaly disease gene. **These mice could also be used in various experiments to study the biochemical pathways in which the WDR62 protein participates,** as these pathways are likely to be similar in mice and humans and would be needed for proper brain development in both species.

chapter 1

Section 1.6

14. Different people may have very different perspectives about their interest in obtaining the DNA sequence of their genome. Genome sequences may be helpful in treating diseases, in making reproductive decisions, and in providing clues about ancestry. At the present time, only a small fraction of the information in genome sequences can be interpreted by scientists because many traits are influenced in very complicated ways by large networks of genes. In some cases, individuals may have excellent reasons for NOT wanting to learn about their genetic predispositions to certain traits. For example, many people whose parents have Huntington disease, a neurodegenerative condition that tends to affect people late in life, can know for certain whether or not they will develop the disease by analysis of the base sequence of a single gene. Some people may wish not to know they will eventually develop this disease because that knowledge may affect their current quality of life.

Your own perspectives about this issue may well change as your understanding of genetics increases.

chapter 2

Mendel's Principles of Heredity

Synopsis

Chapter 2 covers the basic principles of inheritance that can be summarized as Mendel's Laws of Segregation (for one gene) and Independent Assortment (for more than one gene).

Key terms

- **genes** and **alleles** of genes – A gene determines a trait; and there are different alleles or forms of a gene. The color gene in peas has two alleles: the yellow allele and the green allele.

- **genotype** and **phenotype** – Genotype is the genetic makeup of an organism (written as alleles of specific genes), while phenotype is what the organism looks like.

- **homozygous** and **heterozygous** – When both alleles of a gene are the same, the individual is homozygous for that gene (or *pure-breeding*). If the two alleles are different, the organism is heterozygous (also called a *hybrid*).

- **dominant** and **recessive** – The dominant allele is the one that controls the phenotype in the heterozygous genotype; the recessive allele controls the phenotype only in a homozygote.

- **monohybrid** or **dihybrid cross** – a cross between individuals who are both heterozygotes for one gene (monohybrid) or for two genes (dihybrid)

- **testcross** – performed to determine whether or not an individual with the dominant trait is homozygous or heterozygous; an individual with the dominant phenotype but unknown genotype is crossed with an individual with the recessive phenotype

Key ratios

3:1 – Ratio of progeny phenotypes in a cross between monohybrids
  [$Aa \times Aa \rightarrow$ 3 $A-$ (dominant phenotype) : 1 aa (recessive phenotype)]

1:2:1 – Ratio of progeny genotypes in a cross between monohybrids
  ($Aa \times Aa \rightarrow$ 1 AA : 2 Aa : 1 aa)

1:1 – Ratio of progeny genotypes in a cross between a heterozygote and a recessive homozygote
  ($Aa \times aa \rightarrow$ 1 Aa : 1 aa : 1 aa)

1:0 – All progeny have the same phenotype. Can result from several cases:
  [$AA \times$ – – $\rightarrow$ $A-$ (all dominant phenotype)]
  [$aa \times aa \rightarrow$ aa (all recessive phenotype)]

9:3:3:1 – Ratio of progeny phenotypes in a dihybrid cross
  ($Aa\ Bb \times Aa\ Bb \rightarrow$ 9 $A-\ B-$: 3 $A-\ bb$: 3 $aa\ B-$: 1 aa)

chapter 2

Problem Solving

The essential component of solving most genetics problems is to DIAGRAM THE CROSS in a consistent manner. In most cases you will be given information about phenotypes, so the diagram would be:

Phenotype of one parent × phenotype of the other parent → phenotype(s) of progeny

The goal is to assign genotypes to the parents and then use these predicted genotypes to generate the genotypes, phenotypes, and ratios of progeny. If the predicted progeny match the observed data you were provided, then your genetic explanation is correct.

The points listed below will be particularly helpful in guiding your problem solving:

- Remember that **there are two alleles of each gene when describing the genotypes of individuals.** But if you are describing gametes, remember that **there is only one allele of each gene per gamete.**
- **You will need to determine whether a trait is dominant or recessive.** Two main clues will help you answer this question.
    - First, if the parents of a cross are true-breeding for the alternative forms of the trait, look at the phenotype of the F_1 progeny. Their genotype must be heterozygous, and their phenotype is thus controlled by the dominant allele of the gene.
    - Second, look at the F_2 progeny (that is, the progeny of the F_1 hybrids). The 3/4 portion of the 3:1 phenotypic ratio indicates the dominant phenotype.
- You should **recognize the need to set up a testcross** (to establish the genotype of an individual showing the dominant phenotype by crossing this individual to a recessive homozygote).
- You must **keep in mind the basic rules of probability:**
    - *Product rule:* If two outcomes must occur together as the result of independent events, the probability of one outcome AND the other outcome is the product of the two individual probabilities.
    - *Sum rule:* If there is more than one way in which an outcome can be produced, the probability of one OR the other occurring is the sum of the two mutually exclusive individual probabilities.
- Remember that **Punnett squares are not the only means of analyzing a cross; branched-line diagrams and calculations of probabilities according to the product and sum rules are more efficient ways of looking at complicated crosses** involving more than one or two genes.
- **You should be able to draw and interpret pedigrees.** When the trait is rare, look in particular for vertical patterns of inheritance characteristic of dominant traits, and horizontal patterns that typify recessive traits. Check your work by assigning genotypes to all individuals in the pedigree and verifying that these make sense.
- The vocabulary problem (the first problem in the set) is a useful gauge of how well you know the terms most critical for you understanding of the chapter.

chapter 2

Vocabulary

1.

a.	phenotype	4.	observable characteristic
b.	alleles	3.	alternate forms of a gene
c.	independent assortment	6.	alleles of one gene separate into gametes randomly with respect to alleles of other genes
d.	gametes	7.	reproductive cells containing only one copy of each gene
e.	gene	11.	the heritable entity that determines a characteristic
f.	segregation	13.	the separation of the two alleles of a gene into different gametes
g.	heterozygote	10.	an individual with two different alleles of a gene
h.	dominant	2.	the allele expressed in the phenotype of the heterozygote
i.	F_1	14.	offspring of the P generation
j.	testcross	9.	the cross of an individual of ambiguous genotype with a homozygous recessive individual
k.	genotype	12.	the alleles an individual has
l.	recessive	8.	the allele that does not contribute to the phenotype of the heterozygote
m.	dihybrid cross	5.	a cross between individuals both heterozygous for two genes
n.	homozygote	1.	having two identical alleles of a given gene

Section 2.1

2. Prior to Mendel, **people held two basic misconceptions about inheritance.** First was the common idea of blended inheritance: that the parental traits become mixed in the offspring and forever changed. Second, many thought that one parent contributes the most to an offspring's inherited features. (For example, some people thought they saw a fully formed child in a human sperm.)

   In addition, **people who studied inheritance did not approach the problem in an organized way.** They did not always control their crosses. They did not look at traits with clear-cut alternative phenotypes. They did not start with pure-breeding lines. They did not count the progeny types in their crosses. For these reasons, they could not develop the same insights as did Mendel.

3. **Several advantages exist to using peas for the study of inheritance:**
    (1) Peas have a fairly rapid generation time (at least two generations per year if grown in the field, three or four generations per year if grown in greenhouses.
    (2) Peas can either self-fertilize or be artificially crossed by an experimenter.
    (3) Peas produce large numbers of offspring (hundreds per parent).
    (4) Peas can be maintained as pure-breeding lines, simplifying the ability to perform subsequent crosses.
    (5) Because peas have been maintained as inbred stocks, two easily distinguished and discrete forms of many traits are known.
    (6) Peas are easy and inexpensive to grow.

    In contrast, **studying genetics in humans has several disadvantages:**
    (1) The generation time of humans is very long (roughly 20 years).
    (2) There is no self-fertilization in humans, and it is not ethical to manipulate crosses.
    (3) Humans produce only a small number of offspring per mating (usually only one) or per parent (almost always fewer than 20).
    (4) Although people who are homozygous for a trait do exist (analogous to pure-breeding stocks), homozygosity cannot be maintained because mating with another individual is needed to produce the next generation.
    (5) Because human populations are not inbred, most human traits show a continuum of phenotypes; only a few traits have two very distinct forms.
    (6) People require a lot of expensive care to "grow".

    There is nonetheless one major advantage to the study of genetics in humans: Because many inherited traits result in disease syndromes, and because the world's population now exceeds 6 billion people, a very large number of people with diverse, variant phenotypes can be recognized. These variations are the raw material of genetic analysis.

Section 2.2

4. a. **Two phenotypes are seen in the second generation of this cross: normal and albino.** Thus, only one gene is required to control the phenotypes observed.

    b. Note that the phenotype of the first generation progeny is normal color, and that in the second generation, there is a ratio of 3 normal : 1 albino. Both of these observations show that **the allele controlling the normal phenotype (A) is dominant to the allele controlling the albino phenotype (a).**

    c. In a test cross, an individual showing the dominant phenotype but that has an unknown genotype is mated with an individual that shows the recessive phenotype and is therefore homozygous for the recessive allele. The male parent is albino, so **the male parent's genotype is aa.** The normally colored offspring must receive an A allele from the mother, so **the genotype of the normal offspring is Aa.** The albino offspring must receive an a allele from the mother, so **the genotype of the albino offspring is aa. Thus, the female parent must be heterozygous Aa.**

chapter 2

5. Because two different phenotypes result from the mating of two cats of the same phenotype, the short-haired parent cats must have been heterozygous. The phenotype expressed in the heterozygotes (the parent cats) is the dominant phenotype. Therefore, **short hair is dominant to long hair.**

6. a. Two affected individuals have an affected child and a normal child. This outcome is not possible if the affected individuals were homozygous for a recessive allele conferring piebald spotting, and if the trait is controlled by a single gene. Therefore, **the piebald trait must be the dominant phenotype.**

   b. If the trait is dominant, the piebald parents could be either homozygous (*PP*) or heterozygous (*Pp*). However, because the two affected individuals have an unaffected child (*pp*), **they both must be heterozygous (*Pp*).** A diagram of the cross follows:

   piebald × piebald → 1 piebald : 1 normal
   Pp   *Pp*   *Pp*   *pp*

   Note that although the apparent ratio is 1:1, this is not a testcross but is instead a cross between two monohybrids. The reason for this discrepancy is that only two progeny were obtained, so this number is insufficient to establish what the true ratio would be (it should be 3:1) if many progeny resulted from the mating.

7. **You would conduct a testcross between your normal-winged fly (*W*–) and a short-winged fly that must be homozygous recessive (*ww*).** The possible results are diagrammed here; the first genotype in each cross is that of the normal-winged fly whose genotype was originally unknown.

   WW × *ww* → all *Ww* (normal wings)

   Ww × *ww* → ½ *Ww* (normal wings) : ½ *ww* (short wings)

8. First diagram the crosses:

   closed × open → F$_1$ all open → F$_2$ 145 open : 59 closed

   F$_1$ open × closed → 81 open : 77 closed

   The results of the crosses fit the pattern of inheritance of a single gene, with the open trait being dominant and the closed trait recessive. The first cross is similar to those Mendel did with pure-breeding parents, although you were not provided with the information that the starting plants were true-breeding. **The phenotype of the F$_1$ plants is open, indicating that open is dominant. The closed parent must be homozygous for the recessive allele. Because only one phenotype is seen among the F$_1$ plants, the open parent must be homozygous for the dominant allele.** Thus, the parental cucumber plants were indeed true-breeding homozygotes.

   The result of the self-fertilization of the F$_1$ plants shows a 3:1 ratio of the open : closed phenotypes among the F$_2$ progeny. **The 3:1 ratio in the F$_2$ shows that a single gene controls the phenotypes and that the F$_1$ plants are all hybrids (that is, they are heterozygotes).**

The final cross verifies the F₁ plants from the first cross are heterozygous hybrids because **this testcross yields a 1:1 ratio of open: closed progeny.** In summary, all the data are consistent with the trait being determined by one gene with two alleles, and open being the dominant trait.

9. **The dominant trait (short tail) is easier to eliminate from the population by selective breeding.** The reason is you can recognize every animal that has inherited the short tail allele, because only one such dominant allele is needed to see the phenotype. If you prevent all the short-tailed animals from mating, then the allele would become extinct.

   On the other hand, the recessive dilute coat color allele can be passed unrecognized from generation to generation in heterozygous mice (who are *carriers*). The heterozygous mice do not express the phenotype, so they cannot be distinguished from homozygous dominant mice with normal coat color. You could prevent the homozygous recessive mice with the dilute phenotype from mating, but the allele for the dilute phenotype would remain among the carriers, which you could not recognize.

10. The problem already states that only one gene is involved in this trait, and that the dominant allele is dimple (*D*) while the recessive allele is nondimple (*d*).

    a. Diagram the cross described in this part of the problem:

       nondimple ♂ × dimpled ♀ → proportion of F₁ with dimple?

       Note that the dimpled woman in this cross had a *dd* (nondimpled) mother, so the dimpled woman MUST be heterozygous. We can thus rediagram this cross with genotypes:

       dd (nondimple) ♂ × *Dd* (dimple) ♀ → ½ *Dd* (dimpled) : ½ *dd* (nondimpled)

       One half of the children produced by this couple would be dimpled.

    b. Diagram the cross:

       dimple (*D?*) ♂ × nondimpled (*dd*) ♀ → nondimple F₁ (*dd*)

       Because they have a nondimple child (*dd*), the husband must have a *d* allele to contribute to the offspring. **The husband is thus of genotype *Dd*.**

    c. Diagram the cross:

       dimple (*D?*) ♂ × nondimpled (*dd*) ♀ → eight F₁, all dimpled (*D–*)

       The *D* allele in the children must come from their father. The father could be either *DD* or *Dd*, but **it is most probable that the father's genotype is *DD***. We cannot rule out completely that the father is a *Dd* heterozygote. However, if this was the case, the probability that all 8 children would inherit the *D* allele from a *Dd* parent is only $(1/2)^8 = 1/256$.

11. a. The only unambiguous cross is:

       homozygous recessive × homozygous recessive → all homozygous recessive

       The only cross that fits this criteria is: dry × dry → all dry. Therefore, **dry is the recessive phenotype (*ss*) and sticky is the dominant phenotype (*S–*).**

chapter 2

b. A 1:1 ratio comes from a testcross of heterozygous sticky (*Ss*) × dry (*ss*). However, **the sticky x dry matings here include both the *Ss* × *ss* AND the homozygous sticky (*SS*) × dry (*ss*).**

A 3:1 ratio comes from crosses between two heterozygotes, *Ss* × *Ss*, but the sticky individuals are not only *Ss* heterozygotes but also *SS* homozygotes. Thus the sticky x sticky matings in this human population are a mix of matings between two heterozygotes (*Ss* × *Ss*), between two homozygotes (*SS* × *SS*) and between a homozygote and heterozygote (*SS* × *Ss*). **The 3:1 ratio of the heterozygote cross is therefore obscured by being combined with results of the two other crosses.**

12. Diagram the cross:

    black × red → 1 black : 1 red

    No, you cannot tell how coat color is inherited from the results of this one mating. In effect, this was a test cross – a cross between animals of different phenotypes resulting in offspring of two phenotypes. This does not indicate whether red or black is the dominant phenotype. To determine which phenotype is dominant, remember that an animal with a recessive phenotype must be homozygous. Thus, **if you mate several red horses to each other and also mate several black horses to each other, the crosses that always yield only offspring with the parental phenotype must have been between homozygous recessives.** For example, if all the black × black matings result in only black offspring, black is recessive. Some of the red × red crosses (that is, crosses between heterozygotes) would then result in both red and black offspring in a ratio of 3:1. To establish this point, you might have to do several red × red crosses, because some of these crosses could be between red horses homozygous for the dominant allele. You could of course ensure that you were sampling heterozygotes by using the progeny of black × red crosses (such as that described in the problem) for subsequent black × black or red × red crosses.

13. a. **1/6** because a die has 6 different sides.

    b. There are three possible even numbers (2, 4, and 6). The probability of obtaining any one of these is 1/6. Because the 3 events are mutually exclusive, use the sum rule: 1/6 + 1/6 + 1/6 = 3/6 = **1/2**.

    c. You must roll either a 3 or a 6, so 1/6 + 1/6 = 2/6 = **1/3**.

    d. Each die is independent of the other, thus the product rule is used: 1/6 × 1/6 = **1/36**.

    e. The probability of getting an even number on one die is 3/6 = 1/2 (see part [b]). This is also the probability of getting an odd number on the second die. This result could happen either of 2 ways – you could get the odd number first and the even number second, or *vice versa*. Thus the probability of both occurring is 1/2 × 1/2 × 2 = **1/2**.

    f. The probability of any specific number on a die = 1/6. The probability of the same number on the other die =1/6. The probability of both occurring at same time is 1/6 x 1/6 = 1/36. The same probability is true for the other 5 possible numbers on the dice. Thus the probability of any of these mutually exclusive situations occurring is 1/36 + 1/36 + 1/36 + 1/36 + 1/36 + 1/36 = 6/36 = **1/6**.

g. The probability of getting two numbers both over four is the probability of getting a 5 or 6 on one die (1/6 + 1/6 = 1/3) and 5 or 6 on the other die (1/3). The results for the two dice are independent events, so 1/3 × 1/3 = **1/9**.

14. **The probability of drawing a face card = 0.231** (= 12 face cards / 52 cards). **The probability of drawing a red card = 0.5** (= 26 red cards / 52 cards). **The probability of drawing a red face card = probability of a red card × probability of a face card = 0.231 × 0.5 = 0.116.**

15. a. The *Aa bb CC DD* woman can produce 2 genetically different eggs that vary in their allele of the first gene (*A* or *a*). She is homozygous for the other 3 genes and can only make eggs with the *b C D* alleles for these genes. Thus, using the product rule (because the inheritance of each gene is independent), she can make 2 × 1 × 1 × 1 = **2 different types of gametes:** (*A b C D* and *a b C D*).

    b. Using the same logic, an *AA Bb Cc dd* woman can produce 1 × 2 × 2 × 1 = **4 different types of gametes:** *A* (*B* or *b*) (*C* or *c*) *d*.

    c. A woman of genotype *Aa Bb cc Dd* can make 2 × 2 × 1 × 2 = **8 different types of gametes:** (*A* or *a*) (*B* or *b*) *c* (*D* or *d*).

    d. A woman who is a quadruple heterozygote can make 2 × 2 × 2 × 2 = **16 different types of gametes:** (*A* or *a*) (*B* or *b*) (*C* or *c*) (*D* or *d*). This problem (like those in parts (a-c) above) can also be visualized with a branched-line diagram.

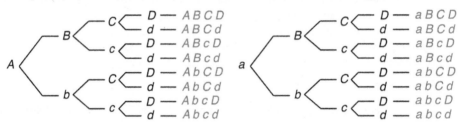

16. a. The probability of any phenotype in this cross depends only on the gamete from the heterozygous parent. The probability that a child will resemble the quadruply heterozygous parent is thus 1/2*A* × 1/2*B* × 1/2*C* × 1/2*D* = 1/16. The probability that a child will resemble the quadruply homozygous recessive parent is 1/2*a* × 1/2*b* × 1/2*c* × 1/2*d* = 1/16. **The probability that a child will resemble either parent is then 1/16 + 1/16 = 1/8.** This cross will produce 2 different phenotypes for each gene or 2 × 2 × 2×2 = **16 potential phenotypes.**

    b. The probability of a child resembling the recessive parent is 0; the probability of a child resembling the dominant parent is 1 × 1 × 1 × 1 = 1. **The probability that a child will resemble one of the two parents is 0 + 1 = 1. Only 1 phenotype is possible in the progeny (dominant for all 4 genes),** as $(1)^4 = 1$.

    c. The probability that a child would show the dominant phenotype for any one gene is 3/4 in this sort of cross (remember the 3/4 : 1/4 monohybrid ratio of phenotypes), so **the probability of resembling the parent for all four genes is $(3/4)^4 = 81/256$.** There are 2 phenotypes possible for each gene, so $(2)^4 = $ **16 different kinds of progeny.**

d. **All progeny will resemble their parents** because all of the alleles from both parents are identical, **so the probability = 1.** There is only 1 phenotype possible for each gene in this cross; because $(1)^4 = 1$, the child can have only **one possible phenotype** when considering all four genes.

17. a. The combination of alleles in the egg and sperm allows only one genotype for the zygote: ***aa Bb Cc DD Ee.***

    b. Because the inheritance of each gene is independent, you can use the product rule to determine the number of different types of gametes that are possible: 1 x 2 x 2 x 1 x 2 = **8 types of gametes.** To figure out the types of gametes, consider the possibilities for each gene separately and then the possible combinations of genes in a consistent order. For each gene the possibilities are: *a*, (*B* : *b*), (*C* : *c*), *D*, and (*E* : *e*). The possibilities can be determined using the product rule. Thus for the first 2 genes [*a*] × [*B* : *b*] gives [*a B* : *a b*] × [*C* : *c*] gives [*a B C* : *a B c* : *a b C* : *a b c*] × [*D*] gives [*a B C D* : *a B c D* : *a b C D* : *a b c D*] × [*E* : *e*] gives [*a B C D E* : *a B C D e* : *a B c D E* : *a B c D e* : *a b C D E* : *a b C D e* : *a b c D E* : *a b c D e*].

    This problem can also be visualized with a branched-line diagram:

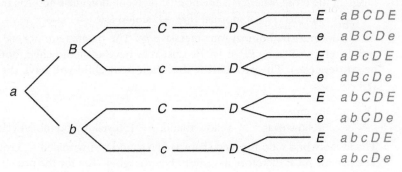

18. The first two parts of this problem involve the probability of occurrence of two independent traits: the sex of a child and galactosemia. The parents are heterozygous for galactosemia, so there is a 1/4 chance that a child will be affected (that is, homozygous recessive). The probability that a child is a girl is 1/2. The probability of an affected girl is therefore $1/2 \times 1/4 = 1/8$.

    a. Fraternal (non-identical) twins result from two independent fertilization events and therefore the probability that both will be girls with galactosemia is the product of their individual probabilities (see above); $1/8 \times 1/8 =$ **1/64**.

    b. For identical twins, one fertilization event gave rise to two individuals. The probability that both are girls with galactosemia is **1/8**.

    For parts c-g, remember that each child is an independent genetic event. The sex of the children is not at issue in these parts of the problem.

c. Both parents are carriers (heterozygous), so the probability of having an unaffected child is 3/4. The probability of 4 unaffected children is 3/4 x 3/4 x 3/4 x 3/4 = **81/256**.

d. The probability that at least one child is affected is all outcomes except the one mentioned in part (c). Thus, the probability is 1 - 81/256 = **175/256**. Note that this general strategy for solving problems, where you first calculate the probability of all events except the one of interest, and then subtract that number from 1, is often useful for problems where direct calculations of the probability of interest appear to be very difficult.

e. The probability of an affected child is 1/4 while the probability of an unaffected child is 3/4. Therefore 1/4 ×1/4 × 3/4 × 3/4 = **9/256**.

f. The probability of 2 affected and 1 unaffected in any one particular birth order is 1/4 × 1/4 × 3/4 = 3/64. There are 3 mutually exclusive birth orders that could produce 2 affecteds and 1 unaffected – unaffected child first born, unaffected child second born, and unaffected child third born. Thus, there is a 3/64 + 3/64 + 3/64 = **9/64** chance that 2 out of 3 children will be affected.

g. The phenotype of any particular child is independent of all others, so the probability of an affected child is **1/4**.

19. Diagram the cross, where P is the normal pigmentation allele and p is the albino allele:

normal ($P?$) × normal ($P?$) → albino (pp)

An albino must be homozygous recessive pp. The parents are normal in pigmentation and therefore could be PP or Pp. Because they have an albino child, **both parents must be carriers (Pp). The probability that their next child will have the pp genotype is 1/4**.

20. Diagram the cross:

yellow round × yellow round → 156 yellow round : 54 yellow wrinkled

The monohybrid ratio for seed shape is 156 round : 54 wrinkled = 3 round : 1 wrinkled. The parents must therefore have been heterozygous (Rr) for the pea shape gene. All the offspring are yellow and therefore have the Yy or YY genotype. **The parent plants were $Y- Rr$ × $YY Rr$** (that is, you know at least one of the parents must have been YY).

21. Diagram the cross:

smooth black ♂ × rough white ♀ → F₁ rough black
→ F₂ 8 smooth white : 25 smooth black : 23 rough white : 69 rough black

a. Since only one phenotype was seen in the first generation of the cross, we can assume that the parents were true breeding, and that the F₁ generation consists of heterozygous animals. **The phenotype of the F₁ progeny indicates that rough and black are the dominant phenotypes. Four phenotypes are seen in the F₂ generation so there are two genes controlling the phenotypes in this cross. Therefore, R = rough, r = smooth; B = black, b = white.** In the F₂ generation, consider each gene separately. For the coat texture, there were 8 + 25 = 33 smooth

: 23 + 69 = 92 round, or a ratio of ~1 smooth : ~3 round. For the coat color, there were 8 + 23 = 31 white : 25 + 69 = 94 black, or about ~1 white : ~3 black, so the F_2 progeny support the conclusion that the F_1 animals were heterozygous for both genes.

b. An F_1 male is heterozygous for both genes, or *Rr Bb*. The smooth white female must be homozygous recessive; that is, *rr bb*. Thus, *Rr Bb* × *rr bb* → 1/2 *Rr* (rough) : 1/2 *rr* (smooth) and 1/2 *Bb* (black) : 1/2 *bb* (white). The inheritance of these genes is independent, so apply the product rule to find the expected phenotypic ratios among the progeny, or **1/4 rough black : 1/4 rough white : 1/4 smooth black : 1/4 smooth white**.

22. Diagram the cross:
    $YY\ rr$ × $yy\ RR$ → all $Yy\ Rr$ → 9/16 Y– R– (yellow round) : 3/16 Y– rr (yellow wrinkled) : 3/16 $yy\ R$– (green round) : 1/16 $yy\ rr$ (green wrinkled).

    Each F_2 pea results from a separate fertilization event. The probability of 7 yellow round F_2 peas is $(9/16)^7$ = 4,782,969/268,435,456 = **0.018**.

23. a. First diagram the cross, and then figure out the monohybrid ratios for each gene:
    $Aa\ Tt$ × $Aa\ Tt$ → 3/4 *A*– (achoo) : 1/4 *aa* (non-achoo) and 3/4 *T*– (trembling) : 1/4 *tt* (non-trembling).

    The probability that a child will be *A*– (and have achoo syndrome) is independent of the probability that it will lack a trembling chin, so the probability of a child with achoo syndrome but without trembling chin is 3/4 *A*– × 1/4 *tt* = **3/16**.

    b. The probability that a child would have neither dominant trait is 1/4 *aa* × 1/4 *tt* = **1/16**.

24. The F_1 must be heterozygous for all the genes because the parents were pure-breeding (homozygous). The appearance of the F_1 establishes that the dominant phenotypes for the four traits are tall, purple flowers, axial flowers and green pods.

    a. From a heterozygous F_1 × F_1, both dominant and recessive phenotypes can be seen for each gene. Thus, you expect 2 × 2 × 2 × 2 = 16 different phenotypes when considering the four traits together. The possibilities can be determined using the product rule with the pairs of phenotypes for each gene, because the traits are inherited independently. Thus: [tall : dwarf] × [green : yellow] gives [tall green : tall yellow : dwarf green : dwarf yellow] × [purple : white] gives [tall green purple : tall yellow purple : dwarf green purple : dwarf yellow purple : tall green white : tall yellow white : dwarf green white : dwarf yellow white] × [terminal : axial] which gives **tall green purple terminal : tall yellow purple terminal : dwarf green purple terminal : dwarf yellow purple terminal : tall green white terminal : tall yellow white terminal : dwarf green white terminal : dwarf yellow white terminal : tall green purple axial : tall yellow purple axial : dwarf green purple axial : dwarf yellow purple axial : tall green white axial : tall yellow white axial : dwarf green white axial : dwarf yellow white axial**. The possibilities can also be determined using the branch method shown on the next page, which might in this complicated problem be easier to track

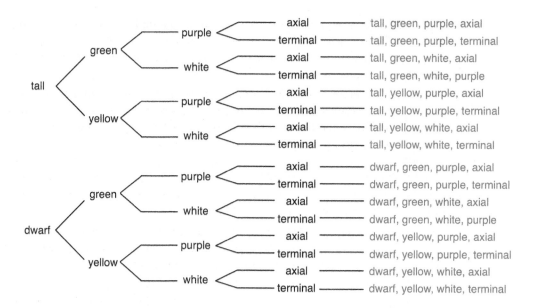

b. Designate the alleles: T = tall, t = dwarf; G = green; g = yellow; P = purple, p = white; A = axial, a = terminal. The cross $Tt\ Gg\ Pp\ Aa$ (an F_1 plant) × $tt\ gg\ pp\ AA$ (the dwarf parent) will produce 2 phenotypes for the tall, green and purple genes, but only 1 phenotype (axial) for the fourth gene or 2 × 2 × 2 × 1 = **8 different phenotypes**. The first 3 genes will give a 1/2 dominant : 1/2 recessive ratio of the phenotypes (for example 1/2 T : 1/2 t) as this is in effect a test cross for each gene. Thus, **the proportion of each phenotype in the progeny will be 1/2 × 1/2 × 1/2 × 1 = 1/8**.

Using either of the methods described in part (a), **the progeny will be 1/8 tall green purple axial : 1/8 tall yellow purple axial : 1/8 dwarf green purple axial : 1/8 dwarf yellow purple axial : 1/8 tall green white axial : 1/8 tall yellow white axial : 1/8 dwarf green white axial : 1/8 dwarf yellow white axial.**

25. For each separate cross, determine the number of genes involved. Remember that 4 phenotypic classes in the progeny means that 2 genes control the phenotypes. Next, determine the phenotypic ratio for each gene separately. A 3:1 monohybrid ratio tells you which phenotype is dominant and that both parents were heterozygous for the trait; in contrast, a 1:1 ratio results from a testcross where the dominant parent was heterozygous.

   a. There are 2 genes in this cross (4 phenotypes). One gene controls purple : white with a monohybrid ratio of 94 + 28 = 122 purple : 32 + 11 = 43 white or ~3 purple : ~1 white. The second gene controls spiny : smooth with a monohybrid ratio of 94 + 32 =126 spiny : 28 + 11 = 39 smooth or ~3 spiny : ~1 smooth. Thus, designate the alleles P = **purple**, p = **white**; S = **spiny**, s = **smooth**. This is a straightforward dihybrid cross: $Pp\ Ss \times Pp\ Ss \rightarrow$ 9 $P-\ S-$: 3 $P-\ ss$: 3 $pp\ S-$: 1 $pp\ ss$.

chapter 2

b. The 1 spiny : 1 smooth ratio indicates a test cross for the pod shape gene. Because all progeny were purple, at least one parent plant must have been homozygous for the *P* allele of the flower color gene. **The cross was either *PP Ss* × *P– ss* or *P– Ss* × *PP ss*.**

c. This is similar to part (b), but here all the progeny were spiny so at least one parent must have been homozygous for the *S* allele. The 1 purple : 1 white test cross ratio indicates that **the parents were either *Pp S–* × *pp SS* or *Pp SS* × *pp S–*.**

d. Looking at each trait individually, there are 89 + 31 = 120 purple : 92 + 27 = 119 white. A 1 purple : 1 white monohybrid ratio denotes a test cross. For the other gene, there are 89 + 92 = 181 spiny : 31 + 27 = 58 smooth, or a 3 spiny : 1 smooth ratio indicating that the parents were both heterozygous for the *S* gene. **The genotypes of the parents were *pp Ss* × *Pp Ss*.**

e. There is a 3 purple : 1 white ratio among the progeny, so the parents were both heterozygous for the *P* gene. All progeny have smooth pods so the parents were both homozygous recessive *ss*. **The genotypes of the parents are *Pp ss* × *Pp ss*.**

f. There is a 3 spiny : 1 smooth ratio, indicative of a cross between heterozygotes (*Ss* × *Ss*). All progeny were white so the parents must have been homozygous recessive *pp*. **The genotypes of the parents are *pp Ss* × *pp Ss*.**

26. Three characters (genes) are analyzed in this cross. While we can usually tell which alleles are dominant from the phenotype of the heterozygote, we are not told the phenotype of the heterozygote (that is, the original pea plant that was selfed). Instead, use the monohybrid phenotypic ratios to determine which allele is dominant and which is recessive for each gene. Consider height first. There are 272 + 92 + 88 + 35 = 487 tall plants and 93 + 31 + 29 + 11 = 164 dwarf plants. This is a ratio of ~3 tall : ~1 dwarf, indicating that **tall is dominant**. Next consider pod shape, where there are 272 + 92 + 93 + 31 = 488 inflated pods and 88 + 35 + 29 + 11 = 163 flat pods, or approximately 3 inflated : 1 flat, so **inflated is dominant**. Finally, consider flower color. There were 272 + 88 + 93 + 29 + 11 = 493 purple flowers and 92 + 35 + 31 + 11 = 169 white flowers, or ~3 purple : ~1 white. Thus, **purple is dominant.**

27. Diagram each of these crosses, remembering that you were told that tiny wings = *t*, normal wings = *T*, narrow eye = *n*, and oval (normal) eye = *N*. You thus know that one gene determines the wing trait and one gene determines the eye trait, and you further know the dominance relationship between the alleles of each gene.

    In cross 1, all of the parents and offspring show the tiny wing phenotype so there is no variability in the gene controlling this trait, and all flies in this cross are *tt*. Note that the eye phenotypes in the offspring are seen in a ratio of 3 oval : 1 narrow. This phenotypic monohybrid ratio means that both parents are heterozygous for the gene (*Nn*). Thus **the genotypes for the parents in cross 1 are: *tt Nn* ♂ × *tt Nn* ♀.**

    In cross 2 consider the wing trait first. The female parent is tiny (*tt*) so this is a test cross for the wings. The offspring show both tiny and normal in a ratio of 82 : 85 or a ratio of 1 tiny : 1 normal. Therefore the normal male parent must be heterozygous for this gene (*Tt*). For eyes the narrow parent is homozygous recessive (*nn*) so again this is a test cross for this gene. Again both eye phenotypes are seen in the offspring in a ratio of

1 oval : 1 narrow, so the oval female parent is a *Nn* heterozygote. Thus **the genotypes for the parents in cross 2 are: *Tt nn* ♂ × *tt Nn* ♀**.

Consider the wing phenotype in the offspring of cross 3. Both wing phenotypes are seen in a ratio of 64 normal flies : 21 tiny or a 3 normal : 1 tiny. Thus both parents are *Tt* heterozygotes. The male parent is narrow (*nn*), so cross 3 is a test cross for eyes. Both phenotypes are seen in the offspring in a 1 normal : 1 narrow ratio, so the female parent is heterozygous for this gene. **The genotypes of the parents in cross 3 are: *Tt nn* ♂ × *Tt Nn* ♀**.

When examining cross 4 you notice a monohybrid phenotypic ratio of 3 normal : 1 tiny for the wings in the offspring. Thus both parents are heterozygous for this gene (*Tt*). Because the male parent has narrow eyes (*nn*), this cross is a test cross for eyes. All of the progeny have oval eyes, so the female parent must be homozygous dominant for this trait. Thus **the genotypes of the parents in cross 4 are: *Tt nn* ♂ × *Tt NN* ♀**.

28. a. Analyze each gene separately: *Tt* × *Tt* will give 3/4 *T–* (normal wing) offspring. The cross *nn* x *Nn* will give 1/2 *N–* (normal eye) offspring. To calculate the probability of the normal offspring apply the product rule to the normal portions of the monohybrid ratios by multiplying these two fractions: 3/4 *T–* × 1/2 *N–* = 3/8 *T– N–*. Thus **3/8 of the offspring of this cross will have normal wings and oval eyes.**

    b. Diagram the cross:
    Tt nn ♀ × *Tt Nn* ♂ → ?
    Find the phenotypic monohybrid ratio separately for each gene in the offspring. Then multiply these monohybrid ratios to find the phenotypic dihybrid ratio. A cross of *Tt* × *Tt* → 3/4 *T–* (normal wings) : 1/4 *tt* (tiny wings). For the eyes the cross is *nn* × *Nn* → 1/2 *N–* (oval) : 1/2 *nn* (narrow). Applying the product rule gives 3/8 *T– N–* (normal oval) : 3/8 *T– nn* (normal narrow) : 1/8 *tt N–* (tiny oval) : 1/8 *tt nn* (tiny narrow). When you multiply each fraction by 200 progeny you will see **75 normal oval : 75 normal narrow : 25 tiny oval : 25 tiny narrow**.

29. a. **The protein specified by the pea color gene is an enzyme called Sgr,** which is required for the breakdown of the green pigment chlorophyll. (See Fig. 2.20b on p. 29.)

    b. **The *y* allele could be a null allele** because it does not specify the production of any of the Sgr enzyme.

    c. The *Y* allele is dominant because in the heterozygote, **the single *Y* allele will lead to the production of some Sgr enzyme,** even if the *y* allele cannot specify any Sgr. **The amount of the Sgr enzyme made in heterozygotes is sufficient for yellow color.**

    d. In *yy* **peas, the green chlorophyll cannot be broken down,** so this pigment stays in the peas, which remain green in color.

    e. If the amount of Sgr protein is proportional to the number of functional copies of the gene, then *YY* homozygotes should have twice the amount of Sgr protein as do *Yy* heterozygotes. Yet both *YY* and *Yy* peas are yellow. These observations suggest that **half the normal amount of Sgr enzyme is sufficient for the pea to break down enough chlorophyll that the pea will still be yellow.**

f. Just as was seen in part (e), for many genes (including that for pea color), half the amount of the protein specified by the gene is sufficient for a normal phenotype. Thus, in most cases, even if the gene is essential, heterozygotes for null alleles will survive. **The advantage of having two copies of essential genes is then that even if one normal allele becomes mutated (changed) so that it becomes a null allele, the organism can survive because half the normal amount of gene product is usually sufficient for survival.**

g. **Yes, a single pea pod could contain peas with different phenotypes** because a pod is an ovary that contains several ovules (eggs), and each pea represents a single fertilization event involving one egg and one sperm (from one pollen grain). If the female plant was *Yy*, or *yy*, then it is possible that some peas in the same pod would be yellow and others green. For example, fertilization of a *y* egg with *Y* pollen would yield a yellow pea, but if the pollen grain was *y*, the pea would be green. However, a pea pod could not contain peas with different phenotypes if the female plant was *YY*, because all the peas produced by this plant would be yellow.

h. **Yes, it is possible that a pea pod could be different in color from a pea growing within it.** One reason is that, as just seen in part (g), a single pod can contain green and yellow peas. But a more fundamental reason is that one gene controls the phenotype of pea color, while a different gene controls the separate phenotype of pod color.

30. If the alleles of the pea color and pea shape genes inherited from a parent in the P generation always stayed together and never separated, then the gametes produced by the doubly heterozygous F_1 individuals in Fig. 2.15 on p. 25 would be either *Y R* or *y r*. (Note that only two possibilities would exist, and these would be in equal frequencies.) On a Punnett square (male gametes shaded in blue, female gametes in red):

	Y R ½	*y r* ½
Y R ½	*YY RR* ¼	*Yy Rr* ¼
y r ½	*Yy Rr* ¼	*yy rr* ¼

Thus **the genotypic ratios of the F_2 progeny would be ¼ *YY RR*, ½ *Yy Rr*, and ¼ *yy rr*. The phenotypic ratios among the F_2 progeny would be ¾ yellow round and ¼ green wrinkled.** These results make sense because if the alleles of the two genes were always inherited as a unit, you would expect the same ratios as in a monohybrid cross.

31. Similar to what you saw in Fig. 2.20 on p. 29, **the most likely biochemical explanation is that the dominant allele *L* specifies functional G3βH enzyme, while the recessive allele *l* is incapable of specifying any functional enzyme** (in

2-15

chapter 2

nomenclature you will see in later chapters, *l* is a *null allele*). The functional enzyme can synthesize the growth hormone gibberellin, so plants with the *L* allele are tall. **Even half the normal amount of this enzyme is sufficient for the tall phenotype,** explaining why *Ll* heterozygotes are tall.

32. *Note:* Your copy of the text might be missing the key figure referred to in this Problem. That figure follows here:

a. As in Problem 31 above, **the dominant allele *P* most likely specifies a functional product (in this case, the protein bHLH), while the recessive *p* allele cannot specify any functional protein.** The fact that the hybrid is purple (as shown on Fig. 2.8 on p. 19) indicates that half the normal amount of active bHLH protein is sufficient for purple color.

b. **Yes, flower color could potentially be controlled by genes specifying the enzymes DFR, ANS, or 3GT in addition to the gene specifying the bHLH protein.** Alleles specifying functional enzymes would yield purple color, while those that could not produce functional enzymes would cause white color. It is likely that the alleles for purple would be dominant.

Section 2.3

33. a. **Recessive** - two unaffected individuals have an affected child (*aa*). Therefore the parents involved in the consanguineous marriage must both be carriers (*Aa*).

    b. **Dominant** - the trait is seen in each generation and every affected person (*A–*) has an affected parent. Note that III-3 is unaffected (*aa*) even though both his parents are affected; this would not be possible for a recessive trait. The term "carrier" is not applicable, because everyone with a single *A* allele shows the trait.

    c. **Recessive** - two unaffected, carrier parents (*Aa*) have an affected child (*aa*), as in part (a).

34. a. Cutis laxa must be a recessive trait because affected child II-4 has normal parents. Because II-4 is affected she must have received a disease allele (*CL*) from both parents. The mother (I-3) and the father (I-4) are both heterozygous (*CL*$^+$ *CL*). **The trait is thus recessive.**

    b. You are told that this trait is rare, so unrelated people in the pedigree, like I-2, are almost certainly homozygous normal (*CL*$^+$ *CL*$^+$). Diagram the cross that gives rise to II-2: *CL CL* (I-1) × *CL*$^+$ *CL*$^+$ (I-2) → *CL*$^+$ *CL*. Thus **the probability that II-**

chapter 2

2 is a carrier is very close to 100%. (In Chapter 21 you will find the definition of a term called the *allele frequency*; if the value of the allele frequency in the population under study is known, you can calculate the very low likelihood that II-2 is a carrier.)

c. As described in part (a) both parents in this cross are carriers: $CL^+ CL \times CL^+ CL$. II-3 is <u>not</u> affected so he cannot be the $CL\ CL$ genotype. Therefore there is a 1/3 probability that he is the $CL^+ CL^+$ genotype and a **2/3 probability that he is a carrier ($CL^+ CL$)**.

d. As shown in part (b), II-2 must be a carrier ($CL^+ CL$). In order to have an affected child II-3 must also be a carrier. The probability of this is 2/3 as shown in part (c). The probability of two heterozygous parents having an affected child is 1/4. Apply the product rule to these probabilities: 1 probability that II-2 is $CL^+ CL \times 2/3$ probability that II-3 is $CL^+ CL \times 1/4$ probability of an affected child from a mating of two carriers = 2/12 = **1/6**.

35. Diagram the cross! In humans this is usually done as a pedigree. Remember that the affected siblings must be $CF\ CF$.

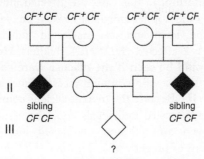

a. **The probability that II-2 is a carrier is 2/3**. Both families have an affected sibling, so both sets of parents (that is, all the people in generation I) must have been carriers. Thus, the expected genotypic ratio in the children is 1/4 affected : 1/2 carrier : 1/4 homozygous normal. II-2 is NOT affected, so she cannot be $CF\ CF$. Of the remaining possible genotypes, 2 are heterozygous. There is therefore a 2/3 chance that she is a carrier.

b. **The probability that II-2 × II-3 will have an affected child** is 2/3 (the probability that the mother is a carrier as seen in part [a]) × 2/3 (the probability the father is a carrier using the same reasoning) × 1/4 (the probability that two carriers can produce an affected child) = **1/9**.

c. The probability that both parents are carriers and that their child will be a carrier is 2/3 × 2/3 × 1/2 = 2/9 (using the same reasoning as in part [b], except asking that the child be a carrier instead of affected). However, it is also possible for $CF^+ CF^+ \times CF^+ CF$ parents to have children that are carriers. Remember that there are two possible ways for this particular mating to occur: homozygous father × heterozygous mother or vice versa. Thus the probability of this sort of mating is 2 × 1/3 (the probability that a particular parent is $CF^+ CF^+$) × 2/3 (the probability that the other parent is $CF^+ CF \times 1/2$ (the probability such a mating could produce a carrier child) = 2/9. **The probability that a child could be carrier** from either of these two scenarios (where both parents are carriers or where only one parent is a carrier) is the sum of these mutually exclusive events, or 2/9 + 2/9 **= 4/9**.

36. a. Because the disease is rare the affected father is most likely to be heterozygous (HD HD^+). **There is a 1/2 chance that the son** inherited the HD allele from his father and **will develop the disease.**

    b. **The probability of an affected child is:** 1/2 (the probability that Joe is HD HD^+) × 1/2 (the probability that the child inherits the HD allele if Joe is HD HD^+) = **1/4**.

37. **The trait is recessive** because pairs of unaffected individuals (I-1 × I-2 as well as II-3 × II-4) had affected children (II-1, III-1, and III-2). There are also two cases in which an unrelated individual must have been a carrier (II-4 and either I-1 or I-2), so **the disease allele appears to be common in the population.**

38. a. **The inheritance pattern seen in Fig. 2.22 on p. 32 could be caused by a rare dominant mutation.** In this case, the affected individuals would be heterozygous ($HD^+ HD$) and the normal individuals would be $HD^+ HD^+$. Any mating between an affected individual and an unaffected individual would give 1/2 normal (HD^+HD^+) : 1/2 affected ($HD^+ HD$) children. **However, the same pattern of inheritance could be seen if the disease were caused by a common recessive mutation.** In the case of a common recessive mutation, all the affected individuals would be HD HD. Because the mutant allele is common in the population, most or even all of the unrelated individuals could be assumed to be carriers ($HD^+ HD$). Matings between affected and unaffected individuals would then also yield phenotypic ratios of progeny of 1/2 normal ($HD^+ HD$) : 1/2 affected (HD HD).

    b. Determine the phenotype of the 14 children of III-6 and IV-6. If the disease is due to a recessive allele, then III-6 and IV-6 must be homozygotes for this recessive allele, and all their children must have the disease. If the disease is due to a dominant allele, then III-6 and IV-6 must be heterozygotes (because they are affected but they each had one unaffected parent), and 1/4 of their 14 children would be expected to be unaffected.

       Alternatively, you could look at the progeny of matings between unaffected individuals in the pedigree such as III-1 and an unaffected spouse. If the disease were due to a dominant allele, these matings would all be homozygous recessive × homozygous recessive and would never give affected children. If the disease is due to a recessive mutation, then many of these individuals would be carriers, and if the trait is common then at least some of the spouses would also be carriers, so such matings could give affected children.

39. Diagram the cross by drawing a pedigree:

chapter 2

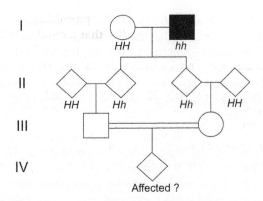

Affected ?

a. Assuming the disease is very rare, the first generation is *HH* unaffected (I-1) × *hh* affected (I-2). Thus, both of the children (II-2 and II-3) must be carriers (*Hh*). Again assuming this trait is rare in the population, those people marrying into the family (II-1 and II-4) are homozygous normal (*HH*). Therefore, the probability that III-1 is a carrier is 1/2; III-2 has the same chance of being a carrier. **Thus the probability that a child produced by these two first cousins would be affected** is 1/2 (the probability that III-1 is a carrier) × 1/2 (the probability that III-2 is a carrier) × 1/4 (the probability the child of two carriers would have an *hh* genotype) **= 1/16 = 0.0625.**

b. If 1/10 people in the population are carriers, then the probability that II-1 and II-4 are *Hh* is 0.1 for each. In this case an affected child in generation IV can only occur if III-1 and III-2 are both carriers. III-1 can be a carrier as the result of 2 different matings: (i) II-1 homozygous normal × II-2 carrier or (ii) II-1 carrier × II-2 carrier. (Note that whether I-1 is *HH* or *Hh*, II-2 must be a carrier because of the normal phenotype (II-2 cannot be *hh*) and the fact that one parent was affected.) The probability of III-1 being a carrier is thus the probability of mating (i) × the probability of generating a *Hh* child from mating (i) + the probability of mating (ii) × the probability of generating an *Hh* child from mating (ii) = 0.9 (the probability II-1 is *HH*, which is the probability for mating [i]) × 1/2 (the probability that III-1 will inherit *h* in mating [i]) + 0.1 (the probability II-1 is *H*, which is the probability for mating [ii]) × 2/3 (the probability that III-1 will inherit *h* in mating [ii]; remember that III-1 is known not to be *hh*) = 0.45 + 0.067 = 0.517. The chance that III-2 will inherit *h* is exactly the same. **Thus, the probability that IV-1 is *hh* =** 0.517 (the probability III-1 is *Hh*) × 0.517 (the probability that III-2 is *Hh*) × 1/4 (the probability the child of two carriers will be *hh*) **= 0.067. This number is slightly higher than the answer to part (a), which was 0.0625, so the increased likelihood that II-1 or II-4 is a carrier makes it only slightly more likely that IV-1 will be affected.**

40. a. Both diseases are known to be rare, so normal people marrying into the pedigree are assumed to be homozygous normal. **Nail-patella (*N*) syndrome is dominant** because all affected children have an affected parent. **Alkaptonuria (*a*) is recessive** because the affected children are the result of a consanguineous mating between 2

unaffected individuals (III-3 × III-4). Because alkaptonuria is a rare disease, it makes sense to assume that III-3 and III-4 inherited the same *a* allele from a common ancestor. Genotypes: **I-1 *nn Aa*; I-2 *Nn AA* (or I-1 *nn AA* and I-2 *Nn Aa*); II-1 *nn AA*; II-2 *nn Aa*; II-3 *Nn A–*; II-4 *nn A–*; II-5 *Nn Aa*; II-6 *nn AA*; III-1 *nn AA*; III-2 *nn A–*; III-3 *nn Aa*; III-4 *Nn Aa*; III-5 *nn A–*; III-6 *nn A–*; IV-1 *nn A–*; IV-2 *nn A–*; IV-3 *Nn A–*; IV-4 *nn A–*; IV-5 *Nn aa*; IV-6 *nn aa*; IV-7 *nn A–*.**

b. The cross is *nn A–* (IV-2) × *Nn aa* (IV-5). The ambiguity in the genotype of IV-2 is due to the uncertainty of her father's genotype (III-2). His parents' genotypes are *nn AA* (II-1) × *nn Aa* (II-2) so there is a 1/2 chance III-2 is *nn AA* and a 1/2 chance he is *nn Aa*. Thus, for each of the phenotypes below you must consider both possible genotypes for IV-2. For each part below, calculate the probability of the child inheriting the correct gametes from IV-2 × the probability of obtaining the correct gametes from IV-5 to give the desired phenotype. If both the possible IV-2 genotypes can produce the needed gametes, you will need to sum the two probabilities.

For the child to have both syndromes (*N– aa*), IV-2 would have to contribute an *n a* gamete. This could only occur if IV-2 were *nn Aa*. The probability IV-2 is *nn Aa* is 1/4: For IV-2 to be *nn Aa*, III-2 would have had to be *nn Aa* and would also have had to give an *n a* gamete to IV-2. The probability of each of those events is 1/2, so the chance of both of them occurring is 1/2 × 1/2 = 1/4. (Note that we can assume that II-2 is *nn Aa* because III-3 must have given two of her children an *a* allele. Therefore, both II-2 and III-3 must be *nn Aa*.) If IV-2 is *nn Aa*, the chance that he would give a child an *n a* gamete is 1/2. The probability that IV-5 would supply an *N a* gamete is also 1/2. Thus, **the probability that the child would have both syndromes is 1/4 × 1/2 × 1/2 = 1/16.** There is no need to sum probabilities in this case because IV-2 cannot produce an *n a* gamete if his genotype is *nn AA*.

For the child to have only nail-patella syndrome (*N– A–*), IV-2 would have to provide an *n A* gamete and IV-5 an *N a* gamete. This could occur if IV-2 were *nn Aa*; the probability is 1/4 (the probability IV-2 is *Aa*) × 1/2 (the probability of an *A* gamete if IV-2 is *Aa*) × 1/2 (the probability of an *N a* gamete from IV-5] = 1/16. This could also occur if IV-2 were *nn AA*. Here, the probability is 3/4 (the probability IV-2 is *nn AA*) × 1 (the probability of an *n A* gamete if IV-2 is *nn AA*) × 1/2 (the probability of an *N a* gamete from IV-5) = 3/8. Summing the probabilities for the two mutually exclusive IV-2 genotypes, **the probability that the child of IV-2 and IV-5 would have only nail-patella syndrome is 1/16 + 3/8 = 7/16.**

For the child to have just alkaptonuria (*nn aa*), IV-2 would have to contribute an *n a* gamete. This could only occur if IV-2 were *nn Aa*. The probability IV-2 is *nn Aa* is 1/4, and the probability of receiving an *n a* gamete from IV-2 if he is *nn Aa* is 1/2. The probability that IV-5 would supply an *n a* gamete is also 1/2. Thus, **the probability that the child of IV-2 and IV-5 would have only alkaptonuria is 1/4 × 1/2 × 1/2 = 1/16.** There is no need to sum probabilities in this case because IV-2 cannot produce an *n a* gamete if his genotype is *nn AA*.

chapter 2

The probability of neither defect is 1 − (sum of the first 3) = 1 - (1/16 + 7/16 + 1/16) = 1 - 9/16 = **7/16.** You can make this calculation because there are only the four possible outcomes and you have already calculated the probabilities of three of them.

41. Diagram the cross(es):

    midphalangeal × midphalangeal → 1853 midphalangeal : 209 normal
    M? × M? → M? : mm

    The following crosses are possible:

    MM × MM → all MM
    Mm × MM → all M-
    MM × Mm → all M-
    Mm × Mm → 3/4 M- : 1/4 mm

    The 209 normal children must have arisen from the last cross, so approximately 3 × 209 = 630 children should be their M- siblings. Thus, about 840 of the children or **~40% came from the last mating** and the other 60% of the children were the result of one or more of the other matings. This problem illustrates that much care in interpretation is required when the results of many matings in mixed populations are reported (as opposed to the results of matings where individuals have defined genotypes).

42. a. An equally likely possibility exists that any child produced by this couple will be affected (A) or unaffected (U). For two children, the possibilities are: AA, AU, UA, UU. The case in which only the second child is affected is UA; this is one of the four possibilities so **the probability that only the second child is affected is 1/4.**

    b. From the list just presented in part (a), you can see that there are two possibilities in which only one child is affected: AU and UA. The probability that either of these two mutually exclusive possibilities will occur is the sum of their independent probabilities: 1/4 + 1/4 = **1/2.**

    c. From the list just presented in part (a), you can see that there is only one possibility in which no child is affected: UU. The probability of this event is **1/4.**

    d. If this family consisted of 10 children, **the case in which only the second child out of 10 is affected** (that is, UAUUUUUUUU) **has a probability of** $1/2^{10}$ = 1/1024 = **~0.00098.** This probability is based on the facts that each birth is an independent event, and that the chance of U and A are each 1/2. We thus use the product rule to determine the chance that each of those 10 independent events will occur in a particular way − a particular birth order.

    In a family of ten children, 10 different outcomes (birth orders) exist that satisfy the criterion that only 1 child has the disease. Only the first child could have the disease, only the second child, only the third child, etc. :

1. AUUUUUUUUU
2. UAUUUUUUUU
3. UUAUUUUUUU
4. UUUAUUUUUU
5. UUUUAUUUUU
6. UUUUUAUUUU
7. UUUUUUAUUU
8. UUUUUUUAUU
9. UUUUUUUUAU
10. UUUUUUUUUA

We have already calculated that the chance of one of these outcomes in particular (#2) is 1/1024. As each of the 10 possibilities has the same probability, **the probability that only one child is affected would be 10 x (1/1024) = 10/1024 = ~0.0098.**

Only one possibility exists in which no child would be affected (UUUUUUUUUU), and just like any other specific outcome, this one has a probability of 1/1024 = ~0.00098.

e. One way to determine the probability that four children in a family of ten will have the disease is to write down all possible outcomes for the criterion, as we did above for the second answer in part (d). Then, also as we did above, sum their individual probabilities, each of which is $(1/2)^{10}$ just as before. If you start to do this......

1. AAAAUUUUUU
2. AAAUAUUUUU
3. AAAUUAUUUU
4. AAAUUUAUUU
5. AAAUUUUAUU
6. AAAUUUUUAU
7. AAAUUUUUUA
8. AAUAAUUUUU
9. AAUAUAUUUU
10. AAUAUUAUUU
    etc.

.....you will realize fairly quickly that writing down every possible birth order in this case is quite a difficult task and you are likely to miss some outcomes. In short – this is not a good way to find the answer! For questions like this, it is far preferable to use a mathematical tool called the **binomial theorem** in order to determine the number of possible outcomes that satisfy the criterion. The binomial theorem looks like this:

chapter 2

> P (**X** will occur **s** times, and **Y** will occur **t** times, in **n** trials) =
>
> $$\frac{n!}{s! \times t!} (p^s \times q^t)$$
>
> P = the probability of what is in parentheses
> **p** = P(**X**)
> **q** = P(**Y**)
>
> **X** and **Y** are the only two possibilities, so **p + q = 1**.
> Also, **s + t = n**.
>
> Remember that **!** means *factorial*: for example, 5! = 5 × 4 × 3 × 2 × 1.

To apply the binomial theorem to the question at hand (assuming you can still remember what the question was!), we'll let **X** = a child has the disease (A), and **Y** = a child does not have the disease (U). Then, **s** = 4, **t** = 6, **n** = 10, **p** = ½, and **q** = ½. The answer to the question is then:

P (4 A and 6 U children out of 10) = (10! / 4! × 6!) (1/2^4 × 1/2^6).

Notice that ($p^s \times q^t$) = (1/2^4 × 1/2^6) = 1/2^{10}. This factor of the binomial theorem equation is the probability of each single birth order, as we saw previously in part (d) above. To get the answer to our question, we need to multiply this factor (the probability of each single birth order) by the number of different birth orders that satisfy our criterion. From the equation in the box above, this second factor is [n!/(s! × t!)] = (10! / 4! × 6!) = 210. Thus, **the probability (P) of only 4 children having the disease in a family of 10 children is 1/2^{10} × 210 ≈ 21%**.

43. **In the case of cystic fibrosis, the alleles causing the disease do not specify active protein** (in this case, the cystic fibrosis transmembrane receptor [CFTR]). Some *CF* disease alleles specify defective CFTR proteins that do not allow the passage of chloride ions, while other *CF* disease alleles do not specify any CFTR protein at all. As you will learn in a later chapter, such alleles are called *loss-of-function* alleles. In a heterozygote, the normal *CF*$^+$ allele still specifies active CFTR protein, which allows for the passage of chloride ions. Because the phenotype of the heterozygote is unaffected, the amount of active CFTR protein allows passage of enough chloride ions for the cells to function normally. Again as you will see, most loss-of-function alleles are recessive to normal alleles for similar reasons. (But it is important to realize that important exceptions are known in which loss-of-function mutations are actually dominant to normal alleles.)

In the case of Huntington disease, the disease-causing allele is dominant. The reason is that **the huntingtin protein specified by this *HD* allele has, in addition to its normal function** (which is not entirely understood), **a second function that is toxic to nerve cells.** This makes the *HD* disease allele a *gain-of-function* allele. The reason *HD* is dominant to *HD*$^+$ is that the protein specified by the disease allele will be toxic to cells even if the cells have normal huntingtin specified by the normal allele. Most (but again not all) gain-of-function mutations are dominant for similar reasons.

chapter 3

Extensions to Mendel's Laws

Synopsis

In Chapter 3, we see that the relationship between genotype and phenotype can be more complicated than envisaged by Mendel. Alleles do not have to be completely dominant or recessive with respect to each other. Not all genotypes are equally viable. Genes can have more than two alleles in a population. One gene can govern more than one phenotype. A single phenotype can be influenced by more than one gene, and these genes can interact in a variety of ways.

Despite these complications, the alleles of individual genes still segregate according to Mendel's Law of Segregation, and different pairs of genes still usually behave as dictated by Mendel's Law of Independent Assortment.

Key terms

wild-type alleles – alleles with a frequency of greater than 1% in the population. Colloquially, wild-type alleles are the "normal" alleles found most commonly in the population.

mutant alleles – rare alleles with a frequency of less than 1% in the population

monomorphic gene – a gene with only one common, wild-type allele

polymorphic gene – a gene with many wild-type alleles. The wild-type alleles of a polymorphic gene are often called **common variants.**

incomplete dominance and **codominance** – cases in which the phenotype of heterozygotes is different than that of either type of homozygote. Incomplete dominance describes alleles where the heterozygote has a phenotype in between that of either homozygote, while heterozygotes for codominant alleles have both of the phenotypes associated with each homozygote. Usually in incomplete dominance one allele is nonfunctional or only partially functional, while in codominance both alleles are fully functional.

recessive lethal allele – an allele (usually a loss-of-function allele) of an **essential gene** necessary to the survival of the individual. A zygote homozygous for a recessive lethal allele cannot survive and thus is not detected among the progeny of a cross.

dominance series of multiple alleles – Although each individual has only two alleles of a gene, many alleles of the gene may exist in the population. These alleles may be completely dominant, incompletely dominant, or codominant with respect to each other as determined by the phenotype of heterozygotes for the particular pair.

pleiotropy – A gene may affect more than one phenotype.

complementary gene action – Function of two different genes is required to produce a phenotype. Nonfunctional recessive alleles of either gene can produce the same abnormal phenotype in homozygotes. Such a phenotype that can be caused by nonfunctional alleles of more than one gene is called a **heterogeneous trait.**

epistasis – An allele of one gene hides the effects of different alleles at a second gene.

redundant genes – Two or more genes provide the same function.

penetrance – the fraction of individuals with a particular genotype who display the genotype's characteristic phenotype

expressivity – the degree to which an affected individual displays the phenotype associated with that individual's genotype. Expressivity of a genotype can vary due to environment, chance, and alleles of other genes (**modifier genes**).

conditional mutation – a change in the base sequence of a gene that affects gene function only under specific environmental conditions

continuous (quantitative) trait – a trait whose phenotype varies over a wide range of values that can be measured. Continuous traits are **polygenic** – they are controlled by the combined activities of many genes.

Exceptions to the 3:1 Mendelian monohybrid ratio

1:2:1 – Ratio of progeny <u>genotypes and phenotypes</u> in a cross between hybrids when there is **incomplete dominance** or **codominance**:

$(Aa \times Aa \rightarrow 1\,AA : 2\,Aa : 1\,aa)$

Note that in incomplete dominance and codominance, a new (third) phenotype will appear in the hybrids (Aa) of the F_1 generation. In the F_2 generation, this same phenotype must be the largest component of the 1:2:1 monohybrid ratio.

2:1 – Ratio of progeny phenotypes observed in a cross between hybrids when one allele is a **recessive lethal allele** that has a dominant effect on a visible phenotype:

$(Aa \times Aa \rightarrow 1\,AA : 2\,Aa : 1\,aa)$

Note that in this case, homozygotes for the recessive lethal allele A die (*red color*), but Aa heterozygotes have a phenotype different from aa homozygotes.

Interactions of two genes

You should be able to recognize traits influenced by two genes as variations on the 9:3:3:1 ratio of genotypic classes resulting from a dihybrid cross. For your convenience, an abbreviated version of Table 3.2 summarizing these gene interactions is presented at the top of the next page. It is particularly useful to understand the concepts of complementary gene action, epistasis, and redundancy.

If you are given the details of a biochemical pathway, you should be able to work out the ratios of phenotypes expected among the progeny of a cross. Note that you cannot

chapter 3

> go in the opposite direction: A particular ratio does not tell much about the underlying biochemistry. Thus, you should NOT try to memorize specific examples relating particular ratios to specific biochemical pathways.

F₂ Genotypic Ratios from an F₁ Dihybrid Cross

Gene interaction	A- B-	A- bb	aa B-	aa bb	F₂ Phenotypic Ratio
None: Four distinct F₂ phenotypes	9	3	3	1	9:3:3:1
Complementary: One dominant allele of each of two genes is necessary to produce the phenotype	9	3	3	1	9:7
Recessive epistasis: Homozygous recessive allele of one gene masks both alleles of another gene	9	3	3	1	9:3:4
Dominant epistasis I: Dominant allele of one gene hides effects of both alleles of another gene	9	3	3	1	12:3:1
Dominant epistasis II: Dominant allele of one gene hides effects of dominant allele of another gene	9	3	3	1	13:3
Redundancy: Only one dominant allele of either of two genes is necessary to produce phenotype	9	3	3	1	15:1

Problem Solving

In Chapter 2, the major goal was to determine which allele of a gene is dominant and which is recessive, and then to ascribe genotypes to various individuals or classes of individuals based on the ratio of progeny types seen in a cross. The challenges become more difficult in this chapter, but the first step in problem solving remains the same: You need to **DIAGRAM THE CROSS** in a consistent manner. The next steps are to answer the following questions:

How many genes are involved in determining the phenotype?
How many alleles of each gene are present?
What phenotypes are associated with which genotypic classes? (The answer to this last question will help you understand the dominance relationships between the alleles of each gene and the interactions between alleles of traits determined by more than one gene.)

The points listed below will be particularly helpful in guiding your problem solving:

- To distinguish between one gene and two gene traits, look for the number of phenotypic classes in the F₂ generation and the ratios in the F₂s among those classes. If a single gene is involved, there will be either two classes (3:1 or 2:1 if an allele is a recessive lethal) or three classes (1:2:1 in the cases of codominance or incomplete

chapter 3

dominance). If two genes are involved, you could see two classes (9:7, 13:3, or 15:1) or three classes (9:3:4 or 12:3:1) or four classes (9:3:3:1). [*Note:* These ratios require that the P generation is true-breeding and that the F_1 crosses examined are between hybrids.]

- Understand that when there is codominance or incomplete dominance, a novel phenotype will appear in the F_1 generation. In the F_2 generation, this same phenotype <u>must</u> be the largest component of the 1:2:1 monohybrid ratio.

- If you see a series of crosses involving different phenotypes for a certain trait like coat color, and each cross gives a monohybrid ratio (3:1 or 1:2:1), then all the phenotypes are controlled by one gene with many alleles that form an **allelic series.** You should write out the dominance hierarchy for this series (e.g., $a = b > c$) to keep track of the relationships among the alleles.

- Lethal alleles are almost always recessive because a zygote with a dominant lethal allele could not grow into an adult. (The only exceptions to this rule involve **conditional lethal alleles** that survive in some environments but not others.) On the basis of what you have learned in this chapter, you can recognize recessive lethal alleles if they are pleiotropic and show a dominant visible phenotype such that the monohybrid phenotypic ratio is 2 (dominant phenotype) : 1 (recessive phenotype).

- Remember that the 9:3:3:1 dihybrid ratio and its variants represent various combinations of the genotypic classes 9 *A– B–* : 3 *A– bb* : 3 *aa B–* : 1 *aa bb,* where the dash indicates either a dominant or recessive allele. Based on the observed ratios, you should be able to tell which genotypic classes correspond to which phenotypes. Although you should <u>not</u> memorize the table on the previous page displaying these variants of 9:3:3:1, you should be able to consider whether particular biochemical explanations fit the ratios seen.

- Don't forget to use the product rule of probability to determine the proportions of genotypes or phenotypes for independently assorting genes.

Vocabulary

1.

a. epistasis   2. the alleles of one gene mask the effects of alleles of another gene

b. modifier genes   5. genes whose alleles alter phenotypes produced by the action of other genes

c. conditional lethal   10. a genotype that is lethal in some situations (for example, high temperature) but viable in others

d. permissive conditions   7. environmental conditions that allow conditional lethals to live

e. reduced penetrance   6. less than 100% of the individuals possessing a particular genotype express it in their phenotype

f.	multifactorial trait	8. a trait produced by the interaction of alleles of at least two genes or from interactions between gene and environment
g.	incomplete dominance	11. the heterozygote resembles neither homozygote
h.	codominance	3. both parental phenotypes are expressed in the F$_1$ hybrids
i.	mutation	4. a heritable change in a gene
j.	pleiotropy	1. one gene affecting more than one phenotype
k.	variable expressivity	9. individuals with the same genotype have related phenotypes that vary in intensity

Section 3.1

2. The problem states that the intermediate pink phenotype is caused by incomplete dominance for the alleles of a single gene. We suggest that you employ genotype symbols that can show the lack of complete dominance; the obvious R for red and r for white does not reflect the complexity of this situation. In such cases we recommend using a base letter as the gene symbol and then employing superscripts to show the different alleles. To avoid any possible misinterpretations, it is always advantageous to include a separate statement making the complexities of the dominant/recessive complications clear. Designate the two alleles f^r = red and f^w = white, so the possible genotypes are $f^r f^r$ = red; $f^r f^w$ = pink; and $f^w f^w$ = white. Note that the phenotypic ratio is the same as the genotypic ratio in incomplete dominance.

   a. Diagram the cross: $f^r f^w \times f^r f^w \rightarrow 1/4\, f^r f^r$ (red) : $1/2\, f^r f^w$ (pink) : $1/4\, f^w f^w$ (white).

   b. $f^w f^w \times f^r f^w \rightarrow 1/2\, f^r f^w$ (pink) : $1/2\, f^w f^w$ (white).

   c. $f^r f^r \times f^r f^r \rightarrow 1\, f^r f^r$ (red).

   d. $f^r f^r \times f^r f^w \rightarrow 1/2\, f^r f^r$ (red) : $1/2\, f^r f^w$ (pink).

   e. $f^w f^w \times f^w f^w \rightarrow 1\, f^w f^w$ (white).

   f. $f^r f^r \times f^w f^w \rightarrow 1\, f^r f^w$ (pink).

   The cross shown in **part (f)** is the most efficient way to produce pink flowers, because all the progeny will be pink.

3. **In Mendel's *Pp* heterozygotes, the amount of enzyme leading to purple pigment is sufficient to produce purple color as intense as the purple color in *PP* homozygotes.** Presumably the heterozygote has enough enzyme P so that the maximal level of purple pigment is produced; more enzyme cannot make more purple pigment.

   In the snapdragons in Fig. 3.3 on p. 47, the amount of red pigment in the *Aa* heterozygote is less than that in the *AA* homozygote. Presumably here, the amount of enzyme A catalyzing the production of the red pigment in the heterozygotes is insufficient to produce the maximum level of the red pigment seen in the AA homozygote. That is, in this case **the intensity of the phenotype is proportional to the dosage of functional alleles (1 in the Aa heterozygote, 2 in the AA homozygote).**

4. a. Diagram the cross:

    $e^+e^+ \times e^+e \rightarrow 1/2\ e^+e^+ : 1/2\ e^+e$.

    The trident marking is only found in the heterozygotes, so the probability is **1/2**.

    b. The offspring with the trident marking are e^+e, so the cross is $e^+e \times e^+e \rightarrow 1/4\ ee : 1/2\ e^+e : 1/4\ e^+e^+$. Therefore, of 300 offspring, **75 should have ebony bodies, 150 should have the trident marking and 75 should have honey-colored bodies.**

5. Diagram the cross:

    yellow × yellow → 38 yellow : 22 red : 20 white

    Three phenotypes in the progeny show that the yellow parents are not true breeding. The ratio of the progeny is close to 1/2 : 1/4 : 1/4. This is the result expected for crosses between individuals heterozygous for incompletely dominant genes. Thus:

    $c^r c^w \times c^r c^w \rightarrow 1/2\ c^r c^w$ (yellow) $: 1/4\ c^r c^r$ (red) $: 1/4\ c^w c^w$ (white).

6. A cross between individuals heterozygous for incompletely dominant alleles of a gene give a ratio of 1/4 (one homozygote) : 1/2 (heterozygote with the same phenotype as the parents) : 1/4 (other homozygote). Because the problem already states which genotypes correspond to which phenotypes, you know that the color gene will give a monohybrid phenotypic ratio of 1/4 red : 1/2 purple : 1/4 white, while the shape gene will give a monohybrid phenotypic ratio of 1/4 long : 1/2 oval : 1/4 round.

    Because the inheritance of these two genes is independent, use the product rule to generate all the possible phenotype combinations (note that there will be 3 × 3 = 9 classes) and their probabilities, thus generating the dihybrid phenotypic ratio for two incompletely dominant genes: **1/16 red long : 1/8 red oval : 1/16 red round : 1/8 purple long : 1/4 purple oval : 1/8 purple round : 1/16 white long : 1/8 white oval : 1/16 white round**. As an example, to determine the probability of red long progeny, multiply 1/4 (probability of red) × 1/4 (probability of long) = 1/16. If you have trouble keeping track of the 9 possible classes, it may be helpful to list the classes in the form of a branch diagram or table as follows:

chapter 3

Phenotype	Probability of phenotype
red, long	1/4 × 1/4 = 1/16
red, oval	1/4 × 1/2 = 1/8
red, round	1/4 × 1/4 = 1/16
purple, long	1/2 × 1/4 = 1/8
purple, oval	1/2 × 1/2 = 1/4
purple, round	1/2 × 1/4 = 1/8
white, long	1/4 × 1/4 = 1/16
white, oval	1/4 × 1/2 = 1/8
white, round	1/4 × 1/4 = 1/16

7. The cross is: white long × purple short → 301 long purple : 99 short purple : 612 long pink : 195 short pink : 295 long white : 98 short white.

    Deconstruct this dihybrid phenotypic ratio for two genes into separate constituent monohybrid ratios for each of the 2 traits, flower color and pod length. For flower color note that there are 3 phenotypes: 301 + 99 purple : 612 +195 pink : 295 + 98 white = 400 purple : 807 pink : 393 white = 1/4 purple : 1/2 pink : 1/4 white. This is a typical monohybrid ratio for incompletely dominant alleles, so **flower color is caused by incompletely dominant alleles of a gene, with c^P giving purple when homozygous, c^W giving white when homozygous, and the $c^P c^W$ heterozygotes giving pink.**

    For pod length, the phenotypic ratio is (301 + 612 + 295) long : (99 + 195 + 98) short = 1208 long : 392 short = 3/4 long : 1/4 short. This 3:1 ratio is that expected for a cross between individuals heterozygous for a gene in which one allele is completely dominant to the other, so **pod shape is controlled by a single gene with the long allele (L) completely dominant to the short allele (l).**

8. a. A person with sickle-cell anemia is a homozygote for the sickle-cell allele: $Hb\beta^S Hb\beta^S$.

    b. The child must be homozygous $Hb\beta^S Hb\beta^S$ and therefore must have inherited a mutant allele from each parent. Because the parent is phenotypically normal, he/she must be a **carrier with genotype $Hb\beta^S Hb\beta^A$**.

    c. Each individual has two alleles of every gene, including the β-globin gene. If an individual is heterozygous, he/she has two different alleles. Thus, if each parent is heterozygous for different alleles, there are **four possible alleles** that could be found in the five children. This is the maximum number of different alleles possible (barring the very rare occurrence of a new, novel mutation in a gamete that gave rise to one of the children). If one or both of the parents were homozygous for any one allele, the number of alleles distributed to the children would of course be less than four.

9. Remember that the gene determining ABO blood groups has 3 alleles and that $I^A = I^B > i$.

   a. The O phenotype means the girl's genotype is ii. Each parent contributed an i allele, so her parents could be ii **(O)** or $I^A i$ **(A)** or $I^B i$ **(B)**.

   b. A person with the B phenotype could have either genotype $I^B I^B$ or genotype $I^B i$. The mother is A and thus could not have contributed an I^B allele to this daughter. Instead, because the daughter clearly does not have an I^A allele, the mother must have contributed the i allele to this daughter. The mother must have been an $I^A i$ heterozygote. The father must have contributed the I^B allele to his daughter, so he could be **either $I^B I^B$, $I^B i$, or $I^B I^A$**.

   c. The genotypes of the girl and her mother must both be $I^A I^B$. The father must contribute either the I^A or the I^B allele, so there is only one phenotype and genotype which would exclude a man as her father - **the O phenotype (genotype ii)**.

10. To approach this problem, look at the mother/child combinations to determine what alleles the father must have contributed to each child's genotype.

    a. The father had to contribute I^B, N, and Rh⁻ alleles to the child. The only male fitting these requirements is **male d** whose phenotype is B, MN, and Rh⁺ (note that the father must be Rh⁺Rh⁻ because the daughter is Rh⁻).

    b. The father had to contribute i, N, and Rh⁻ alleles. The father could be either male c (O MN Rh⁺) or male d (B MN Rh⁺). As we saw previously, male c is the only male fitting the requirements for the father in part (a). Assuming one child per male as instructed by the problem, the father in part (b) must be **male c**.

    c. The father had to contribute I^A, M, and Rh⁻ alleles. Only **male b** (A M Rh⁺) fits these criteria.

    d. The father had to contribute either I^B or i, M, and Rh⁻. Three males have the alleles required: these are male a, male c, and male d. However, of these three possibilities, only **male a** remains unassigned to a mother/child pair.

11. Designate the alleles: p^m (marbled) > p^s (spotted) = p^d (dotted) > p^c (clear).

    a. Diagram the crosses:

       1. $p^m p^m$ (homozygous marbled) × $p^s p^s$ (spotted) → $p^m p^s$ (marbled F₁)

       2. $p^d p^d$ × $p^c p^c$ → $p^d p^c$ (dotted F₁)

       3. $p^m p^s$ × $p^d p^c$ → 1/4 $p^m p^d$ (marbled) : 1/4 $p^m p^c$ (marbled) : 1/4 $p^s p^d$ (spotted dotted) : 1/4 $p^s p^c$ (spotted) = **1/4 spotted dotted : 1/2 marbled : 1/4 spotted**.

    b. The F₁ from cross 1 are **marbled ($p^m p^s$)** from the first cross **and dotted ($p^d p^c$)** from the second cross as shown in part (a).

chapter 3

12. Suppose, as maintained by your fellow student, that spotting is due to the action of one gene with alleles S (spotting) and s (no spots), and that dotting is due to the action of a second gene with alleles D (dotting) and d (no spots). The cross series shown in Fig. 3.4a on p. 48, starting with true-breeding spotted and true-breeding dotted strains, could be diagrammed as:

    $SS\ dd \times ss\ DD \rightarrow Ss\ Dd$ (spotted and dotted F$_1$) $\rightarrow$ F$_2$ consisting of 9 $S-\ D-$ (spotted and dotted) : 3 $S-\ dd$ (spotted, not dotted) : 3 $ss\ D-$ (not spotted, dotted) : 1 $ss\ dd$ (not spotted, not dotted)

    Thus, the alternative hypothesis suggested by your fellow student would predict that some lentils would be found in the F$_2$ generation that would be neither spotted nor dotted. The results shown in Fig. 3.4 do not include any such lentils. **If you counted a large number of F$_2$ individuals and you failed to see lentils that were neither spotted nor dotted, you would be able to exclude the hypothesis that two genes were involved.**

13. a. All of the crosses have results that can be explained by one gene - either a 3:1 phenotypic monohybrid ratio showing that one allele is completely dominant to the other, or a 1:1 ratio showing that a testcross was done for a single gene, or all progeny with the same phenotype as the parents. You can thus conclude that all of the coat colors are controlled by the alleles of **one gene, with chinchilla (C) > himalaya (c^h) > albino (c^a)**.

    b. 1. $c^h c^a \times c^h c^a$
       2. $c^h c^a \times c^a c^a$
       3. $Cc^h \times C(c^h$ or $c^a)$
       4. $CC \times c^h (c^h$ or $c^a)$
       5. $Cc^a \times Cc^a$
       6. $c^h c^h \times c^a c^a$
       7. $Cc^a \times c^a c^a$
       8. $c^a c^a \times c^a c^a$
       9. $Cc^h \times c^h(c^h$ or $c^a)$ or $Cc^a \times c^h c^h$
       10. $Cc^a \times c^h c^a$.

    c. Cc^h (from cross 9) $\times$ Cc^a (from cross 10) $\rightarrow$ 1/4 CC (chinchilla) : 1/4 Cc^a (chinchilla) : 1/4 Cc^h (chinchilla) : 1/4 $c^h c^a$ (himalaya) = **3/4 chinchilla : 1/4 himalaya**, or Cc^a (cross 9) $\times$ Cc^a (cross 10) $\rightarrow$ **3/4 $C-$ chinchilla : 1/4 $c^a c^a$ albino.**

14. Designate the gene p (for pattern). There are 7 alleles, p^1-p^7, with p^7 being the allele that specifies absence of pattern and $p^1 > p^2 > p^3 > p^4 > p^5 > p^6 > p^7$.

    a. There are **7 different patterns** possible. These are associated with the following genotypes: p^1-, $p^2 p^a$ (where $p^a = p^2, p^3, p^4 \ldots p^7$), $p^3 p^b$ (where $p^b = p^3, p^4, p^5 \ldots p^7$),

p^4p^c (where $p^c = p^4, p^5, p^6,$ and p^7), p^5p^d (where $p^d = p^5, p^6,$ and p^7), p^6p^e (where $p^e = p^6$ and p^7), and p^7p^7.

b. The phenotype dictated by the allele p^1 has the greatest number of genotypes associated with it = **7** (p^1p^1 p^1p^2 p^1p^3, etc.). **The absence of pattern** is caused by just one genotype, p^7p^7.

c. This finding suggests that **the allele determining absence of pattern (p^7) is very common** in these clover plants with the p^7p^7 genotype is the most frequent in the population. The other alleles are present, but are much less common in this population.

15. a. This ratio is approximately **2/3 Curly : 1/3 normal**.

    b. The expected result for this cross is: $Cy^+Cy \times Cy^+Cy \rightarrow$ 1/4 $CyCy$ (?) : 1/2 Cy^+Cy (Curly) : 1/4 Cy^+Cy^+ (normal). If the Cy/Cy genotype **is lethal** then the expected ratio will match the observed data.

    c. The cross is $Cy^+Cy \times Cy^+Cy^+ \rightarrow$ 1/2 Cy^+Cy : 1/2 Cy^+Cy^+, so there would be approximately **90 Curly winged and 90 normal winged flies**.

16. There are two keys to this problem: (1) Pollen grains and ovules are gametes that have only one copy of the S incompatibility gene, while the stigma (the part of the female plant on which the pollen grains land has two copies of this gene. (2) Pollen with a particular S gene allele cannot fertilize any ovules in a plant whose stigma has the same S allele, because the pollen will not grow a tube allowing it to fertilize an ovule.

    a. In the cross $S_1S_2 \times S_1S_2$ all of the pollen grains (whether they are S_1 or S_2) will land on the stigmas of plants that have the same alleles, and therefore no progeny would be produced at all.

    b. The way this cross was written is ambiguous because the male and female parents were not specified. The cross could be ♂ $S_1S_2 \times$ ♀ S_2S_3, or ♀ $S_1S_2 \times$ ♂ S_2S_3, or both (because the flowers of these plants have both male and female parts). For the cross ♂ $S_1S_2 \times$ ♀ S_2S_3, the pollen grains would be S_1 or S_2. The S_2 pollen could not fertilize the female plant, but the S_1 pollen could. **The progeny of the ♂ $S_1S_2 \times$ ♀ S_2S_3 cross would thus be S_1S_2 and S_1S_3 (in a 1:1 ratio)**. For the cross written the opposite way (♀ $S_1S_2 \times$ ♂ S_2S_3), the pollen would be S_2 or S_3. The S_2 pollen would not produce any progeny, but the S_3 pollen could produce both S_1S_3 progeny and S_2S_3 progeny. **The progeny of the ♀ $S_1S_2 \times$ ♂ S_2S_3 cross would thus be a 1:1 ratio of S_1S_3 and S_2S_3**.

    c. Because of the ambiguity in the way the cross was written, there are again two possibilities: The cross could be ♂ $S_1S_2 \times$ ♀ S_3S_4, or ♀ $S_1S_2 \times$ ♂ S_3S_4, or both. Regardless of these possibilities, all pollen grains would be able to fertilize all ovules, because the pollen grains do not share any alleles with the female parent. As a result, **any of these crosses could produce four types of progeny in equal numbers: S_1S_3, S_1S_4, S_2S_3, and S_2S_4**.

chapter 3

d. **This mechanism would prevent plant self-fertilization because any pollen grain produced by any plant would land on a stigma sharing the same allele.** For example, if an S_1 pollen grain produced by an S_1S_2 plant lands on a stigma from the same plant, the stigma would have the same allele and no pollen tube would be able to grow to allow fertilization. The same would be true for a S_2 pollen grain from the same plant. (Of interest, tomato plants in the wild cannot self-fertilize because of this incompatibility mechanism; they proliferate only through cross-fertilization. However, many domesticated cultivars of tomatoes can self-fertilize because they were selected for varieties that have mutations causing the failure of the incompatibility mechanism.)

e. **Plants with functioning incompatibility systems must be heterozygotes because a pollen grain cannot fertilize a female plant sharing the same allele of the S incompatibility gene.** For example, an S_1 pollen grain cannot fertilize successfully any female plant that also has an S_1 allele. No way thus exists to create S_1S_1 homozygous progeny.

f. **Peas cannot be governed by this mechanism because you already saw in Chapter 2 that Gregor Mendel routinely self-fertilized his peas in the F_1 generation to produce the F_2 generation.**

g. **The larger the number of different alleles of the S gene that are present in the population, the more likely it is that any given pollen grain of any genotype would land on the stigma of a flower that did not share the same allele, and the less likely that the pollen will interact unproductively with flowers that share the same allele.** Within the population, the proportion of matings that could produce progeny would increase with a greater variety in S gene alleles; this would clearly increase the fertility (and thus the average evolutionary fitness) of the population as a whole.

17. a. The 2/3 montezuma : 1/3 wild type phenotypic ratio, and the statement that montezumas are never true-breeding, together suggest that there is a **recessive lethal allele** of this gene. When there is a recessive lethal, crossing two heterozygotes results in a 1:2:1 genotypic ratio, but one of the 1/4 classes of homozygotes do not survive. **The result is the 2:1 phenotypic ratio as seen in this cross. Both the montezuma parents were therefore heterozygous, *Mm*.** The *M* allele must confer the montezuma coloring in a dominant fashion, but homozygosity for *M* is lethal.

b. Designate the alleles: *M* = montezuma, *m* = greenish; *F* = normal fin, *f* = ruffled. Diagram the cross: *Mm FF* × *mm ff* → expected monohybrid ratio for the *M* gene alone: 1/2 *Mm* (montezuma) : 1/2 *mm* (wild type); expected monohybrid ratio for the *F* gene alone: all *Ff*. The expected dihybrid ratio = **1/2 *Mm Ff* (montezuma) : 1/2 *mm Ff* (greenish, normal fin)**.

c. *Mm Ff* × *Mm Ff* → expected monohybrid ratio for the *M* gene alone: 2/3 montezuma (*Mm*) : 1/3 greenish (*mm*); expected monohybrid ratio for the *F* gene alone: 3/4 normal fin (*F-*) : 1/4 ruffled (*ff*). The expectations when considering both

chapter 3

genes together is: **6/12 montezuma normal fin : 2/12 montezuma ruffled fin : 3/12 green normal fin : 1/12 green ruffled fin.**

Section 3.2

18. The cross is: walnut × single → F_1 walnut × F_1 walnut → 93 walnut : 29 rose : 32 pea : 11 single

    a. How many genes are involved? The four F_2 phenotypes means there are 2 genes, *A* and *B*. Both genes affect the same structure, the comb. The F_2 phenotypic dihybrid ratio among the progeny is close to 9:3:3:1, so there is no epistasis. Because walnut is the most abundant F_2 phenotype, it must be the phenotype due to the *A– B–* genotype. Single combs are the least frequent class, and are thus *aa bb*. Now assign genotypes to the cross. If the walnut F_2 are *A– B–*, then the original walnut parent must have been *AA BB*:

    AA BB × *aa bb* → *Aa Bb* (walnut) → 9/16 *A– B–* (walnut) : 3/16 *A– bb* (rose) : 3/16 *aa B–* (pea) : 1/6 *aa bb* (single).

    b. Diagram the cross, recalling that the problem states the parents are homozygous:

    AA bb (rose) × *aa BB* (pea) → *Aa Bb* (walnut) → 9/16 *A– B–* (walnut) : 3/16 *A– bb* (rose) : 3/16 *aa B–* (pea) : 1/6 *aa bb* (single). Notice that this F_2 is in identical proportions as the F_2 generation in part (a).

    c. Diagram the cross: *A– B–* (walnut) × *aa B–* (pea) → 12 *A– B–* (walnut) : 11 *aa B–* (pea) : 3 *A– bb* (rose) : 4 *aa bb* (single). Because there are pea and single progeny, you know that the walnut parent must be *Aa*. The 1 *A–* : 1 *aa* monohybrid ratio in the progeny also tells you the walnut parent must have been *Aa*. Because some of the progeny are single, you know that both parents must be *Bb*. In this case, the monohybrid ratio for the *B* gene is 3 *B–* : 1 *bb*, so both parents were *Bb*. The original cross must have been *Aa Bb* × *aa Bb*. You can verify that this cross would yield the observed ratio of progeny by multiplying the probabilities expected for each gene alone. For example, you anticipate that 1/2 the progeny would be *Aa* and 3/4 of the progeny would be *Bb*, so 1/2 × 3/4 = 3/8 of the progeny should be walnut; this is close to the 12 walnut chickens seen among 30 total progeny.

    d. Diagram the cross: *A– B–* (walnut) × *A– bb* (rose) → all *A– B–* (walnut). The progeny are all walnut, so **the walnut parent must be *BB*.** No pea progeny are seen, so **both parents cannot be *Aa*, so one of the two parents must be *AA*.** This could be either the walnut or the rose parent or both.

19. black × chestnut → F_1 bay → F_2 black : bay : chestnut : liver

    Four phenotypes in the F_2 generation means there are two genes determining coat color. The F_1 bay animals produce four phenotypic classes, so they must be doubly heterozygous, *Aa Bb*. Crossing a liver colored horse to either of the original parents resulted in the parent's phenotype. The liver horse's alleles do not affect the phenotype, suggesting the recessive genotype *aa bb*. Though it is probable that the original black mare was *AA bb* and the chestnut stallion was *aa BB*, each of these animals only produced 3 progeny, so it cannot be concluded definitively that these animals were homozygous for the dominant allele they carry. Thus, **the black mare was *A– bb*, the chestnut stallion was *aa B–*, and the F₁ bay animals are *Aa Bb*.** The F₂ horses were: bay (*A– B–*), liver (*aa bb*), chestnut (*aa B–*), and black (*A– bb*).

20. a. Because unaffected individuals had affected children, the **trait is recessive**. From affected individual II-1, you know the mutant allele is present in this generation. The trait was passed on through II-2 who was a carrier. All children of affected individuals III-2 x III-3 are affected, as predicted for a recessive trait. However, generation V seems inconsistent with recessive inheritance of a single gene. This result is consistent with two different genes involved in hearing with a defect in either gene leading to deafness: The trait is polymorphic. **The two family lines shown contain mutations in two separate genes, and the mutant alleles of both genes determining deafness are recessive.**

    b. Individuals in generation V are doubly heterozygous (*Aa Bb*), having inherited a dominant and recessive allele of each gene from their parents (*aa BB* × *AA bb*). The people in generation V are not affected because **the product of the dominant allele of each gene is sufficient for normal function.** This is an example of *complementation:* The gamete from each parent provided the dominant allele that the gamete from the other parent lacked. (See Fig. 3.21 on p. 65).

21. green × yellow → F₁ green → F₂ 9 green : 7 yellow

    a. Two phenotypes in the F₂ generation could be due to one gene or to two genes with epistasis. If this is one gene, then *GG* × *gg* → *Gg* → 3/4 *G–* (green) : 1/4 *gg* (yellow). The actual result is a 9:7 ratio, not a 3:1 ratio. The 9:7 ratio is a variant of the 9:3:3:1 phenotypic dihybrid ratio, and 9:7 ratios indicate the occurrence of epistasis. Thus, **there are 2 genes controlling color.** The genotypes are:

    AA BB (green) × *aa bb* (yellow) → F₁ *Aa Bb* (green) → F₂ 9/16 *A– B–* (green) : 3/16 *A– bb* (yellow) : 3/16 *aa B–* (yellow) : 1/16 *aa bb* (yellow).

    b. F₂ *Aa Bb* × *aa bb* → 1/4 *Aa Bb* (green) : 1/4 *aa Bb* (yellow) : 1/4 *Aa bb* (yellow) : 1/4 *aabb* (yellow) = 1/4 green : 3/4 yellow.

22. a. white × white → F₁ white → F₂ 126 white : 33 purple

    At first glance this cross seems to involve only one gene, as true-breeding white parents give white F₁s. However, if this were true, then the F₂ MUST be totally white as well. The purple F₂ plants show that this cross is NOT controlled by 1 gene. These results may be due to 2 genes. To determine if this is the case, it makes sense to ask: Does a ratio of 126 : 33 represent a variant of the 9:3:3:1 dihybrid ratio? Usually when you are given raw numbers of individuals for the classes, you divide through by the smallest number, yielding in this case 3.8 white : 1 purple. This is neither a recognizable monohybrid nor dihybrid ratio. Dividing through by the smallest class is NOT the correct way to convert raw numbers to a ratio. Assuming that the F₁ in this case are dihybrids, there must have been 16 different equally likely fertilization events that produced the F₂ progeny (16 boxes in the 4 ×

4 Punnett square), even though the phenotypes may not be distributed in the usual 9/16 : 3/16 : 3/16 : 1/16 ratio. If the 159 F_2 progeny are divided equally into 16 fertilization types, then there are 159/16 = ~10 F_2 plants/fertilization type. The 126 white F_2s therefore represent 126/10 = ~13 of these fertilizations. Likewise the 33 purple plants represent 33/10 = ~3 fertilization types. **The F_2 phenotypic ratio is thus 13 white : 3 purple. The data fit the hypothesis that two genes control color, and that the F_1 are dihybrids.**

You can now assign genotypes to the parents in the cross. Because the parents are homozygous (true-breeding) and there are 2 genes controlling the phenotypes, there are two possible ways to set up the genotypes of the parents so that the F_1 dihybrids are heterozygous for dominant and recessive alleles of each gene. One option is: *AA BB* **(white)** × *aa bb* **(white)** → *Aa Bb* **(white, same as *AA BB* parent)** → 9 *A– B–* **(white)** : 3 *A– bb* **(unknown phenotype)** : 3 *aa B–* **(unknown phenotype)** : 1 *aa bb* **(white, same as *aa bb* parent)**. If you assume that *A– bb* is white and *aa B–* is purple (or *vice versa*), then this is a match for the observed data presented in the cross above [(9 + 3 + 1) = 13 white : 3 purple].

Alternatively, you could try to diagram the cross as *AA bb* (white) × *aa BB* (white) → *Aa Bb* (whose phenotype is unknown as this is NOT a genotype seen in the parents) → 9 *A– B–* (same unknown phenotype as in the F_1) : 3 *A– bb* (white like the *AA bb* parent) : 3 *aa B–* (white like the *aa BB* parent) : 1 *aa bb* (unknown phenotype). Such a cross cannot give an F_2 phenotypic ratio of 13 white : 3 purple. The only F_2 classes that could be purple are *A– B–*, but this is impossible because this class is larger than the number of purple plants observed and because the F_1 plants must then have been purple; or the *aa bb* class which is smaller than the number of purple plants observed. Therefore, the first set of possible genotypes (written in bold above) is the best fit for the observed data.

Assume that *A– bb* plants are white, and *aa B–* plants are purple. Our model above states that in order to be purple, a plant must have a *B* allele and no *A* allele. Thus, we can say that *A* is epistatic to *B*.

b. white F_2 × white F_2 (self) → 3/4 white : 1/4 purple. Assume again that the *aa B–* class is purple in part (a) above. A 3:1 monohybrid ratio means the parents are both heterozygous for one gene with purple due to the recessive allele. The second gene is not affecting the ratio, so both parents must be homozygous for the same allele of that gene. Thus the cross must be: *Aa BB* (white) × *Aa BB* (white self cross) → 3/4 *A– BB* (white) : 1/4 *aa BB* (purple).

c. purple F_2 × self → 3 purple : 1 white. Again, the selfed parent must be heterozygous for one gene and homozygous for the other gene. Because purple is *aa B–*, the genotypes of the purple F_2 plants must be *aa Bb*.

d. white F_2 × white F_2 (not a self cross) → 1/2 purple : 1/2 white. The 1:1 monohybrid ratio means a test cross was done for one of the genes. The second gene is not altering the ratio in the progeny, so the parents must be homozygous for that gene. If purple is *aa B–*, then the genotypes of the parents must be *aa bb* (white) × *Aa BB* (white) → 1/2 *Aa Bb* (white) : 1/2 *aa Bb* (purple).

3-14

chapter 3

23. Dominance relationships are between <u>alleles of the same gene</u>. **Only one gene is involved when considering dominance relationships. Epistasis involves two genes.** The alleles of one gene affect the phenotypic expression of the second gene.

24. a. The cross is between two normal flies that carry *H* and *S*. These individuals cannot be homozygous for *H* or for *S*, because we are told that both are lethal in homozygotes. Thus, the mating described is a dihybrid cross: *Hh Ss* × *Hh Ss*. The genotypic classes among the progeny zygotes should be 9 *H– S–*, 3 *H– ss*, 3 *hh S–*, and 1 *hh ss*. However, all zygotes that are *HH* or *SS* or both will die before they hatch into adult flies. You could do this problem as a branched-line diagram as shown in the following figure, in which the progeny should be 2/3 *Hh* and 1/3 *hh* (considering the *H* gene alone) and 2/3 *Ss* and 1/3 *ss* (considering the *S* gene alone). As can be seen from the diagram, **7/9 of the adult progeny will be normal, and 2/9 will be hairless.**

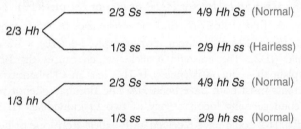

b. As just seen in the diagram, the hairless progeny of the cross in part (a) are *Hh ss*, and these are mated with parental flies that are *Hh Ss*. You could again portray the results of this cross as a branched-line diagram. For the *H* gene, again 2/3 of the viable adult progeny will be *Hh* and 1/3 will be *hh*. The cross involving the *S* gene is a testcross, and all the progeny will be viable, so 1/2 the progeny will be *Ss* and 1/2 will be *ss*. As seen in the diagram that follows, **2/6 = 1/3 of the progeny will be hairless and the remaining 2/3 will be normal.**

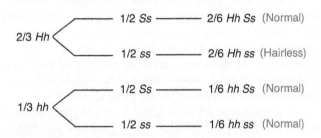

25. $I^A I^B$ *Ss* × $I^A I^A$ *Ss* → expected monohybrid ratio for the *I* gene of 1/2 $I^A I^A$: 1/2 $I^A I^B$; expected ratio for the *S* gene considered alone of 3/4 *S–* : 1/4 *ss*. Use the product rule to generate the phenotypic ratio for both genes considered together and then assign phenotypes, remembering that all individuals with the *ss* genotype look like type O. The phenotypic ratio for both genes is: 3/8 $I^A I^A$ *S–* : 3/8 $I^A I^B$ *S–* : 1/8 $I^A I^A$ *ss* : 1/8 $I^A I^B$ *ss* = 3/8 A : 3/8 AB : 1/8 O : 1/8 O = **3/8 Type A : 3/8 Type AB : 2/8 Type O.**

26. The difference between traits determined by a single pleiotropic gene and traits determined by several genes would be seen if crosses were done using pure-breeding plants (wild type × mutant), then selfing the F$_1$ progeny. If **several genes** were involved there **would be several different combinations of the petal color, markings and stem position phenotypes** in the F$_2$ generation. If all 3 traits were due to an allele present at **one gene**, the **three phenotypes would always be inherited together** and the F$_2$ plants would be either yellow, dark brown, erect OR white, no markings and prostrate.

27. a. blood types: **I-1 AB; I-2 A; I-3 B; I-4 AB; II-1 O; II-2 O; II-3 AB; III-1 A; III-2 O**.

    b. genotypes: **I-1** Hh $I^A I^B$; **I-2** Hh $I^A i$ (or $I^A I^A$); **I-3** $H-$ $I^B I^B$ (or $I^B i$); **I-4** $H-$ $I^A I^B$; **II-1** $H-$ ii; **II-2** hh $I^A I^A$ (or $I^A i$ or $I^A I^B$); **II-3** Hh $I^A I^B$; **III-1** Hh $I^A i$; **III-2** hh $I^A I^A$ (or $I^A I^B$ or $I^A i$ or $I^b i$ or $I^B I^B$)

    At first glance, you find inconsistencies between expectations and what could be inherited from a parent. For example, I-1 (AB) × I-2 (A) could not have an O child (II-2). The epistatic h allele (which causes the Bombay phenotype) could explain these inconsistencies. If II-2 has an O phenotype because she is hh, her parents must both have been Hh. The Bombay phenotype would also explain the second seeming inconsistency of two O individuals (II-1 and II-2) having an A child. II-2 could have received an I^A allele from one of her parents and passed this on to III-1 together with one h allele. Parent II-1 would have to contribute the H allele so that the I^A allele would be expressed; the presence of H means that II-1 must also be ii in order to be type O. A third inconsistency is that individuals II-2 and II-3 could not have an ii child since II-3 has the $I^A I^B$ genotype, but III-2 has the O phenotype. This could also be explained if II-3 is Hh and III-2 is hh.

28. a. Diagram one of the crosses:

    white-1 × white-2 → red F$_1$ → 9 red : 7 white

    Even though there are only 2 phenotypes in the F$_2$, color is not controlled by one gene - the 9:7 ratio is a variation of 9:3:3:1, so there are 2 genes controlling these phenotypes. Individuals must have at least one dominant allele of each gene in order to get the red color; this is an example of *complementary gene action*. Thus the genotypes of the two pure-breeding white parents in this cross are *aa BB* × *AA bb*. The same conclusions hold for the other 2 crosses.

    If white-1 is *aa BB* and white-2 is *AA bb*, then white-3 must be *AA BB cc*. The reason is that if white-3 had the same genotype as white-1 or white-2, then one of the three crosses would have produced an all white F$_1$ (no complementation). Because none of the crosses had an all white F$_1$, we can conclude that **three genes are involved**.

    b. White-1 is *aa BB CC*; white-2 is *AA bb CC* and white-3 is *AA BB cc*.

    c. *aa BB CC* (white-1) × *AA bb CC* (white-2) → *Aa Bb CC* (red) → 9/16 *A– B– CC* (red) : 3/16 *A– bb CC* (white) : 3/16 *aa B– CC* (white) : 1/6 *aa bb CC*

chapter 3

(white). Red color requires a dominant, functional allele of each of the three genes (*A– B– C–*).

29. Diagram the cross. Figure out an expected monohybrid ratio for each gene separately, then apply the product rule to generate the expected dihybrid ratio.

    $A^y A\ Cc \times A^y A\ cc \rightarrow$ monohybrid ratio for the *A* gene alone: 1/4 $A^y A^y$ (dead) : 1/2 $A^y A$ (yellow) : 1/4 *AA* (agouti) = 2/3 $A^y A$ (yellow) : 1/3 *AA* (agouti); monohybrid ratio for the *C* gene: 1/2 *Cc* (non-albino) : 1/2 *cc* (albino).

    Overall there will be 2/6 $A^y A\ Cc$ (yellow) : 2/6 $A^y A\ cc$ (albino) : 1/6 *AA Cc* (agouti) : 1/6 *AA cc* (albino) = **2/6 $A^y A\ Cc$ (yellow) : 3/6 –– *cc* (albino) : 1/6 *AA Cc* (agouti)**. Note that the $A^y A\ cc$ animals must be albino because the albino parent had exactly the same genotype; this indicates that *cc* is epistatic to all alleles of gene *A*. Although you were not explicitly told that the *AA cc* animals are also albino, this makes sense because in Fig. 3.23 on p. 67, *cc* is epistatic to all alleles of another gene *B* in animals that must have been *AA*. (Another way to think of this is that the *cc* albino color must be epistatic to alleles of all genes that confer color because no pigments are produced.)

30. a. **No**, a single gene cannot account for this result. While the 1:1 ratio seems like a testcross, the fact that **the phenotype of one class of offspring (linear) is not the same as either of the parents** argues against this being a testcross.

    b. The appearance of four phenotypes means **two genes** are controlling the phenotypes.

    c. The 3:1 ratio suggests that **two alleles of one gene** determine the difference between the wild-type and scattered patterns.

    d. The true-breeding wild-type fish are homozygous by definition, and the scattered fish have to be homozygous recessive according to the ratio seen in part (c), so the cross is: *bb* (scattered) × *BB* (wild type) → F₁ *Bb* (wild type) → F₂ 3/4 *B–* (wild type) : 1/4 *bb* (scattered).

    e. The inability to obtain a true-breeding nude stock suggests that the nude fish are heterozygous (*Aa*) and that the *AA* genotype dies. Thus *Aa* (nude) × *Aa* (nude) → 2/3 *Aa* (nude) : 1/3 *aa* (scattered).

    f. Going back to the linear cross from part (b), the fact that there are four phenotypes led us to propose two genes were involved. The 6:3:2:1 ratio looks like an altered 9:3:3:1 ratio in which some genotypes may be missing, as predicted from the result in part (e) that *AA* animals do not survive. The 9:3:3:1 ratio results from crossing double heterozygotes, so **the linear parents are doubly heterozygous *Aa Bb*. The lethal phenotype associated with the *AA* genotype produces the 6:3:2:1 ratio. The phenotypes and corresponding genotypes of the progeny of the linear × linear cross are: 6 linear, *Aa B–* : 3 wild-type, *aa B–* : 2 nude, *Aa bb* : 1 scattered, *aa bb*.** Note that the *AA BB*, *AA Bb*, *AA Bb*, and *AA bb* genotypes are missing due to lethality.

31. You know about the alleles of the *A* gene (*A* for agouti, a^t for black/yellow, *a* for black, and A^y for recessive lethal yellow) from Fig. 3.7 on p. 51 and Fig. 3.8 on p. 52. You know about the interaction of the *B* and *C* genes governing the variation in black, brown, and albino colors from Fig. 3.23 on p. 67.

   a. The yellow parent must have an A^y allele, but we don't know the second allele of the *A* gene (A^y–). We don't know at the outset what alleles this yellow mouse has at the *B* gene, so we'll leave these alleles for the time being as *??*. Since this mouse does show color we know it is not *cc* (albino), so it must have at least one *C* allele (*C*–). The brown agouti parent has at least one *A* allele (*A* -); it must be *bb* at the *B* gene; and since there is color it must also be *C*–. The mating between these two can thus be represented as A^y– *?? C*– × *A*– *bb C*–.

      Now consider the progeny. Because one pup was albino (*cc*), the parents must both be *Cc*. A brown pup (*bb*) indicates that both parents had to be able to contribute a *b* allele, so we now know the first mouse (the yellow parent) must have had at least one *b* allele. The fact that this brown pup was non-agouti means both parents carried an *a* allele. The black agouti progeny tells us that the first mouse must have also had a *B* allele. This latter fact also clarifies that A^y is epistatic to *B* because this parent was yellow rather than black. The complete genotypes of the mice are therefore: **A^ya Bb Cc x Aa bb Cc**.

   b. Think about each gene individually, then the effect of the other genes in combination with that phenotype. *C*– leads to a phenotype with color; *cc* gives albino (which is epistatic to all colors determined by the other genes because no pigments are produced). The possible genotypes of the progeny of this cross for the *A* gene are A^y*A*, A^y*a*, *Aa* and *aa*, giving yellow, yellow, agouti and non-agouti phenotypes, respectively. Since yellow (A^y) is epistatic to *B*, non-albino mice with A^y will be yellow regardless of the genotype of the *B* gene. *Aa* is agouti; with the *aa* genotype there is no yellow on the hair (non-agouti). The type of coloration depends on the *B* gene. For *B* the offspring could be *Bb* (black) or *bb* (brown). In total, **six different coat color phenotypes are possible**: albino (– – – – *cc*), yellow (A^y(*A* or *a*) – – *C*–), brown agouti (*A*– *bb C*–), black agouti (*A*– *B*– *C*–), brown (*aa bb C*–), and black (*aa B*– *C*–). [*Note:* Although A^y (yellow color) is in fact epistatic to *B* (black) or *bb* (brown) colors governed by the *B* gene, you were not explicitly told this. Thus, based on the information provided, you might have included an additional color phenotype if you considered that A^y(*A* or *a*) *bb C*– had a lighter color than the yellow of A^y(*A* or *a*) *Bb C*– animals.]

32. In Figure 3.28b on p. 72 the A^1 and B^1 alleles each have the same effect on the **phenotype** (plant height in this example), while the A^0 and B^0 alleles are non-functional. Thus, the shortest plants are $A^0A^0B^0B^0$, and the tallest plants are $A^1A^1B^1B^1$. **The phenotypes are determined by the total number of A^1 and B^1 alleles in the genotype**. Thus, $A^1A^0B^0B^0$ plants are the same phenotype as $A^0A^0B^0B^1$. In total there will be 5 different phenotypes: 4 '0' alleles (total = 0); 1 '1'

chapter 3

allele + 3 '0' alleles (total = 1); 2 '1' alleles + 2 '0' alleles (total = 2); 3 '1' alleles + 1 '0' allele (total = 3); and 4 '1' alleles (total = 4).

In Figure 3.22 on p. 65 the a allele = b allele = no function (in this case no color = white). If the A allele has the same level of function as a B allele then you would see 5 phenotypes as was the case for Figure 3.28b. But since there are a total of 9 phenotypes, this cannot be true so $A \neq B$. Notice that $aa\ Bb$ is lighter than $Aa\ bb$ even though both genotypes have the same number of dominant alleles. Thus, **in Fig. 3.22 an A allele has more effect on coloration than a B allele**. If you assume, for example, that $B = 1$ unit of color and $A = 1.5$ unit of color, then 16 genotypes lead to 9 phenotypes.

33. a. $Aa\ Bb\ Cc \times Aa\ Bb\ Cc \rightarrow$ 9/16 A– B– × 3/4 C– : 9/16 A– B– × 1/4 cc : 3/16 A– bb × 3/4 C– : 3/16 A– bb x 1/4 cc : 3/16 $aa\ B$– × 3/4 C– : 3/16 $aa\ B$– × 1/4 cc : 1/16 $aa\ bb$ × 3/4 C– : 1/16 $aa\ bb$ × 1/4 cc = 27/64 A– B– C– (wild type) : 9/64 A– B– cc : 9/64 A– $bb\ C$– : 3/64 A– $bb\ cc$: 9/64 $aa\ B$– C– : 3/64 $aa\ B$– cc : 3/64 $aa\ bb\ C$– : 1/64 $aa\ bb\ cc$ = **27/64 wildtype : 37/64 mutant**.

   b. Diagram the crosses:
      1. unknown male × $AA\ bb\ cc \rightarrow$ 1/4 wild type (A– B– C–) : 3/4 mutant
      2. unknown male × $aa\ BB\ cc \rightarrow$ 1/2 wild type (A– B– C–) : 1/2 mutant
      3. unknown male × $aa\ bb\ CC \rightarrow$ 1/2 wild type (A– B– C–) : 1/2 mutant

   The 1:1 ratio in test crosses 2 and 3 is expected if the unknown male is heterozygous for one of the genes that are recessive in the testcross parent. The 1 wild type : 3 mutant ratio arises when the male is heterozygous for two of the genes that are homozygous recessive in the testcross parent. (If you apply the product rule to 1/2 B– : 1/2 bb and 1/2 C– : 1/2 cc in the first cross, then you find 1/4 B– C–, 1/4 B– cc, 1/4 $bb\ C$–, and 1/4 $bb\ cc$. Only B– C– will be wild type, the other 3 classes will be mutant). Thus the unknown male must be $Bb\ Cc$. In testcross 1 the male could be either AA or aa. Crosses 2 and 3 show that the male is only heterozygous for one of the recessive genes in each case: gene C in testcross 2 and gene B in testcross 3. In order to get wild-type progeny in both crosses, the male must be AA. **Therefore the genotype of the unknown male is $AA\ Bb\ Cc$**.

34. a. For all 5 crosses, determine the number of genes involved in the trait and the dominance relationships between the alleles. Cross 1: 1 gene, red>blue. Cross 2: 1 gene, lavender>blue. Cross 3: 1 gene, codominance/incomplete dominance (1:2:1), bronze is the phenotype of the heterozygote. Cross 4: 2 genes with epistasis (9 red : 4 yellow : 3 blue). Cross 5: 2 genes with epistasis (9 lavender : 4 yellow : 3 blue). **In total there are 2 genes. One gene controls blue (c^b), red (c^r) and lavender (c^l) where $c^r = c^l > c^b$. The second gene controls the yellow phenotype: Y seems to be colorless (or has no effect on color), so the phenotype is determined by the alleles of the c gene. The y allele makes the flower yellow, and is epistatic to all alleles of the c gene.**

   b. cross 1: $c^r c^r\ YY$ (red) × $c^b c^b\ YY$ (blue) $\rightarrow$ $c^r c^b\ YY$ (red) $\rightarrow$ 3/4 c^r– YY (red) : 1/4 $c^b c^b\ YY$ (blue)

      cross 2: $c^l c^l\ YY$ (lavender) × $c^b c^b\ YY$ (blue) $\rightarrow$ $c^l c^b\ YY$ (lavender) $\rightarrow$ 3/4 c^l– YY (lavender) : 1/4 $c^b c^b\ YY$ (blue) cross 3: c^lc^l YY (lavender) × c^rc^r YY (red) → c^lc^r YY (bronze) → 1/4 c^lc^l YY (lavender) : 1/2 c^lc^r YY (bronze) : 1/4 c^rc^r YY (red)

cross 4: c^rc^r YY × c^bc^b yy (yellow) → c^rc^b Yy (red) → 9/16 c^r– Y– (red) : 3/16 c^r– yy (yellow) : 3/16 c^bc^b Y– (blue) : 1/16 c^bc^b yy (yellow)

cross 5: c^lc^l yy (yellow) × c^bc^b YY (blue) → c^lc^b Yy (lavender) → 9/16 c^l– Y– (lavender) : 3/16 c^l– yy (yellow) : 3/16 c^bc^b Y– (blue) : 1/16 c^bc^b yy (yellow)

   c. c^rc^r yy (yellow) × c^lc^l YY (lavender) → c^rc^l Yy (bronze) → monohybrid ratio for the *c* gene is 1/4 c^rc^r : 1/2 c^rc^l : 1/4 c^lc^l and monohybrid ratio for the *Y* gene is 3/4 Y– : 1/4y. Using the product rule, these generate a dihybrid ratio of **3/16 c^rc^r Y– (red) : 3/8 c^rc^l Y– (bronze) : 3/16 c^lc^l Y– (lavender) : 1/16 c^rc^r yy (yellow) : 1/8 c^rc^l yy (novel genotype) : 1/16 c^lc^l yy (yellow).** You expect the c^rc^l yy genotype to be yellow as *y* is normally epistatic to the *c* gene. However, you have no direct evidence from the data in any of these crosses that this will be the case, so it is possible that this genotype could cause a different and perhaps completely new phenotype.

35. a. Analyze each cross by determining how many genes are involved in the phenotypes and the relationships between the alleles of these genes. In cross 1, there are 2 genes because there are 3 classes in the F_2 showing a modified 9:3:3:1 ratio (12:1:3), and LR is the doubly homozygous recessive class. In cross 2, only 1 gene is involved because there are 2 phenotypes in a 3:1 ratio; WR>DR. In cross 3, there is again only 1 gene involved (2 phenotypes in a 1:3 ratio); DR>LR. In cross 4, there is 1 gene (2 phenotypes, with a 3:1 ratio); WR>LR. In cross 5, there are again 2 genes (and as in cross 1, there is a 12:1:3 ratio of three classes); LR is the double homozygous recessive. In total, **there are 2 genes controlling these phenotypes in foxgloves**.

   b. Remember that all four starting strains are true-breeding. In cross 1 the parents can be assigned the following genotypes: *AA BB* (WR-1) × *aa bb* (LR) → *Aa Bb* (WR) → 9 *A– B–* (WR) : 3 *A– bb* (WR; this class displays the epistatic interaction) : 3 *aa B–* (DR) : 1 *aa bb* (LR). The results of cross 2 suggested that DR differs from WR-1 by one gene, so **DR is *aa BB*;** cross 3 confirms these genotypes for DR and LR. Cross 4 introduces WR-2, which differs from LR by one gene and differs from DR by 2 genes, so **WR-2 is *AA bb*.** Cross 5 would then be *AA bb* (WR-2) × *aa BB* (DR) → *Aa Bb* (WR) → 9 *A– B–* (WR) : 3 *A– bb* (WR) : 3 *aa B–* (DR) : 1 *aa bb* (LR) = 12 WR : 3 DR : 1 LR.

   c. WR from the F_2 of cross 1 LR → 253 WR : 124 DR : 123 LR. Remember from part (b) that LR is *aa bb* and DR is *aa B–* while WR can be either *A– B–* or *A– bb* = *A– ??*. The experiment is essentially a testcross for the WR parent. The observed monohybrid ratio for the *A* gene is 1/2 *Aa* : 1/2 *aa* (253 *Aa* : 124 + 123 *aa*), so the WR parent must be *Aa*. The DR and LR classes of progeny show that the WR parent is also heterozygous for the *B* gene (DR is *Bb* and LR is *bb* in these progeny). Thus, the cross is *Aa Bb* (WR) × *aa bb* (LR).

chapter 3

36. The hairy × hairy → 2/3 hairy : 1/3 normal cross described in the first paragraph of the problem tells us that the hairy flies are heterozygous, that the hairy phenotype is dominant to normal, and that the homozygous hairy progeny are lethal (that is, hairy is a recessive lethal). Thus, hairy is *Hh*, normal is *hh*, and the lethal genotype is *HH*. Normal flies therefore should be *hh* (**normal-1**) and a cross with hairy (*Hh*) would be expected to always give 1/2 *Hh* (hairy) : 1/2 *hh* (normal) as seen in cross 1.

   In cross 2, the progeny MUST for the same reasons be 1/2 *Hh* : 1/2 *hh*, yet they ALL appear normal. This suggests the normal-2 stock has another mutation that suppresses the hairy wing phenotype in the *Hh* progeny. The hairy parent must have the recessive alleles of this suppressor gene (*ss*), while the normal-2 stock must be homozygous for the dominant allele (*SS*) that suppresses the hairy phenotype. Thus cross 2 is *hh SS* (**normal-2**) × *Hh ss* (hairy) → 1/2 *Hh Ss* (normal because hairy is suppressed) : 1/2 *hh Ss* (normal).

   In cross 3, the normal-3 parent is heterozygous for the suppressor gene: *hh Ss* (**normal-3**) × *Hh ss* (hairy) → the expected ratios for each gene alone are 1/2 *Hh* : 1/2 *hh* and 1/2 *Ss* : 1/2 *ss*, so the expected ratio for the two genes together is 1/4 *Hh Ss* (normal) : 1/4 *Hh ss* (hairy) : 1/4 *hh Ss* (normal) : 1/4 *hh ss* (normal) = 3/4 normal : 1/4 hairy.

   In cross 4 you see a 2/3 : 1/3 ratio again, as if you were crossing hairy x hairy. After a bit of trial-and-error examining the remaining possibilities for these two genes, you will be able to demonstrate that this cross was *Hh Ss* (**normal-4**) × *Hh ss* (hairy) → expected ratio for the individual genes are 2/3 *Hh* : 1/3 *hh* and 1/2 *Ss* : 1/2 *ss*, so the expected ratio for the two genes together from the product rule is 2/6 *Hh Ss* (normal) : 2/6 *Hh ss* (hairy) : 1/6 *hh Ss* (normal) : 1/6 *hh ss* (normal) = 2/3 normal : 1/3 hairy.

37. This problem shows that gene interactions producing variations of the 9:3:3:1 ratio in addition to those shown in Table 3.2 on p. 64 are possible though rare.

    a. Using the information provided, it is clear that one of the pure-breeding white strains must be homozygous for recessive alleles of gene *A* and the other pure-breeding white strain must be homozygous for recessive alleles of gene *B*. That is, the cross was *AA bb* (**white**) × *aa BB* (**white**) → F₁ *Aa Bb* (all blue).

    b. In the F₂ generation produced by self-mating of the F₁ plants, you would find a genotypic ratio of 9 *A– B–* : 3 *A– bb* : 3 *aa B–* : 1 *aa bb*. The *A– B–* plants would have blue flowers because colorless precursor 1 would be converted into blue pigment. (Colorless precursor 2 would not produce blue pigment in these flowers because the pathway is suppressed by the proteins specified by the dominant alleles of the two genes. The *A– bb* plants would be white because the first pathway could not produce blue pigment in the absence of the protein specified by *B*, while the second pathway would be shut off by the protein specified by *A*. The *aa B–* plants would be white because the first pathway could not produce blue pigment in the absence of the protein specified by *A*, while the second pathway would be shut off by the protein specified by *B*. Interestingly, the *aa bb* plants would be blue because even though the first pathway would not function, the second would as it is not suppressed. **You would thus expect in the F₂ generation a ratio of 10 blue (9 *A– B–* +1 *aa bb*) : 6 white (3 *A– bb* + 3 *aa B–*).**

chapter 3

38. The answers are presented in the table below. Different colors in the table represent different phenotypes; these colors are chosen arbitrarily and do not signify anything. The numbers in parentheses indicate the compounds that are present to produce the colors.

Part	9 A– B–	3 A– bb	3 aa B–	1 aa bb	Ratio
a	(2 + 4)	(2 + 3)	(1 + 4)	(1 + 3)	9:3:3:1
b	(2)	(2)	(2)	(1)	15:1
c	(3)	(2)	(1)	(1)	9:3:4
d	(2)	(1)	(1)	(1)	9:7
e	(2 + 3)	(2)	(3)	(1)	9:3:3:1
f	(2 + 4) = (2)	(2 + 3) = (2)	(1 + 4)	(1 + 3)	12:3:1
g	(3)	(2) = (1)	(1) = (2)	(1) = (2)	9:7
h	(2)	(1)	(2)	(2)	13:3

[*Note:* In part (e), you could have interpreted a limitless supply of compound 1 to mean that compound 1 would be present in all the phenotypes (for example, the A– B– genotype would have compounds 1 + 2 + 3, etc.). The ratio would still be the same under this assumption as the ratio listed in the table.]

39. **A particular phenotypic ratio does not allow you to infer the operation of a specific biochemical mechanism because** as can be seen from the answers to Problem 38, **different biochemical mechanisms can produce the same ratio of phenotypes** (for example, the pathways in parts [d] and [g] are different yet both yield 9:7 ratios). The particular ratio seen in a cross may nonetheless provide information about types of biochemical pathways you could exclude from consideration because those pathways could not produce the observed ratio.

   In contrast, **if you know the biochemical mechanism behind a gene interaction** and you also know the dominance relationships of the alleles, you can then trace out the consequences of each genotypic class and thus **you can predict the ratios of phenotypes you would see among the F_2 progeny.**

40. a. The mutant plant lacks the function of all three genes, so **its genotype must be *aa bb cc*.**

   b. Considering each gene separately, 3/4 of the F_2 progeny will have at least one dominant allele, whereas 1/4 will be homozygous for the recessive allele. As just seen in part (a), mutant plants must be triply homozygous recessive. The chance that a plant will have the *aa bb cc* genotype is 1/4 x 1/4 x 1/4 = 1/64. All other F_2 plants will be normal for this phenotype, so **the fraction of normal plants = 1 – 1/64 = 63/64.**

   c. The most likely explanation for redundant gene function is that in the relatively recent past, a single gene became duplicated (or in this case, triplicated). The three copies of the *SEP* gene are nearly identical to each other and thus fulfill the same function. Only if the functions of all three genes are lost does a mutant phenotype result. In fact, these kinds of gene duplication events occur often enough in nature that redundant gene function is a common phenomenon.

chapter 3

41. a. There are actually **two different phenotypes** mentioned in this problem. One phenotype is the shape of the erythrocytes. All people with the genotype SPH^- SPH^- have **spherical erythrocytes**. Therefore this phenotype **is fully penetrant and shows no variation in expression**. The second **phenotype is anemia. Here the expressivity among anemic patients varies from severe to mild**. There are even **some people with the SPH^- SPH^- genotype (150/2400) with no symptoms of anemia at all. Thus the penetrance of the anemic phenotype is 2250/2400 or 0.94**.

   b. The disease causing phenotype is the anemia and the severity of the anemia is greatly reduced when the spleen functions poorly and does not "read" the spherical erythrocytes. Therefore **treatment might involve removing the spleen (an organ which is not essential to survival). The more efficiently the spleen functions the earlier in a patient's life it should be removed**. SPH^- SPH^- individuals with no symptoms of anemia should not be subjected to this drastic treatment.

42. a. The most likely mode of inheritance is a **single gene with incomplete dominance** such that $f^n f^n$ = normal (<250 mg/dl), $f^n f^a$ = intermediate levels of serum cholesterol (250-500 mg/dl) and $f^a f^a$ homozygotes = elevated levels (>500 mg/dl). Some of the individuals in the pedigrees do not fit this hypothesis. In two of the families (Families 2 and 4), two normal parents have a child with intermediate levels of serum cholesterol. **One possibility is that in each family, at least one of these normal parents (I-3 and/or I-4 in Family 2; I-1 and/or I-2 in Family 4) was actually a $f^n f^a$ heterozygote who did not have elevated cholesterol in excess of 250 mg/dl. In this scenario, familial hypercholesterolemia is a trait with incomplete penetrance,** so that some unaffected people have a genotype that causes the disease in other people. It is also possible that the affected children of these parents do not have an f^a allele associated with elevated serum cholesterol, but they show the trait for other reasons such as diet, level of exercise, or other genes. This explanation is reasonable, but perhaps less likely because multiple children would have to have the trait but not the f^a allele.

   b. **Familial hypercholesterolemia also shows variable expressivity,** meaning that people with the same genotype have the condition, but to different extents. This suggests that factors other than just the genotype are involved in the expression of the phenotype. Such factors could again include diet, level of exercise, and other genes.

43. a. The pattern in both families **looks like a recessive trait** since unaffected individuals have affected progeny and the trait skips generations. For example, in the Smiths II-3 must be a carrier, but in order for III-5 to be affected II-4 must also be a carrier. **If the trait is rare (as is this one) you wouldn't expect two heterozygotes to mate by chance as many times as required by these pedigrees. The alternative explanation is that the trait is dominant but not 100% penetrant.**

b. Assuming this is a dominant but not completely penetrant trait, **individuals II-3 and III-6 in the Smiths' pedigree individual and II-6 in the Jeffersons' pedigree** must carry the dominant allele but not express it in their phenotypes.

c. If the trait were common, **recessive inheritance** is the more likely mode of inheritance.

d. **None**; in cases where two unaffected parents have an affected child, both parents would be carriers of the recessive trait.

44. If polycystic kidney disease is dominant, then the child is *Pp* and inherited the *P* disease allele from one parent or the other, yet phenotypically the parents appear to be *pp*. Perhaps one of the parents is indeed *Pp*, but this parent does not show the disease phenotype for some reason. Such situations are not uncommon: the unexpressed dominant allele is said to have **incomplete penetrance** in these cases. Alternatively, it could be that both parents are indeed *pp* and the *P* allele inherited by the child was due to a **spontaneous mutation** during the formation of the gamete in one of the parents; we will discuss this topic in Chapter 7. It is also possible that the **father of the child is not the male parent of the couple**. In this case the biological father must have the disease.

45. The Black Lab: A solid black dog must: make eumelanin (*E*-); not be brown (*B*-); have no pheomelanin striping in the hairs [(*K^b*- and any *A* gene alleles) or (*aa* and any *K* gene alleles)]; must not be diluted to gray (*D*–); no spotting (*S*–); and no merle (M^2M^2). Because Labs always breed true for solid colors (black, brown, or some light yellow color), the black Lab cannot be heterozygous at any gene for recessive alleles that specify non-solid colors. So **the Black Lab is most likely: *EE* or *Ee*, *BB*, (K^bK^b and any gene *A* alleles) or (K^bk^y *aa*), *DD*, *SS*, M^2M^2**.

The Chocolate Lab: **A solid chocolate brown dog would be the same genotype as solid black, except *bb*.**

The Yellow Lab: A solid yellow dog must not make eumelanin (*ee*). Any alleles of gene *B* are possible, and as above, the dog must be *DD*, *SS*, and M^2M^2. The same considerations for genes *A* and *K* apply as for the other Labs above. The yellow dog pictured cannot be *aa*, however, or it would be white. Therefore, **yellow labs are *ee*, *B*– or *bb*, K^bK^b, any gene *A* alleles except *aa*, *DD*, *SS*, M^2M^2.**

chapter 4

The Chromosome Theory of Inheritance

Synopsis

Chapter 4 is critical for understanding genetics because it connects Mendel's laws for the transmission of genes with the behavior of chromosomes during meiosis. Genes are located on chromosomes and travel with them during cell division, gamete formation, and fertilization. Mendel's first law of segregation is a consequence of the fact that homologous chromosomes segregate from each other (move to opposite spindle poles) during the first meiotic division in germ cells. Mendel's second law of independent assortment results from the independent alignment of each pair of homologous chromosomes during the first meiotic division.

Many of the experiments that verified the Chromosome Theory of Inheritance relied on the facts that in many organisms the two sex chromosomes are morphologically distinct and do not carry the same genes. As a result, the inheritance patterns of X-linked genes in fruit flies and humans differ from those of autosomal genes, and scientists were able to correlate these unusual inheritance patterns with the presence of particular sex chromosomes.

Key terms

haploid (*n*) – cells such as gametes that contain only one copy of each chromosome pair

diploid (2*n*) – cells that contain both copies of an homologous chromosome pair

sister chromatids – the two copies of a single chromosome that result after chromosome replication during S phase. The sister chromatids are connected to each other at the **centromere**.

metacentric/acrocentric – terms that describe the position of the centromere on a chromosome. In *metacentric chromosomes,* the centromere is near the center so the chromosome contains two long *arms*. In *acrocentric chromosomes,* the centromere is near (but not quite at) one of the ends.

homologous chromosomes (homologs) – chromosomes that match in shape, size, and banding because they have the same set of genes. Normally, in a pair of homologous chromosomes, one chromosome is of maternal origin, while the other is of paternal origin. As a result, homologous chromosomes may carry different alleles of genes.

nonhomologous chromosomes – chromosomes that do not match because they have different sets of genes

karyotype – a representation of an organism's genome made by arranging micrographs of each chromosome as homologous pairs of decreasing size.

chapter 4

sex chromosomes – the chromosomes that determine an individual's sex; in humans and *Drosophila*, these are the **X chromosome** and the **Y chromosome**

SRY – the gene found on the Y chromosome in humans and other mammals whose presence determines maleness. In rare cases of **sex reversal,** the *SRY* gene has shifted from the Y chromosome to the X chromosome.

autosomes – chromosomes that do not determine sex and that are found as homologous pairs

pseudoautosomal regions (PARs) – regions containing about 30 genes that are found on both the X and Y chromosomes in humans. The presence of PARs allows the X and Y chromosomes to pair with each other during the first meiotic division.

cell cycle – repeating pattern of cell growth and division. **Interphase** (the period between cell divisions) consists of **G1, S** (the time of chromosome replication), and **G2.** The cells divide during **M phase,** and the **daughter cells** separate from each other through **cytokinesis** at the end of M phase. (See Fig. 4.9 on p. 94 for details.)

spindle – cellular structure made of **microtubules** emanating from two **spindle poles** that forms during M phase and guides chromosome movements

kinetochores – structures found at centromeres that connect the chromosomes to the microtubule fibers of the spindle

mitosis – cell division that preserves the number and kinds of chromosomes. Successive rounds of mitosis starting from a fertilized zygote create a multicellular organism whose **somatic cells** are genetically identical.

prophase/prometaphase/metaphase/anaphase/telophase – successive stages of M phase (see Fig. 4.10 on p. 95 for details)

germ cells – cells destined for a specialized role in the production of gametes. In animals, the germ cells are set aside as the **germ line** early in embryogenesis.

meiosis – cell divisions of germ cells that result in the formation of gametes. Meiosis consists of two rounds of division (Meiosis I and Meiosis II) that follow a single round of chromosome replication; as a result, the gametes are haploid.

synapsis – the pairing of homologous chromosome during Meiosis I. Synapsis is mediated by a protein structure called the **synaptonemal complex.** The paired chromosomes during prophase and metaphase of meiosis I are called **bivalents** or **tetrads.**

crossing-over – exchange of genetic material between homologous chromosomes that takes place during prophase of meiosis I at sites called **chiasmata.** Crossing-over results in the **recombination** of genetic material between nonsister chromatids.

nondisjunction – rare mistakes that occur during cell division in which chromosomes or chromatids that should separate from each other fail to separate from each other

gametogenesis/oogenesis/spermatogenesis – processes that produce gametes

oogonia/spermatogonia – germ-line cells in the ovaries or testes that undergo mitotic divisions

chapter 4

- **oocytes/spermatocytes** – germ-line cells in the ovaries or testes that undergo meiosis. **Primary oocytes/spermatocytes** are the cells that undergo meiosis I; **secondary oocytes/spermatocytes** undergo meiosis II.
- **polar bodies** – small cells produced at the end of meiosis I or meiosis II in females that disintegrate and cannot become eggs
- **ovum** – the haploid egg cell produced at the end of meiosis II in females
- **spermatid** – haploid cells produced at the conclusion of meiosis II in males that will undergo additional development to become **sperm**
- **X linkage** – inheritance pattern seen for genes located on the X chromosome but not the Y chromosome. X-linked genes may exhibit **crisscross inheritance** in which the sons look like their mothers and the daughters resemble their fathers. Males are **hemizygous** for X-linked genes because they have only one copy of these genes.
- **dosage compensation** – processes that ensure that cells in males and females produce the same amount of protein encoded by X-linked genes. In humans, dosage compensation occurs due to **X chromosome inactivation:** when an individual has two or more X chromosomes, all of the X chromosomes but one are "turned off" because they turn into inactive **Barr bodies**.
- **sex-limited traits** – traits that affect a structure or process found in one sex but not the other. **Sex-influenced traits** can be seen in both sexes, but the expression differs between the sexes. The terms *sex-limited* and *sex-influenced* do not say anything about a gene's location, because both sex-linked and autosomal genes may affect the two sexes differently.

The Keys to Mitosis and Meiosis

The figures at the bottom of the next page depict metaphase of mitosis and meiosis I and summarize how chromosomes are transmitted during these types of cell division. You should note the following points:

- Just as in Fig. 4.3 on p. 88, sister chromatids are shown in the same shade of the same color, homologous chromosomes are shown in different shades of the same color, and nonhomologous chromosomes are shown in different colors. This scheme reflects the facts that sister chromatids have identical DNA sequences, homologous chromosomes are nearly the same (they have the same genes but may have different alleles of those genes), and nonhomologous chromosomes have different genes.
- In metaphase of mitosis, the sister chromatids are connected through their kinetochores to microtubules emanating from the opposite spindle poles. The chromosomes congress to the metaphase plate because the microtubules exert forces on the sister chromatids in opposite directions, but these counterbalance each other because the sister chromatids are connected at the centromere. During anaphase, the centromeric connection is dissolved, so the sister chromatids can move to the opposite poles.

- In metaphase of meiosis I, the two sister chromatids of each chromosome attach to microtubules from the same spindle pole. It is the homologous chromosomes that attach to microtubules from opposite spindle poles; at anaphase of meiosis I, it is the homologous chromosomes that go to opposite spindle poles. During anaphase of meiosis I, the centromeres remain intact, so the sister chromatids remain together as the homologous chromosome go to opposite spindle poles. The separation of homologous chromosomes during meiosis I explains Mendel's first law of allele segregation.

- The reason homologous chromosomes connect to opposite spindle poles and can congress to the metaphase plate during meiosis I, is that the homologs are attached to each other by crossing-over at chiasmata (*not shown*). Crossing-over is thus critical for ensuring that the homologous chromosomes pair so that they can separate from each other at anaphase of meiosis I when the chiasmata are dissolved. In the absence of crossing-over, homologous chromosomes are likely to misbehave and go to the same spindle pole at anaphase of meiosis I; this is one cause of *nondisjunction*. In Chapter 5, you will learn more about how crossing-over ensures proper chromosome behavior during meiosis I.

- During metaphase of meiosis I, two different pairs of homologous chromosomes can connect to the spindle in two equally likely ways, explaining Mendel's second law of independent assortment.

- (*Not shown*). Metaphase of meiosis II is similar to metaphase of mitosis, but the cells have only half the number of chromosomes. For example, cells in metaphase of meiosis II would have only one *blue* chromosome and one *red* chromosome.

Metaphase of Mitosis

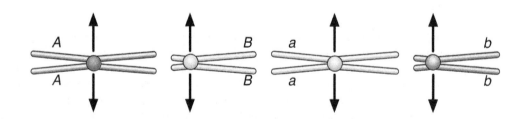

Metaphase of Meiosis I

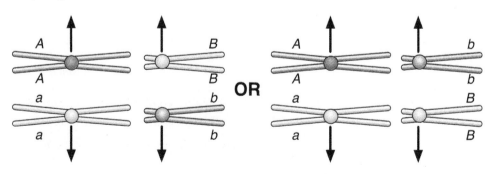

OR

4-4

chapter 4

Problem Solving

For problems involving mitosis and meiosis, you should make sure you can draw from memory the two figures on the previous page and understand the forces that pull sister chromatids (mitosis and meiosis II) or homologous chromosomes (meiosis I) to opposite spindle poles and the structures (centromeres in mitosis and meiosis II; the chiasmata resulting from crossover during meiosis I) that keep the sister chromatids or homologous chromosomes together at the metaphase plate until the centromeres or chiasmata are dissolved. Remember that in this book, a chromosome consists of all chromatids connected through a common centromere. Thus, a single chromosome during mitotic metaphase has two sister chromatids, but then these chromatids become two chromosomes during anaphase of mitosis.

Remember that you should start each problem involving the outcomes of crosses by diagramming the crosses, determining the number of genes involved and the phenotypes associated with each of the alleles. Based on what you learned in this chapter, you can see that you need to consider the possibility that a gene controlling a phenotype is X-linked. You can determine whether a gene involved in the cross is X-linked by looking for a clear phenotypic difference between the sexes in one generation's progeny of a cross. This is NOT a difference in the absolute numbers of males and females of a certain phenotype, but instead a trait that is present in one sex and totally absent in the other sex. This difference between the sexes will be seen in <u>either</u> the F_1 generation <u>or</u> the F_2 generation, but **not** in both generations in the same cross series. It is not possible to make a definitive conclusion about X linkage based on just one generation of a cross - you **must** see the data from **both** the F_1 and F_2 progeny. If the sex difference is seen in the F_1 generation, then the female parent had the X-linked trait. If the sex difference is seen in the F_2 generation, then the male parent had the X-linked trait. X-linked genes usually show a 1:1 monohybrid ratio.

Once you have generated a model to explain the data, use your model to re-diagram the cross by assigning genotypes to the parental (P) generation. Then follow the cross through, figuring out the expected genotypes and phenotypes in the F_1 and F_2 generations. Remember that your genotype-phenotype assignments must always agree with those initially assigned to the parents. Next, compare your predicted results to the observed data you were given. If the two sets of information match, then your initial genotypes were correct! In many cases there may seem to be two possible sets of genotypes for the parents. If your predicted results do not match the data given, try the other possible set of genotypes for the parents.

Vocabulary

1.

    a. meiosis   13. one diploid cell gives rise to four haploid cells

    b. gametes   7. haploid germ cells that unit at fertilization

    c. karyotype   11. the array of chromosomes in a given cell

chapter 4

d.	mitosis	10.	one diploid cell gives rise to two diploid cells
e.	interphase	12.	the part of the cell cycle during which the chromosomes are not visible
f.	syncytium	8.	an animal cell containing more than one nucleus
g.	synapsis	9.	pairing of homologous chromosomes
h.	sex chromosomes	1.	X and Y
i.	cytokinesis	6.	division of the cytoplasm
j.	anaphase	15.	the time during mitosis when sister chromatids separate
k.	chromatid	3.	one of the two identical halves of a replicated chromosome
l.	autosomes	2.	chromosomes that do not differ between the sexes
m.	centromere	16.	connection between sister chromatids
n.	centrosomes	4.	microtubule organizing centers at the spindle poles
o.	polar body	14.	cell produced by meiosis that does not become a gamete
p.	spermatocytes	5.	cells in the testes that undergo meiosis

Section 4.1

2. A diploid number of 46 means there are 23 homologous pairs of chromosomes.

   a. A child receives **23 chromosomes from the father**.

   b. Each somatic cell has **44 autosomes** (22 pairs) and **2 sex chromosomes** (1 pair).

   c. **A human ovum (female gamete) contains 23 chromosomes** - one of each homologous pair (22 autosomes and one X chromosome).

   d. **One sex chromosome (an X chromosome) is present in a human ovum.**

Section 4.2

3. a. **7 centromeres** are shown.

   b. There are **7 chromosomes** in the diagram. Note that by definition the number chromosomes is equal to the number of centromeres.

   c. There are **14 chromatids**.

   d. There are **3 pairs** of homologous chromosomes and one chromosome without a partner.

   e. There are **4 metacentric** chromosomes and **3 acrocentric** chromosomes.

chapter 4

    f. This is an organism in which males are XO, so it is most likely that **females would be XX. The karyotype of females would be the same as the one shown, except that there would be 2 copies of the left-most chromosome instead of one copy.**

4. Sex-reversed XX individuals are males because they have a functional *SRY* gene whose presence activates testes development in embryos. The embryonic testes secrete hormones that trigger the development of other male sex organs. However, **sex-reversed XX males lack the male fertility genes found in the MSY region of the Y chromosome, so they are infertile** (compare Figs. 4.7 and 4.8 on p. 91).

5. a. If the purpose of the SRY protein is to activate the *Sox9* gene, then male development takes place when *Sox9* is active, but female development is the default state that occurs when *Sox9* is inactive. Thus, **an XY individual who is homozygous for a nonfunctional mutant allele of *Sox9* would develop as a female.** Even though the SRY protein is made, it will not be able to activate copies of *Sox9* that are inherently nonfunctional.

    b. In the experiment described in the Fast Forward Box on p. 93, researchers added an *SRY* gene to the genomes of XX mice, and these mice developed as males. Suppose now that instead of *SRY*, you added a *Sox9* transgene to an XX mouse. **This XX mouse with a *Sox9* transgene would develop as a female, not a male.** The reason is that the *Sox9* transgene could not be activated in the absence of the SRY protein. (In fact, XX mice already have two copies of *Sox9* even without the transgene because *Sox9* is autosomal.)

Section 4.3

6. Mitosis maintains the chromosome number. Thus, **mitosis produces 2 daughter cells each with 14 chromosomes ($2n$, diploid).**

7. a. iii. In anaphase, the sister chromatids separate and move to opposite spindle poles. Note in the figure that each chromatid assumes a V-shaped configuration because the chromatid is being pulled towards the pole along the microtubules that connect the pole to the kinetochore in the centromeric region. As a result, the centromeric region "leads" the chromatid towards the pole, and the telomeres at the ends of the arms lag behind as the chromatid moves towards the poles.

    b. i. In prophase, the chromosomes have already replicated so that they consist of two sister chromatids. These chromosomes have not yet attached to the spindle because the nuclear envelope is still intact.

    c. iv. In metaphase, the chromosomes (consisting at this stage of two sister chromatids connected to each other at the centromere) are found at the metaphase plate at the center of the cell. There is no nuclear envelope because it broke down previously, allowing the kinetochores of the two sister chromatids to connect to spindle fibers emanating from opposite spindle poles. (These spindle fiber attachments are what allow the chromosomes to congress to the metaphase plate.)

d. ii. During G_2 of the cell cycle, the chromosomes, which have not yet condensed, are still contained in an intact nucleus.

e. v. During telophase, the sister chromatids, which already moved to opposite spindle poles during anaphase, are incorporated into the nuclei of the two daughter cells. Cytokinesis is the process that separates the cytoplasms of these two daughter cells.

8. a. G_1, S, G_2 and M (see Fig. 4.9 on p. 94).

    b. G_1, S and G_2 are all part of interphase.

    c. G_1 is the time of major cell growth that precedes DNA synthesis and chromosome replication. Chromosome replication occurs during S phase. G_2 is another phase of cell growth after chromosome replication during which the cell synthesizes many proteins needed for mitosis.

9.

	G_1	S	G_2	Prophase	Metaphase	Anaphase	Telophase
a. # of chromatids/ chromosome	1	1 → 2	2	2	2	1	1
b. nucleolus?	yes	yes	yes	yes → no	no	no	no → yes
c. spindle?	no	no	no	no → yes	yes	yes	yes → no
d. nuclear membrane?	yes	yes	yes	yes → no	no	no	no → yes

10. **No**, mitosis can (and does) occur in haploid cells. This is because each chromosome aligns independently at the metaphase plate; there is no requirement for homologous chromosome pairing.

Section 4.4

11. **Meiosis produces 4 cells (*n*, haploid), each with 7 chromosomes** (one half the number of chromosomes of the starting cell).

12. In order to answer this question count the number of independent centromeres. If two sister chromatids are connected by one centromere this counts as one chromosome. **Meiosis I is the only division that reduces the chromosome number by half, so it is a reductional division.** A cell undergoing meiosis II or mitosis has the same number of centromeres (and thus chromosomes) as the each daughter cell. Therefore, the **meiosis II and mitotic divisions are both equational.**

13. a. **mitosis, meiosis I, meiosis II**

    b. **mitosis, meiosis I**

    c. **mitosis**

chapter 4

d. Mitosis is obviously excluded, but the rest of the answer depends on whether your definition of ploidy counts chromosomes or chromatids. Meiosis I in a diploid organism produces daughter cells with n chromosomes but $2n$ chromatids; meiosis II produces haploid daughter cells with n chromosomes or n chromatids. Thus there are **two possible answers: meiosis II (if you count chromatids or chromosomes) and meiosis I (if you count chromosomes but not if you count chromatids).** To avoid potential confusion, geneticists usually use the terms "n", "$2n$", and "ploidy" only to describe cells with unreplicated chromosomes, so they don't often use these terms when talking about the cells produced at the end of meiosis I.

e. **meiosis I**

f. **none**

g. **meiosis I**

h. **meiosis II, mitosis**

i. **mitosis, meiosis I**

14. Remember that the problem states that all cells are from the same organism. This fact is needed to allow the designation of mitosis, meiosis I and II. The stage of the cell cycle can be inferred from the morphology of the spindle, the presence or absence of the nuclear membrane, and whether homologous chromosomes (or sister chromatids) are paired (or connected through a centromere) or separated. **The n number is 3 chromosomes** (this is the number of chromosomes in the haploid gametes).

    a. **anaphase of meiosis I**

    b. **metaphase of mitosis** (<u>not</u> meiosis II because there are 6 chromosomes or $2n$ in this cell)

    c. **telophase of meiosis II**

    d. **anaphase of mitosis**

    e. **metaphase of meiosis II**

15. a. The cell is in **metaphase/early anaphase of meiosis I in a male**, assuming the heterogametic sex (that is, the sex with two different sex chromosomes) in *Tenebrio* is male. Note the *heteromorphic* chromosome pair (with non-identical chromosomes) in the center of the cell.

    b. It is not possible to distinguish **centromeres, telomeres or sister chromatids**, among other structures.

    c. $n = 5$

16. In the following figures, chromosomes are drawn in the same style as in <u>Fig. 4.3</u> on p. 88: Nonhomologous chromosomes are in different colors (here, *blue* and *red*); homologous chromosomes are represented in different shades (*dark* or *light*) of the same color, while sister chromatids are in the same shades of the same color. In this scheme, the relative degree of similarity in color and shade illustrates the degree of nucleotide sequence similarity. Nonhomologous chromosomes are unrelated and do not have the

same genes. Homologous chromosomes have DNA sequences that are about 99.9% identical (they have the same genes, but different alleles of those genes). Sister chromatids have identical DNA sequences, excepting very rare new mutations.

To further your understanding of chromosome behaviors during the various types of cell division, the figures also have arrows showing the direction in which the chromatids or chromosomes will move during the subsequent anaphase. (This was not requested in the instructions to the problem.)

a. Metaphase of mitosis:

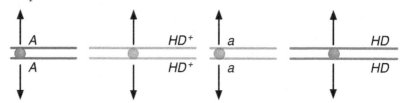

b. Metaphase of meiosis I: (Note the pairing of homologous chromosomes and the two possible alignments of the 2 non-homologous chromosome pairs.)

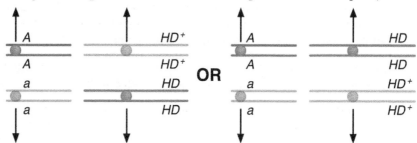

c. Metaphase of meiosis II: Shown below is only one of the two products of meiosis I from the cell diagrammed at the left in part (b); the other product of this meiosis I would have chromosomes bearing *a* and *HD*. The other alignment in meiosis I (shown at the right in part [b]) will give two *A HD*$^+$ and two *a HD*$^+$ daughter cells.

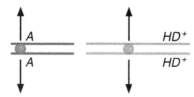

17. Using the assumptions given, each person can produce 2^{23} genetically different gametes. Thus, the couple could potentially produce $2^{23} \times 2^{23} = 2^{46}$ or **70,368,744,177,664 different zygotic combinations**. That is 70 trillion, 368 billion, 744 million, 177 thousand, 664 genetically different children.

It is **very realistic to assume that homologous chromosomes carry different alleles of some genes**. As we will see later in the book there are about 3 million differences between the DNA sequences in the two haploid human genomes in any one human being, or on average about 130,000 differences between any two homologous chromosomes. In contrast, **crossing-over almost always occurs between homologous chromosomes in any meiosis**; you will remember that crossing-over is needed to allow the homologous chromosomes to segregate properly during meiosis I. Thus **the second assumption is much less realistic**.

The fact that crossing-over occurs between homologous chromosomes means that even though the 2^{46} number just calculated is enormous, it is nonetheless a gross underestimate of the number of different zygotes two people could produce. As you will see in the next chapter, crossing-over leads to *genetic recombination,* meaning that a chromosome in a gamete is a pastiche of pieces from the chromosomes the parent obtained from his or her parents. Different gametes from any one parent have different mixtures of genomic segments from that person's mother and father. So, when someone says they have never known anyone genetically like you, she is exactly right (unless you are an identical twin or triplet)!

18. The diploid sporophyte contains 7 homologous pairs of chromosomes. One chromosome in each pair came from the male gamete and the other from the female gamete. When the diploid cell undergoes meiosis one homolog of each pair ends up in the gamete. For each homologous pair, the probability that the gamete contains the homolog inherited from the father = 1/2. **The probability that a gamete contains only the homologs inherited from the father is $(1/2)^7 = 0.78\%$.**

19. Absolutely. **Meiosis requires the pairing of homologous chromosomes during meiosis I, so meiosis cannot occur in a haploid organism.**

20. a. **The diagram at left is correct. In this diagram the connection between sister chromatids is maintained in the region between the chiasma and the telomeres**; in the diagram on the right this connection is incorrectly shown as it occurs between the chromatids of homologous chromosomes. If you trace out the chromatids you will notice that the sister chromatid cohesion in the region between the chiasma and the telomeres provides a physical connection that joins homologous chromosomes, allowing them to stay together at the metaphase plate during meiosis I. This shows why recombination is so important for proper meiotic chromosome behavior. Problem 5.17 in Chapter 5 presents this same idea about the physical connection between homologous chromosomes during meiosis I in a different format.

   b. **The cohesin complexes at the centromere must be different than those along the chromosome arms, at least during meiosis I. At anaphase I the cohesin complexes along the arms must be destroyed in order to allow the homologous chromosomes to separate. However, the cohesin at the centromeres must remain because sister chromatids stay together until anaphase of meiosis II.** Recent research indicates that this difference between the cohesin complexes occurs because centromeric cohesin is specifically protected from destruction at anaphase I by a protein called *shugoshin* (Japanese for *guardian spirit*). We will discuss cohesins and shugoshins in more detail in Chapter 11.

21. Most likely, **some kind of crossing-over event occurred between a normal X chromosome and a normal Y chromosome,** as shown by the *red* "X" at the left of the figure below. This crossing-over event would then move the *SRY* gene from the Y onto the X chromosome, making an *SRY*⁺ X chromosome and an *SRY*⁻ Y chromosome. A person with a normal X chromosome and an *SRY*⁺ X chromosome would be a sex-reversed male. A person with a normal X chromosome and an *SRY*⁻ Y chromosome would be a sex-reversed female.

    This crossing-over event must have occurred in a male, and it must have occurred between the centromere of the Y chromosome and the *SRY* gene in order to move the gene from the Y to the X. [You should note that this crossing-over event must be "illegitimate" (meaning that it is a rare mistake in meiosis) because the X and Y chromosome are not homologous (do not share the same DNA sequences) in the region where the crossing-over event occurred]. In contrast, crossing-over does occur at least once in each meiosis between either the PAR1 regions on the X and Y and/or the PAR2 regions on the X and Y; it is crossing-over in the PAR regions that insures the X and Y chromosomes pair and then disjoin properly during meiosis I in normal males. See Problem 5-16 in Chapter 5 for a view of a crossover in the PAR region.

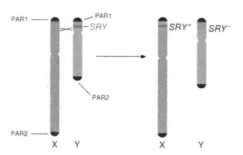

Section 4.5

22. Remember that this textbook uses the convention that the number of chromosomes is equal to the number of separate centromeres. Thus, after DNA synthesis each chromosome has replicated and so has 2 chromatids held together at the replicated by attached centromere. This structure is one chromosome with two chromatids. The centromeres separate at anaphase in mitosis or at anaphase II in meiosis; at this point, the two chromatids now become two chromosomes. For an overview of egg formation in humans see Fig. 4.18 on p. 107; for an overview of sperm formation in humans see Fig. 4.19 on p. 108.

    a. **96 chromosomes** with 1 chromatid each = **96 chromatids**;

    b. **48 chromosomes** with 2 chromatids each = **96 chromatids**;

    c. **48 chromosomes** with 1 chromatid each = **48 chromatids**;

    d. **48 chromosomes** with 1 chromatid each (unreplicated in G_1) = **48 chromatids**;

    e. **48 chromosomes** with 2 chromatids each (replicated in G_2) = **96 chromatids**;

chapter 4

    f. **48 chromosomes** (unreplicated in G_1 preceding meiosis) = **48 chromatids**;

    g. **48 chromosomes** with 2 chromatids each = **96 chromatids**;

    h. **48 chromosomes** with 1 chromatid each (unreplicated before S) = **48 chromatids**;

    i. **24 chromosomes** with 1 chromatid each = **24 chromatids**;

    j. **48 chromosomes** with 2 chromatids each = **96 chromatids**;

    k. **24 chromosomes** with 2 chromatids each = **48 chromatids**;

    l. **24 chromosomes** with 1 chromatid each = **24 chromatids**;

    m. **24 chromosomes** with 1 chromatid each = **24 chromatids**.

23. a. **400 sperm** are produced from 100 primary spermatocytes;

    b. **200 sperm** are produced from 100 secondary spermatocytes;

    c. **100 sperm** are produced from 100 spermatids;

    d. **100 ova** are formed from 100 primary oocytes. Remember that although each primary oocyte will produce three or four meiotic products (depending on whether the first polar body undergoes meiosis II), only one will become an egg (ovum);

    e. **100 ova** develop from 100 secondary oocytes – the other 100 haploid products are polar bodies;

    f. **No ova** are produced from polar bodies.

24. The primary oocyte is arrested in prophase I, so it contains a duplicated set of the diploid number of chromosomes (46 chromosomes and 92 chromatids). During meiosis I, the homologous chromosomes segregate into two separate cells, so the chromosome carrying the *A* alleles will segregate into one cell while the chromosome carrying the *a* alleles will segregate into the other cell. (This outcome assumes that no crossing-over has occurred between the *A* gene and the centromere of the chromosome that carries this gene.) One of these cells becomes the secondary oocyte (containing 23 chromosomes each with 2 chromatids and more of the cytoplasm) and the other becomes the polar body. **The genotype of the dermoid cyst that develops from a secondary oocyte could be either *AA* or *aa*.** Note that the attached sister chromatids in the secondary oocyte must separate from each other before the first mitosis leading to cyst formation.

25. Remember that in turkeys, ZW is ♀, ZZ is ♂ and WW is lethal.

    a. The ZW eggs would give rise to **only ZW females.**

    b. Cells resulting from meiosis will be 1/2 Z : 1/2 W. Upon chromosomal duplication they would become ZZ or WW. ZZ cells develop into males and WW is lethal, so **only males** are produced by this mechanism.

    c. After eggs have gone through meiosis I, they will contain either a replicated Z or a replicated W chromosome. If the sister chromatids separate to become the chromosomes you will have 1/2 ZZ : 1/2 WW. Again **only ZZ males** are produced since WW cells are inviable.

4-13

d. Meiosis of a ZW cell produces 4 haploid products, 2 Z : 2 W, one of which is the egg and the other 3 are the polar bodies. If the egg is Z then it has 1/3 chance of fusing with a Z polar body and 2/3 chance of fusing with a W polar body = 1/3 ZZ (♂) : 2/3 ZW (♀). If the egg is W then the fusion products will be 1/3 WW (lethal): 2/3 ZW (♀). Because these 2 types of eggs are mutually exclusive, you add these 2 probabilities: 1/2 (probability of Z egg) x (1/3 ZZ : 2/3 ZW) + 1/2 (probability of W egg) x (1/3 WW : 2/3 WZ) = 1/6 ZZ : 2/6 ZW + 1/6 WW : 2/6 WZ = 1/6 ZZ (♂) : 4/6 ZW (♀) : 1/6 WW (lethal) = **1/5 ZZ (♂) : 4/5 ZW (♀)**.

Section 4.6

26. In birds, males are ZZ, females are ZW. Diagram the crosses between true-breeding birds:

yellow ♀ × brown ♂ → brown ♀ and ♂; brown ♀ × yellow ♂ → yellow ♀ and brown ♂
These two crosses are reciprocal crosses, and the progeny show a sex-associated difference in phenotypes – both sexes are the same phenotype (brown) in the first cross but different phenotypes in the second cross (brown ♀ and yellow ♂). This finding indicates the **trait is sex-linked**; a single sex-linked gene is sufficient to explain the results. Also, the second cross shows crisscross inheritance - brown females × yellow males → brown sons and yellow daughters. Criss-cross inheritance is also characteristic of a sex-linked trait. Because the parents are true-breeding, the first cross shows that brown > yellow. **Therefore, the alleles of the gene are Z^B (brown allele on Z chromosome) and Z^b (yellow allele on Z chromosome); the W chromosome does not carry the feather color gene.** The first cross was $Z^bW \times Z^BZ^B \rightarrow Z^BW$ (brown females) and Z^BZ^b (brown males). The second cross was $Z^BW \times Z^bZ^b \rightarrow Z^BZ^b$ (brown males) and Z^bW (yellow females).

27. a. Ivory eyes is the recessive phenotype and brown is the dominant phenotype; females have 2 alleles of every gene and males have only one. Diagram the cross: ivory ♀ (*bb*) × brown ♂ (*B*) → fertilized eggs are *Bb* ♀ **(brown)**; unfertilized eggs are *b* ♂ **(ivory)**.

b. The cross is *Bb* F$_1$ ♀ × *B* ♂ → fertilized eggs are **1/2 *Bb* ♀ (brown) : 1/2 *BB* ♀ (brown) = all brown ♀ progeny**; unfertilized eggs are **1/2 *B* ♂ (brown) : 1/2 *b* ♂ (ivory)**.

28. Brown eye color (*bw*) and scarlet (*st*) are autosomal recessive mutations while vermilion (*v*) is an X-linked recessive trait. The genes interact such that both *bw v* double mutants and *bw st* double mutants are white-eyed. When diagramming a cross involving more than one gene you <u>must</u> start with a genotype for each parent that includes information on <u>both genes</u>. Then figure out the genotype of the F$_1$ progeny. In order to predict the F$_2$ results, find the expected ratio for each gene separately, and then cross-multiply the ratios to generate the F$_2$. Diagram the following crosses. [*Note:* In this answer as well as those to several problems that follow, X-linked genes will not be written as a superscript of the X symbol. Instead only the gene symbol will be used, as is done for

chapter 4

the autosomal genes. You can determine that the gene in question is X-linked because it will be paired with the Y chromosome in the male genotypes.]

a. vermilion ♀ ($v\ v\ bw^+bw^+$) × brown ♂ ($v^+Y\ bw\ bw$) → **F$_1$ $v\ v^+\ bw^+bw$ (wild type females) × $vY\ bw^+bw$ (vermilion males)** → F$_2$ ratio for brown alone = 3/4 bw^+– : 1/4 $bw\ bw$; the ratio for vermilion alone in both F$_2$ females and males = 1/2 v^+ : 1/2 v (the other sex chromosome will be either v or Y); **the ratio for both genes in both F$_2$ females and males = 3/8 $v^+\ bw^+$– (wild type) : 3/8 $v\ bw^+$– (vermilion) : 1/8 $v^+\ bw\ bw$ (brown) : 1/8 $v\ bw\ bw$ (white).**

b. brown ♀ ($v^+v^+\ bw\ bw$) × vermilion ♂ ($vY\ bw^+bw^+$) → **F$_1$ $vv^+\ bw^+bw$ (wild-type females) × $v^+Y\ bw^+bw$ (wild type males)** → F$_2$ ratio for brown alone = 3/4 bw^+– : 1/4 $bw\ bw$; ratio for vermilion alone in the F$_2$ females = 1 v^+– and in the males = 1/2 v^+ : 1/2 v. The ratio for both genes in the **F$_2$ females is 3/4 v^+– bw^+– (wild type) : 1/4 v^+– $bw\ bw$ (brown)** and the dihybrid ratio in the **F$_2$ males is 3/8 $v^+Y\ bw^+$– (wild type) : 3/8 $vY\ bw^+$– (vermilion) : 1/8 $v^+Y\ bw\ bw$ (brown) : 1/8 $vY\ bw\ bw$ (white).**

c. scarlet ♀ ($bw^+bw^+\ st\ st$) × brown ♂ ($bw\ bw\ st^+st^+$) → **F$_1$ $bw^+bw\ st^+st$ (wild type, males and females are the same because both genes are autosomal)** → F$_2$ monohybrid ratio for scarlet = 3/4 st^+– : 1/4 $st\ st$ and for brown = 3/4 bw^+– : 1/4 $bw\ bw$. **The F$_2$ dihybrid ratio (which hold for both sexes) = 9/16 st^+– bw^+– (wild type) : 3/16 st^+– $bw\ bw$ (brown) : 3/16 $st\ st\ bw^+$– (scarlet) : 1/6 $st\ st\ bw\ bw$ (white).**

d. brown ♀ ($bw\ bw\ st^+st^+$) × scarlet ♂ ($bw^+bw^+\ st\ st$) → **F$_1$ $bw^+bw\ st^+st$ (wild type)** → **F$_2$ as in part (c) above.**

29. In birds, females are the heterogametic sex (ZW); Z^B represents the Z chromosome with the barred allele, Z^b is non-barred.

    a. Z^BW (barred hen) × Z^bZ^b (non-barred rooster) → **Z^bW (non-barred females) and Z^BZ^b (barred males).**

    b. Z^BZ^b F$_1$ × Z^bW → **Z^BW (barred) and Z^bW (non-barred) females and Z^BZ^b and Z^bZ^b (barred and non-barred) males.**

30. The answer to this question depends upon when the nondisjunction occurs. Nondisjunction in meiosis I in the male results in one secondary spermatocyte that contains both sex chromosomes (the X and the Y) and the other secondary spermatocyte that lacks sex chromosomes. If the red-eyed males are $X^{w+}Y$, then the sperm that are formed after nondisjunction in meiosis I are $X^{w+}Y$ and nullo (O) sex chromosome. The X^wX^w female makes X^w eggs, so fertilization will produce **X^wO (white-eyed, sterile male) and $X^{w+}X^wY$ (red-eyed female)** progeny. If nondisjunction occurred in meiosis II in the male, the sperm would be $X^{w+}X^{w+}$ or

YY and nullo (O) sex chromosome. After fertilization of the X^w eggs, the zygotes would be $X^{w+}X^{w+}X^w$ (lethal) or **$X^w YY$ (fertile white-eyed males) and $X^w O$ (sterile white-eyed males)**. Notice that the normal progeny of this cross will be $X^{w+}X^w$ (red-eyed females) and $X^w Y$ (white-eyed males), which are indistinguishable from the nondisjunction progeny in terms of their eye colors.

31. It is most likely that **the bag-winged females have one mutation on the X chromosome that has a dominant effect on wing structure and that also causes lethality in homozygous females or hemizygous males**. This is analogous to the A^y allele in mice discussed in Chapter 3, except here the bag gene is on the X chromosome. Thus, all of the bag winged flies are heterozygous for this mutation.

    Diagram the cross: $Bg\ Bg^+$ (bag-winged female) × $Bg^+\ Y$ (normal male) → 1/4 $Bg\ Bg^+$ (bag-winged females) : 1/4 $Bg^+\ Bg^+$ (normal females) : 1/4 $Bg\ Y$ (dead) : 1/4 $Bg^+\ Y$ (normal males). The ratio of the surviving progeny is 1/3 bag-winged females : 1/3 normal females : 1/3 normal males.

32. a. Eosin is an X-linked recessive mutation. The cream colored variants were found occasionally in the eosin true-breeding stock. This suggests that the cream flies have the eosin genotype and <u>also</u> have a second mutation that further lightens the eosin phenotype. Diagram the cross:

    wild type ♀ × cream ♂ → F₁ wild type → F₂ 104 wild type ♀ : 52 wild type ♂ : 44 eosin ♂ : 14 cream ♂ (8:4:3:1 ratio)

    How many genes are controlling the phenotype? There are 3 phenotypes seen in the F₂ males. This number of phenotypes could be due to one gene with codominance/incomplete dominance, but the phenotypes are not present in a 1:2:1 ratio. Instead the F₂ phenotypic ratio seems to be an epistatic modification of 9:3:3:1. Therefore the eye colors must be controlled by 2 genes whose mutant alleles are recessive. The fact that the cream eye color initially arose in the eosin mutant stock suggests that cream is a modifier of eosin. This sort of modifier only alters the mutant allele of the gene it is modifying and has no effect on the wild-type allele. The F₁ shows you that the wild-type phenotype is dominant. As expected, the difference in the phenotypes between the F₂ males and females shows that eosin (*e*) is X-linked, but the cream modifier (*cr*) is probably not on the X chromosome (that is, it is autosomal) because some F₂ animals are eosin but not cream. These considerations suggest that the crosses observed were as follows:

    $e^+e^+\ cr^+cr^+$ ♀ × $eY\ crcr$ ♂ → F₁ $e^+e\ cr^+cr$ ♀ × $e^+Y\ cr^+cr$ ♂ → F₂ ratios for each gene alone = 1/2 e^+– ♀ : 1/4 e^+Y : 1/4 eY and 3/4 cr^+– : 1/4 $crcr$. When these are cross multiplied, the **F₂ ratio for both genes = 3/8 e^+– cr^+– (wild type)** ♀ : 1/8 e^+– $crcr$ (wild type) ♀ : 3/16 $e^+Y\ cr^+$– (wild type) ♂ : 1/16 $e^+Y\ crcr$ (wild type) ♂ : 3/16 $eY\ cr^+$– (eosin) ♂ : 1/16 $eY\ crcr$ (cream) ♂ = 8 wild type ♀ : 4 wild type ♂ : 3 eosin ♂ : 1 cream ♂.

    b. $eY\ cr^+cr^+$ (eosin ♂) × $ee\ crcr$ (cream ♀) → F₁ $ee\ cr^+cr$ ♀ × $eY\ cr^+cr$ ♂ → F₂ ratios for each gene alone = 1/2 ee ♀ : 1/2 eY ♂ and 3/4 cr^+– : 1/4 $crcr$. The F₂

chapter 4

ratios for both genes = 3/8 *ee cr⁺–* (eosin ♀) : 3/8 *eY cr⁺–* (eosin ♂) : 1/8 *ee crcr* (cream ♀) : 1/8 *eY crcr* (cream ♂).

c. *ee cr⁺cr⁺* (eosin ♀) × *eY crcr* (cream ♂) → **F₁ *ee cr⁺cr* ♀ × *eY cr⁺cr* ♂** →
F₂ ratios for each gene alone = 1/2 *ee* ♀ : 1/2 *eY* ♂ and 3/4 *cr⁺–* : 1/4 *crcr*. **F₂ ratios for both genes = 3/8 *ee cr⁺–* (eosin ♀) : 3/8 *eY cr⁺–* (eosin ♂) : 1/8 *ee crcr* (cream ♀) : 1/8 *eY crcr* (cream ♂)**. This is the same result as in part (b).

33. wild type ♀ × yellow vestigial ♂ → F₁ wild type ♀ and ♂ → F₂ 16 yellow vestigial ♂ : 48 yellow ♂ : 15 vestigial ♂ : 49 wild type ♂ : 31 vestigial ♀ : 97 wild type ♀

    There are 4 different phenotypes in the F₂, so 2 genes involved. One gene determines body color, and the F₁ shows that wild type is dominant to yellow. The other gene determines wing length, and the F₁ shows that wild type is dominant to vestigial. This conclusion is reinforced by the F₂ progeny where there are 48 + 49 + 97 normal wing individuals : 16 + 15 + 31 vestigial winged flies = 194 normal : 61 vestigial = 3:1. In the F₂, there are wild-type and vestigial males and females in a 3:1 ratio, so wing shape is an autosomal trait; in the F₂ there are yellow males but no yellow females, so body color is an X-linked trait. Thus, **vestigial is the recessive allele of an autosomal gene and yellow is the recessive allele of an X-linked gene**. The original cross was thus $X^{y+}X^{y+} vg^+vg^+$ ♀ × $X^yY\ vgvg$ ♂.

34. a. The white eye allele is an X-linked, recessive mutation. Diagram the cross:

    white ♂ × purple ♀ → F₁ wild type eye color → F₂ 3/8 wild type ♀ : 1/8 purple ♀ : 3/16 wild type ♂ : 1/4 white ♂ : 1/16 purple ♂.

    How many genes are involved? The white and purple eye colors cannot be caused by two different alleles of the same gene because then the F₁ males would be purple due to criss-cross inheritance. Because white and purple are mutant eye colors, the parental cross can also be considered to be a complementation test; the wild-type phenotype of the F₁ heterozygotes means there are 2 different genes controlling eye color (review Fig. 3.12 on p. 58). The mutant allele of one of the genes causes white eye color, the mutant allele of the other gene causes purple eye color. The wild-type alleles of both genes contribute to wild-type eye color. **In both cases the mutant allele is recessive ($w < w^+$ and $p < p^+$)** as demonstrated by the wild-type phenotype of the F₁ animals. **You also expect that white (the absence of pigment) should be epistatic to eyes with red or purple pigmentation.** Although you are already told that **white is X-linked; purple is an autosomal trait** because it shows no difference between the sexes in either the F₁ or the F₂ generation - there is a 3 $p^+–$: 1 pp ratio in both males and females in the F₂. The original cross was thus: $wY\ p^+p^+$ (white ♂) × $w^+w^+\ pp$ (purple ♀).

    b. $ww\ p^+p^+$ (white ♀) × $w^+Y\ pp$ (purple ♂) → **F₁ $w^+w\ p^+p$ (wild-type eye color ♀) × $wY\ p^+p$ (white ♂)** → F₂ ratio for white alone = 1/2 w^+w : 1/2 ww females and 1/2 w^+Y : 1/2 wY males; for the purple gene alone the ratio is 3/4 $p^+–$

: 1/4 pp. The F$_2$ ratio for both genes in females = 3/8 w^+– p^+– (wild type) : 1/8 w^+– pp (purple) : 3/8 ww p^+– (white) : 1/8 ww pp (white) = 3/8 wild type : 1/2 white : 1/8 purple. **The ratio for the males is the same**. Note that in this reciprocal cross there is a phenotypic difference between the sexes in the F$_1$ generation and no difference in the F$_2$ generation.

Section 4.7

35. Color-blindness in this family is an X-linked recessive condition, because two unaffected parents have affected children, and because two unrelated individuals would have to carry rare alleles if the trait were autosomal. Since the males are hemizygous with only one allele for this trait, their phenotype directly represents their genotype. **II-2 and III-3 are affected, so they must be $X^{cb}Y$**. Now consider the parents of II-2. **I-2 is normal and therefore $X^{CB}Y$. I-1 must be a carrier so her genotype is $X^{CB}X^{cb}$. II-1 can be either $X^{CB}X^{CB}$ or $X^{CB}X^{cb}$** (but she is more likely to be a normal homozygote if the trait is rare). **II-3 must be a carrier ($X^{CB}X^{cb}$) because she had an affected son. II-4 is $X^{CB}Y$; III-1 must be $X^{CB}X^{cb}$** (because she has normal color vision yet she must have received X^{cb} from her father); **III-2 is either $X^{CB}X^{CB}$ or $X^{CB}X^{cb}$; III-4 is an unaffected male and therefore must be $X^{CB}Y$**.

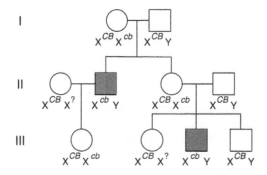

36. Pedigrees 1-4 show examples of each of the four modes of inheritance.

    a. **Pedigree 1** represents a recessive trait because two unaffected individuals have affected children. If the trait were X-linked, then I-1 would have to be affected in order to have an affected daughter. The trait is **autosomal recessive**. **Pedigree 2** represents recessive inheritance [see part (a)]. This is **X-linked recessive** inheritance as autosomal recessive inheritance was already accounted for in part (a). This conclusion is supported by the data showing that the father (I-1) is unaffected and only sons show the trait in generation II, implying that the mother must be a carrier. **Pedigree 3** shows the inheritance of a dominant trait because affected children always have an affected parent (remember that all four diseases are rare). The trait must be **autosomal dominant** because the affected father transmits it to a son. **Pedigree 4** represents an **X-linked dominant** trait as characterized by the transmission from affected father to all of his daughters but none of his sons.

b. **Pedigree 1** - both parents are carriers for this autosomal recessive trait so there is a **1/4 chance** that the child will be affected (*aa*). **Pedigree 2** – individual I-2 is a carrier for this X-linked trait. The probability is 1/2 that she will pass on X^a to her daughter II-5. The unaffected father (I-1) contributes a normal X^A chromosome, so the probability that is II-5 is a carrier = 1/2. The **probability of an affected son** = 1/2 (probability II-5 is a carrier) × 1/2 (probability II-5 contributes X^a) × 1/2 (probability of Y from father II-6) = **1/8**. The **probability of an affected daughter = 0** because II-6 must contribute a normal X^A. **Pedigree 3** - for an autosomal dominant trait there is a **1/2 chance** that the heterozygous mother (II-5) will pass on the mutant allele to a child of either sex. **Pedigree 4** - the father (I-1) passes on the mutant X chromosome to all his daughters and none of his sons. Therefore II-6 does not carry the mutation, as shown by his normal phenotype. The **probability of an affected child = 0**.

37. a. Unaffected individuals have affected children, so albinism is **recessive**.
    b. If albinism were X-linked, then I-9 would have to be an affected hemizygote in order to have an affected daughter. As this is not the case, albinism is **autosomal**.
    c. *aa*
    d. *Aa*
    e. *Aa*
    f. *Aa*
    g. *Aa*
    h. *Aa*.

38. a. Draw the pedigree:

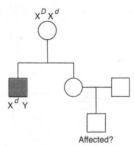

Affected?

The grandmother (I-1) of the child in question must have been heterozygous for the *d* allele causing the disease because her son (II-1) was affected. The probability that II-2 will have an affected son = 1/2 (the probability that II-1 inherited the X^d chromosome from I-1) × 1/2 (the probability that II-1's son receives the X^d chromosome from her if she is $X^D X^d$) = **1/4**.

b.

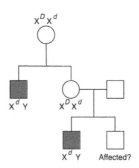

We now know that II-2 is in fact a carrier since she had an affected son. Therefore the probability is **1/2** that she will pass on the X^d chromosome to III-2.

c.

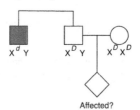

The mother of these two men was a carrier. She passed on the X^d chromosome to the affected son and she passed on the X^D chromosome to the unaffected son. There is **no chance** that the unaffected man will pass the disease allele to his children.

d.

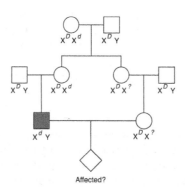

The probability of IV-1 being an affected son = 1/2 (probability that II-3 is a carrier) × 1/2 (probability that III-2 inherits X^d) × 1/2 (probability that IV-1 inherits X^d) × 1/2 (probability IV-1 inherits the Y from III-1) = **1/16**. **The probability that IV-1 is an affected girl** = 1/2 (probability that II-3 is a carrier) × 1/2 (probability that III-2 inherits X^d) × 1/2 (probability that IV-1 inherits X^d) × 1/2 (probability IV-1 inherits X^d from III-1) = **1/16**. **The chance that IV-1 is not**

affected = 1 - (1/16 probability of an affected male + 1/16 probability of an affected female) = 1 - 2/16 = **7/8**.

e.

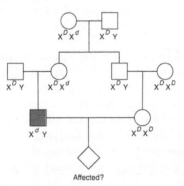

Affected?

II-2 must be a carrier in order for her son to inherit the disease. II-3 is the brother and he is normal, so his genotype must be X^DY. Therefore III-2, his daughter, must be homozygous normal. The probability that the fetus (diamond) will be an **affected boy = 0**; the probability of an **affected daughter = 0**; the probability of an **unaffected child = 100%**.

39. **Yes, it is possible that a woman heterozygous for the disease-causing allele of the *DMD* gene could have symptoms in some parts of the body and not others.** From Fig. 4.25 on p. 116, you can see that in some cells of a female, one X chromosome is inactivated, while in other cells, the other X chromosome is inactivated. The decision of which chromosome to inactivate is made at random early in development, when the embryo has only 500-1000 cells.

If the X chromosome that is inactivated carries the wild-type allele, the cell would not have the normal function of the *DMD* gene because the *DMD* allele on the active X chromosome is the nonfunctional disease-causing allele. If the chromosome that is inactivated is that carrying the *DMD* mutant allele, then the cell would have normal function of the *DMD* gene from the X chromosome with the wild-type allele. From the perspective of the patient, the severity of the symptoms will depend upon which particular cells in the body possess or lack *DMD* gene function.

40. a. In contrast with the situation described in the previous problem (Problem 39), **it is very unlikely that females heterozygous for the hemophilia disease allele would have the condition in some parts of the body but not others.** It is true that some bone marrow cells (those that inactivate the X chromosome with the nonfunctional alleles of the gene encoding clotting factor VIII) will have normal function of the gene, while other bone marrow cells (those that inactivate the X chromosome with the wild-type allele of this gene) will not be able to produce the clotting factor. However, all bone marrow cells will "dump" the clotting factor VIII they make into the blood serum. Thus, the serum as a whole would be expected to have less (roughly half) of the clotting factor VIII than does the serum in a woman homozygous for the wild-type allele. The phenotype is determined by the level of this clotting factor in the serum, not the amount that is present in particular cells.

b. Because the serum in a woman heterozygous for a hemophilia-causing mutant allele has a lower concentration of clotting factor VIII than does the serum in a normal homozygote, **you would expect the blood in the heterozygote to coagulate at a lower rate** (assuming that the rate of coagulation is directly related to the amount of this protein).

    It is conceivable that the rate of clotting may vary considerably among heterozygous women, depending upon what proportion of the bone marrow cells can make clotting factor VIII and what proportion cannot. These proportions are determined by X chromosome-inactivating events early in embryogenesis that randomly inactivate either the maternal or paternal X chromosome; once made, the "decision" is inherited by the descendants of the embryonic cell. Because of the random nature of the decision early in development, it is impossible to predict which bone marrow cells in the adult will have inactivated maternal or paternal X chromosome. On average, heterozygous women would have about 50% of the normal levels of clotting factor VIII, but some heterozygous women could have less and others more.

41. a. Hypertrichosis is almost certainly an **X-linked dominant trait**.

    b. **The trait is dominant because every affected child has an affected parent**. If the trait were recessive then several of the people marrying into the family would have to be carriers, yet this trait is rare. Hypertrichosis is highly likely to be **X-linked because two affected males (II-4 and IV-3) pass the trait on to all of their daughters but none of their sons**.

    We can actually calculate the probability that the trait is autosomal dominant rather than X-linked dominant. If the trait is an autosomal dominant, then the fathers would have to be heterozygous *Hh* because they have unaffected children; the mothers of these children are unaffected so the mothers would have to be *hh*. The probability of an unaffected male child from such matings = 1/2 (probability of inheriting *h*) × 1/2 (probability the child will be male) = 1/4. The probability of an affected daughter = 1/4. The probability of having 9 unaffected sons and 5 affected daughters = $(1/4)^9 \times (1/4)^5$ = $(1/4)^{14}$ = **3.7× 10⁻⁹**. In other words, the probability that the trait is autosomal dominant is only about 4 in a billion. (The 9 unaffected sons and 4 of the 5 affected daughters are the children of II-4, while the last affected daughter is the progeny of IV-3.)

    c. **III-2 had 4 mates and III-9 had 6 mates!**

42. When nondisjunction occurs in meiosis I (MI), there are two types of affected gametes. Half of the gametes receive no sex chromosomes (nullo) while the rest of the gametes receive one copy of each of the sex chromosomes in the parent. Thus MI nondisjunction in a male will give XY sperm and nullo sperm. If the nondisjunction occurs in MII, then half of the affected gametes are nullo while the rest receive 2 copies of one of the original sex chromosomes. MII nondisjunction in a male will therefore give either XX sperm or YY sperm and nullo sperm with no sex chromosomes.

chapter 4

a. The boy received both an X^B and a Y from his father. Because both the X and Y segregated into the same gamete, the nondisjunction occurred in **meiosis I in the father**.

b. The son could have received both the X^A and X^B from the mother and the Y from the father, or alternatively the X^B from the mother and the X^A and Y from the father. You can determine that the nondisjunction occurred in **meiosis I, but you cannot determine in which parent**.

c. The son has two X^A chromosomes and a Y. Therefore the nondisjunction occurred in the **mother, but because she is homozygous X^AX^A you cannot determine during which meiotic division**.

43. a. The father contributed the i or I^A, Rh^+, M or N, and $Xg^{(a-)}$ alleles. Only **alleged father #3** fits these criteria.

b. Alleged father #1 was ruled out based on his X chromosome genotype. He is $Xg^{(a+)}$ and cannot have an XX daughter who is $Xg^{(a-)}$. If the daughter was XO (Turner), she could have inherited the one X from either mother or father. If the X came from her mother (who must have been heterozygous for Xg), then her father did not contribute an X chromosome. Therefore **alleged fathers #1 and #3** both fit the criteria for paternity.

44. Remember that white tigers are shown by <u>unshaded</u> symbols. Test each possibility by assigning genotypes based on the mode of inheritance.

a. **No**. Y linked genes are only expressed in males. Since there are white females, the trait could not be Y-linked.

b. **No**. White males would have to have white daughters but not white sons. Mohan's daughter is not white and Tony has a normally colored daughter, Kamala, and a white son, Bim.

c. **Yes**. The information in the pedigree is consistent with dominant autosomal inheritance.

d. **Yes**. The information in the pedigree is consistent with a recessive X-linked trait.

e. **Yes**. There is no data in the cross scheme to allow you to rule out recessive autosomal inheritance in this highly inbred family.

45. a. **Individual III-5** was not related to anyone in the previous generation. Therefore the cancer with which he was afflicted is not likely to be caused by the rare mutation in *BRCA2*. Instead his cancer may have been due to a mutation in another gene that can predispose to cancer or exposure to environmental influences like carcinogens.

b. There is a vertical pattern of inheritance of the cancers, so **the *BRCA2* mutation has a dominant effect on causing cancer**. Most of the affected children in this pedigree have an affected parent, although there are a couple of cases where this is not seen (example V-1 and V-8). It is very unlikely that the mutant *BRCA2* allele is recessive because then many unrelated people would have to be carriers of the rare mutant allele.

c. The *BRCA2* mutation is not Y-linked as affected males (for example III-9 and IV-6) pass the trait to their daughters. **The data do not clearly distinguish between X-linked and autosomal inheritance**. The X-linked hypothesis is supported by the fact that all of the daughters of men who could transmit the mutant allele are affected (for example the five daughters of IV-6). *BRCA2* is actually on autosome 13.

d. **The penetrance of the cancer phenotype is incomplete.** If the mutant allele is dominant then women II-4 and III-8 should have had cancer if the mutant disease-causing allele is dominant. It is critical to note that the presence of the *BRCA2* mutation does not always cause cancer. Instead the mutation increases the predisposition to cancer. The reasons for this will be explained in Chapter 17.

e. **The expressivity is variable.** Some of the women who have inherited the mutant cancer causing allele do not have breast cancer but instead have ovarian or other cancers. This suggests that the *BRCA2* mutation can predispose to cancers other than breast cancer, even though breast tumors are the most likely outcome.

f. **Ovarian cancer is sex-limited** since males don't have ovaries. Although the data are fragmentary, **the penetrance of breast cancer may be sex-influenced**. Of the five males who should be carrying the *BRCA2* mutant allele (II-5, III-9, IV-2, IV-9, IV-4), only one has breast cancer (III-9), giving a rate of 20% affected men. Of the 21 women who must be carrying the mutant *BRCA2* allele 16 have breast cancer - a rate of 80% affected women. Perhaps the hormonal environment or the amount of breast tissue helps determine whether a person with the *BRCA* mutation will develop breast cancer.

g. The absence of cancer in the first two generations **could be explained by the low penetrance of the cancer phenotype, particularly among men**. Three individuals in generations I and II must have carried the mutant allele of *BRCA2* but only one of these was a woman. Other explanations are also possible. One interesting hypothesis has to do with the fact that cancer is a disease whose frequency increases with age and life expectancies have improved only in recent times. Thus the individuals in the first two generations of this pedigree may have died early of other causes before they had the chance to develop cancer.

46. a. If the trait is an autosomal dominant, then the fathers must be heterozygous *Rr* because they have unaffected daughters. The mothers would then be homozygous normal, *rr*. The probability of an affected male child from such matings = 1/2 (probability of inheriting R) × 1/2 (probability the child will be male) = 1/4. The probability of an unaffected daughter = 1/4. The probability of having 6 affected sons and 5 unaffected daughters = $(1/4)^6 \times (1/4)^5 = $ **$(1/4)^{11} = 2 \times 10^{-7}$** or extremely unlikely! (The reasoning here is the same as that used previously in Problem 41b above.)

b. **This trait could be an example of Y-linked inheritance or it could represent sex-limited expression of the mutant allele** (that is, the allele is dominant in

chapter 4

47. a. Heterozygous *Oo* female cats have tortoiseshell coats. In some patches of cells the chromosome with the *O* allele is inactivated so the coat is black as determined by the *o* allele. In other patches of cells the chromosome with the *o* allele is inactivated so the fur is orange since the *O* allele is still functioning. **Crosses that could yield *Oo* females include *OO* × *o*Y (orange females x black males), *oo* × *O*Y (black females x orange males), *Oo* × *o*Y (tortoiseshell females × black males), and *Oo* × *O*Y (tortoiseshell females × orange males).**

    b. **Male tortoiseshell cats could be XXY Klinefelter males who are heterozygous *Oo*.** One of their X chromosomes would be inactivated in some patches of cells; the other X chromosome would be inactivated in other patches of cells, just as for the tortoiseshell females in part (a).

    c. Coat color in calico cats requires the action of another gene to produce white fur, but the genotype needed for white color cannot be completely epistatic to the *orange* gene, because the black and orange patches would not be visible if this was true. One possibility is that the *Oo* or *Oo*Y cats are also heterozygous for an X-linked white coat color gene. However if this were the case, the white coat allele would have to be on either the X chromosome with *O* or that with *o*. These cats would be either white + black or white + orange, but instead we see that calico cats have all three colors. It turns out that **an autosomal gene called the *white-spotting* or *piebald* gene causes the white spotting - a dominant allele of this gene causes white fur, but in heterozygotes this allele has variable expressivity so some patches have a color dictated by the functional alleles of the *orange* gene.**

48. a. All of the progeny will have **mutant coat color** since the wild-type allele is on the inactivated paternal X chromosome.

    b. All of the progeny will have **wild-type coat color** since the mutant allele is on the inactivated paternal X chromosome.

    c. To characterize alleles as recessive or dominant, you need to examine the phenotype of heterozygotes under conditions where both alleles are expressed. Here, **the heterozygotes are all females whose phenotype was determined by the allele received by the mother.**

    d. In tortoiseshell cats the maternal X chromosome is inactivated in some embryonic cells and their descendants, whereas the paternal X chromosome is inactivated in other embryonic cells and their descendants. **In marsupials, the paternal X chromosome is always inactivated.**

49. a. **The *SHOX* allele that causes LWD is dominant.** You can easily see this because of the vertical inheritance pattern in which the rare disease is transmitted from parent to child in every generation.

chapter 4

b. **An X-linked dominant would travel from mother to daughter and son, and from father to daughter. In this pedigree, you can see transmission from father (III-3) to both daughter (III-7) and son (III-5).** Even though the *SHOX* gene is located on the sex chromosomes, the disease shows an inheritance pattern similar to that of a condition caused by an autosomal dominant allele. This is the reason PAR1 is called a *pseudoautosomal region*.

c. The *SHOX* disease allele is located on the indicated sex chromosomes in the following affected individuals: **I-2 X, II-2 X, II-3 X, III-3 X, III-5 Y, III-7 X, IV-1 X, IV-2 X.**

d. The unusual observation is the transmission of the disease allele from father (II-3) to both his son (III-5) and his daughter (III-7). The father (II-3) got the *SHOX* disease allele on an X chromosome from his mother. He gives his X chromosome (with the *SHOX* disease allele) to his daughter, and he also gives his son a Y chromosome with the *SHOX* disease allele. Because the *SHOX* gene is in PAR1, the disease allele can be transferred from the X to the Y by crossing-over during meiosis.

e. The crossing-over event that switches the *SHOX* alleles between the X and Y chromosomes is shown on the following diagram. This figure focuses only on the short arms of the two sex chromosomes that include the *SHOX* gene in PAR1 and (on the Y) the *SRY* gene specifying male development. (The diagram does not show the centromeres nor the long arms of these chromosomes; for a more complete picture, see the answer to Problem 21 above.)

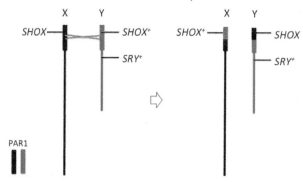

You should note an important difference between the crossing-over event depicted here and the crossover shown in the answer to Problem 21 earlier in this chapter that leads to the phenomenon of sex reversal. In this case, the crossing-over that shifts the *SHOX* gene alleles between the X and Y chromosomes occurs in PAR1. This region is homologous in the X and Y chromosomes, and thus crossing-over occurs regularly during almost every meiosis in PAR1. (Such crossing-over is necessary for the X and Y chromosomes to pair and then separate from each other during meiosis I in males.) In contrast, the crossing-over between the X and Y that causes sex reversal is a mistake (an illegitimate recombination) because sex reversal requires a crossover in the interval between the Y chromosome centromere and the

SRY gene (see the answer to Problem 21). The X and Y chromosomes do not share DNA sequences in this region, so crossing-over does not usually occur there.

chapter 5

Linkage, Recombination, and the Mapping of Genes on Chromosomes

Synopsis

In Chapter 5, we learn the consequences of the fact that hundreds or even thousands of genes can be found on one chromosome. If two genes lie close together on a chromosome, their alleles do not assort independently; instead, the genes are *linked* and parental classes of gametes will be formed more frequently than recombinant classes. The *Recombination Frequency (RF)* increases to a maximum of 50% as the distance between the two genes increases. The formation of recombinant classes depends on the occurrence of physical crossovers between nonsister chromatids that take place during prophase of the first meiotic division, after the chromosomes have replicated and homologous chromosomes have paired with each other. The greater the distance between the genes, the more likely the chance of a crossover and the greater the RF; the 50% limit to RF occurs when genes are sufficiently far apart that at least one crossover occurs between them in every meiosis. By measuring RF, geneticists can map the location of genes along the chromosomes.

In most organisms, RF can be tracked only by looking at individual progeny of a cross involving at least one parent who is a heterozygote for the two genes in question. But in the fungi *S. cerevisiae* and *N. crassa*, all the products of a meiosis remain together in a sac called an *ascus*. As a result, researchers can infer more details about the locations and kinds of crossovers that took place during the meiosis that produced that ascus.

Finally, Chapter 5 discusses what happens when recombination takes place during mitosis - rather than meiosis. In contrast with meiotic recombination, which is a programmed event that must occur at least once on every chromosome during every meiosis, mitotic recombination reflects rare mistakes that occur (usually in response to DNA damage). Mitotic recombination provides a valuable tool to researchers because it allows them to create *mosaic* organisms in which different cells have different genotypes.

Key terms

syntenic genes – genes located on the same chromosome

parental types/classes – gametes whose alleles are in the same combinations as in the gametes that gave rise to previous generation; or progeny generated by parental gametes

recombinant types/classes – gametes whose alleles are in different combinations than in the gametes that gave rise to the previous generation; or progeny generated by recombinant gametes. Although recombinant types can also form from independent assortment if the two genes are on nonhomologous chromosomes (see Chapter 4), in this chapter the term refers exclusively to gametes or progeny resulting from a meiosis in which a crossover took place between the two genes in question.

linkage – when two genes are located close enough to each other on the same chromosome so that a doubly heterozygous parent makes more parental type than recombinant type gametes

recombination frequency (RF) – the proportion of the total number of gametes or progeny that are recombinant types (recombinants/total)

map unit (mu)/centimorgan (cM) – a measure of genetic distance between linked genes; 1 mu = 1 cM = 1% RF.

locus (singular)/**loci** (plural) – a specific location on a particular chromosome

interference – a phenomenon in which a crossover event on a chromosome prevents or limits the occurrence of a second crossover event nearby on the same chromosome. Interference is thought to take place as a way to help ensure that at least one crossover occurs on each bivalent, so that homologous chromosomes segregate accurately during the first meiotic division.

coefficient of coincidence (c.o.c.) – the ratio of the number of double crossovers observed in a cross to the number of double crossovers expected if the two crossovers were independent of each other. The *interference* = 1 – c.o.c. This makes sense because the amount of interference reflects a decrease in the frequencies of double crossovers (DCOs).

chi-square test – a statistical device used to measure the likelihood that a particular set of observed experimental results could have been obtained as chance deviations from the expectations from a particular hypothesis to be tested (the **null hypothesis**)

significant result – experimental results that are highly unlikely to be explained by the null hypothesis. A significant result allows you to reject the null hypothesis. An **insignificant result** does <u>not</u> allow you to reject the null hypothesis. Note that these definitions mean that the chi-square test can never prove a null hypothesis, but can only reject the null hypothesis.

degrees of freedom (d.o.f) – In the chi-square test for the goodness of fit, the d.o.f. = number of classes – 1. The d.o.f. is required for interpreting the chi-square test.

***p* value** – the probability that the observed results could be obtained by chance deviations from the expectations of the null hypothesis. If the *p* value is low (usually below 0.05 for genetic analysis), you can reject the null hypothesis.

ascus – in fungi, a sac that contains all four products of a single meiosis. These haploid products are the **ascospores** (or simply **spores**). A **tetrad** is the collection of the 4 spores in one ascus. (In *Neurospora*, the ascus is an **octad** containing 8 spores due to the occurrence of one round of mitosis after meiosis is complete.)

ordered versus **unordered tetrad** – In ordered tetrads (*N. crassa*), the geometry of the spores in the ascus reflects the movement of chromatids during the two meiotic divisions. In unordered tetrads (*S. cerevisiae*), the four ascospores are arranged at random within the ascus.

chapter 5

parental ditype (PD) – a tetrad in which all four spores are parental types (two spores of each of the two parental types possible)

nonparental ditype (NPD) – a tetrad in which all four spores are recombinant types (two spores of each of the reciprocal recombinant types)

tetratype (T) – a tetrad with four spores all of different genotypes (the two reciprocal parental types and the two reciprocal recombinant types)

first-division segregation pattern (MI) and **second-division segregation pattern (MII)** – arrangements of spores in ordered tetrads (octads in *Neurospora*). In an MI pattern, if an imaginary line is drawn through the center of the ascus, all the spores with one allele are on one side of the line and all those with the other allele are on the other side of the line (4:4). In an MII pattern, pairs of spores with both alleles are found on either side of the imaginary line at the ascus middle (2:2:2:2).

twin spots – adjacent patches of tissues that are phenotypically and genotypically different from each other, and phenotypically and genotypically different from the surrounding cells. Twin spots can result from **mitotic crossing-over** depending on the arrangement of alleles and the sites of the mitotic crossovers.

mosaic – an organism harboring cells of different genotypes. For example, a fruit fly with twin spots is a mosaic.

sector – in yeast, a region of a colony of cells in which the cells have a different phenotype and genotype from the other cells in the same colony

Key Equations

- Recombination frequency (RF) = # recombinant progeny / total # progeny. (Multiply × 100 to express the RF as %.)
- 1% RF = 1 mu or 1 cM.
- Coefficient of coincidence (c.o.c) = # DCOs observed / # DCOs expected. To calculate the # DCOs expected, multiply the crossover frequency in interval 1 × crossover frequency in interval 2, and then multiply this value by the total number of progeny.
- Interference = 1 – c.o.c.
- $X^2 = \Sigma [(\text{\# observed} - \text{\# expected})^2 / \text{\# expected}]$.
  (In other words, X^2 is the sum over all classes of $[(O-E)^2 / E]$ calculated for each class.)
- Degrees of freedom (d.o.f.) = # classes -1.
- In X^2 analysis, the ultimate goal is to determine the probability p that the observed results could have been obtained by chance if the null hypothesis is correct. To determine p, you need to calculate X^2 and d.o.f., and then you must consult a table of critical values such as Table 5.2 on p. 147.
- See the equations in the box Three Easy Rules for Tetrad Analysis that follows.

… chapter 5

> **Three Easy Rules for Tetrad Analysis**
>
> To understand the derivation of the rules below, study Fig. 5.21 on p. 152 (two genes on different chromosomes) and Fig. 5.23 on p. 153 (two linked genes). You should note that the kind of tetrad analysis discussed in this chapter concerns the mapping of two genes at a time; no three point cross analysis is involved.
>
> Rule #1 - If the number of PD tetrads is about equal to the number of NPD tetrads, then the genes are unlinked. Genes are unlinked if they are on nonhomologous chromosomes or if the genes are far apart on the same chromosome. If the # PD tetrads >>> # NPD tetrads, the genes are linked.
>
> Rule #2 - Distance between linked genes is determined by calculating recombination frequency using the equation:
>
> RF = [(NPD + ½T)/ total # tetrads] × 100.
>
> Rule #3 – If a crossover occurs between a gene and the centromere, the two alleles separate from each other at the second meiotic division (MII segregation) instead of during the first meiotic division (MI segregation). Centromere distance can be measured in fungi with ordered asci such as *Neurospora* using the equation:
>
> gene ↔ centromere distance = [(½ × # MII tetrads) / total # tetrads] × 100.

Problem Solving

For many of the problems in this chapter, you will still need to start by answering the same questions you dealt with in the first four chapters:

How many genes are involved in the cross?
For each of these genes, what is the dominance relationship between the alleles?
For each of these genes, is it X-linked or autosomal?

In the problems in this chapter, you now have to consider the additional possibility that genes may be syntenic (on the same chromosome) and that they are genetically linked.

When thinking about genes that are linked, you must write the genotypes in a way that represents the linkage. Remember that there is one allele per homolog, so *aa* becomes *a / a*. It is usually advantageous to indicate the homologous chromosomes with horizontal, parallel lines so that you can track which alleles of which genes were present on which chromosomes. Thus, $a^+ b^+ c / a b c^+$ would be diagrammed as:

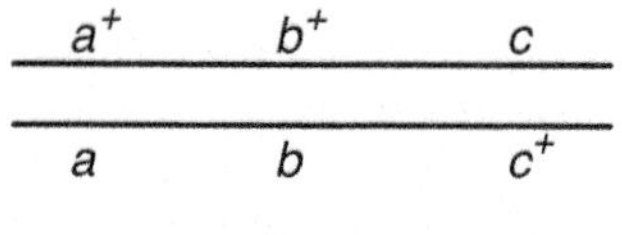

```
   a⁺      b⁺      c
 ─────────────────────
   a       b       c⁺
```

Although the two lines in the diagram above represent the two homologous chromosomes, you should always keep in mind that recombination actually occurs at the four-strand stage, after each chromosome has replicated into two sister chromatids. This fact plays an important role in understanding why there is a 50% limit of the RF measured between two genes, and it also helps you visualize what is occurring when you are analyzing tetrads in fungi.

chapter 5

Tips for two-point crosses:

- The minimum requirement for detecting recombination is that one parent must be heterozygous for 2 genes. The recombination events that can be detected are the ones that occur between the 2 genes, giving recombinant gametes instead of parental gametes.
- It is easiest to detect the parental versus recombinant gametes if you do a testcross.
- If the genes are assorting independently, in a test cross of $aa^+ \, bb^+ \times aa \, bb$, the expected phenotypic frequencies and classes of progeny are $1 \, a^+- \, b^+- : 1 \, aa \, bb : 1 \, a^+- \, bb : 1 \, aa \, b^+-$. But the genes are genetically linked if you see more parental than recombinant progeny.
- Recombination frequency (RF) = # recombinant progeny / total # progeny. (Multiply × 100 to express the RF as %.) 1% RF = 1 mu or 1 cM.
- Genes on the X chromosome can be mapped without a testcross. Just use the hemizygous male progeny as in Problem 5-6.

Tips for three-point crosses:

- In a three-point cross, a parent heterozygous for the three genes generates the progeny. Therefore, all classes (parental, etc.) will occur as reciprocal pairs of progeny. These reciprocal pairs will be both genetic reciprocals and numerically equivalent.
- Designate the different gametes or offspring as noncrossover (NCO; parental), single crossover (SCO) or double crossover (DCO). The NCO classes are those classes of progeny who have one of the intact, nonrecombinant homologs from the parent. The NCO classes will be represented by the reciprocal pair with the greatest numbers of offspring. There will be SCOs occurring between the gene on the left and the gene in the middle (two reciprocal classes), and SCOs occurring between the gene in the middle and the gene on the right (another two reciprocal classes). The DCO classes will be represented by the reciprocal classes with the smallest numbers of offspring (Fig. 5.12 on p. 138). Thus, in a three-point cross there will usually be 8 classes of progeny. However, sometimes one or both double crossover classes are missing because they are rare.
- By examining the pattern of data seen in a problem, you can often start solving the problem with a basic understanding of the linkage relationships of the genes. Some of the more common patterns of data are:
    - 3 unlinked genes give 8 classes of data that occur as 4 genetically reciprocal pairs, but all classes are seen in a 1:1:1:1:1:1:1:1 ratio;
    - 3 linked genes give 8 classes of data that occur as 4 reciprocal pairs genetically and numerically unless one or both of the DCO classes are missing, in which case you will see 6 classes as 3 reciprocal pairs or 7 classes as 3 reciprocal pairs plus an additional unpaired class;
    - 2 linked genes plus one unlinked gene will yield 8 classes of data. 4 of these classes will be numerical equal to each other, while the other 4 classes will also be numerically equal to each other. The group of 4 classes with larger numbers will consist of the reciprocal parental classes for the linked genes

together with either allele of the unlinked gene. The group of 4 classes with the smaller numbers will be the reciprocal recombinant classes of the linked genes together with either allele of the unlinked gene.

- Begin the process of mapping the genes by ordering the genes. To figure out which gene is in the middle of a group of three genes, choose one of the double crossover classes. Compare it to the most similar parental class of progeny where two of the three genes will have the same combination of alleles. The gene that differs is the gene in the middle. See Problem 5-22c for further explanation.

- The last step is to determine the distance between the genes on each end and the gene in the middle. Use the formula RF = # recombinants between the 2 genes / total # of progeny. Remember for each interval that the # recombinants is the number of progeny in the reciprocal SCO classes representing crossovers in that interval, plus the number of recombinants in the reciprocal DCO classes.

Tips for crosses involving 4 genes:

- A few problems in this chapter deal with crosses involving 4 genes. You should realize that to provide you with easily solvable problems, the arrangements of the 4 genes were carefully chosen such that the data patterns immediately suggest rather simple solutions. If you had 4 linked genes in a group, you would get 16 classes of progeny in 8 reciprocal pairs, but this would be extremely difficult to analyze because you would have to consider many types of double crossovers as well as triple crossovers, in addition to the SCOs in each interval. You should thus be on the lookout for the following patterns with simpler solutions:
    - 4 unlinked genes give 16 classes of progeny in a 1:1:1:1:1:1:1:1:1:1:1:1:1:1:1:1 ratio;
    - 3 linked genes and 1 gene assorting independently gives 16 classes of data occurring as 8 reciprocal pairs genetically and 4 groups of 4 numerically.
    - 4 linked genes that show only 8 classes of data indicate that two of these genes are so tightly linked to each other that they never separate by recombination. You can thus deal with this as a three-point cross.

Tips for tetrad analysis problems:

- Remember that the designations PD, NPD, and T refer to the set of four spores in one ascus, NOT to individual spores.

- Although tetrad analysis may seem daunting, it is actually in many ways straightforward, particularly if you follow the Three Easy Rules for Tetrad Analysis presented above.

- Ordered tetrads such as those in *Neurospora* allow you to determine the distances between genes and centromeres. (This is normally not possible in unordered tetrads, except in one special and useful case described in Problem 5.46f in which one of the genes is located right at the centromere.) To compute gene ↔ centromere distances, you need to count the number of tetrads showing MI and MII segregation patterns. To do this, draw an imaginary line through the middle of the ascus and assess for

chapter 5

each gene whether all the spores in the ascus are the same allele (MI pattern) or have different alleles (MII pattern).

Tips for mitotic recombination problems:

- For mitotic recombination in a heterozygous parental cell to produce daughter cells homozygous for either allele, mitotic recombination must take place between the gene and the centromere.

- Any mitotic recombination between a gene and the centromere will automatically make homozygous any other gene that is even further from the centromere.

Vocabulary

1.

a.	recombination	8.	formation of new genetic combinations by exchange of parts between homologs
b.	linkage	4.	when two loci recombine in less than 50% of gametes
c.	chi-square test	1.	a statistical method for testing the fit between observed and expected results
d.	chiasma	11.	structure formed at the spot where crossing-over occurs between homologs
e.	tetratype	2.	an ascus containing spores of four different genotypes
f.	locus	5.	the relative chromosomal location of a gene
g.	coefficient of coincidence	6.	the ratio of observed double crossovers to expected double crossovers
h.	interference	3.	one crossover along a chromosome makes a second nearby crossover less likely
i.	parental ditype	10.	an ascus containing only two nonrecombinant kinds of spores
j.	ascospores	12.	fungal spores contained in a sac
k.	first-division segregation	9.	when the two alleles of a gene are segregated into different cells at the first meiotic division
l.	mosaic	7.	individual composed of cells with different genotypes

Section 5.1

2. a. Diagram the cross. WT = wild type.

scabrous ♂ ($scsc\ j^+j^+$) × javelin ♀ ($sc^+sc^+\ jj$) → F$_1$ WT ♀ ($scsc^+\ j^+j$) × scabrous javelin ♂ ($scsc\ jj$) → 1/4 scabrous (P) : 1/4 javelin (P) : 1/4 WT (R) : 1/4 scabrous javelin (R).

This F$_1$ female will make four different types of gametes in equal frequency – 1/4 $sc^+\ j$: 1/4 $sc\ j^+$: 1/4 $sc^+\ j^+$: 1/4 $sc\ j$. Because this is a testcross, the male parent will always provide the recessive alleles of the genes. Thus the phenotypes of the progeny will be determined by the gamete they receive from the heterozygous F_1 female. (Recall that in *Drosophila*, meiotic recombination does not occur in the male germ line, and so testcrosses are almost always set up like the one here.)

Of course these genes may be linked. In this case the cross would be diagrammed as follows:

scabrous ♂ ($sc\ j^+ / sc\ j^+$) × javelin ♀ ($sc^+ j / sc^+ j$) → F_1 WT ♀ ($sc\ j^+ / sc^+ j$) × scabrous javelin ♂ ($sc\ j / sc\ j$) → scabrous ($sc\ j^+$) (P) : javelin ($s^+ j$) (P) : WT ($sc^+ j^+$) (R) : scabrous javelin ($sc\ j$) (R).

Again, this F_1 female will make 4 different types of gametes. **There will be the two Parental (P) types and two Recombinant (R) types - $s^+ j$ (P) : $sc\ j^+$ (P) : $sc^+ j^+$ (R) : $sc\ j$ (R)**. The two Parental types must be of equal frequencies, and the two Recombinant types must be equal to each other. The relative proportion of Parentals to Recombinants will depend on the genetic distance (recombination frequency = RF) between the genes on the chromosome. The recombination frequency could be anything from 0 mu (the genes are very closely linked) to 50 mu (the genes are far apart on the same chromosome). In the latter case P = R and the proportions of the four types of gametes will also be 1 : 1 : 1 : 1.

b. In these results the Parentals (77 scabrous + 74 javelin) are equal frequency to the Recombinants (76 wild type + 73 scabrous javelin). **Thus the genes assort independently.**

c. F_1 WT ♀ ($sc\ sc^+\ j^+ j$) × WT ♂ ($sc^+ sc^+\ j^+ j^+$) → all wild-type (WT) progeny. The F_1 female will still make four types of gametes in equal frequency. However the male parent in this cross can only contribute $sc^+ j^+$ gametes so **all the progeny will be phenotypically wild type. Thus you could not determine the frequency of parental and recombinant gametes from the F_1 female.**

d. Diagram the cross.

javelin ♀ ($sc^+ sc^+\ jj$) × scabrous javelin ♂ ($scsc\ jj$) → F_1 javelin ♀ ($sc\ sc^+\ jj$).

This F_1 female will make $sc^+\ j$ and $sc\ j$ parental gametes. **Her recombinant gametes will be the same genotypes as her parental gametes thus making it impossible to detect crossing-over.** In order to detect crossing-over (or independent assortment), the parent must be heterozygous for two genes or markers. The crossovers detected will occur between the two genes.

3. a. Diagram this cross.
   $B_1B_2\ D_1D_4$ ♂ × $B_3B_3\ D_2D_3$ ♀ → $B_1B_3\ D_1D_3$ ♂ (John).
   Thus John received a $B_1\ D_1$ gamete from his father and a $B_3\ D_3$ gamete from his mother. John will produce **parental gametes of these types: $B_1\ D_1$ and $B_3\ D_3$**.

   b. The **recombinant gametes produced by John will be $B_1\ D_3$ and $B_3\ D_1$**.

   c. All 100 sperm have the parental genotypes. There are no recombinant sperm. Therefore **the B and D loci are linked and are 0 mu apart.**

chapter 5

4. a. The Punnett square that follows illustrates the outcomes expected for a dihybrid cross where the F$_1$ individuals are *A B / a b*. This Punnett square is adjusted so that the width and height of the boxes reflect approximately the proportions of progeny of each class you would expect to see in the F$_2$ generation. The boxes with the F$_2$ genotypes are not all equal in size because the 4 types of gametes are not produced in equal amounts (as would be the case if the two genes were unlinked). Instead, there are more Parental gametes *(A B* and *a b)* made by each parent than Recombinant gametes *(A b* and *a B)*. Because 80% of the gametes made by the F$_1$ individuals are Parental, and because the two Parental classes are equal to each other, 40% of the gametes will be *A B*, 40% will be *a b*, 10% will be *A b*, and 10% will be *a B*. (The two Recombinant classes are also equal to each other because they are the reciprocal products of the same crossovers.) The color-coding in the Punnett square boxes represents the phenotypic classes, and the numbers in the boxes represent the proportions of the F$_2$ in the given class. **Among the F$_2$s, 66% will be *A– B–*, 9% will be *aa B–*, 9% will be *A– bb*, and 16% will be *aa bb*.**

Gametes	A B 0.40	a b 0.40	A b 0.10	a B 0.10
A B 0.40	AA BB 0.16	Aa Bb 0.16	AABb 0.04	AaBB 0.04
a b 0.40	Aa Bb 0.16	aa bb 0.16	Aabb 0.04	aaBb 0.04
A b 0.10	AA Bb 0.04	Aa bb 0.04	AAbb 0.01	AaBb 0.01
a B 0.10	Aa BB 0.04	aa Bb 0.04	AaBb 0.01	aaBB 0.01

b. The Punnett square that follows shows the expected outcomes of a dihybrid cross if the F$_1$ generation is *A b / a B* and 80% of the gametes are the two Parental classes. The color-coding of the boxes is the same as in part (a). **Among the F$_2$s, 51% will be *A– B–*, 24% will be *aa B–*, 24% will be *A– bb*, and 1% will be *aa bb*.**

Gametes	A b 0.40	a B 0.40	A B 0.10	a b 0.10
A b 0.40	AA bb 0.16	Aa Bb 0.16	AABb 0.04	Aabb 0.04
a B 0.40	Aa Bb 0.16	aa BB 0.16	AaBB 0.04	aaBb 0.04
A B 0.10	AA Bb 0.04	Aa BB 0.04	AABB 0.01	AaBb 0.01
a b 0.10	Aa bb 0.04	aa Bb 0.04	AaBb 0.01	aabb 0.01

Section 5.2

5. a. The parental females are heterozygous for the X-linked dominant Greasy fur (*Gs*) and Broadhead (*Bhd*) genes. They are crossed to homozygous recessive males. The male progeny of this cross will show four phenotypes, which will fall into two genetic reciprocal pairs. One pair will be Gs Bhd and wild type (*Gs Bhd* and *Gs$^+$ Bhd$^+$*) while the other pair will be Gs and Bhd (*Gs Bhd$^+$* and *Gs$^+$ Bhd*). If the genes are far apart on the X chromosome, then they will assort independently and all four classes will show equal frequency. If the genes are linked, then there will be a more frequent reciprocal pair (the parental pair) and a less frequent pair (the recombinant pair). There are 49 *Gs Bhd$^+$* and 48 *Gs$^+$ Bhd* male progeny. Thus, this is the parental pair and the genotype of the female is *Gs Bhd$^+$ / Gs$^+$ Bhd*. **The cross can be diagrammed:**

    Gs Bhd$^+$ / Gs$^+$ Bhd ♀ × *Gs$^+$ Bhd$^+$ / Y* ♂ → 49 *Gs Bhd$^+$* ♂ : 48 *Gs$^+$ Bhd* ♂ : 2 *Gs Bhd* ♂ : 1 *Gs$^+$ Bhd$^+$* ♂.
    The distance between these two genes: RF = 2 + 1 / 100 = 0.03 = 3 mu.

   b. The daughters in this cross must inherit the *Gs$^+$ Bhd$^+$* X chromosome from their fathers. This chromosome carries the recessive alleles of both genes, so this is a true testcross. Thus the **genotypes, phenotypes and frequencies of the female progeny would be the same as their brothers**.

6. The parents are from true-breeding stocks. Diagram the cross:

    raspberry eye color ♂ × sable body color ♀ → F$_1$ wild-type eye and body color ♀ × sable body color ♂ (if no mention is made of the eye color, then it is assumed to be wild type) → F$_2$ 216 wild-type ♀ : 223 sable ♀ : 191 sable ♂ : 188 raspberry ♂ : 23 wild-type ♂ : 27 raspberry sable ♂

    The phenotypes seem to be controlled by 2 genes, one for eye color and the other for body color. The F$_1$ female progeny show that the wild-type allele is dominant for body color (*s$^+$ > s*) and the wild-type allele is also dominant for eye color (*r$^+$ > r*). Sable body color is seen in the F$_1$ males but not the F$_1$ females, so the *s* gene is X-linked. In the F$_2$ generation, the raspberry eye color is seen in the males but not in the females, so *r* is also an X-linked gene. We can now assign genotypes to the true-breeding parents in this cross:

    r s$^+$ / Y (raspberry ♂) × *r$^+$ s / r$^+$ s* ♀ → F$_1$ *r s$^+$ / r$^+$ s* ♀ (wild type) × *r$^+$ s / Y* (sable ♂) → [the heterozygous F$_1$ female can make the following gametes: (parentals) *r s$^+$, r$^+$ s* and (recombinants) *r s, r$^+$ s$^+$*; the F$_1$ male can make Y and *r$^+$ s* gametes] → F$_2$ will be *r s$^+$ / r$^+$ s* (wild-type females), *r$^+$ s / r$^+$ s* (sable females), *r s / r$^+$ s* (sable females), *r$^+$ s$^+$ / r$^+$ s* (wild-type females), *r$^+$ s / Y* (sable males), *r s$^+$ / Y* (raspberry males), *r s / Y* (raspberry sable males), *r$^+$ s$^+$ / Y* (wild-type males)

    The F$_1$ female is heterozygous for both genes and will therefore make parental and recombinant gametes. The F$_1$ male is not a true testcross parent, because he does not carry the recessive alleles for both of the X-linked genes. However, this sort of cross can be used for mapping, because F₂ sons receive only the Y chromosome from the F₁ male and are hemizygous for the X chromosome from the F₁ female. Thus the phenotypes in the F₂ males represent the array of parental and recombinant gametes generated by the F₁ female as well as the frequencies of these gametes. Using the F₂ males, the **RF between sable and raspberry = 23 (wild-type males) + 27 (raspberry sable males) / 429 (total males) = 0.117 x 100 = 11.7 cM**.

7. To determine the probability that a child will have a particular genotype, look at the gametes that can be produced by the parents. In this example, *A* and *B* are 20 mu apart, so in a doubly heterozygous individual 20% of the gametes will be recombinant and the remaining 80% will be parental. The *a b / a b* homozygous man can produce only *a b* gametes. The doubly heterozygous woman, with a genotype of *A B / a b*, can produce 40% *A B*, 40% *a b*, 10% *A b* and 10% *a B* gametes. (Total of recombinant classes = 20%.) **The probability that a child receives the *A b* gamete from the female (and is therefore *A b / a b*) = 10%**.

8. a. Diagram the cross:

   $CC\ DD \times cc\ dd \rightarrow F_1\ C\ D\ /\ c\ d \times cc\ dd \rightarrow$ 997 *Cc Dd*, 999 *cc dd*, 1 *Cc dd*, 3 *cc Dd*.

   Because the gamete from the homozygous recessive parent is always *c d*, we can ignore one *c* and one *d* allele (the *c d* homolog) in each class of the F₂ progeny. The remaining homolog in each class of F₂ is the one contributed by the doubly heterozygous F₁, the parent of interest when considering recombination. In the F₂ the two classes of individuals with the greatest numbers represent parental gametes (*C D* or *c d* from the heterozygous F₁ parent combining with the *c d* gamete from the homozygous recessive parent). The other two types of progeny result from a recombinant gametes (*C d* or *c D* combining with the *c d* gamete from the homozygous recessive parent). The number of recombinants divided by the total number of offspring × 100 gives the map distance: (1 + 3)/(997 + 999 + 1 + 3) = 4/2000 = 0.002 × 100 = 0.2% RF or **0.2 map units (mu) or 0.2 cM**.

   b. $CC\ dd \times cc\ DD \rightarrow F_1\ C\ d\ /\ c\ D \times c\ d\ /\ c\ d \rightarrow$? As determined in part (a), *c* and *d* are 0.2 cM apart. Thus the gametes produced by the heterozygous F₁ will be 49.9% *C d*, 49.9% *c D*, 0.1% *C D*, 0.1% *c d*. After fertilization with *c d* gametes, there would be **49.9% *Cc dd*, 49.9% *cc Dd*, 0.1% *Cc Dd*, 0.1% *cc dd***. Because this is a testcross, the gametes from the doubly heterozygous F₁ parent determine the phenotypes of the progeny.

   c. Because the RF is so low (0.2%), very few crossovers occur between genes *C* and *D*. A single crossover in a single meiosis would produce 2 Recombinant progeny. With an RF = 0.002, you would have 2 Recombinant progeny out of 1000 total progeny. 1000 progeny could be obtained in 1000/4 = 250 meioses. Thus, only 1 out of every 250 meioses would have a crossover between genes *C* and *D*; in any one meiosis, the chance of a crossover occurring between the genes would be 1/250 = 0.004 = 0.04%. Thus, **in a typical meiosis, no crossovers would occur between genes *C* and *D***.

   d. In this case, you see equal frequencies of all progeny classes, so the number of Recombinants equals the number of Parentals. In other words, the genes are syntenic but so far apart that they are genetically unlinked; the RF is ~50%. From the discussion on p. 136 of the text, you should realize that when genes are unlinked, every meiosis must have at least one recombination event occurring between the two genes in question (in other words, when RF = 50%, no meiosis can be an NCO). However, we cannot determine the average number of crossovers between the two genes other to say that this number must be greater than 1. The reason is that as the physical distance between syntenic but unlinked genes increases, the number of double crossovers (DCOs) and triple or higher level crossovers increases yet the RF still remains ~50%. **Thus, in a typical meiosis, at least one crossover occurs between genes C and D.**

9. Designate the alleles of the genes: A = normal pigmentation and a = albino allele; $Hb\beta^A$ = normal globin and $Hb\beta^S$ = sickle allele.

   a. Because both traits are rare in the population, we assume that the parents are homozygous for the wild-type allele of the gene dictating their normal traits (that is, they are not carriers). Diagram the cross: $a\ Hb\beta^A\ /\ a\ Hb\beta^A$ (father) × $A\ Hb\beta^S\ /\ A\ Hb\beta^S$ (mother) → $a\ Hb\beta^A\ /\ A\ Hb\beta^S$ (son). Given that the genes are separated by 1 map unit, parental gametes = 99% and recombinant gametes = 1% of the gametes. **The son's gametes will consist of: 49.5% $a\ Hb\beta^A$, 49.5% $A\ Hb\beta^S$, 0.5% $a\ Hb\beta^S$ and 0.5% $A\ Hb\beta^A$.**

   b. In this family, the cross is $A\ Hb\beta^A\ /\ A\ Hb\beta^A$ (father) × $a\ Hb\beta^S\ /\ a\ Hb\beta^S$ (mother) → $a\ Hb\beta^S\ /\ A\ Hb\beta^A$ (daughter). **The daughter's gametes will be: 49.5% $a\ Hb\beta^S$, 49.5% $A\ Hb\beta^A$, 0.5% $a\ Hb\beta^A$ and 0.5% $A\ Hb\beta^S$.**

   c. The cross is: $a\ Hb\beta^A\ /\ A\ Hb\beta^S$ (son) × $a\ Hb\beta^S\ /\ A\ Hb\beta^A$ (daughter). **The probability of an $a\ Hb\beta^S\ /\ a\ Hb\beta^S$ child (sickle cell and anemic) = 0.005** (probability of $a\ Hb\beta^S$ from son) x 0.495 (probability of $a\ Hb\beta^S$ from daughter) = 0.0025.

10. a. Designate the alleles: H = Huntington allele, h = normal allele; B = brachydactyly, b = normal fingers. **John's father is $bb\ Hh$; his mother is $Bb\ hh$.**

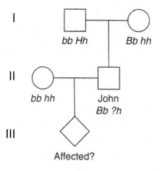

   b. We know the John is Bb because he has brachydactyly. **His complete genotype could be $Bb\ Hh$ or $Bb\ hh$.**

   In order to answer part (c) below, it would be helpful here to calculate the probability that John is Hh rather than hh. This calculation is a bit tricky – it requires you to use *conditional probability*, something that you have done in previous problems in the book, but those examples were simpler. For example, in Chapter 2 on p. 38, to answer Solved Problem III you need to calculate the probability that the daughter of two *Tt* parents is *Tt*. If you had no other information, the probability would be 1/2, as the progeny of a monohybrid cross appear in the ratio 1 *TT* : 2 *Tt* : 1 *tt*. However, a *condition* exists in the problem: We are told that the daughter in question does not have Tay-Sachs disease, which means that she is not *tt*. This fact means that the "1 *tt*" can be eliminated from the set of her possible genotypes, leaving 1 *TT* : 2 *Tt* and thus a 2/3 chance that she is *Tt*.

    Now back to John. If we had no information about John, the probability that John is *Hh* because he inherited *H* from his *Hh* father would be 1/2. However, we do have some information about John - he is 50 and has no symptoms of Huntington disease; this fact - or *condition* – as well as the information that 2/3 of people who inherit *H* have symptoms of the disease by age 50 - has to be used to calculate the chance that John inherited the *H* allele from his father. The easiest way to think about this conditional probability is the following:

    John's father or anyone else who is *Hh* could have two classes of children: those who inherit the *H* allele (*Hh*), and those who inherit the *h* allele (*hh*). As we said previously, with no other knowledge, the ratio of John's progeny would be 1 *Hh* : 1 *hh*. However, we know that John is 50 and has no symptoms. All of the *hh* progeny will have no symptoms at 50, and also 1/3 of the *Hh* progeny will have no symptoms. These facts change the genotypic ratio, because we are considering the symptom-free people only; for symptom-free people, the ratio is 1/3 *Hh* : 1 *hh*. Another way to express the same ratio is 1/3 *Hh* : 3/3 *hh*, or 1 *Hh* : 3 *hh*. Thus, **the chance that John is *Bb Hh* is 1/4.**

c. In order for the child to express both diseases, the child must be *Bb Hh*, and so John must be *Bb Hh*. We just calculated in part (b) that the probability John is *Bb Hh* is 1/4. The chance that the child is *Bb Hh* is 1/4 (the chance that John has the *H* allele) × 1/2 (the chance that the child inherits *H* from John) × 1/2 (the chance that the child inherits *B* from John) = 1/16. The chance that a *Bb Hh* child will express brachydactyly is 90%, and the chance such a child will have Huntington disease by age 50 is 2/3. Therefore, **the probability that John and his wife's child will express both diseases is 1/16 × 0.90 × 2/3 = 0.0375 = 3.75%.**

d. If the two loci are linked, the alleles on each of John's homologs will be either *B h / b H* or *B h / b h*. The *B H* gamete could only be produced if John is *B h / b H* and recombination occurs. The probability of that specific recombinant gamete = 10% (the other 10% of the recombinant gametes are *b h*). As shown in part (b), the probability that John's genotype is *B h / b H* = 1/4. **The probability that John's child will have both Huntington and brachydactyly = 1/4 (probability that John is *B h / b H*) × 1/10 (probability of child inheriting *B H* recombinant gamete from John) × 9/10 (probability of expressing brachydactyly) × 2/3 (probability of expressing Huntington's by age 50) = 0.015 = 1.5%.**

11. Diagram the cross:

    CC bb (brown rabbits) × *cc BB* (albinos) → F₁ *Cc Bb* × *cc bb* → 34 black : 66 brown : 100 albino.

a. If the genes are unlinked, the F$_1$ will produce *C B*, *c b*, *C b* and *c B* gametes in equal proportions. A mating to animals that produce only *c b* gametes will produce four genotypic classes of offspring: 1/4 *Cc Bb* (black) : 1/4 *cc Bb* (albino) : 1/4 *Cc bb* (brown) : 1/4 *cc BB* (albino); that is, a ratio of **1/4 black : 1/2 albino : 1/4 brown**.

b. If gene *C* and gene *B* are linked, then the genotype of the F$_1$ class is *C b / c B* and you would expect the parental type *C b* and *c B* gametes to predominate; *c b* and *C B* are the recombinant gametes present at lower levels. The parental gametes are represented in the F$_2$ by the *Cc bb* (brown) and *cc Bb* (albino) classes. Since we cannot distinguish between the albinos resulting from fertilization of recombinant gametes (*c b*), and those resulting from parental gametes (*c B*), we have to use the proportion of the *C B* recombinant class and assume that the other class of recombinants (*c b*) is present in equal frequency. Because crossing-over is a reciprocal exchange, this assumption is reasonable. There were 34 black progeny; assuming that 34 of the 100 albino progeny were the result of recombinant gametes, **the genes are (34 + 34) recombinant / 200 total progeny = 34% RF = 34 cM apart**.

12. Diagram the cross:

    blue smooth × yellow wrinkled → 1447 blue smooth, 169 blue wrinkled, 186 yellow smooth, 1510 yellow wrinkled.

    a. To determine if genes are linked, first predict the results of the cross if the genes are unlinked. In this case, a plant with blue, smooth kernels (*A– W–*) is crossed to a plant with yellow, wrinkled kernels (*aa ww*). As there are four classes of progeny, the parent with blue, smooth kernels must be heterozygous for both genes (*Aa Ww*). From this cross, **we would predict equal numbers of all four phenotypes in the progeny if the genes were unlinked. Since the numbers are very skewed, with the smaller classes representing recombinant offspring, the genes are linked. RF = (169 + 186) / (1447 + 169 + 186 + 1510) = 355/3312 = 10.7% = 10.7 cM**.

    b. The genotype of the blue smooth parent was *Aa Ww*. The arrangement of alleles in the parent is determined by looking at the phenotypes of the largest classes of progeny (the parental reciprocal pair). Since blue, smooth and yellow wrinkled are found in the highest proportion, *A W* must be on one homolog and *a w* on the other = *A W / a w*.

    c. The genotype of the blue, wrinkled progeny is *A w / a w*. This genotype can produce *A w* or *a w* gametes in equal proportions. Recombination *cannot* be detected here, because this genotype is heterozygous for only one gene. Recombination between these homologs yields the same two combinations of alleles (*A w* and *a w*) as the parental, so each type of gamete is expected 50% of the time. The yellow smooth progeny plant has a genotype of *a W / a w*. Again since recombination cannot be detected, the frequency of each type is 50%. Thus the cross is *A w / a w* (blue wrinkled) × *a W / a w* (yellow smooth). **Four types of offspring are expected in equal proportions: 1/4 *A w / a W* (blue smooth) : 1/4 *A w / a w* (blue wrinkled) : 1/4 *a w / a W* (yellow smooth) : 1/4 *a w / a w* (yellow wrinkled)**.

13. a. *AA BB* × *aa bb* → F$_1$ *A B / a b* (*A B* on one homolog and *a b* on the other homolog). The F$_1$ progeny will produce parental type gametes *A B* and *a b* and the

recombinant types *A b* and *a B*. The genes are 40 cM apart, so **the recombinants will make up 40% of the gametes (20% *A b* and 20% *a B*). The remaining 60% of the gametes are parental: 30% *A B* and 30% *a b*.** Set up a Punnett square to calculate the frequency of the 4 phenotypes in the F_2 progeny. In the Punnett square the phenotypes associated with the genotypes in each box are indicated with the same colors as was used previously in Problem 4 above. **The F_2 phenotypic ratio is: 0.59 *A– B–* : 0.16 *A– bb* : 0.16 *aa B–* : 0.09 *aa bb*.**

Gametes	A B 0.30	a b 0.30	A b 0.20	a B 0.20
A B 0.30	AA BB 0.09	Aa Bb 0.09	AA Bb 0.06	Aa BB 0.06
a b 0.30	Aa Bb 0.09	aa bb 0.09	Aa bb 0.06	aa Bb 0.06
A b 0.20	AA Bb 0.06	Aa bb 0.06	AA bb 0.04	Aa Bb 0.04
a B 0.20	Aa BB 0.06	aa Bb 0.06	Aa Bb 0.04	aa BB 0.04

b. If the original cross was *AA bb* × *aa BB*, the allele combinations in the F_1 would be *A b / a B*. Parental gametes in this case are 30% *A b* and 30% *a B* and the recombinant gametes are 20% *A B* and 20% *a b*. Set up a Punnett square, as in part (a). **The F_2 phenotypic ratio is: 0.54 *A– B–* : 0.21 *A– bb* : 0.21 *aa B–* : 0.04 *aa bb*.**

Gametes	A b 0.30	a B 0.30	A B 0.20	a b 0.20
A b 0.30	AA bb 0.09	Aa Bb 0.09	AA Bb 0.06	Aa bb 0.06
a B 0.30	Aa Bb 0.09	aa BB 0.09	Aa BB 0.06	aa Bb 0.06
A B 0.20	AA Bb 0.06	Aa BB 0.06	AA BB 0.04	Aa Bb 0.04
a b 0.20	Aa bb 0.06	aa Bb 0.06	Aa Bb 0.04	aa bb 0.04

chapter 5

14. Notice that you are asked for the <u>number of different *kinds* of phenotypes</u>, not the number of individuals with each of the different phenotypes.

   a. **2** (*A–* and *aa*);

   b. **3** (*AA*, *Aa*, and *aa*);

   c. **3** (*AA*, *Aa*, and *aa*);

   d. **4** (*A– B–*, *A– bb*, *aa B–*, and *aa bb*);

   e. **4** [*A– B–*, *A– bb*, *aa B–*, and *aa bb*; because the genes are linked, the frequency of the four classes will be different than that seen in part (d)];

   f. **Nine** phenotypes in total. There are three phenotypes possible for each gene. The total number of combinations of phenotypes is $(3)^2 = 9$ [(*AA*, *Aa*, and *aa*) × (*BB*, *Bb*, and *bb*)].

   g. Normally there are four phenotypic classes, as in parts (d) and (e): (*A– B–*, *A– bb*, *aa B–*, and *aa bb*). In this case, there is recessive epistasis, so two of the classes (for example, *A– bb* and *aa bb*) have the same phenotype, **giving 3 phenotypic classes**.

   h. Two genes means four phenotypic classes (*A– B–*, *aa B–*, *A– bb* and *aa bb*). Because gene function is redundant, the first three classes are all phenotypically equivalent in that they have function, and only the *aa bb* class will have a different phenotype, being without function. Thus there are only **2 phenotypic classes**.

   i. There is 100% linkage between the two genes. The number of phenotypic classes will depend on the arrangement of alleles in the parents. **If the parents are *A B / a b* × *A B / a b*, the progeny will be 3/4 *A– B–* : 1/4 *aa bb* and there will be two phenotypic classes. If the parents are *A b / a B* × *A b / a B*, the progeny will be 1/4 *AA bb* : 1/2 *Aa Bb* : 1/4 *aa BB* and there will be three phenotypic classes in the offspring.**

15. V^1, V^2, and V^3 are codominant alleles of the DNA variant marker locus, while *D* and *d* are alleles of the disease gene. The marker and the disease locus are linked. Diagram the cross:

   $V^1 D / V^2 d$ (father) × $V^3 d / V^3 d$ (mother) → $V^2 ? / V^3 d$.

   The fetus <u>must</u> get a $V^3 d$ homolog from the mother. The fetus also received the V^2 allele of the marker. The father could have given a $V^2 d$ (nonrecombinant) or a $V^2 D$ (recombinant) gamete.

   a. If the *D* locus and the V^1 allele of the marker are 0 mu apart (there is no recombination between them; they appear to be the same gene), **the probability that the child, who received the V^2 allele of the marker, has the *D* allele = 0**.

   Parts (b)-(d) require thinking in terms of *conditional probability* – similar to Problem 10 above. In this case, the *condition* is the fact that you already know that the child received the V^2 marker allele. This fact alters the otherwise simple expectations based on recombination frequency and equal segregation of alleles into gametes.

chapter 5

b. If the distance between the disease locus and marker is 1 mu, 1% of the father's gametes are recombinant between *D* and the marker locus, and 99% of his gametes are parentals. Half of the recombinant gametes will be V^2 *D,* and the other half will be V^1 *d*; half of the parental gametes will be V^1 *D,* and the other half will be V^2 *d*. If we consider 200 of his gametes, there will be 1 V^2 *D* : 1 V^1 *d* : 99 V^1 *D* : 99 V^2 *d.* As we already know that the child inherited V^2, and the ratio of gametes with the V^2 allele is 1 V^2 *D* : 99 V^2 *d,* you can see that **the chance that the fetus has the *D* allele is 1/100 = 1%.** Parts (b) – (d) are solved using the same logic.

c. **5%.**

d. **10%.**

e. **50%.**

16. a. You would know that this figure represents meiosis I in a mouse primary spermatocyte (as opposed to a human primary spermatocyte) **because you can see that there are 20 bivalents, each bivalent being represented by one *red* line showing the synaptonemal complex** [with the exception of the X-Y pair, see part (d) below.] **In humans, there would be 23 bivalents because there are 23 pairs of chromosomes.**

    b. **Because of a printing error, it is very hard to discern the *blue* color in this micrograph, so you may not be able to answer this problem based on what you see. (Our apologies.)** If you could see the *blue* color, it would be located at one end of each red string of synaptonemal complex. (At the end of this answer is a similar photograph properly showing the *blue* staining.) **This fact indicates that all mouse chromosomes are acrocentric** because the centromere staining in *blue* is at one chromosomal end. Interestingly, you know from various karyotypes you have already seen in this book that many human chromosomes are metacentric. Thus, similar preparations from human primary spermatocytes would show in some cases *blue* staining in the middle of a *red* strand. The location of the centromeres is thus another criterion by which you could be sure you are looking at mouse rather than human primary spermatocytes.

    c. In mice, *n* = 20 so 2*n* = 40. The figure legend indicates that these are primary spermatocytes in mid-prophase of meiosis I (which makes sense because this is the time at which the synaptonemal complexes are present). At this stage of meiosis, the chromosomes have already replicated and each chromosome is made of two sister chromatids. Thus, the cell contains 2 × 2*n* chromatids, or **80 chromatids.**

    d. **The X-Y bivalent is indicated by the *white arrow* in Fig. 5.7a on p. 134.** The reason you know this to be the case is that the two chromosomes are associated at only one end. The problem tells you that there is only one PAR on the mouse X and Y chromosomes, so this is where they associate. If you look hard at the point of association (just at the arrowhead), you can see a small *green* spot, indicating that a recombination event (represented by a recombination nodule) is occurring between the PARs on the X and the Y. This recombination event is absolutely required for proper disjunction of the X and Y chromosomes during meiosis I [see part (e) below]. (*Note:* If you could see the centromeres in *blue,* you would find that they are located at the ends of the X and the Y chromosome that do not contain the PARs; see the additional figure at the end of this answer.)

In case you are curious, the reason the red color extends along the length of the X and Y chromosomes, even though pairing occurs only at the PARs, is that the synaptonemal complex includes some proteins that extend along the chromosome length (these form so-called *lateral elements*) as well as proteins that are found only at the sites of pairing (forming so-called *central elements*). The protein stained in *red* is part of the lateral elements. A figure summarizing the structure of the X-Y bivalent is shown below. The colors are as in Fig. 5.7a, including the *blue* centromeres if they were visible at the ends of these two acrocentric chromosomes.

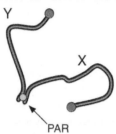

e. In order for the X and Y chromosomes to segregate faithfully so that each sperm contains either an X or a Y but not both, the X and Y need to pair with each other. The X and the Y also need to stay together so that they can migrate to the metaphase plate during meiosis I before they separate at anaphase of meiosis I. The pseudoautosomal regions of the X and the Y have nearly identical DNA sequences. This allows the X and Y chromosomes to pair with each other. Crossing-over within the pseudoautosomal regions holds the X-Y pair together until anaphase of meiosis I.

Special note: The photograph that follows shows a different mouse primary spermatocyte prepared in the same way, but in this case the *blue* centromere staining demonstrating that mouse chromosomes are all acrocentric is clearly visible. The arrow points to the PAR region of the X-Y bivalent (note again the *green* recombination nodule in the PAR region); the longer, somewhat more brightly red staining strand in this bivalent is the X.

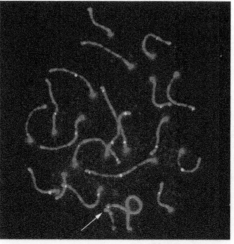

© Dr. Paula Cohen & Dr. Kim Holloway, The Cohen Lab, Center for Reproductive Genomics, Cornell University

chapter 5

17. **The figure that follows shows clearly that the cohesin complexes located distal to the chiasmata (that is, further away from the centromere than the site of recombination) are the critical cohesin complexes for maintaining the homologous chromosomes together in a bivalent.** The two centromeres are being pulled towards the opposite spindle poles. The *left* panel of the figure below shows what would occur if only the cohesin complexes distal to the chiasmata remained (and none proximal to the chiasmata existed). The cohesin complexes distal to the centromere are sufficient to prevent the homologous chromosomes from separating. As seen at the *center* and *right*, the cohesin complexes proximal to the chiasmata have no such function; in the absence of the distal complexes, there is nothing to prevent the homologous chromosomes from being separated and being pulled to opposite spindle poles (*right*).

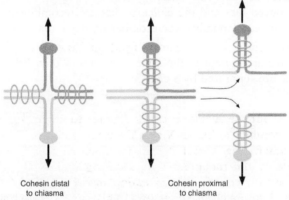

Cohesin distal to chiasma   Cohesin proximal to chiasma

Careful observation of the drawing will also illustrate some additional points of interest. (i) Note that these are acrocentric mouse chromosomes, so the centromeres are found at one chromosome end. The centromeres (*dark green* and *light green ovals*) are made of many cohesin complexes that are not shown. The spindle fibers exerting poleward forces (*black arrows*) are connected to the kinetochores (*not shown*), which are found at the centromeres but which involve proteins other than the cohesin complexes. (ii) At anaphase of meiosis I, the cohesin complexes along the arms (both proximal and distal to the chiasmata) dissolve, allowing the homologous chromosomes to be pulled to opposite spindle poles. (iii) Note that the cohesin complexes are rings inside of which are the two sister chromatids. In other words, the rings include two *light green* chromatids or two *dark green* chromatids, but never a *light green* and a *dark green* chromatid. The reason for this is straightforward: The cohesin rings are made immediately after chromosomes replicate into sister chromatids, before the homologous chromosomes pair with each other.

Section 5.3

18. Diagram the cross:

wild type ♀ × reduced cinnabar ♂ → F₁ ♀ × F₁ ♂ → 292 wild type, 9 cinnabar, 7 reduced, 92 reduced cinnabar.

Two genes are involved in this cross, but the frequencies of the phenotypes in the second generation offspring do not look like frequencies expected for a cross between double heterozygotes for two independently assorting genes (9:3:3:1). The genes must be linked. Designate the alleles: cn^+ = wild type, cn = cinnabar; rd^+ = wild type, rd = reduced. The cross is:

$cn^+\ rd^+ / cn^+\ rd^+\ ♀ \times cn\ rd / cn\ rd\ ♂ \rightarrow F_1\ cn^+\ rd^+ / cn\ rd$.

Recombination occurs in *Drosophila* females but not in males. Thus males can produce only the parental $cn^+\ rd^+$ or $cn\ rd$ gametes. The females produce both the parental gametes and the recombinant gametes $cn^+\ rd$ and $cn\ rd^+$. These gametes can combine as shown in the Punnett square that follows. The phenotypes in the boxes are indicated with various colors. You do not need to fill in the numbers in the Punnett square to solve this problem, but they are added as a check of the final answer described below the table.

The reduced flies and the cinnabar flies are recombinant classes. However, there should be an equal number of recombinant types that have a wild-type phenotype because they got a $cn^+\ rd^+$ gamete from the male parent. If we assume that these recombinants are present in the same proportions, then RF = 2 × (7+9)/400 = 9%. **The genes are separated by 8 cM.**

Eggs \ Sperm	$cn^+\ rd^+$ 0.50	$cn\ rd$ 0.50
$cn^+\ rd^+$ Parental 0.46	$cn^+\ rd^+ / cn^+\ rd^+$ (wild type) 0.23	$cn^+\ rd^+ / cn\ rd$ (wild type) 0.23
$cn\ rd$ Parental 0.46	$cn^+\ rd^+ / cn\ rd$ (wild type) 0.23	$cn\ rd / cn\ rd$ (cinnabar reduced) 0.23
$cn^+\ rd$ Recombinant 0.04	$cn^+\ rd^+ / cn^+\ rd$ (wild type) 0.02	$cn^+\ rd / cn\ rd$ (reduced) 0.02
$cn\ rd^+$ Recombinant 0.04	$cn^+\ rd^+ / cn\ rd^+$ (wild type) 0.02	$cn\ rd^+ / cn\ rd$ (cinnabar) 0.02

19. a. There are four equally frequent phenotypic classes in the progeny; these must include a reciprocal pair of parentals and a reciprocal pair of recombinants. The parental reciprocal pair could either be [(wild-type wings, wild-type eyes) and (dumpy wings, brown eyes)] or [(wild-type wings, brown eyes) and (dumpy wings, wild-type eyes)]. The first possibility means the genotype of the heterozygous female was $dp^+\ bw^+ / dp\ bw$ while the second pair means the genotype of the

female was *dp bw+ / dp+ bw*. In either case the recombination frequency is 50% (178 + 181 / 716 or 185 + 172 / 716) so **the two genes are assorting independently**.

b. In this cross the heterozygous parent is the male and there is no recombination in *Drosophila* males. If dumpy and brown were on separate chromosomes then they must assort independently and there would be four classes of progeny as in part (a). We can tell that **the two genes are on the same chromosome** because the lack of recombination in the male parent then explains why only two (parental) classes of progeny are seen. **The genotype of the heterozygous male must have been *dp+ bw+ / dp bw*.**

c. If the genes are far enough apart on the same chromosome they will assort independently in the first cross. This is because (1) **recombination occurs at the four strand stage of meiosis; and (2) in every meiosis, at least one crossover will occur between the two genes**. Independent assortment does not occur in the second cross because there is no recombination in *Drosophila* males. In *Drosophila*, therefore, **it is a simple matter to decide if genes are syntenic** (located on the same chromosome), **even if the alleles of these genes assort independently**. A cross between a male heterozygous for the two genes of interest and a recessive female will clearly distinguish between genes on separate chromosomes and syntenic genes. In the case of genes on separate chromosomes there will be four equally frequent classes of progeny. In the case of genes on the same chromosome (even genes that are very far apart on the same chromosome) there will only be two classes of progeny – the parental classes.

d. The *dp* ↔ *bw* distance cannot be measured in any two-point cross. The measured recombination frequency can be no higher than 50% when looking at data involving a single pair of genes. **Large genetic distances can be measured accurately only by summing up the values obtained for smaller distances separating other genes in between those at the ends**.

20. a. A testcross is the best way to find the map distance between two genes. One parent must be heterozygous for both genes and the other parent must homozygous recessive. Thus the cross would be ***Bb Cc* × *bb cc***. Because these genes are syntenic the heterozygous parent may be either *B C / b c* or *B c / b C*.

b. There are two possible orders for these genes, shown in the following figure:

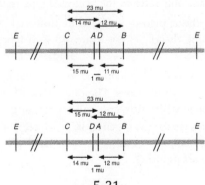

c. Because the map positions of *A* and *D* are so close, **the relative order of *A* and *D* is not clear. Another uncertainty is the location of gene *E*.** This gene must be on this chromosome because we are told all of these genes are syntenic. However the *E* gene is genetically unlinked to any of the other genes (*A*, *B*, *C* and *D*).

d. To figure out the actual order of the genes you must **do a three-point cross with either *B A D* or *A D C* and order the genes** by finding the double crossover class, comparing this to the parental class and ascertaining the order of the genes. This procedure is described in the Tips for Three-Point Crosses in the Problem Solving section at the beginning of the chapter. Also see Problem 5-22c for a further explanation of this method of determining the gene order.

The location of the *E* gene can be determined by finding new genes that are genetically linked to the left of *C* and to the right of *B* in the maps seen in part (b). This **extension of the linkage group** will eventually allow the placement of gene *E* when it shows genetic linkage to one of these new genes.

21. The shortest distances are the most accurate. Therefore begin assembling a map using the genes that are closest together. MAT-LEU2 are 16 cM apart and MAT-THR4 are 20 cM apart. This gives two possible maps: MAT - (16 cM) - LEU2/THR4 or LEU2 - (16 cM) - MAT - (20 cM) - THR4. Because THR4 is 35 cM from LEU2 the second map must be the correct one. Note that the two smaller distances do not sum to the longer distance because these recombination frequencies are based on two-point crosses. The HIS4 gene is 23 cM from LEU2. This initially leads to two possible maps: HIS4 - (23 cM) - LEU2 - MAT - THR4 or LEU2 - (23 cM) - HIS4/MAT - THR4. The first map is the most likely because HIS4 is 37 cM from MAT instead of being very close to MAT. So **the order of the genes is HIS4 - LEU2 - MAT - THR4** (or the inverse).

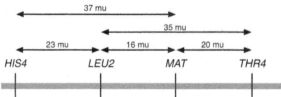

22. In foxgloves, wild-type flower color is red and the mutant color is white; the mutation *peloria* causes the flowers at the apex of the stem to be very large; normal foxgloves are very tall, and *dwarf* affects the plant height. When you describe the phenotype of an individual you usually only refer to the non-wild type traits. The cross is:

white flowered × dwarf peloria → F₁ white flowered × dwarf peloria → 172 dwarf peloria, 162 white, 56 dwarf peloria white, 48 wild type, 51 dwarf white, 43 peloria, 6 dwarf, 5 peloria white.

a. Because there is only 1 phenotype of F₁ plant, the parents must have been homozygous for all three genes. The phenotype of the F₁ heterozygote indicates the **dominant alleles: white flowers, tall stems, and normal-sized flowers.**

b. Designate the alleles for the 3 genes: *W* = white, *w* = red; *P* = normal-sized flowers, *p* = peloria; *T* = tall, *t* = dwarf. The cross is: ***WW PP TT* (white flowered)** × ***ww dd pp* (dwarf peloria).**

c. Note that all 3 of these genes are genetically linked. There are only 2 classes of parental progeny as defined both phenotypically and numerically. In order to draw a map of these genes, organize the testcross data (see the Table that follows), figure out which of the 3 genes is in the middle and calculate the recombination frequencies in regions 1 and 2. By definition, the parentals are the class with the same phenotype as the original parents of the cross, and the DCO class is the least frequent reciprocal pair of progeny. At this point, arbitrarily one of the remaining 2 reciprocal pairs of progeny is SCO 1 and the last remaining pair is SCO 2.

Classes of gametes	Genotype	Numbers
Parental (P)	W P T	162
	w p t	172
SCO 1	W p t	56
	w P T	48
SCO 2	W P t	51
	w p T	43
DCO	w P t	6
	W p T	5

Compare the DCO class with the parental class. Remember that a DCO comes from a meiosis with a simultaneous crossover in regions 1 and 2. As a result, the allele of the gene in the middle switches with respect to the alleles of the genes on the ends. In this data set, take one of the DCO classes (for example *W p T*) and compare it to the most similar parental gamete (*W P T*). Two alleles out of three are in common; the one that differs (*p* in this case) is the gene in the middle. Thus, the order is *W P T* (or *T P W*).

Once you know the order of the 3 genes, you must calculate the 2 shortest distances - from the end to the center (*W↔P*) and from the center to the other end (*P↔T*). The total number of progeny in this testcross = 543. RF *W↔P* = (56 + 48 + 6 + 5)/543 = 115/543 = 21.2 cM; RF *P↔T* = (51 + 43 + 6 + 5)/543 = 19.3 cM.

The map is:

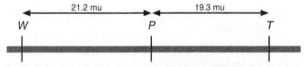

d. Interference (I) = 1 - coefficient of coincidence (coc)
coc = frequency of observed DCO / frequency of expected DCO

The expected percentage of double crossovers is the product of the RF in each interval: (0.193) × (0.212) = 0.0409. The observed DCO frequency = 11/543 = 0.0203; **coc = 0.0203/0.0409 = 0.496; I = 1 - 0.496 = 0.504.**

23. Although this problem considers three different genes, each cross only considers two at a time. You are thus asked to construct a map of the three genes based on 3 two-point crosses. Note the unusual usage of a comma to separate the genes in the genotypes written for the parents of each cross. This is meant to show that we don't know if any

of the genes are on the same chromosome. If the two genes in each cross are assorting independently then the four phenotypic classes in the testcross progeny must occur in equal frequencies. This is clearly not the case in any of the three crosses. Therefore the *mb*, *e* and *k* genes are all linked. The genotypes of the parents in the first cross may be correctly written as $mb^+ \, e^+ / mb^+ \, e^+ \times mb \, e / mb \, e$.

In the first cross, the $mb^+ \, e^+$ and $mb \, e$ classes are the parentals. This is based both on the known genotypes of the true breeding parental generation and on the fact that this reciprocal pair is the most frequent. The recombinants are the $mb^+ \, e$ and the $mb \, e^+$ flies, so the RF = 11 + 15 / 250 = 0.104 = 10.4 mu. In the second cross the recombinant reciprocal pair is $k^+ \, e^+$ and $k \, e$, so the recombination frequency is 11 + 7 / 312 = 0.058 = 5.8 mu. The $k \, mb^+$ and $k^+ \, mb$ classes are the recombinant reciprocal pair for the third cross. In this case the RF = 11 + 15 / 422 = 0.062 = 6.2 mu. The *mb* and *e* genes in the first cross are the furthest apart, so the *k* gene must be in between them. Thus the map of these data is:

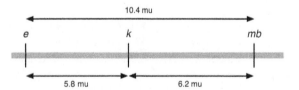

Note that the distance calculated between *e* and *mb* in cross one (10.4 mu) does NOT equal the sum of the two shorter distances (12.0 mu). The distance calculated from cross one is less accurate because it is based on single crossovers between *e* and *mb* and does not include any of the double crossovers that occurred in the *e*↔*k* and *k*↔*mb* regions simultaneously. Thus, **the best map of these genes is:**

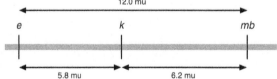

24. Diagram the cross:

    pink petals, black anthers, long stems × pink petals, black anthers, long stems → the progeny listed in the table.

    a. Looking at flower color alone, the cross is pink × pink → (78 + 26 + 44 + 15) red : (39 + 13 + 204 + 68) pink : (5 + 2 + 117 + 39) white = 163 red : 324 pink : 163 white. The appearance of two new phenotypes (red and white) suggests that flower color shows incomplete dominance. This idea is confirmed by the 1:2:1 monohybrid ratio seen in the self-cross progeny. **The pink flowered plants are *Pp*, red are *PP* and white are *pp*.**

    b. The expected ratio of red: pink : white would be 1:2:1. Calculating for the 650 plants, this equals **162.5 red, 325 pink, and 162.5 white**.

    c. The monohybrid ratio of the black and tan anthers = (78+26+39+13+5+2) tan : (44 + 15 + 204 + 68 + 117 + 39) black = 163 tan : 487 black = ~ **1 tan : 3 black. Therefore black (*B*) is dominant to tan (*b*)**. The monohybrid ratio for the stem

5-24

chapter 5

length = 487 long stems : 163 short stems. = ~ **3 long : 1 short, so long (L) is dominant to short** *(l)*.

d. Designate the alleles: *PP* = red, *Pp* = pink, *p* = white; *B* = black, *b* = tan, *L* = long, *l* = short. Because all 3 monohybrid phenotypic ratios are characteristic of heterozygous crosses, **the genotype of the original plant is *Pp Bb Ll*.**

e. If the stem length and anther color genes assort independently, the 9:3:3:1 phenotypic ratio should be seen in the progeny. Totaling all the progeny in each of the classes: 365 long black : 122 short black : 122 long tan : 41 short tan. The observed dihybrid ratio is close to a 9:3:3:1 ratio, so **the genes for anther color (*B*) and stem length (*L*) are unlinked.**

The expected monohybrid ratio for flower color is 1 red : 2 pink :1 white, while that for stem length is 3 long :1 short. If the two genes are unlinked, the expected dihybrid ratio can be calculated using the product rule to give 3/8 long pink : 3/16 long red : 3/16 long white : 1/8 short pink : 1/16 short red : 1/16 short white, or a 6 *L– Pp* : 3 *L– PP* : 3 *L– pp* : 2 *ll Pp* : 1 *ll PP* : 1 *ll pp* ratio. The observed numbers are: 243 long pink : 122 long red : 122 long white : 81 short pink : 41 short red : 41 short white. This observed ratio is close to the predicted ratio, so **the genes for petal color (*P*) and stem length (*L*) are unlinked.**

The same analysis is done for flower color and anther color. The expected dihybrid ratio here is also 6:3:3:2:1:1. The observed numbers are: 272 black pink : 59 black red : 156 black white : 52 tan pink : 104 tan red : 7 tan white. Because these numbers do not fit a 6:3:3:2:1:1 ratio, we can conclude that **flower color (*P*) and anther color (*B*) are linked genes.**

f. It is clear that the original snapdragon was heterozygous for *Pp* and *Bb*. However the genotype of this heterozygous plant could have been either *P B / p b* or *P b / p B*. If it is the former and the genes are closely linked, then the *pp bb* phenotype will be very frequent (almost 1/4 of the progeny because it is nonrecombinant). Instead the *pp bb* class is very infrequent, accounting for only about 1% (7/650) of the progeny. Thus, **the parental genotype must have been *P b / p B*. In this case, the infrequent *pp bb* class received a *p b* recombinant gamete from each parent. The frequency of the *pp bb* genotype = (frequency of the *p b* recombinant gamete)². The recombination frequency (RF) between the *P* gene and the *B* gene = frequency of *P B* + *p b* gametes = 2(√(#*pp bb*/# total progeny) = 2(√(7/650)) = ~20 map units.**

25. a. See Problem 5-29 for a detailed explanation of the methodology. Diagram the cross:

$a^+ b^+ c^+ / a\ b\ c\ ♀ \times a\ b\ c / a\ b\ c\ ♂ \rightarrow ?$.

Recombination occurs in *Drosophila* females, so the female parent will make the following classes of gametes. Because this is a testcross, these gametes will determine the phenotypes of the progeny. The Table that follows indicates the kinds of gametes that the female parent will produce and their frequencies.

Classes of gametes	Genotype	Frequency of reciprocal pair	Numbers	Frequency of each class	# of progeny freq x 1,000
Parental (P)	$a^+ b^+ c^+$ $a\ b\ c$	1 - all else	1 - (0.18 + 0.08 + 0.02) = 0.72	0.36 0.36	360 360
SCO 1	$a^+ b\ c$ $a\ b^+ c^+$	RF in region 1 = SCO 1 + DCO	SCO 1 = 0.2 - 0.02 = 0.18	0.09 0.09	90 90
SCO 2	$a^+ b^+ c$ $a\ b\ c^+$	RF in region 2 = SCO 2 + DCO	SCO 2 = 0.1 - 0.02 = 0.08	0.04 0.04	40 40
DCO	$a^+ b\ c^+$ $a\ b^+ c$	(RF in region 1) x (RF in region 2)	(0.2) x (0.1) = 0.02	0.01 0.01	10 10

b. Diagram the cross: $a^+ b^+ c^+ / a\ b\ c\ \male \times a\ b\ c / a\ b\ c\ \female \rightarrow$?. Here, the heterozygous parent is the male. Recombination does not occur in male *Drosophila*. Thus, the heterozygous parent will generate only 2 types of gametes, the parental types. If you score 1,000 progeny of this cross, you will find **500 $a^+ b^+ c^+$ and 500 $a\ b\ c$ progeny**.

26. First re-write the data as genotypes grouping the genetic reciprocal pairs.

Classes of gametes	Genotype	Numbers
Parental (P)	th h st + + +	432 429
Recombinant	th h + + + st	37 33
Recombinant	th + st + h +	35 34

The *th h st* and wild-type (+ + +) classes are defined as parental because they are the most frequent reciprocal pair.

a. Each class of the parental reciprocal pair corresponds to one homolog of the heterozygous *Drosophila* female that was testcrossed to the homozygous recessive male. Thus the genotype of this female is ***th h st* / $th^+ h^+ st^+$**.

b. This is a testcross with three linked genes, but there are only 6 classes of data (3 reciprocal pairs) instead of 8 classes. The missing reciprocal pair is *th + +* and *+ h st*. This pair must be the least frequent DCO class. Comparing the *th + +* DCO class to the *+ + +* parental class (the most similar) shows that *th* is the gene that differs, so *th* must be the gene in the middle. The *h* ↔ *th* distance is 35 + 34 / 1000 = 0.069 = 6.9 mu. The *th* ↔ *st* distance is 37 + 33 / 1000 = 0.07 = 7 mu. **The best map is:**

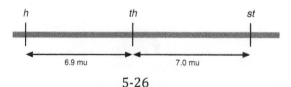

chapter 5

c. I = 1 − coefficient of coincidence; coefficient of coincidence = observed DCO / expected DCO. Thus coefficient of coincidence = 0 / (0.069)(0.07) = 0 and I = 1 − 0 = 1. **The interference is complete.**

27. a. wild type ♀ × scute echinus crossveinless black ♂ → 16 classes of data

Among the 16 classes there are wild type for all traits and scute, echinus, crossveinless, and black. This tells you that the parental female was heterozygous for all 4 traits. The fact the parental female is wild type also tells you that the wild-type alleles of all 4 genes are dominant. Remember that if you do a testcross with a female that is heterozygous for 3 linked genes, the data shows a very specific pattern. Because each type of meiosis (no recombination, DCO, etc.) gives a pair of gametes, you will see 8 classes of data that will occur in 4 pairs, both genetically and numerically. Thus, if you begin a cross with a female that is heterozygous for 4 linked genes, you should see 16 classes of progeny in a pattern of 8 genetic and numeric pairs. Although we have 16 classes of data, they do not occur in numeric pairs - instead we see numeric groups of 4. If one (or more) of the genes instead assorts independently of the others, then in a testcross you must see numeric groups of 4. For example, the most frequent classes will be parental, and there will be a 1:1:1:1 ratio of the 4 parental types.

Which gene is assorting independently relative to the other 3 genes? To answer this question, list the genotypes of the gametes that came from the heterozygous parent in the largest group of 4. This group should include the parental classes for the 3 linked genes; there are 4 genotypes here to account for the independent assortment of the unlinked gene. Then choose one of the 4 genes, and remove the allele of that gene from all 4 groups. When you do this with the gene that is assorting independently of the rest, you will find there are only 2 reciprocal classes of data left, which are the parental classes for the linked genes. If you choose one of the linked genes to remove, you will still have 4 different phenotypic classes left.

Try removing the *b* gene first, as shown in the *third column* of the Table that follows. You see that when the *b* allele is removed, only 2 classes are left: *s e c* and + + +. But when this analysis is repeated removing the allele of the *s* gene there are still 4 different phenotypes left (see the *right column* of the Table).

Genotype	Numbers	remove b	remove s
b s e c	653	s e c	b e c
+ s e c	670	s e c	+ e c
+ + + +	675	+ + +	+ + +
B + + +	655	+ + +	b + +

The *b* gene is therefore assorting independently of the other 3 genes. When the *b* gene is removed from all 16 classes of data, you can see that the data reduces to 8 classes that form 4 genetic and numeric reciprocal pairs, just as in any three-point cross [see answer to part (b) below]. Thus, the genotype of the parental female is:

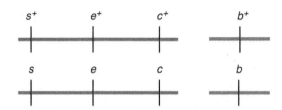

b. Write the classes out as reciprocal pairs.

Classes of gametes	Genotype	Numbers
Parental (P)	+ + +	1323
	s e c	1330
SCO 1	s + +	144
	+ e c	147
SCO 2	s e +	171
	+ + c	169
DCO	s + c	2
	+ e +	2

Compare the DCO $s + c$ to the parental $s\ e\ c$; this shows that e is in the middle. Calculate RF $s \leftrightarrow e$ = (144 + 147 + 2 + 2)/3288 = 9.0 cM; RF $e \leftrightarrow c$ = (171 + 169 + 2 + 2)/3288 = 10.5 cM.

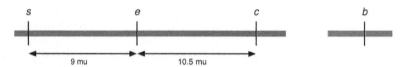

c. Interference = 1 - coefficient of coincidence
coc = observed DCO frequency / expected DCO frequency = (4/3288) / (0.09 × 0.105) = 0.001 / 0.009 = 0.11.
I = 1 – 0.11 = 0.89; yes, there is interference.

28. a. Recombination does not occur in male *Drosophila*. Therefore, in the cross $A\ b\ /\ a\ B$ ♀ × $A\ b\ /\ a\ B$ ♂ → F$_1$, the females will make 4 types of gametes (parental $A\ b$ and $a\ B$; recombinant $A\ B$ and $a\ b$). The frequencies of the 2 parental gametes will be equal to each other, and the frequency of the $A\ B$ recombinant will be equal to the frequency of its reciprocal recombinant, $a\ b$. Males, however, will make only two (parental) gamete types in equal numbers: $A\ b$ and $a\ B$. The Punnett square that follows shows how this will play out; the relative sizes of the boxes do not affect the ultimate answer.

chapter 5

Sperm Eggs	A b	a B
A b Parental	A b / A b (A– bb)	A b / a B (A– B–)
a B Parental	a B / A b (A– B–)	a B / a B (aa B–)
A B Recombinant	A B / A b (A– B–)	A B / a B (A– B–)
A b Recombinant	A b / A b (A– bb)	a b / a B (aa B–)

There is a ratio of 1/4 *A– bb* (*orange* boxes): 1/2 *A– B–* (*blue* boxes): 1/4 *aa B–* (*purple* boxes) for the progeny receiving the parental gametes AND the same ratio among the progeny receiving the recombinant gametes. Thus, **the overall phenotypic dihybrid ratio will always be 1/4 *A– bb* : 1/2 *A– B–* : 1/4 *aa B–*, independent of the recombination frequency between the *A* and *B* genes.**

   This will **not** be true of the cross *A B / a b* ♀ × *A B / a b* ♂ (see Table on the following page). The male will make the parental gametes, *A B* and *a b*, while the female will make 4 types of gametes: parental *A B* and *a b*; recombinant *A b* and *a B*. In this case, half of the progeny will look *A– B–* because they received the *A B* gamete from the male parent irrespective of the gamete from the female parent. When the male parent donates the *a b* gamete, then it is the gamete from the female parent that determines the phenotype of the offspring. The progeny that are *A– bb* and *aa B–* have received a recombinant gamete from the female parent (and the *a b* gamete from the male). These classes of progeny can then be used **to estimate the recombination frequency between the *A* and *B* genes: RF = 2(# of *A– bb* + # of *aa B–*)/total progeny.** (The factor of 2 is included because you cannot see half of the recombinants.)

chapter 5

Eggs \ Sperm	A B	a b
A B Parental	A B / A B (A– B–)	A B / a b (A– B–)
a b Parental	a b / A B (A– B–)	a b / a b (aa bb)
A b Recombinant	A B / A B (A– B–)	A b / a b (A– bb)
a B Recombinant	A b / A B (A– B–)	a B / a b (aa B–)

b. In mice, recombination occurs in both females and males. Therefore, in the cross $A\ b\ /\ a\ B\ ♀\ \times\ A\ b\ /\ a\ B\ ♂$ both sexes will make the same array of gametes (parental $A\ b$ and $a\ B$; recombinant $A\ B$ and $a\ b$). In this case, the only phenotype of progeny with a singular genotype will be $aa\ bb$. In the cross under consideration here, the $aa\ bb$ phenotype can only arise if both parents donate the $a\ b$ recombinant gamete. Of course, the other recombinant gamete, $A\ B$, occurs with equal frequency. The probability of the $aa\ bb$ genotype = (probability of an $a\ b$ recombinant gamete)^2. If you know the frequency of the $aa\ bb$ phenotypic class in the progeny, **recombination frequency (frequency of recombinant products) between the A and B genes = $2(\sqrt{\#aa\ bb / \#}$ total progeny)**.

In the case of the $A\ B\ /\ a\ b\ ♀\ \times\ A\ B\ /\ a\ b\ ♂$ cross in mice, both sexes are making the same parental gametes ($A\ B$ and $a\ b$) and recombinant gametes ($A\ b$ and $a\ B$) gametes. Again, the only phenotype of progeny with a singular genotype will be $aa\ bb$. In this example, this phenotype is the result of the fusion of the $a\ b$ parental gamete from each parent. The frequency of the $aa\ bb$ phenotype = (the frequency of the $a\ b$ gamete) × (the frequency of the $a\ b$ gamete). If you know the frequency of the $aa\ bb$ phenotypic class in the progeny, **the frequency of nonrecombinant products between the A and B genes = $2(\sqrt{\#aa\ bb / \#}$ total progeny)**. Recombination frequency = 1 - frequency of nonrecombinant products between the A and B genes.

chapter 5

29. Diagram the cross:

    MCS/MCS (Virginia strain) $\times$ mcs/mcs (Carolina strain) $\rightarrow$ F_1 MCS/mcs $\times$ mcs/mcs $\rightarrow$?

    Assume no interference, and remember the map of these 3 genes:

    You are being asked to calculate the proportion of the testcross progeny that will have the Virginia parental phenotype. In Chapter 3 we could answer this question by calculating the monohybrid ratios for each pair of alleles ($M : m$, $C : c$ and $S : s$) in the heterozygous parent. The testcross parent can provide only the recessive alleles for each gene, so the probabilities of the various phenotypes can be determined by applying the product rule. Unfortunately, this method of arriving at the probability of the progeny phenotypes only works when the genes under discussion are assorting independently. When the 3 genes are linked, as in this problem, to calculate the frequency of a parental class we must calculate first calculate the frequencies of *all* of the phenotypes expected in the testcross progeny.

    A parent that is heterozygous for 3 genes will give 8 classes of gametes. In a test cross, the gamete from the heterozygous parent determines the phenotype of the progeny. When the 3 genes are genetically linked, these 8 classes will be found as 4 reciprocal pairs. In other words, each meiosis in the heterozygous parent must give 2 reciprocal products that occur at equal frequency. For instance, a meiosis with no recombination will produce the parental gametes, MCS and mcs at equal frequency. This particular reciprocal pair will also be the most likely event and so the most frequent pair of products. The least probable meiotic event is a double crossover which is a recombination event in the region between M and C (region 1) and simultaneously in the region between C and S (region 2). The remaining gametes are produced by a single crossover in region 1 (the reciprocal pair known as single crossovers in region 1 or SCO 1) and a single crossover in region 2 (the reciprocal pair known as single crossovers in region 2 or SCO 2).

    The numbers shown on the map of this region of the chromosome represent the recombination frequencies in the gene-gene intervals. There are 6 mu between the M and C genes, so 6% (RF = 0.06) of the progeny of this cross will have had a recombination event in region 1. This recombination frequency includes all detectable recombination events between these 2 genes - both SCO in region 1 *and* DCO. Likewise, 17% of all the progeny will be the result of a recombination event in region 2. The DCO class is the result of a simultaneous crossover in region 1 and region 2. Recombination in two separate regions of the chromosome should be independent of each other, so we can apply the product rule to calculate the expected frequency of DCOs. Thus, the frequency of DCO = (0.06) $\times$ (0.17) = 0.01. Remember that the recombination frequency between M and C (region 1) includes both SCO 1 and DCO. Thus, 0.06 = SCO 1 + 0.01; solve for SCO 1 = 0.06 - 0.01 = 0.05. The same calculation for region 2 shows that SCO 2 = 0.16. The parental class = 1 - (SCO 1 + SCO 2 + DCO). Also remember that each class of gametes (parental, etc.) is made up of a

5-31

reciprocal pair. If the frequency of the DCO class is 0.01, then the frequency of the *M c S* gamete is half of that = 0.005. This logic is summarized in the table below.

Classes of gametes	Genotype	Frequency of reciprocal pair	Numbers	Frequency of each class
Parental (P)	M C S m c s	1 - all else	1 - (0.05 + 0.16 + 0.01) = 0.78	0.39 0.39
SCO 1	M c s m C S	RF in region 1 = SCO 1 + DCO	SCO 1 = 0.06 - 0.01 = 0.05	0.025 0.025
SCO 2	M C s m c S	RF in region 2 = SCO 2 + DCO	SCO 2 = 0.17 - 0.01 = 0.16	0.08 0.08
DCO	M c S m C s	(RF in region 1) x (RF in region 2)	(0.06) x (0.17) = 0.01	0.005 0.005

a. The proportion of backcross progeny resembling Virginia (parental, *M C S*) = 0.39.

b. Progeny resembling *m c s* (P) = 0.39.

c. Progeny with *M c S* (DCO) = 0.005.

d. Progeny with *M C s* (SCO 2) = 0.8.

30. On the basis of a recombination frequency (RF) of 5%, the physical distance between the *HD* gene and the *G8* marker was estimated to be 5 million bp, but when the human genome was sequenced, the actual physical distance between the gene and the marker was found to be only 500,000 bp. Why is the actual distance much lower than the estimated distance? The reason is that the estimate of the distance on the basis of the RF assumes that recombination is uniform over the entire genome. This assumption is not true. Although over the whole genome the *average* relationship is that 1% RF corresponds to 1 million bp everywhere in the genome, this value hides a great deal of variation in the relationship in different regions of the genome.

In this particular case, it appears that about 10× more crossovers occur in the interval between the *HD* gene and the *G8* marker than in the average 500,000 bp interval in the genome. Clearly then, **the genomic DNA between *HD* and *G8* contains a recombination hotspot.**

31. In cross #1 the criss-cross inheritance of the recessive alleles for *dwarp* and *rumpled* from mother to son tells you that all these genes are X-linked. In cross #2, you see the same pattern of inheritance for *pallid* and *raven*, so these genes are X-linked as well. The fact that the F_1 females in both crosses were wild type tells you that the wild-type allele of all four genes is dominant to the mutant allele. Designate alleles: dwp^+ and *dwp* for the *dwarp* gene, rmp^+ and *rmp* for the *rumpled* gene, pld^+ and *pld* for the *pallid* gene, and rv^+ and *rv* for the *raven* gene. Assign the genes an arbitrary order to write the genotypes. If

you keep the order the same throughout the problem, it is sufficient to represent the wild-type allele of a gene with +.

Cross 1: *dwp rmp + + / dwp rmp + +* × *+ + pld rv* / Y → *dwp rmp + + / + + pld rv* (wild-type females) and *dwp rmp + +* / Y (dwarp rumpled males).

Cross 2: *+ + pld rv / + + pld rv* × *dwp rmp + +* / Y → *+ + pld rv / dwp rmp + +* (wild-type females) and *+ + pld rv* / Y (pallid raven males).

Final cross: *dwp rmp + + / + + pld rv* (cross 1 F₁ females) × *dwp rmp pld rv* / Y → 428 *+ + pld rv*, 427 *dwp rmp + +*, 48 *+ rmp pld rv*, 47 *dwp + + +*, 23 *+ rmp pld +*, 22 *dwp + + rv*, 3 *+ + pld +*, 2 *dwp rmp + rv*.

All 4 genes <u>might</u> be genetically linked, as they are all on the X chromosome. If so, you expect 16 classes of progeny (2 × 2 × 2 × 2) in 8 genetic and numeric pairs. Notice that there are only 8 classes of progeny. The data that is seen shows <u>exactly</u> the pattern you expect for a female that is heterozygous for 3 linked genes. If 2 of the 4 genes <u>never</u> recombine then you would expect the pattern of data that is seen. If 2 genes never recombine, they will always show the parental configuration of alleles. Thus, examine the various pairs of genes for the presence or absence of recombinants. Note that you <u>never</u> see recombinants between *pallid* and *dwarp*: all the progeny are either pallid or dwarp, but never pallid and dwarp nor wild type for both traits. This suggests that the two genes are so close together that there is essentially no recombination between the loci. If a much larger number of progeny were examined, you might observe recombinants. Treat *dwp* and *pld* as 2 genes at the same location, so one of them (*dwp*, for instance) can be ignored, and this problem becomes a three-point cross between *pld*, *rv* and *rmp*.

Classes of gametes	Genotype	Numbers
Parental (P)	dwp rmp + +	427
	+ + pld rv	428
SCO 1	+ rmp pld rv	48
	dwp + + +	47
SCO 2	dwp + + rv	22
	+ rmp pld +	23
DCO	+ + pld +	3
	dwp rmp + rv	2

When the *rmp + rv* DCO class is compared to the *rmp + +* parental class, you can see that *rv* is in the middle. The *rmp* ↔ *rv* RF = (48 + 47 + 3 + 2)/1000 = 10 cM; the *pld* ↔ *rv* RF = (22 + 23 + 3 + 2)/1000 = 5 cM. I = 1 - coc; coc = observed frequency of DCO/expected frequency of DCO. Thus, coc = (5/1000)/(0.05)(0.1) = 0.005/0.005 = 1; I = 1 - 1 = 0, so **there is no interference**.

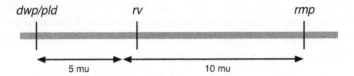

chapter 5

32. a. Interference is a phenomenon by which the occurrence of one crossover prevents the occurrence of a second crossover in a nearby region of the same chromosome. Scientists think that the existence of interference helps ensure that each bivalent has at least one crossover, which is necessary in turn to ensure that homologous chromosomes segregate from each other properly during the first meiotic division. One way of thinking about the utility of interference is the following: If you assume that only a limited amount of recombination enzymes are available in primary spermatocytes or primary oocytes, the limited number of crossovers that can occur will be apportioned among all the bivalents.

    Fig. 5.7a on p. 134 shows a preparation of chromosomes from a primary spermatocyte. Note that each bivalent shows at least one *green* recombination nodule (a few may be hard to see, but they are present). **The existence of interference is suggested by the fact that even on the bivalents that show more than one spot of recombination nodules, these nodules are far away from each other.** That is, one recombination event prevents the occurrence of a nearby recombination event on the same bivalent.

    b. Figure 5.7a merely suggests that two crossover events do not occur often at nearby positions on the same bivalent. To show that this conclusion about the existence of interference actually hold up from this type of evidence, you would have to look at many meiotic figures from many spermatocytes and/or oocytes. **You would then have to ask: If recombination nodules can occur at any position in the genome with equal likelihood (that is, if interference did not exist), how likely would it be that two recombination nodules would occur in a given specified distance on one bivalent. (You could calculate this probability by totaling the average number of recombination nodules from many meiotic figures and dividing by the total length of all the chromosomes to find the average distance between nodules. You could then see if the space between nodules can be graphed as a *normal* (Gaussian) distribution around this average.** If the data did not resemble the expected bell-shaped curve, then the distribution is not random. If interference did exist, you would expect to see the data shifted over from a bell-shaped curve so as to favor larger-than-average distances between adjacent nodules.

33. To answer this question, you need first to count the number of recombination nodules visible in the whole genome depicted in Fig. 5.7a on p. 134. There is at least one nodule per bivalent, and there are 20 bivalents in mice. By our own count, we see 29 nodules, but you may see (or imagine you see) a few more or less. We will take 29 as the proper count here. In this meiotic figure, the 29 nodules are distributed among 1,386 cM in the male genetic map. If the same mouse genome corresponds to 1,817 cM in the female genetic map, this must be because more recombination events take place in the average primary oocyte than in the average primary spermatocyte. So you would multiply 29 times the ratio of the cM in the female : male genetic maps (1,817 / 1,386) = ~**38 recombination nodules in the average prophase I primary oocytes.** [Your final answer may vary depending on how many nodules you counted in Fig. 5.7a, but your answer must be more than 20 × (1,817 / 1,386) = 26.]

chapter 5

Section 5.4

34. The null hypothesis is that there is independent assortment of 2 genes yielding a dihybrid phenotypic ratio of 9/16 R– Y– : 3/16 R– yy : 3/16 rr Y– : 1/16 rr yy. Use the chi square (X^2) test to compare Mendel's observed data with the 9:3:3:1 ratio expected for two genes that assort independently.

Genotypes	Observed #	Expected #	X^2 square equation	Sum of X^2
R– Y–	315	9/16 (556) = 313	$(315 – 313)^2/313$	0.01
R– yy	108	3/16 (556) = 104	$(108 – 104)^2/104$	0.15
rr Y–	101	3/16 (556) = 104	$(101 – 104)^2/104$	0.09
rr yy	32	1/16 (556) = 35	$(32 – 35)^2/35$	0.26
				0.51

The number of classes is 4, so the degrees of freedom is 4–1 or 3. Using Table 5.2 on p. 147 of the text, the probability of having obtained this level of deviation by chance alone is between 0.9 and 0.99 (90 - 99%). Thus we *cannot* reject the null hypothesis. In other words, **the data are consistent with independent assortment and we therefore conclude that Mendel's data could indeed result from the independent assortment of the 2 genes.**

You should note, however, that although Mendel's data are consistent with independent assortment, this chi square analysis does not prove that these genes assort independently. This analysis did not exclude any hypothesis. We could not formulate the null hypothesis as stating that the genes do not assort independently; we cannot predict the expected number of progeny under this hypothesis given that linked genes can have RF anywhere from 0% to just under 50%.

35. a. Diagram the cross.

orange (O– bb) × black (oo B–) → F_1 brown (O– B–) → 100 brown (O– B–) : 25 orange (O– bb) : 22 black (oo B–) : 13 albino (oo bb)

Because the F_1 snakes were all brown, we know that the orange snake could not have contributed an *o* allele, or there would have been some black snakes. The orange snake must be *OO bb*. The black snake could not have contributed a *b* allele or there would have been some orange snakes, so the black parent must be *oo BB*. Therefore **the F_1 snakes must be *Oo Bb*.**

b. The F_1 snakes are heterozygous for both genes (*Oo Bb*). If the two loci assort independently, we expect the F_2 snakes to show a 9 brown : 3 orange : 3 black : 1 albino ratio. The total number of F_2 progeny is 160. **We expect 90 (160 × 9/16) of these progeny to be brown, 30 (160 × 3/16) to be orange, 30 to be black and 10 (160 × 1/16) to be albino.**

c.

Genotypes	Observed #	Expected #	X^2 square equation	Sum of X^2
O– B–	100	90	$(100-90)^2/90$	1.11
O– bb	25	30	$(25-30)^2/30$	0.83
Oo B–	22	30	$(22-30)^2/30$	2.13
oo bb	13	10	$(13-10)^2/10$	0.9
				4.97

There are three degrees of freedom (4 classes – 1) and the p value is between 0.5 and 0.1. **The observed values do not differ significantly from the expected.**

d. There is a 10% - 50% probability that these results would have been obtained by chance if the null hypothesis were true; this is simply another way of writing the meaning of the p value.

36. a. The cross is:

    normal (*DD*) × dancer (*dd*) → F$_1$ normal (*Dd*)   F$_2$ 3/4 *D–* (normal) : 1/4 *dd* (dancer)

    1/4 of the F$_2$ mice will be dancers if the trait is determined by a single gene with complete dominance.

    b. Diagram the cross:

    normal (*AA BB*) × dancer (*aa bb*) → F$_1$ normal (*Aa Bb*)   F$_2$ 15/16 normal (*A– B–* + *A– bb* + *aa B–*) : 1/16 dancer (*aa bb*)

    1/16 of the mice would be expected to be dancers given the second hypothesis that dancing mice must be homozygous for the recessive alleles of two genes.

    c. Calculate the chi square values for each situation. Null hypothesis #1: Dancing is caused by the homozygous recessive allele of one gene, so 1/4 of the F$_2$ mice should be dancers. Calculating the expected numbers, 1/4 × 50 mice or 13 should have been dancers, 37 should have been nondancers.

Genotypes	Observed #	Expected #	X^2 square equation	Sum of X^2
nondancers	42	(3/4) × 50 = 37	$(42-37)^2/37$	0.68
dancers	8	(1/4) × 50 = 13	$(8-13)^2/13$	1.92
				2.60

With one degree of freedom the p value is between 0.5 and 0.1 and we cannot reject the null hypothesis. The hypothesis that dancing is caused by the homozygous recessive allele of one gene is therefore a good fit with the data. Null hypothesis #2: Dancing is caused by being homozygous for the recessive alleles of two genes (*aa bb*), so 1/16 of the F$_2$ mice should be dancers.

chapter 5

Genotypes	Observed #	Expected #	X² square equation	Sum of X²
nondancers	42	47	$(42-47)^2/47$	0.53
dancers	8	3	$(8-3)^2/3$	8.33
				8.86

With one degree of freedom, the p value is < 0.005, so the null hypothesis that two genes control the dancer phenotype is not a good fit; in fact, the hypothesis can be rejected by the criteria employed by most geneticists. **The one gene hypothesis is a better fit with the data.**

Section 5.5

37. a. The number of meioses represented here is the total of the number of asci = **334**. Each ascus contains the 4 products of one meiosis.

   b. Diagram the cross: $a + c \times + b +$

   To map these genes, use the Three Easy Rules for Tetrad Analysis. First designate the type of asci represented. This has to be done for each pair of loci as PD (P), NPD (N) and T refer exclusively to the relationship between two genes. In the table below, the top row shows the designations for all three pairs: the a-b comparison is at the lower left, the b-c comparison is at the lower right and the a-c comparison is at the top of the pyramid.

	P		P		T		T		N		P	
P		P	N	N	T	P	T	N	N	P	T	T
a + c		a b c	+ + c	+ b c	a b +	a + c
a + c		a b c	a + c	a b c	a b +	a b c
+ b +		+ + +	+ b +	+ + +	+ + c	+ + +
+ b +		+ + +	a b +	a + +	+ + c	+ b +
137		141	26	25	2	3
I I I		I I I	II I I	II I I	I I I	I II I

   Rule #1: For genes a and b PD = NPD, so these two genes are not linked. For genes b and c PD = NPD, so genes b and c are not linked. For genes a and c, PD>>NPD, so the genes are linked. Calculate RF between a and c = 2 + (1/2)(26 + 25)/334 = 8.2 cM (Rule #2).

   Gene-centromere distances can be calculated in *Neurospora* (Rule #3), so analyze the data for MI and MII segregation patterns for the alleles of each gene in each ascus type. This analysis is done separately for each gene, unlike the gene↔gene analysis done above. The designation for each gene is presented under that gene at the bottom row of the table (I - MI, II = MII). Rule #3 shows that the distance between a and the centromere = (1/2)(26 + 25)/334 = 7.6 mu; the distance between b and the centromere = (1/2)(3)/334 = 0.4 mu; the distance between c and the centromere = (1/2)(0)/334 = 0.

Now compile all of these pieces of data into one map. Rule #1 shows that gene *b* is 0.4 mu from the centromere and is on a different chromosome from genes *a* and *c*. Genes *a* and *c* are on the same chromosome, so the *a*↔centromere distance and the *c*↔centromere distance refer to the same centromere. Gene *c* is 0 mu from the centromere. As you can see from the map, there are 2 slightly different distances for the gene *a*↔gene *c* region. The gene↔gene distance is 8.2 mu and the *a*↔centromere distance is 7.6 mu. In this case the longer gene↔gene distance is more accurate as it includes the SCOs between *a* and *c* as well as some of the DCOs between *a* and *c* (the 4-strand DCO, Fig. 5.23 on p. 153).

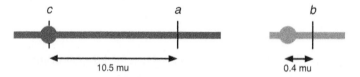

c. Carefully consider the information you have for the different chromosomes in the ascus type chosen (the group with 3 members). The *a* and *c* genes show PD segregation, which can mean either no crossing-over between them or a 2-strand DCO. Both genes show MI segregation, which means there haven't been any single crossovers between either gene and the centromere. **Gene *b* shows MII segregation, which means there has been an SCO between the gene and the centromere.** This crossover also means the *a-b* and *c-b* comparisons in this class will show the tetratype pattern (T are due to crossovers between either gene and its centromere when the genes are on separate chromosomes).

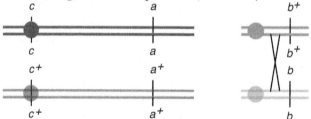

38. a. These data are presented as phenotypes of individual spores. Because the data are presented for 3 genes, they may be analyzed as a 3-point cross. Organize the data into reciprocal pairs of spores.

Classes of gametes	Genotype	Numbers
Parental (P)	a + +	31
	a f g	29
SCO 1	a + g	6
	a f +	6
SCO 2	a + g	13
	a f +	14
DCO	a + +	1
	a f g	1

5-38

The α f g DCO spore type is most similar to the a f g parental spore type. Thus, the mating type (a/α) is the gene in the middle. The distances are: f ↔ a/α = (6 + 6 + 1 + 1)/101 = 13.9 cM and a/α ↔ g = (13 + 14 + 1 + 1)/101 = 28.7 cM.

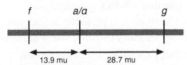

b. This problem says you have an ascus with an α f g spore. This spore is the result of a double crossover [see part (a)]. The reciprocal product would be the a + + spore, but this is not seen in the ascus. Remember that there are 3 different types of double crossovers: 2-strand DCOs, 3-strand DCOs, and 4-strand DCOs. Each of these types of DCOs gives a different array of spores in the resulting ascus (Fig. 5.23 on p. 153). Draw a meiotic figure of this chromosome and try some different types of DCOs. **A 3-strand DCO gives the desired result.**

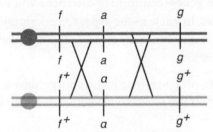

39. Diagram the cross and summarize the data: $met^-\ lys^- \times met^+\ lys^+$

P	T
$met^+\ lys^+$	$met^+\ lys^+$
$met^+\ lys^+$	$met^-\ lys^+$
$met^-\ lys^-$	$met^+\ lys^-$
$met^-\ lys^-$	$met^-\ lys^-$
89	11

a. **The two types of cells in the first group of 89 asci are $met^+\ lys^+$ (could grow on all four types of media) and $met^-\ lys^-$ (require the addition of met and lys to minimal medium).** These asci are parental ditypes (PD). The four types of cells in the second group of 11 asci are $met^+\ lys^+$; $met^+\ lys^-$ (grew on min + lys and on min + lys + met); $met^-\ lys^+$ (grew on min + met and on min + lys + met); and $met^-\ lys^-$. These are tetratype (T) asci.

b. Because the number of PD >>> NPD, **the genes are linked.** The distance between them is:

[NPD + (1/2)T] / total tetrads = [0 + (1/2)(11)]/100 × 100 = **5.5 m.u.**

c. NPD should be seen eventually and would result from four-strand double crossovers. There would be two types of spores: $met^-\ lys^+$ (could grow on min +

met and on min + met + lys) and *met⁺ lys⁻* (could grow on min + lys and on min + met + lys).

40. Use the Three Easy Rules for Tetrad Analysis to help you solve this problem.

    a. In cross 1, the number of PD (parental ditypes) = NPD (nonparental ditypes) so the *ad* gene and the mating type locus assort independently. In cross 2 the number of PD >> NPD so we can conclude the *p* gene and the mating locus are linked; RF *p↔mating type* = [NPD + (1/2)T]/ total tetrads = (3) + (1/2)(27)/54 =16.5/54= .31 × 100 = 31 cM between the two genes.

    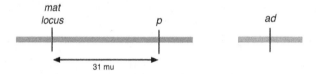

    b. **To calculate gene↔centromere distances you need information on the order of ascospores in each ascus type.** Only with this information can you calculate gene↔centromere distances based on 1/2(# of asci showing MII segregation for the gene)/total asci.

41. a. The genotype of a true-breeding wild-type diploid strain of *Saccharomyces* can be written + / +. If this diploid undergoes meiosis, **all (100%) of the asci will have 4 viable spores.**

    b. The genotype of this strain is + / *n*, where *n* = a null activity allele of an essential gene. The diploid cells will be viable, because they have functional enzyme for this essential gene from the + allele. If this diploid undergoes meiosis, then **all (100%) of the tetrads will have 2 + : 2 n spores (that is, only 2 spores in each tetrad will be viable).**

    c. Gene *a* and gene *b* are different essential genes; *a* and *b* represent temperature-sensitive alleles of these genes. When you diagram a cross you must write complete genotypes for both haploid parents: *a b⁺* (strain a) × *a⁺ b* (strain b) → *aa⁺ bb⁺*. Each haploid parent strain will die when grown under restrictive conditions because they cannot produce a required product, while the diploid cells are viable because they have a wild-type allele for both genes.

    If these genes are unlinked, then after meiosis PD asci = NPD asci (Three Easy Rules for Tetrad Analysis: Rule #1). When the genes are unlinked, also remember that T tetrads (asci) arise from SCOs between either gene and the centromere. Because each of these genes is 0 mu from the centromere, you will not see T asci. Thus, after meiosis, you will have 50% PD (2 *a b⁺* spores : 2 *a⁺ b* spores) : 50% NPD (2 *a⁺ b⁺* : 2 *a b*). None of the spores in the PD tetrads are viable under restrictive conditions, while 2 of the 4 spores in the NPD asci are viable (*a⁺ b⁺*). Thus, **50% of the asci will have 0 viable spores (PD) and 50% of the asci will have 2 viable spores (NPD).**

    d. Again genes *a* and *b* are unlinked essential genes, but now gene *a*↔centromere = 0 mu and gene *b*↔centromere = 10 mu. A SCO between gene *b* and the centromere

chapter 5

will happen in 20% of the asci (because only half of the ascospores of an SCO are recombinant) and will give T asci. A T ascus will have a spore ratio of 1 $a^+ b^+$: 1 $a b$: 1 $a b^+$: 1 $a^+ b$. There is 1 viable spore ($a^+ b^+$) in a T ascus. The remaining 80% of the asci will be divided equally between PD and NPD because the genes are unlinked. Thus, **40% of the asci will have 0 viable spores (PD), 40% of the asci will have 2 viable spores (NPD) and 20% of the asci will have 1 viable spore (T)**.

e. Because both genes are on the same chromosome, you can diagram this cross: $a b^+$ (strain a) × $a^+ b$ (strain b) → $a b^+ / a^+ b$. When the genes are linked, SCO and 3-strand DCO between the genes give T asci and 4-strand DCO between the genes gives NPD asci. The remainder of the asci are PD (no crossover and 2-strand DCO). If the recombination frequency between gene a and gene b = 0 mu, then the only possible result will be PD asci (2 $a b^+$: 2 $a^+ b$) in which none of the spores are viable. Thus, **100% of the asci will have 0 viable spores**.

f. Now the genes are 10 mu apart and there are no DCO events. This means that 20% of the asci produced will be T [see part (d)] and the remaining 80% will be PD. Therefore, **80% of the asci will have 0 viable spores (PD) and 20% of the asci will have 1 viable spore (T)**.

g. A 4-strand DCO gives an NPD ascus. The genotypes of the spores will be 2 $a^+ b^+$: 2 $a b$. Thus, **2 of the 4 spores will be viable**.

42. Genes a, b, and c are all on different chromosomes. [For simplicity, we will refer to the centromere of the chromosome on which a gene is found with the possessive (that is, as the gene's centromere.)] Thus, all crosses between these genes are expected to give an equal number of PD and NPD asci. The fact that the cross involving genes a and b yields no T indicates that **genes a and b are both very close to their respective centromeres**. If two genes are on separate chromosomes, tetratypes arise when a crossover occurs between one of the genes and its centromere, or between the other gene and its centromere. In the crosses involving genes a and c or genes b and c, many T asci are seen. Because genes a and b are tightly centromere-linked, **gene c must be very far from its centromere** in order to generate these T asci.

43. Diagram the yeast cross:

    $his^- lys^+$ × $his^+ lys^-$ → $his^- lys^+ / his^+ lys^-$ → 233 PD, 11 NPD, 156 T.

    a. The **PD** asci must have 2 $his^- lys^+$: 2 $his^+ lys^-$ spores, the **NPD** asci have 2 $his^+ lys^+$: 2 $his^- lys^-$ spores and the **T** asci have 1 $his^- lys^+$: 1 $his^+ lys^-$: 1 $his^+ lys^+$: 1 $his^- lys^-$ spores.

    b. **PD>>NPD so the genes are linked; RF between 2 genes = [11 + (1/2)(156)]/400 = 22.3 mu.**

    c. Because the genes are linked, you know that NPD asci are the result of 4-strand DCOs. This sort of DCO is 1/4 of all the DCO events. The 3-strand DCOs (1/2 of all DCOs) give T while 2-strand DCOs (1/4 of all DCOs) give PD. In the data you see 11 NPD tetrads, or 11 meioses that underwent 4-strand DCOs. There are another 22 tetrads that are the result of 3-strand DCOs (and are Ts), and another 11 asci that underwent 2-strand DCOs (and are PDs), for **a total of (11 + 22 + 11 =) 44 asci that underwent 2 crossovers. The remaining (156 - 22 =) 134 T asci underwent a SCO, or 1 crossover, and the remaining (233 – 11 =) 222 PD asci underwent 0 crossovers (NCOs)**. From this analysis, you can see that a general equation for the number of SCO meioses is SCO = T-2NPD. Likewise, a general equation for the number of DCO meioses is DCO = 4NPD.

d. There are 44 asci that underwent DCOs for a total of 88 crossover events. There were another 134 asci that underwent SCOs. Thus, there were 134 + 88 = 222 crossover events / 400 meioses = **0.555 crossovers/meiosis**. We have just calculated "m", the mean number of crossovers between *his* and *lys* per meiosis. A general equation for m is: m = [(SCO + 2 (DCO)]/total asci.

e. The equation for recombination frequency between 2 genes used in the textbook and in part (b) above only includes 4-strand DCO (NPD). **This formula ignores 3-strand and 2-strand DCOs**. This calculation for recombination frequency assumes that all T asci are due to SCOs. Though this is true for the majority of T asci, some T asci are due to 3-strand DCO events; the larger the distance between the 2 genes in question, the more Ts are due to DCOs. Also, it is an oversimplification to assume that all PD asci are nonrecombinant, because some of them are due to 2-strand DCOs; the fraction of PDs due to DCOs also increases with the distance between the two genes in question. Remember that the 3 types of DCO events occur in a ratio of 1 (2-strand DCOs) : 2 (3-strand DCOs) : 1 (4-strand DCOs)(see Fig. 5.23 on p. 153).

A more accurate formula for RF between two genes that takes into account all crossovers is: RF = m/2, where m is the mean number of crossovers per meiosis, as defined in part (d) above. The reason is that every meiosis that occurs with a single crossover (every SCO) results in 2 recombinant chromosomes and 2 nonrecombinants. You can prove to yourself that this equation makes sense by imagining two theoretical situations. First, suppose that every meiosis were an SCO (m = 1); then RF = 1/2 or 50%, as expected when no NCOs occur. Second, suppose that half of all meioses occurred as SCOs (m = 1/2) and the other half were NCOs; then RF = 25%, which makes sense as 1/4 of all resulting chromosomes would be recombinants. (Note that the equation for the green line in Fig. 5.15 on p. 142 is m/2.)

With this understanding, and also the relationships we determined in parts (c) and (d), we can derive a more accurate formula for recombination frequency: **RF = m/2 = (1/2 SCO + DCO)/total asci = [1/2 (T – 2NPD) + (4NPD)]/total asci = [(1/2)T + 3NPD]/total asci**. Many yeast geneticists use this more accurate formula in preference to the one in your textbook.

f. RF = [(1/2)(156 – 22) + 4(11)]/400 = [(1/2)(134) + 44]/400 = 111/400 = 0.278 = **27.8 mu**. As you can see, this value is somewhat larger than the distance calculated in part (b). As expected, this corrected RF value = m/2, where m = 0.555 [see part (d)].

44. This problem can be solved only if you make assumptions about interference in the 22 m.u. interval between genes *C* and *D*. The reason that interference is important is that the degree of interference determines the number of double crossovers (DCOs) that

chapter 5

occur, and this will of course influence the proportion of tetrads that are PD, NPD, and T. The solution is by far easiest to calculate if interference = 1, because no DCOs occur in this situation. We will thus discuss each of the two parts of this problem first under the assumption that interference = 1. We then consider each part of the problem at the other extreme, in which interference = 0. If interference is greater than 0 but less than 1, the proportions of the three types of tetrads would vary between these extremes.

a1. Situation: Cross in *Saccharomyces cerevisiae*; Interference = 1. **Because there are no double crossovers (DCOs) between genes *C* and *D*; there can be no NPD tetrads.** All the recombinants between the 2 genes are due to single crossovers (SCO). If two genes are linked, then SCOs between the genes give T asci. Remember that the RF between *C* and *D* is (.07 + .15 = 0.22), and that the formula for recombination frequency between 2 genes = [NPD + (1/2)T]/total asci. Solve for T: 0.22 = 0 + (1/2)T so T = 2(0.22) = 0.44. Therefore, **44% of the asci will be Ts and the remaining 56% will be PDs.** In Problem 43 above you just derived a more accurate formula for RF = [3NPD + (1/2)T]/total asci. This new formula accounts for the Ts and NPDs produced by DCOs, and you can see that when there are no DCOs, both formulas are equally accurate.

b1. Situation: Cross in *Neurospora crassa*; Interference = 1. In *Neurospora* the recombination events that underlie the formation of PD, NPD, and T are the same as in yeast. The difference is that the ordered tetrads allow you to distinguish whether a SCO event occurs between *C* and the centromere or between *D* and the centromere. If the SCO occurs between *C* and the centromere, then you will see MII segregation for the alleles of the *C* gene and MI segregation for the alleles of the *D* gene. If the SCO occurs between *D* and the centromere then you will see the reverse - MI segregation for *C* and MII segregation for *D*.

Here, I = 1, so there are no DCOs. Therefore, as in part (a), T = 0.44 and PD = 0.56. However 7/22 of the SCO events occur between gene *C* and the centromere, while the remaining 15/22 of the SCOs occur between *D* and the centromere. **The expected results are summarized in the table below.**

Crossover type and location:	NCO (no crossover)	SCO C↔cent.	SCO D↔cent
ascus type:	PD	T	T
MI or MII gene C:	MI	MII	MI
MI or MII gene D:	MI	MI	MII
frequency:	0.56	7/22 x 0.44 = 0.14	15/22 x 0.44 = 0.30

Before looking at the math for situations where the interference is not equal to 1, you should think about this problem qualitatively. As interference decreases, a higher proportion of meioses will have two or more crossovers. Because NPDs can be produced only by meioses with more than one crossover (in the parts of the problem solved above for interference = 1, there were no NPDs), as interference decreases you will see more NPD tetrads. Because in the corrected equation for RF in tetrad analysis from Problem 43 {where RF = [3NPD + (1/2)T]/total asci}, if the RF was constant, the proportion of T tetrads would have to decrease to compensate for the NPDs. This equation also sets limits on the proportions of the total tetrads that would be T and NPD. Understanding the qualitative relationship of interference with the proportions of tetrads is more important than obtaining numerical values for these proportions, as we next describe.

a2. **Situation:** Cross in *Saccharomyces cerevisiae;* Interference = 0. You should note at the outset that the figure accompanying this problem is somewhat ambiguous: Do the map units shown correspond to the measured recombination frequency (RF) in an experiment, or are they instead corrected map units corresponding to the physical distance between the genes (as in Fig. 5.15 on p. 142)? To match all the other problems in this chapter, we assume that the map units shown on the figure represent RFs measured in an experiment. (This ambiguity does not exist when the interference = 1 because the two definitions are then identical.)

If there is no interference, then the likelihood that crossovers occur anywhere on the chromosome is given by a Poisson distribution. Although mathematical methods exist to predict the frequencies of single, double, and higher level crossovers in chromosomal intervals of given lengths from the Poisson distribution, we will simplify the discussion by using an approximation for the expected frequencies of DCOs based on the idea that the two single crossovers occurred independently. That is, the expected frequency of DCO = (recombination frequency in the $C \leftrightarrow D$ region) × (recombination frequency in the $C \leftrightarrow D$ region) = (0.22) × (0.22) = 0.0484. [You may ask why you could not estimate the expected frequencies of DCOs by multiplying the $C \leftrightarrow$ centromere and centromere$\leftrightarrow D$ distances (0.07) **x** (0.15). The reason is that such a calculation would ignore all the DCOs in which two crossovers take place either in the $C \leftrightarrow$ centromere or in the centromere$\leftrightarrow D$ intervals.]

The RF between $C \leftrightarrow D$ = SCO + DCO (looking at individual spores). [The reason this is true is that even though DCOs have two crossovers, on average, only half of the spores from meioses with double crossovers could be recognized as recombinants.] Thus, 0.22 = SCO + 0.0484; SCO = 0.22 - 0.0484 = 0.1716. If SCO frequency between C and D = 0.1716 then T due to SCO = 0.3432 (see the case above where Interference = 1).

Remember that when analyzing tetrads the three different types of DCOs can be distinguished: 2-strand DCOs, which give PD asci; 3 strand DCOs, which give T asci; and 4-strand DCOs, which give NPD asci (Fig. 5.23b-f on p. 153). These DCOs occur in the ratio of 1/4 (2 strand DCOs) : 1/2 (3-strand DCOs) : 1/4 (4-strand DCOs). If the total DCO frequency = 0.0484 then 1/4 (0.0484) = 0.0121 is the frequency of 4-strand DCOs (NPD), 1/2 (0.0484) = 0.0242 is the frequency of 3-strand DCOs (T) and the remaining 1/4 of the DCOs (0.0121) are 2-strand DCOs, which will be PD tetrads.

In total, **NPD = 0.0121, T = 0.0242 (due to DCO) + 0.3674 (due to SCO) = 0.3795 and the remainder are PD = 0.6205**.

b2. **Situation:** Cross in *Neurospora crassa;* Interference = 0. Here, the proportions of PD, NPD, and T tetrads will be the same as in part (a2) above. However, several patterns of MI and MII segregation patterns for the two genes are possible depending upon where the SCO and DCO events are occurring. Let's first consider the NCO and SCO meiosis (this is the same table as shown in part [b] above but without the frequencies):

crossover type and location:	no crossover	SCO C↔cent.	SCO cent.↔D
ascus type:	PD	T	T
gene C (MI/MII)	MI	MII	MI
gene D (MI/MII)	MI	MI	MII

Next we consider DCO meioses in which both crossovers occur in the same interval: either C↔centromere or centromere↔D. In either case, the DCO could either involve 2, 3, or 4 strands.

crossover type and location:	DCO C↔cent. 2-strand	DCO C↔cent. 3-strand	DCO C↔cent. 4-strand	DCO cent.↔D 2-strand	DCO cent.↔D 3-strand	DCO cent.↔D 4-strand
ascus type:	PD	T	NPD	PD	T	NPD
gene C (MI/MII)	MI	MII	MI	MI	MI	MI
gene D (MI/MII)	MI	MI	MI	MI	MII	MI

Finally, let's look at the DCO meioses in which one crossover occurs in C↔centromere and the other crossover occurs in centromere↔D. Such DCOs could again involve either 2, 3, or 4 strands.

crossover type and location:	DCO C↔cent. cent.↔D 2-strand	DCO C↔cent. cent.↔D 3-strand	DCO C↔cent. cent.↔D 4-strand
ascus type:	PD	T	NPD
gene C (MI/MII)	MII	MII	MII
gene D (MI/MII)	MII	MII	MII
frequency:	1/4 x 0.0105 = 0.003	1/2 x 0.0105 = 0.005	1/4 x 0.0105 = 0.003

Note that in these tables, we have calculated the frequencies only for the DCOs in which one crossover occurs in C↔centromere and the other crossover occurs in centromere↔D. In this situation, the expected number of simultaneous crossovers in both regions is easy to calculate as the product of the map distances in the two intervals (0.07) x (0.15) = 0.0105. The number of meioses with 3-strand DCOs will be twice that of the meioses with either 2-strand or 4-strand DCOs. The frequencies for the classes shown in the two previous tables are complex to calculate (for example, you would have to determine the expected number of DCOs with both crossovers in each interval) and are not worth the effort. What is worthwhile is to look at the patterns of PD, NPD, and T superimposed with the patterns for MI and MII segregation for the two genes, as these patterns tell us what occurred during meiosis in each case.

45. a. If at least one crossover, and often multiple crossovers, occurred between a gene the centromere in every meiosis, the MI or MII configuration of octads would be randomized. As there are 2 different types of MIs (4*a+* : 4*a-* and 4*a-* : 4*a+*), and 4 different types of MII octads (2*a+* : 2*a-* : 2*a+* : 2*a-* and 2*a+* : 2*a-* : 2*a-* : 2*a+* and 2*a-* : 2*a+* : 2*a+* : 2*a-* and 2*a-* : 2*a+* : 2*a-* : 2*a+*), **when MI and MII octads are produced randomly, 2/6 will be MI, and 4/6 will be MII. As the RF between a gene and the centromere is (1/2)(# MII octads)/(total # octads), the maximal RF between a gene the centromere is (1/2)(4/6) = 1/3 = 33%.**

b. **Yes – a gene and the centromere can be unlinked. This simply means that at least one crossover occurs between them in every meiosis, and so the fraction of MII octads is 33% [see part (b)].** Of course a gene is always physically connected to a centromere through a DNA strand; however, it need not be linked to the centromere genetically.

c. A distance of 30 m.u. between a gene and the centromere is close to the maximal RF that can be measured by ordered tetrad analysis, indicating that the gene is far from the centromere. This map distance is unlikely to be accurate; the reason is that DCOs between the gene and the centromere will ignored because they result in MI octads.

46. a. Refer to the three strains as *trp1*, *trp2* and *trp3*. Remember that in *Neurospora* the 4 meiotic products undergo a subsequent mitosis to give 8 spores in the ascus. Consequently, the first 2 spores are identical to each other (they are mitotic products of the same initial spore), the 3d and 4th spores are the same as each other, and so on. For the purposes of discussing the results, assume that the ascus is made up of the 4 original spores prior to this extra mitosis. You cross *trp1-* × wild type → diploid → 2 wild type : 2 *trp1-*. You are seeing a monohybrid ratio of 1:1, which means there is only one gene controlling the Trp- phenotype in strain *trp1*. Each of the strains gives the same result, so **in each haploid strain a single mutant gene is responsible for the Trp- phenotype**.

b. First consider the cross between *trp3-* and wild type. The diploid is *trp3-* / *trp3+*. After meiosis the first 2 spores were either both *trp3+* (could grow on minimal media) or they were both *trp3-* (could not grow on minimal media). Thus, all the asci were 2 *trp3-* : 2 *trp3+*, and the order of the alleles was either - - + + or + + - -. In other words, all the asci showed MI segregation for the *trp3* gene. In the case of the crosses with *trp1-* × wild type and *trp2-* × wild type, the resultant diploids gave some asci that gave a different result - only one spore of the top 2 was trp+ while the other one was trp-. In other words, these asci showed MII segregation for the *trp* gene in question. In summary, **the *trp3* gene is very closely linked to its centromere (no T = no crossovers between the gene and the centromere), while the *trp1* and *trp2* mutations are further from their centromere(s)**. (Note that we are again using the possessive "its centromere" or "their centromeres" to indicate the centromere of the chromosome on which the gene is found.)

c. If two strains have mutations in the same gene, then the resulting diploid ($trpx^-$ / $trpx^-$) would be unable to grow on minimal media and would give asci where all 4 spores were $trpx^-$ (0 spores viable on minimial media). This result is never seen, so each mutant strain must have a mutation in a different gene, for a total of 3 different trp^- genes in the 3 strains. Diagram the cross between *trp1* and *trp2*:

$trp1^- \ trp2^+ \times trp1^+ \ trp2^- \rightarrow trp1^- \ trp1^+ \ trp2^+ \ trp2^-$

When this diploid is allowed to undergo meiosis you see **78 asci** with 0 viable spores (2 $trp1^- \ trp2^+$: 2 $trp1^+ \ trp2^-$ = **PD and 22 asci** with 2/8 or 1/4 viable spores (1 $trp1^+ trp2^+$: 1 $trp1^- trp2^-$: 1 $trp1^- \ trp2^+$: 1 $trp1^+ \ trp2^-$ = **T**).

In the ***trp1 × trp3* cross** you see a new class of asci, those with 2 viable spores (2 $trp1^+ trp3^+$: 2 $trp1^- \ trp2^-$ = NPD). In this cross **there are 46 PD, 48 NPD, and 6 T asci.**

In the last cross, ***trp2 × trp3*, there are 42 PD, 42 NPD and 16 T asci.**

d. In *trp1 × trp2*, PD>>NPD (NPD= 0) so the genes are linked. The T asci are caused by SCOs between the 2 genes. The recombination frequency between *trp1* and *trp2* = [0 + (1/2)(22)]/100 = 0.11 = 11 mu. In *trp1 × trp3* there are 46 PD = 48 NPD so the genes are unlinked; the 6 T asci arise from SCOs between *trp1* and its centromere. None of these T asci can arise from crossovers between *trp3* and its centromere because you found in part (b) that they were very tightly linked. Thus, *trp1* is 3 mu from its centromere. In the cross between *trp2* x *trp3*, PD = NPD so the genes are unlinked. There are 16 T asci, so gene 2 must be 8 mu from its centromere. **The map is shown in the following diagram.**

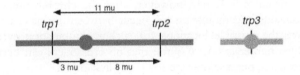

e. Ordered octads show the segregation pattern for each gene separately. In this example, **both mutant genes give the same phenotype (Trp⁻). Thus, it is impossible to determine if a Trp⁻ spore is + - or - - or - +.** If you cannot distinguish these different types of spores, then you cannot determine the segregation pattern (MI vs MII) for the individual genes.

f. **You can calculate gene↔centromere distances because you discovered that one of the genes (*trp3*) is tightly linked to its own centromere**. Therefore, when $trp3^-$ is crossed with either of the other mutants, any T asci must be due to SCO between the other gene and its centromere, and these T asci can then be used to calculate a distance between the other gene and its centromere. This same sort of analysis can also be used in yeast to map gene↔centromere distances if a strain is available with a mutation known to be right at a centromere.

chapter 5

Section 5.6

47. a. The colony is mostly white because the cells are heterozygotes ($ade2^+$ / $ade2^-$) and the $ade2^+$ allele is dominant. Red sectors arise when one cell in the growing colony becomes homozygous $ade2^-$ / $ade2^-$. As all the cells continue to grow, the colony continues to expand and the $ade2^-$ / $ade2^-$ cells form a red sector within the white colony (Fig. 5.29 on p. 159). **Red cells of the $ade2^-$ / $ade2^-$ genotype could arise by mitotic recombination (Fig. 5.28 on p. 158), by loss of the entire chromosome containing the $ade2^+$ allele, by a deletion of the portion of the chromosome containing the $ade2^+$ allele, or by spontaneous mutation of the $ade2^+$ allele to an $ade2^-$ allele.** (*Note:* Using the usual genetic nomenclature in yeast, the $ade2^+$ allele could be written as *ADE2* and the $ade2^-$ allele as *ade2*. These are alternative ways of representing the same things.)

    b. The size of the red sectors depends on when in the formation of the colony the event occurred to form the initial $ade2^-$ / $ade2^-$ cell. If the event occurred early in the formation of the colony there will be a larger red sector than if the event occurred near the end of colony formation. All of the events mentioned in part (a) are rare, so in general they will occur later in colony formation when there are more cells in which they could occur. As a result, **most of the red sectors will be small.** (The colonies in Fig. 5.29 on p. 159 actually contain several very small sectors that are too tiny to show up in the photograph.)

48. In a normal mitosis (with no recombination), all of the daughter cells produced are genotypically *a b c leth d e* / a^+ b^+ c^+ $leth^+$ d^+ e^+ and are phenotypically wild type, like the original cell. Rarely, however, after the chromosomes have replicated into sister chromatids, recombination can occur between nonsister chromatids. The figure below shows the possible locations for such recombination events; these are indicated by Roman numerals I-V. Assume that no crossovers can occur between gene *a* and the centromere because they are so closely linked.

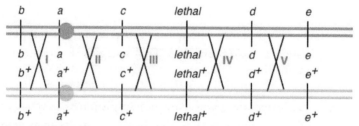

If you review Fig. 5.28 on p. 158, you will see that after mitotic recombination, only one orientation of the chromatids on the homologous chromosomes will segregate to yield progeny genotypically different from the parents, and we will only discuss this orientation here. The important rule to keep in mind is that daughter cells of an originally heterozygous parental cell will become homozygous for an allele only if the mitotic crossover occurs between the centromere and the gene. If the mitotic crossover occurs further away from the centromere than the gene, all daughter cells will remain heterozygotes for that gene (see again Fig. 5.28 on p. 158).

Consider now the results of a crossover in region I of the figure. In the segregation pattern of interest, one daughter cell will be homozygous for the b^+ allele and heterozygous for the rest of the genes on the chromosome. Thus, this cell will be phenotypically wild type and indistinguishable from the nonrecombinant cells surrounding it. However, the reciprocal product of this mitosis will be a daughter cell whose genotype will be homozygous mutant for gene b and heterozygous wild type for everything else. This cell will continue to divide mitotically and yield a patch of b mutant tissue in a sea of wild-type tissue.

Next consider a crossover in region II. In this case, the segregation pattern of interest will yield one wild-type cell and a reciprocal daughter cell which will be b^+ a^+ c $leth$ d e / b a c $leth$ d e. Because this cell is homozygous for the recessive lethal mutation, it will die and you will never see it. A crossover in region III will also give a lethal recombinant product (genotypically b^+ a^+ c^+ $leth$ d e / b a c $leth$ d e) that again you will never see.

A crossover in region IV will give a d e patch of mutant tissue (genotypically b^+ a^+ c^+ $leth^+$ d e / b a c $leth$ d e), while a crossover in region V will give a patch of e mutant tissue (genotypically b^+ a^+ c^+ $leth^+$ d^+ e / b a c $leth$ d e).

Thus, **the only phenotypes that will be found in sectors as a result of mitotic recombination will be those associated with b, d, and e** (representing crossovers in regions I, IV, and V, respectively).

49. The genotype of the female fly is y^+ sn^+ / y sn. A diagram of potential mitotic crossovers in the cells of this animal is shown in the following figure:

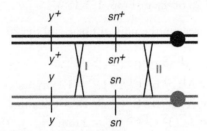

a. **The larger patch of yellow tissue arises from a mitotic crossover in region I.** This mitotic recombination will generate a cell of the genotype y sn^+ / y sn (Problem 5-48). The smaller patch of yellow and singed tissue must have arisen from a second mitotic recombination between the centromere and sn, in region II. This would produce cells that are y sn / y sn. Because the yellow patch is larger, the recombination in region I occurred earlier in development. The yellow, singed cells are a small patch within a larger patch of yellow cells, so the mitotic crossover in region II happened after (later in development than) the crossover in region I, and this second crossover happened in a recombinant daughter cell of the first recombination event. No wonder these patches within patches are rare!

b. If the genotype of the female was $y^+\ sn\ /\ y\ sn^+$, then **a recombination event in the region I would give you a detectably recombinant cell (yellow phenotype) with the genotype $y\ sn\ /\ y\ sn^+$. A subsequent second mitotic crossover in region II in one of these originally recombinant $y\ sn\ /\ y\ sn^+$ cells will give you a patch of $y\ sn$ tissue inside a patch of yellow tissue, as in part (a)**.

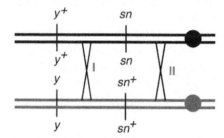

Although the problem did not explicitly ask, you should note that one kind of mitotic crossover can occur which will give a different result in a $y^+\ sn^+\ /\ y\ sn$ female [part (a)] than a $y^+\ sn\ /\ y\ sn^+$ female [part (b)]. After a recombination event in region II the female from part (a) will show a $y\ sn$ patch of tissue. The other recombinant product will be homozygous $y^+\ sn^+$, and this cell and its descendants will be indistinguishable from nonrecombinant cells. The same sort of recombination event in region II in the female from part (b) will yield one daughter cell of genotype $y^+\ sn\ /\ y^+\ sn$ which will give a patch of sn tissue. The reciprocal product of this mitotic recombination will be $y\ sn^+\ /\ y\ sn^+$ and will give an adjacent patch of yellow tissue. This phenomenon is called *twin spots* and was depicted in Figs. 5.27 and 5.28 of the text (pp. 157-158).

50. List the phenotypes seen in the normal tissue and the 20 tumors and their frequency of occurrence:
    - normal tissue = $A^F A^S\ B^F B^S\ C^F C^S\ D^F D^S$
    - tumor type 1 = $A^F\ B^F B^S\ C^F C^S\ D^F$ = 12 tumors
    - tumor type 2 = $A^F\ B^F B^S\ C^F C^S\ D^F D^S$ = 6 tumors
    - tumor type 3 = $A^F\ B^S\ C^F C^S\ D^F$ = 2 tumors

Remember that all of the tumor tissues were also homozygous for $NF1^-$ while the normal tissue is heterozygous $NF1^-\ NF1^+$.

a. **Mitotic recombination** could have caused all 3 types of tumors.

b. Remember that mitotic recombination causes genes further from the centromere to become homozygous, while genes between the recombination event and the centromere, or those on the other side of the centromere, remain heterozygous (review Problem 5-48). Notice that all 3 types of tumor cells are homozygous for $NF1^-$, so by definition all 20 mitotic recombination events that gave rise to the 20 tumors occurred between the centromere and the $NF1$ gene. Notice also that all 20 tumors are homozygous for gene *A*, specifically the A^F allele. Because gene *A* is affected by all of the mitotic recombination events, it must be further from the centromere than the *NF1* gene. Because these tumors become homozygous for the A^F allele, that allele must be on the same homolog with the *NF1⁻* allele. There are 6 tumors that are homozygous for these 2 genes (tumor type 2).

As you work your way along the chromosome from *NF1* toward the centromere the next gene to also become homozygous is the *D* gene (tumor type 1), specifically the D^F allele, then the *B* gene (tumor type 3), the B^S allele. Gene *C* never becomes homozygous, yet we are told it is on this chromosome as well. Thus it could either be on the other side of the centromere from *NF1*, *A*, *D* and *B* or it could be on the same side of the centromere as the rest of the genes but very closely linked to the centromere (like the *a* gene in Problem 5-48). **The order of the genes and coupling of the alleles in normal tissue (prior to mitotic recombination) is shown in the following figure:**

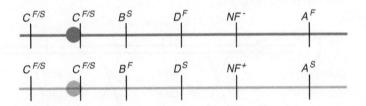

It is possible to use the <u>mitotic</u> recombination frequencies with which the various genotypes of tumors arose as a rough, relative approximation of the distances between the genes. These numbers are NOT to be confused with the meiotic recombination frequency we have been calculating in the other problems in this and other chapters! In tumor type 2 the mitotic recombination event happened between gene *NF1* and gene *D* (*NF1⁻* is homozygous, and so further from the centromere than the recombination event while gene *D* is still heterozygous), and this occurred in 6/20 tumors = 0.3, so **the relative recombination frequency between *NF1* gene *D* is 0.3**. In tumor type 1 the mitotic recombination event happened between gene *B* and gene *D* (*B* is still heterozygous in these tumors while *D* is homozygous D^F), and this happened in 12/20 tumors = 0.6, so **the relative recombination frequency between genes *B* and *D* is 0.6**. In tumor type 3 the mitotic recombination event happened between the centromere and gene *B* (or between gene *C* and gene *B* if you place gene *C* on the same side of the centromere as the NF1 gene). This event happened in 2/10 tumors = 0.1 so **the relative recombination frequency between the centromere and gene *B* is 0.1**.

c. An entire homolog of one chromosome is lost. If the lost homolog were the one with the *NF1⁺* allele, then **the resulting cell would be hemizygous for the *NF1⁻* allele, and would develop into a tumor. This did NOT occur here, because ALL 3 genotypes of tumors are still heterozygous for at least the *C* gene.**

5-51

d. **Yes, deletions of portions of the $NF1^+$ homolog which cause loss of the $NF1^+$ allele** could cause tumors to develop in the resulting $NF1^-$ cells. If these deletions extended to the neighboring genes *A, D,* or *B,* the tumor cells could show the same protein variants as in the problem. Although point mutations which inactivate the $NF1^+$ allele could cause tumors to develop in the resulting $NF1^-$ cells, this mechanism does not seem to be responsible for any of the particular tumors seen, because all of the tumors express only one allele of gene *A* (the A^F allele), and this should not be the case if a point mutation occurred in the *NF* gene.

51. This problem illustrates how scientists studying the development of multicellular organisms can use GFP (green fluorescent protein) and mitotic crossing-over to create *mosaic* animals in which different cells have different genotypes. This technique has the additional virtue that the scientists can recognize these different cells within the body of the animal to see how gene function in various cells influence the organism's development.

    a. See following diagram:

    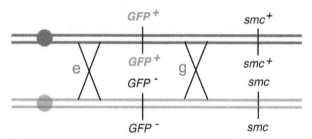

    b. To study the importance of the smc^+ gene, scientists want to study what happens when the gene function is removed; that is, to "knock out" the function of the gene. Because animals homozygous for the loss-of-function *smc* mutation die as early embryos, you cannot obtain any adult mice that completely lack *smc* function. **By using mitotic recombination, researchers can create animals in which most cells are heterozygotes (smc^+/smc), but some cells are smc/smc homozygotes without *smc* function.** As long as the lack of the *smc*-encoded protein in those particular cells does not prevent mice from being born, scientists can then study what happens to the animal if the protein is removed from specific cells.

    c. **The light green cells (most of the cells in the heterozygous animal) are cells with one copy of the GFP gene. The dark green cells are those with two copies of GFP** that are formed from mitotic recombination (these are homozygous for GFP^+). Because these cells have two copies of this gene, they make more GFP protein and fluoresce in green more brightly. **The cells with no GFP^+ gene fail to fluoresce and are shown in white in the figure.** These are the other, reciprocal daughter cells produced from the same mitotic recombination events giving rise to the bright green cells.

    d. As just mentioned in the answer to part (c), **the bright green cells (clone 2) and the white cells (clone 1) are next to each other because they are the reciprocal

chapter 5

products of the same mitotic recombination events. Right after mitosis, these cells are adjacent to each other because they were produced from the same parent cell. If the progeny of these reciprocal daughter cells do not move in the epithelium, they will stay in the same general area and make the *twin spots* that are shown.

e. **See the mitotic recombination event "e" depicted in the diagram shown in the solution to part (a).** The mitotic crossover must occur in the interval between the centromere and the GFP gene so as to create daughter cells homozygous for GFP^+ (the bright green cells) or homozygous for no *GFP* gene (the white cells). Because the *smc* gene is further away from the centromere than the *GFP* gene, the mitotic crossover shown in the diagram will make a daughter cell that is homozygous for both d for bright green and normal sized), and a daughter cell that is homozygous for no *GFP* gene and for the *smc* mutation (no fluorescence, small sized).

f. There are more cells in clone 1 than in clone 2 because **the cells in clone 2 are homozygous for the recessive *smc* mutation that causes these cells to divide prematurely** before they reach normal size, making the cells smaller than the normal cells (like those in clone 2) with one or more copies of the smc^+ allele. There will be more of such small white cells because **in any given period of time, they will undergo more rounds of cell division than the normal sized cells.**

g. **See the mitotic recombination event "g" depicted in the diagram shown in the solution to part (a).** This event occurs in the interval between *GFP* and *smc*. These daughter cells will be light green (one copy of GFP^+) and small (because they are homozygous for *smc*). No twin spots will result from this type of mitotic crossover because the reciprocal daughter cell will be light green (GFP^+ / no *GFP*) and normal sized (smc^+ / smc^+), just like all the surrounding cells.

chapter 6

DNA Structure, Replication, and Recombination

Synopsis

The statement "DNA's genetic functions flow directly from its molecular structure" is a good starting point for comprehending how DNA serves as the fundamental molecule of inheritance. The structure of DNA is not only beautiful in its own right, but the functions of DNA are also inherent in this structure. Make sure you get a good mental image of the DNA molecule and its construction. Understanding where hydrogen and covalent bonds are found, the polarity of the strands of DNA, and why complementarity is important will provide a good basis for appreciating many cellular processes (for example DNA replication, recombination, transcription, and translation).

The most important principle to understand and remember from this chapter is what DNA polymerase needs in order to synthesize DNA (Fig. 6.22 on p. 190). DNA polymerase requires:

- A **template**. The template is stretch of single-stranded DNA to be copied by DNA polymerase through base complementarity.
- A **primer.** The primer is a segment of RNA (or DNA) that hybridizes to the template and that provides a free 3' end. DNA polymerase adds nucleotides to this 3' end in succession; DNA polymerase thus synthesizes new DNA in the 5'-to-3' direction as the moves in the 3'-to-5' direction along the template.
- All four **deoxyribonucleotides** (**dNTPs**). DNA polymerase adds dNTPs onto the 3' end of the primer, using the template to add complementary base pairs one-by-one.

DNA synthesis by DNA polymerases is not only essential for chromosome replication, but also for DNA repair and even for genetic recombination by crossing-over. You will see later in the book that many procedures in recombinant DNA technology that allow modern-day geneticists to manipulate DNA molecules also involve DNA synthesis by DNA polymerase.

Key terms

deoxyribonucleic acid (DNA) – a polymer that contains genetic information made of of four different types of subunits called **nucleotides**. Each nucleotide contains a sugar (**deoxyribose**), a **phosphate group**, and one of four **nitrogenous bases**: either a double-ring **purine** [**adenine (A)** or **guanine (G)**], or a single-ring **pyrimidine** [**cytosine (C)** or **thymine (T)**].

phosphodiester bonds – covalent linkages between adjacent nucleotides in DNA. The phosphate on the 5' carbon on the sugar ring of an incoming nucleotide is linked to the hydroxyl group on the 3' carbon of the sugar of the last nucleotide in the growing polymer.

chapter 6

antiparallel – describes the opposing orientations of the two strands of double-stranded DNA. DNA polymers have direction – **polarity** – because at one end (the **5' end**), the 5' carbon of the sugar on the first nucleotide is not connected to any other nucleotide; while at the other end (the **3' end**), the 3' carbon of the sugar on the final nucleotide is not connected to any other nucleotide.

complementary base pairing – formation of **hydrogen bonds** between specific bases (A-T and G-C) on antiparallel DNA strands. The two DNA strands in double-stranded DNA are complementary; the base sequence of one strand predicts the base sequence of the other strand.

semiconservative DNA replication – each strand of a double-stranded DNA molecule is used as a **template** for synthesis of a complementary antiparallel DNA strand. Thus, after DNA replication, every double-stranded DNA molecule consists of one old strand (the template) and one newly synthesized strand.

replication fork – during DNA replication, a Y-shaped area where the top of the "Y" consists of two unwound DNA strands that are both used as templates for semiconservative DNA replication. The replication fork moves continually in the direction of the stem of the Y as the double-stranded DNA that constitutes the stem continually unwinds.

leading strand and **lagging strand** – During DNA replication, the leading strand is synthesized continuously in the **5'-to-3' direction** toward the unwinding replication fork. The lagging strand has a polarity opposite to that of the leading strand, and so it must be synthesized discontinuously as small **Okazaki fragments** that are ultimately joined into a continuous strand.

ribonucleic acid (RNA) – a polymer of **ribonucleotides**, that is similar in structure to DNA with the following exceptions: (i) in RNA, the sugar molecule is ribose instead of deoxyribose in DNA; (ii) **uracil** is present instead of thymine; and (iii) RNA is usually single-stranded.

RNA primer – during DNA replication, a short stretch of RNA synthesized by **primase** enzyme that initiates ("primes") DNA synthesis

DNA polymerase – an enzyme complex that polymerizes DNA in the 5'-to-3' direction by extending a primer bound to a DNA template. DNA polymerase III extends the RNA primers, and DNA polymerase I fills in the gaps after the RNA primers are removed.

DNA ligase – during DNA replication, the enzyme that stitches Okazaki fragments together by forming phosphodiester bonds between adjacent nucleotides

recombination – the generation of new combinations of alleles by crossing-over and/or independent assortment. (In this chapter, we focus specifically on the generation of new allelic combinations through crossing-over.)

Holliday junctions – interlocked regions of two nonsister chromatids in recombination intermediates. During recombination, **resolvase** enzyme separates the two chromatids by breaking one DNA strand of each nonsister chromatid at each of two Holliday junctions.

chapter 6

> **heteroduplex** – region of DNA where the the two strands originate from different nonsister chromatids. Heteroduplexes form during the process of recombination.
>
> **crossover pathway** and **noncrossover pathway** – after initiation of recombination, crossing-over may occur through Holliday junction formation and resolution (crossover pathway), or it can be aborted by **anticrossover helicase** enzyme (noncrossover pathway), resulting only in the formation of one heteroduplex region.
>
> **gene conversion** – in a heterozygote, change in the base sequence of one allele to that of the other allele as a result of heteroduplex formation and mismatch repair during recombination.

Problem Solving

The problems in this chapter require a thorough understanding of:
- the molecular structure of DNA
- the molecular mechanism of semiconservative DNA replication
- the molecular mechanism of recombination

Unlike the problems in previous chapters, many of the problems in Chapter 6 require you to interpret or devise biochemical (as opposed to genetic) experiments. In some cases these experiments ask you to tag molecules with radioactivity so that you can follow them. Radioactive label can be incorporated into protein or DNA if a cell or organisms is grown in or fed a radioactive precursor that goes specifically into the type of molecule you want to follow. In designing experiments using radioactive labeling, be sure to consider how you can get a unique label into the molecule of interest.

Vocabulary

1.

a.	transformation	6.	Griffith experiment
b.	bacteriophage	11.	a virus that infects bacteria
c.	pyrimidine	9.	a nitrogenous base containing a single ring
d.	deoxyribose	2.	the sugar within the nucleotide subunits of DNA
e.	hydrogen bonds	4.	noncovalent bonds that hold the two strand of the double helix together
f.	complementary bases	8.	two nitrogenous bases that can pair via hydrogen bonds
g.	origin	10.	a short sequence of bases where unwinding of the double helix for replication begins
h.	Okazaki fragments	6.	short DNA fragments formed by discontinuous replication of one of the strands

chapter 6

i.	purine	3.	a nitrogenous base containing a double ring
j.	topoisomerases	13.	enzymes involved in controlling DNA supercoiling
k.	semiconservative replication	5.	Meselson and Stahl experiment
l.	lagging strand	1.	the strand that is synthesized discontinuously during replication
m.	telomeres	7.	structures at the ends of eukaryotic chromosomes

Section 6.1

2. (See Fig. 6.5, p. 177.) The proof that DNA was the transforming principle was **the treatment of the transforming extract with an enzyme (DNase) that degrades DNA. After this treatment, the extract was no longer able to transform rough, nonvirulent strains of *Streptococcus pneumoniae* bacteria into smooth, virulent cells that could kill host mice.** Avery, MacCleod and McCarty also showed that treatments with RNase and proteinase did not abolish the transforming activity of their extracts, indicating that the transforming principle was neither RNA nor protein. These experiments using enzymatic treatments were important because it could be argued that the purified "transforming principle" these investigators isolated as DNA might have contained proteins or other molecules. Even with this extensive evidence, for almost 20 years many scientists still remained unconvinced the DNA carries genes.

3. In DNA transformation, the DNA that enters the cell is in the form of randomly sized fragments, usually generated by mechanical forces that shear the DNA while it is being extracted and prepared from the bacterial cell. Therefore, two genes that are closer together on the chromosome will end up on the same fragment more often than genes that are far apart from each other on the chromosome. A high cotransformation frequency between two genes thus indicates that they are close together. **Gene *a* is closer to gene *c* than it is to gene *b*** because there are many instances when *a* and *c* were cotransformed, but only a few instances when *a* and *b* were cotransformed.

4. **Sulfur is found only in proteins, never in DNA,** while phosphorus is a major constituent of the backbone of the DNA molecule and is found only very rarely in proteins (none of the amino acids in proteins contain phosphorous, though as we will see in Chapter 8 phosphorous can sometimes be added to certain proteins at certain times). Nitrogen and carbon, on the other hand, are found in both proteins and DNA. **Hershey and Chase needed to differentiate between protein and DNA, so they needed to be able to specifically label the proteins and not the DNA and *vice versa*. If they had used labeled nitrogen or carbon there would be no way to differentiate protein and nucleic acid.**

Section 6.2

5. (See Feature Figure 6.11b,c on p. 183 for an overview of DNA structure.) **In Tube #1 all the sugar-phosphate (phosphodiester) bonds are broken. You would in theory**

6-4

chapter 6

see individual pairs of complementary nucleotides held together by hydrogen bonds. The sugars in the nucleotides would not be attached to phosphate groups; instead, the phosphates would be free in solution (Fig. 6.10 on p. 182). In reality, the hydrogen bonds that hold together individual complementary nucleotides (2 for A-T pairs and 3 for G-C pairs) are not very stable. Thermal forces working at room temperature would disrupt the hydrogen bonds between the nucleotides of one pair. You usually need at least 4 nucleotide pairs in order to have DNA that is stably double stranded at room temperature.

In Tube #2 the bonds that attach the bases to the sugars are broken. You would see base pairs (similar to Fig. 6.10 without the 'sugar') and sugar-phosphate chains without the bases (similar to Feature Figure 6.9b without the 'bases').

Tube #3 would contain single strands of DNA since the hydrogen bonds between bases were broken.

6. X-ray diffraction studies yielded a **crosswise pattern of spots, indicating that DNA is a helix** containing repeating units spaced every 3.4 Å. One complete turn of the helix occurs every 34 Å. **The diameter of the molecule is 20 Å, indicating that DNA must be composed of more than one polynucleotide chain** (Feature Figure 6.11b,c on p. 183). The key X-ray diffraction pictures were taken by Rosalind Franklin and Maurice Wilkins in 1951-1952; James Watson and Francis Crick then built models based on the known chemistry of the nucleotide building blocks to fit the X-ray data.

7. a. Human DNA is double stranded. If 30% of the bases are A, and A pairs with an equal amount of T (30%), that leaves 40% to be C + G. Thus, there will be **20% C**.

    b. **30% T** (see answer to part [a] above)

    c. **20% G** (must equal the amount of C)

8. Remember that in double-stranded DNA, the amount of $A = T$ and the amount of $G = C$. But the red amounts do not have to equal the blue amounts. Therefore one of the nucleotides in red + one of the nucleotides in blue must equal the other nucleotide in red plus the other nucleotide in blue.

    a. **True.** $A + C = T + G$

    b. **True.** $A + G = C + T$

    c. **False.** $A + T$ does not equal $G + C$. This fact (that $[A + T] \neq [G + C]$) is actually quite important. First, the DNA of different species can vary a great deal in the proportions of A-T base pairs relative to G-C base pairs. Second, the proportions of A-T base pairs and G-C base pairs can be very different in different regions of the same chromosome. In most organisms, the regions between genes have a higher proportion of A-T base pairs (they are "A-T rich") than the genes themselves.

    d. **False.** $A / G \neq C / T$ (except in genomes where the frequencies of all the bases are 25%).

    e. **True.** $A / G = T / C$ because the frequencies of the *red* nucleotides are equal to each other, and the frequencies of the *blue* nucleotides are equal to each other.

    f. **True.** $(C + A) / (G + T) = 1$. This is another way of writing $A + C = T + G$ as in part (a) above.

chapter 6

9. In double-stranded DNA, adenine pairs with thymine so the percentage of A must equal the percentage of T (this is also true for G and C). Here the amount of A is not equal to the amount of T (nor does C equal G), so **the chromosome of this virus must be single stranded**. (The DNA of single-stranded viruses becomes converted into double-stranded DNA when the virus infects the host cell.)

10. a. **The A-T base pairs have only two hydrogen bonds, so it takes less heat energy to denature these base pairs.** G-C base pairs have three hydrogen bonds holding them together. It thus takes more energy to break the bonds between Cs and Gs. Remember that the DNA of different species can vary a great deal in the proportions of A-T base pairs relative to G-C base pairs. Moreover, this ratio can be very different in different regions of the same chromosome. In most organisms, the regions between genes have a higher proportion of A-T base pairs (they are "A-T rich") than the genes themselves. In the early stages research on genomes, scientists sometimes tried to locate genes by looking for regions of DNA that were more resistant to heat denaturation, and thus had a higher G-C content.

    b. **The denatured single-stranded DNA must contain stretches of nucleotides that are complementary to a nearby sequence but in an inverted orientation.** These *stem loop* structures are regions where the single strand of DNA formed a double-stranded region. The loops and the strings holding the stems together are still single-stranded DNA.

11. 5'...**CAGAATGGTGCTCTGCTAT**...3'. This could also be written backwards as 3'.....TATCGTCTCGTGGTAAGAC.....5'. It is best to indicate polarity by showing the 5' and 3' directions on a strand of DNA or RNA. If no polarity is indicated then the 5' end is assumed to be at the left end of the sequence. **The dots show that this is a short region of a much longer nucleotide chain.**

Section 6.3

12. A note about nomenclature: Depending on the context, the letters A, C, G and T can be used to represent either the nitrogenous bases alone, or the nucleosides (base + sugar) or the nucleotides (base + sugar + phosphate) containing those bases. Although it is not of great importance for most genetics courses, you might find it useful to know that organic chemists distinguish between bases, nucleosides and nucleotides with a complicated nomenclature. The name of the base is altered slightly to indicate a nucleoside (for example adenine becomes adenosine) and the name "deoxy", or more precisely "2-deoxy" is added to the nucleoside name if the sugar is deoxyribose. To name a nucleotide, the number of phosphate groups it contains and the position of their connection to the sugar are also specified. Thus 2'-deoxythymidine 5'-monophosphate signifies a nucleotide containing deoxyribose, the base thymine and a single phosphate group connected to the 5'-carbon atom of deoxyribose. See if you can find this nucleotide in Feature Figure 6.9 on p. 180.

    a. The information in the DNA is contained in the order of its building blocks. There are 3 billion nucleotides in the complete haploid set of 23 human chromosomes. This amount of sequence can potentially provide a huge amount of information. **Although there are only 4 different building blocks, they can be combined**

chapter 6

in a huge number of combinations. For instance, when considering a short 10 nucleotide long piece of DNA there are 4^{10} or 1,048,576 different possible sequence permutations. The information may be recognized by proteins that bind directly to DNA (see Chapters 15 and 16), or, as you will see in Chapter 8, it can be copied into RNA to direct the synthesis of proteins.

b. **Each of the four building blocks is a nucleotide.** Each nucleotide is made of the sugar deoxyribose and one of the four nitrogenous bases (A, C, G or T) and a phosphate group (see Feature Figure 6.9 on p. 180). **Deoxyribose plus a base makes a nucleoside. When phosphate groups are added these become nucleotides.** In a strand of DNA the **adjacent nucleotides are connected by phosphodiester bonds** that link a phosphate group to both the 3'-carbon atom of the deoxyribose of one nucleotide and the 5' carbon atom of the deoxyribose of the next nucleotide in the chain.

c. **Four major differences between DNA and RNA exist:** (i) In RNA the sugar is ribose instead of deoxyribose. (ii) RNA contains the base U instead of T. (iii) Most DNA molecules found in nature are double stranded and most RNA molecules are single stranded, but there are exceptions to both of these cases. (iv) DNA strands can be very long – more than 100,000,000 nucleotides in a human chromosome, for example. The longest naturally occurring RNA molecules are much shorter – about 20,000 nucleotides at most.

13. The complementary DNA would have the complementary sequence with the opposite polarity. Note also the presence of T in DNA in contrast with U in RNA:

    3' GGGAACCTTGATGTTTCGGCTCTAATT 5'

14. Mix RNA from virus type 1 with protein from virus type 2 to reconstitute a "hybrid" virus. In a parallel experiment, mix RNA from virus type 2 with protein from virus type 1. Infect cells with each of these reconstituted hybrid viruses separately, and analyze the protein in the progeny viruses that result from each infection. You will find that **the progeny viruses in each case have the protein that corresponds to the type of RNA in the parent hybrid virus.** The protein in the progeny did not correspond to the protein in the parent hybrid virus.

15. a. The CAP protein recognizes (binds to) a sequence of double-stranded DNA that is 16 base pairs long, but only 10 of these base pairs are specified. If the 4 types of bases are all present at equal frequencies in the DNA, the chance of finding a CAP binding site in a random sequence is **1/4^{10} = 1/1,048,576 base pairs**. (Note that the unspecified Ns don't figure into the calculation because the probability of an N base [that is, any base] appearing is 1.)

    b. **The DNA sequence is rotationally symmetric – or palindromic;** the base sequence is identical when read in the 5'-to-3' direction on each strand. This makes sense if a DNA-binding region on each identical subunit of a dimer recognizes the major groove of DNA in the same way. (The specific amino acids in the DNA-binding part (*domain*) of each CAP monomer make the same noncovalent contacts with the same bases in the major groove.)

    c. **No; DNA helicase is not required for CAP to bind the major groove of DNA.** DNA helicase unwinds the double-stranded DNA molecule into single strands.

DNA-binding proteins like CAP can bind double-stranded DNA (usually making contacts with sites in the major groove) without unwinding. Specific amino acids in the DNA-binding domain of CAP form noncovalent bonds with specific nucleotides in DNA; some atoms of the bases are available to form these bonds even when the bases are paired in double-stranded DNA.

Section 6.4

16. a. **1/4 of the DNA is in an H/H (^{15}N/^{15}N) band near the bottom of the gradient; 3/4 of the DNA is in an L/L (^{14}N/^{14}N) band near the top.**

    b. **1/4 of the DNA is in an L/H (^{14}N/^{15}N) band in the middle of the gradient; 3/4 of the DNA is in an L/L (^{14}N/^{14}N) band near the top.**

    c. After one round, a single L/H band near the middle of the gradient would be expected. In subsequent rounds this band would be expected to spread gradually toward the top of the gradient as the heavy DNA is diluted across all of the new molecules in a random fashion. As a result **you would not expect discrete bands. Instead you would see a broad "smear."**

    d. **1/8 of the DNA is in an H/H (^{15}N/^{15}N) band near the bottom of the gradient; 7/8 of the DNA is in an L/L (^{14}N/^{14}N) band near the top.**

17. It is best to consider the individual strands of DNA when calculating the amount of DNA of different densities. Meselson and Stahl started with H:H double stranded DNA (with nucleotides only containing ^{15}N). After one generation in ^{14}N media (with L or light ^{14}N nitrogen), this original molecule becomes two molecules - H:L and H:L, both of which have an intermediate density. After a second generation these 2 molecules will become 4 - H:L, L:L and H:L, L:L. Thus there will be equal amounts of the intermediate band and the light band, as stated in the problem. After another round of DNA replication in the light media (round 3) these 4 molecules will become 8 - H:L, L:L, L:L, L:L and H:L, L:L, L:L, L:L. Thus **after 3 rounds 1/4 of the total DNA will be intermediate density and 3/4 will be light density**. After the next round of replication (round 4) the 8 molecules will become 16 - H:L, L:L, L:L, L:L, L:L, L:L, L:L, L:L and H:L, L:L, L:L, L:L, L:L, L:L, L:L, L:L. **At this point (after round 4), 1/8 of the double-stranded DNA molecules would be of the intermediate density (^{14}N/^{15}N) and the remaining 7/8 would be light density (^{14}N/^{14}N).**

18. After one S phase, the label would be in one strand of each DNA double helix (call this the round 1 helix), so each chromatid (a double-stranded molecule) contains label on one of its two strands. The labeled ^{3}H-thymidine was removed before the next S phase, so the next set of new strands are not labeled. When the unlabeled strand of each round 1 chromatid is used as a template in round 2, the resulting double stranded chromatid will be unlabeled. When the labeled strand in the round 1 helix is replicated, the resultant chromatid contains this labeled strand and an unlabelled, newly synthesized complementary strand. Thus, after this second round of replication is complete every chromosome would have one labeled chromatid (but the label is only on one strand) and one unlabeled chromatid. Therefore, for each homologous chromosome pair, **one chromatid of each of the two chromosomes would contain label [option (c)].**

chapter 6

The figure below illustrates the experiment. In this figure, each line represents a single strand of DNA. *Red* strands of DNA are labeled with radioactive (^{3}H)-thymidine, whereas *blue* strands of DNA do not have radioactivity. You can see that immediately after S phase in the presence of radioactive (^{3}H)-thymidine, both chromatids of both homologs will have radioactivity. After mitosis is completed and after a second round of S phase (this time in the absence of radioactivity), only one chromatid of each of the two chromosomes will contain label; this is option (c). [*Notes:* (1) After mitosis, only one of the two daughter cells is shown, but the situation is exactly the same in the other daughter cell. (2) The bottom of the figure shows what you would see in the microscope when looking at metaphase chromosomes. The microscope does not have sufficient resolution to distinguish the two strands of DNA that make up a chromatid.]

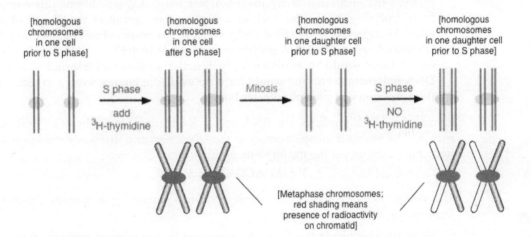

19. In the following figure, *blue* lines indicate parental DNA strands, while *red* lines are newly synthesized DNA. Note that this figure actually shows a complete *replication bubble* with a Y-shaped *replication fork* at each end (the problem asked you to draw only one of these two replication forks). At each replication fork, one new strand of DNA is synthesized continuously (the *leading strand*) while the other strand of DNA must be synthesized discontinuously as small *Okazaki fragments* (small *red* arrows) that need to be stitched together to make a contiguous new strand.

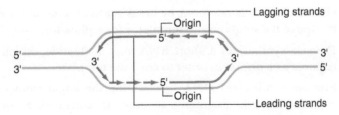

20. a. **DNA polymerase needs to interact with the hydroxyl (–OH) group at the 3' end of the previously incorporated nucleotide and with high energy phosphate groups at the 5' end of the incoming nucleotide triphosphate. From the point of view of the products of the reaction, DNA polymerase allows**

the –OH at the 3' end of the growing strand to "attack" the high-energy bond between the phosphate closest to the 5' end of the incoming nucleotide triphosphate and the next phosphate. In essence, this reaction adds the incoming nucleotide (with one phosphate) to the 3' end of the growing strand via a new phosphodiester bond. This action of the DNA polymerase automatically dictates the 5'-to-3' growth of the DNA chain during replication.

b. DNA ligase also catalyzes the formation of a phosphodiester bond between the nucleotides located at the 3' end of one Okazaki fragment and the 5' end of the next Okazaki fragment. **Because only a single phosphate group is attached to the 5' carbon atom of the nucleotide at the 5' end of an Okazaki fragment after the primer is removed, DNA ligase needs an external source of energy to catalyze the formation of a phosphodiester bond. Ligase obtains this energy by hydrolyzing ATP. This lack of a triphosphate group at the 5' end of an Okazaki fragment also explains why DNA polymerase cannot be used to join Okazaki fragments – DNA polymerase does not hydrolyze ATP and no high energy bond would be available at the junction of Okazaki fragments, so the DNA polymerase enzyme would have no available energy source to catalyze the formation of a phosphodiester bond.**

21. Primers for DNA synthesis are RNA molecules that are made by the "Primase" enzyme. The primer is complementary to the DNA sequence that was shown in bold; furthermore, the primer has the opposite polarity:

    5' UAUACGAAUU 3'

22. A replication bubble is formed by 2 replication forks proceeding in opposite directions from a single origin of replication.

    a. There are 3 bubbles in this figure, so there must be **3 origins of replication** in this DNA molecule.

    b. There are **6 replication forks,** one at each of the two ends of each of the 3 bubbles.

    c. If all replication forks move at the same rate, then the largest bubble was the first one activated. The **smallest bubble (the one in the middle) was the last origin of replication to be activated.**

23. a. **Topoisomerase relieves the stress of the over wound DNA ahead of the replication fork.**

    b. **Helicase unwinds the DNA by breaking hydrogen bonds between base pairs to expose the single strand templates for replication.**

    c. **Primase synthesizes a short RNA oligonucleotide, which DNA polymerase requires as a primer in order to copy the template.**

    d. **Primase synthesizes DNA ligase joins the sugar phosphate backbones of adjacent Okazaki fragments in order to construct a continuous strand of newly synthesized DNA.**

24. During DNA replication, after an RNA primer (*green* in the diagram below) is removed from the 5' end of an Okazaki fragment, the "lost" information in the strand being synthesized (*red*) can be replaced. The replacement occurs when DNA polymerase

chapter 6

extends the 3' end of the preceding Okazki fragment by copying the template strand (*blue*) exposed by primer removal.

However, this replacement of RNA primers cannot occur at the 5' ends of the new DNA strands, because there is no preceding Okazaki fragment that could be extended. DNA polymerase can add nucleotides only to a free –OH group at the 3' end of a primer, but once the RNA is removed no such primer end is available). In other words, **when the RNA primer is removed from the very 5' end of a newly synthesized strand in a linear chromosome, there is no way to replace that primer with DNA. As a result, you would expect that information equal to the length of the removed primer will be lost from the 5' end of the new strand at each end of the chromosome** if there is not some sort of alternate replication methodology at the chromosome ends. This leads to a successive shortening of the chromosomes in each generation of cells when the DNA replicates.

The following figure shows the situation at the right end of a linear chromosome. The same loss of DNA sequence happens at the left end of the chromosome as well. You can picture the events at the other end by rotating this diagram 180º while remembering that the RNA primer on the left end of the chromosome will be found on the pair of strands that is continuous at the right end of the chromosome.

It turns out that linear chromosomes in eukaryotic organisms have special structures called *telomeres* at their ends that allow them to overcome this obstacle to their replication. Chapter 12 discusses the nature of these telomeres in detail.

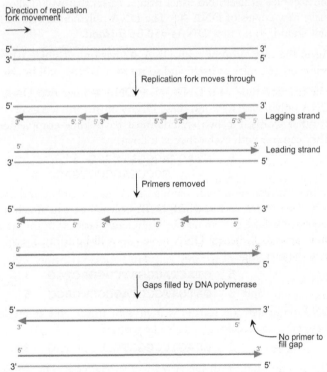

25. **Fig. 6.18 shows both strands of DNA being replicated in the same direction relative to the replication fork.** We now know that each replication fork has a lagging

6-11

strand and a leading strand because DNA polymerase catalyzes the addition of nucleotides only to the 3' end of the growing chain. Watson and Crick's figure also makes it seem as if DNA replication happens spontaneously without the participation of any of the many proteins that are now known to be involved in the process (see Feature Figure 6.23, p. 191). In spite of these simplifications, this drawing was one of the most influential in the history of modern biology. Watson and Crick's simplification made it intuitively obvious how double helical DNA could be passed from one generation to the next. This powerful idea was very important in convincing scientists of the time that the Watson-Crick model for DNA structure was valid.

26. In order to unlock the intertwined helixes, you must break both strands of one of the intertwined molecules. The unbroken strand can then pass through the break in the other DNA molecule. The broken helix must then be rejoined. This breakage/rejoining is mediated by the enzyme *topoisomerase*.

27. Remember that DNA polymerase requires: (1) a primer supplying a free 3' end so that nucleotides can be added to the growing new strand; (2) a single-stranded template DNA to copy; (3) a source of dNTPs (which here were added to the reaction tubes).

   a. The reaction tube with DNA #1 has only a single type of DNA that cannot form base pairs with other copies of itself. As a result, although DNA #1 could potentially act as a template, it cannot simultaneously serve as a primer to help make more copies of itself. No other primer exists that would allow DNA polymerase to make new copies of DNA #1. The DNA polymerase cannot function to produce a new strand, so **no new DNAs will be formed.**

   b. Again the reaction tube with DNA #2 has only a single type of DNA that cannot prime other copies of itself, and so **no new DNAs will be formed.**

   c. The reaction tube with DNA #1 + DNA #2 has two DNA strands. In order for DNA polymerase to work, these DNAs must form a region of double helical DNA; in other words, the two strands must have some complementary base sequences. They can pair with each other as follows:

   5' CTACTACGGATCGGG 3'
   3'   TGCCTAGCCCTGACC 5'

   In this configuration both strands can act as a template and as a primer. Each strand supplies a free 3' end to which DNA polymerase can attach new nucleotides whose sequence will be determined by complementarity with the template supplied by the other strand. Two new DNA molecules will be formed, each 5 nucleotides longer than the two original DNAs:

   5' **CTACTACGGATCGGGACTGG** 3'
   and 3' **GATGATGCCTAGCCCTGACC** 5'

   DNA sequences are usually written with the 5' end at the left, so the second of these DNA molecules could also be written as:

   5' **CCAGTCCCGATCCGTAGTAG** 3'.

   d. The reaction tube with DNA #3 is a special case. This single strand of DNA has two regions that have complementary base sequence, so the DNA can form a so-

chapter 6

called *stem loop* (see Problem 6-10 above). The two regions can base pair with each other as follows:

```
5' AGTAGCCAGTGGGG A A
      GGTCACCCC A A A
```

Note that there is now a free 3' end to which DNA polymerase can add nucleotides as well as a template to guide the addition. The product will therefore be 5 nucleotides longer than the original:

5' AGTAGCCAGTGGGGAAAAACCCCACTGGCTACT 3'

It is also possible for two of these DNA molecules to pair as shown:

```
5' AGTAGCCAGTGGGGAAAAACCCCACTGG
   3' GGTCACCCCAAAAAGGGGTGACCGATGA
```

The 5 As on the two strands (underlined) will not pair. Instead they will form a bubble in the middle of this double stranded molecule. There are now free 3' ends to which DNA polymerase can add nucleotides as well as templates to guide the addition. The product will be the same as that from the template with the stem loop.

Section 6.5

28. The coinfection of bacteria with $c+$ and c bacteriophage allowed recombination to occur between the two genotypes of bacteriophage. How can a single progeny bacteriophage produce both $c+$ and c progeny in the second round of infection? How can one bacteriophage with a single double stranded chromosome contain two kinds of genetic information? Remember that a key intermediate step in recombination is the formation of a *heteroduplex region* (Feature Figure 6.27, pp. 196-197). Normally, the heteroduplex regions are corrected by DNA repair systems. However, **in rare cases the heteroduplex is not corrected and a single bacteriophage particle can be generated with a chromosome that contains the mismatch. One strand of DNA would be $c+$ and the other strand would be c.** When such a bacteriophage infects a new bacterial host cell, DNA replication creates some molecules in which both strands are $c+$ and other DNAs in which both strands are c.

    As an aside, you might be wondering how you can actually isolate a single bacteriophage and look at its progeny. This will become clear in the next chapter, when methods for handling bacteriophage are explored in more detail.

29. **A mutant *spo11⁻* yeast strain will not be able to undergo recombination.** Any event requiring recombination would not occur normally. In addition to not being able to reassort alleles on the same chromosome through crossing-over, the mutant strain would not be able to perform certain types of DNA repair; the latter will be discussed in Chapter 7.

30. The numbers of B and b alleles are not in the 2:2 ratio predicted from meiosis of an $A\ B\ C\ /\ a\ b\ c$ diploid. **The 3:1 ratios indicate that gene conversion occurred**. Gene conversion results from the DNA repair of the nucleotide mismatches of heteroduplex DNA that forms during branch migration in the process of recombination (Fig. 6.28, p.

200). Because the *B* gene shows the 3:1 ratio of alleles, this gene must have been in the heteroduplex region.

Recombination events can be resolved in two ways when the flanking markers are considered - genes *A* and *C* in this case. In one resolution (the noncrossover pathway), the flanking markers do not show recombination (**Feature Figure 6.27: Steps 4'-5', p. 198**). Tetrad I is the noncrossover type (**Fig. 6.29b, p. 200**). The *A* and *C* gene alleles show the nonrecombinant configuration of *AC* and *ac*. In this tetrad the correction of the nucleotide mismatches in the two regions of heteroduplex DNA was to *B* giving spores of the genotypes *A B C*, *A B C*, *a B c* and *a b c* (**Fig. 6.29b, p. 200**).

In the other resolution of the recombination event, the flanking markers do show recombination (**Feature Figure 6.27: Steps 4-6, pp. 197-198**). In tetrad II the flanking markers show tetratype genotypes (*A C*, *A c*, *a C* and *a c*) indicating a single cross over between these genes (**Fig. 6.29a, p. 200**). Again the heteroduplex DNA must have included the region of the *B* gene. In this case the heteroduplexes were both repaired to the *b* allele giving spores of the genotypes *A B C*, *A b c*, *a b C* and *a b c* (**Fig. 6.29a, p. 200**).

31. a. **The octad is neither MI (4:4) nor MII (2:2:2:2)**; instead, it's 3:1:1:3.

b. The diagram requested is shown on the following page. In the figure, each line represents a single strand of DNA. Also for the sake of clarity, this figure does not use the usual convention for differentiating between homologous chromosomes (where they are shown in different shades of different colors); instead, homologous chromosomes are shown here in different colors (*red* and *blue*), whereas sister chromatids are in the same colors. Spores that have the phenotype associated with the a^+ allele are shaded in *yellow*, whereas spores with the phenotype associated with the a^- allele are shaded in *gray* to highlight the 3:1:1:3 pattern.

The figure shows that when crossing-over occurred between two nonsister chromatids, the heteroduplex regions formed on each chromatid (**Feature Figure 6.27: Steps 4-6, pp. 197-198**) included the DNA sequence difference between the a^+ and a^- alleles. The mismatched bases were not repaired (neither base was changed so as to be complementary to the one on the other DNA strand) prior to the mitosis that generates 8 spores form the 4 meiotic products. As mismatch repair is the ultimate cause of gene conversion, no gene conversion occurred: no a^+ alleles were converted to a^-, or *vice versa*, and so 4 a^+ and 4 a^- spores are present in the octad. Yet, the resulting octad is neither MI nor MII because as shown in the figure, heteroduplex formation during crossing-over switched the positions of the two central spores (one a^+ spore and one a^- spore switched positions) resulting in a 3:1:1:3 octad.

c. Yes – it is possible to observe evidence of heteroduplex formation in a *Neurospora* ascus even if gene conversion did not occur; that is precisely what happened in this example. As shown in the diagram above, **the 3:1:1:3 octad was produced as the result of heteroduplex formation during crossing-over and the absence of gene conversion.**

[*Note:* As shown in the following diagram and explained in part (b), the 3:1:1:3 octad was produced by recombination through the crossover pathway (**Feature**

Figure 6.27: Steps 4-6, pp. 197-198). The anticrossover pathway results in the formation of a heteroduplex on only one chromatid (Feature Figure 6.27: Steps 4'-5', p. 198). Thus, without mismatch repair, the anticrossover pathway would produce a 3:1:4 octad (rather than a 3:1:1:3 octad) from the same recombination event.]

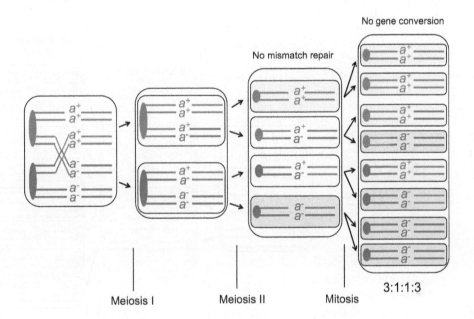

32. a. **Recombination must have initiated (that is, the first double-strand DNA break occurred) between *e* and *g* – either between *e* and *f* or between *f* and *g*.**

b. **Crossing-over did not occur; the alleles of the genes flanking gene *f* are in their parental configurations in the spores (e^+g^+ or e^-g^-).** This meiosis must have involved the anticrossover helicase pathway and the formation of a single heteroduplex region (Feature Figure 6.27: Steps 4'-5', p. 198). Gene *f* must have been included in the heteroduplex region to obtain the resulting octad (because gene *f* does not segregate 4:4).

c. The chromosome that ended up with the f^+f^- mismatch in the heteroduplex region was corrected by mismatch repair to $f^-\ f^-$. As shown in the diagram that follows part (d), **neither the *e* gene nor the *g* gene were within the heteroduplex region. Mismatch repair did not affect either of these two genes, so the e^+ and e^- alleles, as well as the g^+ and g^- alleles, segregated 4:4.** [In this figure, each line represents a single strand of DNA. *Blue* lines are DNA strands from one parent, while *red* lines represent DNA strands from the other parent (on the homologous chromosome).] *Yellow* shading indicates spores with the f^+ phenotype; *gray* shading spores with the f^- phenotype. *Black* type indicates alleles resulting from mismatch repair that eventually caused gene conversion.

d. The octads are MI for gene e and for gene g. For gene f, they are neither MI or MII because gene conversion has resulted in a 2:6 ratio of $f^+ : f^-$ alleles. To be MI or MII, an octad must have to be equal numbers of f^+ and f^- alleles.

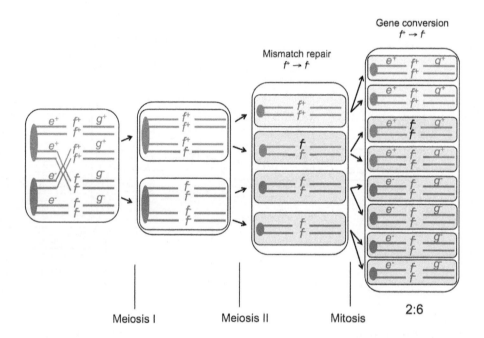

chapter 7

Anatomy and Function of a Gene: Dissection Through Mutation

Synopsis

This chapter covers a lot of ground, but it's all about mutations. Mutations – changes in the base sequence of DNA – are the heart and soul of genetics. Mutations are the basis of genetic variation, the raw material for evolution, and the basic tool of genetic analysis. Now that you have been introduced to the structure of DNA, you can understand the molecular nature of different kinds of mutations, how DNA damage or DNA replication errors can result in mutation, and how various repair mechanisms can prevent the damage or errors from causing mutation.

This chapter describes landmark experiments – particularly Benzer's investigations with T4 phage and Beadle and Tatum's "one-gene-one-enzyme" study in *Neurospora* – that used mutations to reveal key insights into the nature of the gene. In addition, this chapter introduces the idea that the reason mutations in genes affect an organism's phenotype is that most genes encode proteins. Chapter 8 will later expand upon this concept by describing how the information in genes is used to produce specific proteins.

As you review this chapter, be careful not to confuse the critical terms **complementation**, **recombination**, and **supplementation**. These words sound similar but, but their meanings are quite different.

Key terms

mutation – a change in the base pair sequence of DNA. In multicellular organisms, mutations can occur in any cell, but they are only heritable if they are passed through the gametes.

forward mutation and **reverse mutation** – mutation from a wild-type (normal) allele to a non-wild-type (mutant) allele is *forward* mutation. Changing a mutant base pair sequence back to the wild-type sequence is *reverse* mutation or **reversion**; the reverse mutants are called **revertants.**

point mutations – small changes in base pair sequence, such as: base pair substitutions (**transitions** substitute one purine for the other or one pyrimidine for the other, and **transversions** substitute a pyrimidine for a purine or *vice versa*); **small deletions** (removal of one or a few base pairs); and **small insertions** (addition of one or a few base pairs)

somatic cells and **germ-line cells** – In animals, all of the body cells are *somatic cells* except for those that that produce the gametes, which are the *germ-line cells*. Mutations can be inherited only if they occur in germ-line cells.

spontaneous mutation – changes in the base pair sequence that occur naturally in cells. Spontaneous mutations are caused initially either by DNA damage due to by-products of metabolism or the environment, or by rare errors in the process of DNA replication. Regardless of their origin, mutations only become *fixed* in the DNA sequence after both strands of DNA are converted to the mutant sequence by DNA replication. The DNA-damaging processes include: **depurination** (removal of As or Gs from nucleotides); **deamination** (removal of amine groups from bases); **thymine dimerization** (joining together of adjacent Ts); and **oxidation** (oxidized Gs pair with As). Mutation caused directly by errors in the process of DNA replication include incorporation of incorrect bases due to **base tautomerization** (a base in a rare form when the replication fork comes through); and **slipped mispairing** (pausing of DNA polymerase – **stuttering** – at short repeated sequences results in slippage of one of the two DNA strands and consequently either addition or deletion of repeat units).

mutagen – an agent that damages DNA and thereby promotes mutation. Mutagens include **base analogs** (molecules resembling nucleotides that can be incorporated into DNA); **X-rays** (radiation that makes double-strand breaks in DNA); **alkylating agents** (bases can be alkylated); **intercalators** (molecules that insert between adjacent base pairs). The DNA damage caused by mutagens ultimately causes mutations either as a result either of DNA replication with a damaged template, or of error-prone DNA repair.

unstable trinucleotide repeat – a region of DNA containing tandem base triplet repeats. DNA replication through the repeat region can result in mutation by slipped mispairing that increases or decreases the number of repeats. Diseases associated with unstable trinucleotide repeats (e.g., Huntington disease and fragile X syndrome) are usually due to alleles with increased repeat numbers.

base excision repair and **nucleotide excision repair** – DNA repair mechanisms in which a damaged base is removed by a specific enzyme (*base excision repair*), or an entire damaged nucleotide is removed (*nucleotide excision repair*). In both mechanisms, enzyme systems remove several nucleotides on either side of the originally damaged base, and DNA polymerases restore the missing nucleotides using the other (wild-type) DNA strand as a template for DNA synthesis.

double-strand break repair – Broken DNA initiates homologous recombination between sister or nonsister chromatids that results in rejoining of the broken DNA ends.

nonhomologous end-joining (NHEJ) – a DNA repair mechanism where double-strand breaks are protected from degradation and brought together by proteins so that phosphodiester bonds may be formed by DNA ligase.

methyl-directed mismatch repair – correction of mismatched base pairs caused by DNA replication errors. Proteins recognize mismatched base pairs and remove the nucleotide in the mismatch from the newly synthesized (unmethylated) DNA strand; the new strand is repaired using the original template strand as a template for DNA synthesis.

unequal crossing-over – aberrant meiotic recombination between two similar, but different, DNA sequences (for example, two related genes) that are adjacent to each other on a chromosome and that misalign. The result can be gene *duplication*, gene *deletion*, or the formation of *hybrid genes*.

chapter 7

colony – a mound of bacteria growing on agar media in a petri plate that all descend from a single bacterium. Assuming that no mutations occurred, all of the bacteria in the colony have the same genotype.

plaque – a round clearing in a *bacterial lawn* (a culture of bacteria spread on a petri plate too densely to form individual colonies) devoid of bacteria but filled with bacteriophage; all the phage descend from a single phage that infected and lysed a single bacterium, thus initiating rounds of infection and lysis of neighboring bacteria.

biochemical pathway – an ordered set of chemical reactions, each step of which is performed by a protein (an *enzyme*), through which an organism converts molecules it ingests or has made into different molecules that it needs.

auxotroph and **prototroph** – Bacteria that can grow on minimal medium (*prototrophs*) have no mutations in genes for biosynthetic enzymes; *auxotrophs* have a mutation in a gene encoding an enzyme required for biosynthesis of an essential molecule and thus cannot grow on minimal medium unless it is supplemented with that molecule or another compound that the auxotroph can use to synthesize it.

Ames test – an assay that determines whether or not a compound is a mutagen by testing its ability to revert auxotrophic bacteria to prototrophs

complementation test – a genetic assay to determine whether or not two different recessive (to wild-type) loss-of-function mutations reside in the same gene. If an organism with the two mutant DNAs has a wild-type phenotype, the mutants **complement** one another; each mutant DNA provides the function that the other lacks, and so the two mutations must reside in different genes. If a mutant phenotype is observed (no complementation), then the two mutations must be in the same gene; neither mutant DNA supplies the function that the other lacks.

complementation group – a set of recessive (to wild-type) loss-of-function mutations that fail to complement each other because they are all mutant alleles of the same gene

selection and **screen** – A *selection* employs growth conditions in which only organisms with a desired phenotype can survive. In a *screen*, large numbers of organisms are tested individually for a particular phenotype. Selections can detect extremely rare events leading to a particular phenotype, whereas screens cannot.

fine structure mapping – measuring the small recombination frequencies between different mutations in the same gene to determine the relative locations of the mutant bases

supplementation – addition of biosynthesis products (for example, amino acids) or intermediates in biosynthesis pathways to minimal media in order to support growth of auxotrophs

protein (also called a **polypeptide**) – a polymer of **amino acids** linked covalently by **peptide bonds**. The first amino acid in a protein is the **N terminus**, and the final amino acid added to the polypeptide is the **C terminus**. Proteins are important components of most cellular structures, and they constitute most of the enzymes that carry out cellular metabolism.

primary, secondary, tertiary, and **quaternary protein structure** – the sequence of amino acids in a polypeptide is its *primary structure*; some regions of polypeptides are locally ordered into α-helices or β-pleated sheets (*secondary structures*); the polypeptide chain as a whole folds into a characteristic three-dimensional shape (*tertiary structure*); often, separate polypeptides interact to form multisubunit structures (**multimers**) whose three-dimensional configuration is the *quaternary structure*.

Unifying principles of mutation and DNA repair

- **Mutation and DNA replication are intimately associated:**

  (1) Mutations can occur as mistakes in DNA replication.

  (2) If DNA is damaged, mutations can become fixed in the DNA sequence permanently if the damage is not reversed before the next round of DNA replication.

  (3) Because of reasons (1) and (2), the more rounds of DNA replication, the greater the chance that mutations can occur and become fixed in the DNA sequence.

  (4) Important DNA repair mechanisms involve DNA replication (see below). Such DNA repair pathways can cause a mutation to be fixed in the DNA sequence if they use the mutant base in a mismatched pair as the template.

- **DNA repair usually works in one of three general ways:**

  (1) Removal and resynthesis: A damaged or mispaired nucleotide is removed; that DNA strand is degraded on either side of that nucleotide to create a single-stranded gap; and finally DNA polymerase uses the complementary DNA strand as a template to synthesize a new DNA strand containing the correct base.

  (2) Homologous recombination: A double-strand break in a DNA molecule can initiate homologous recombination with a sister or nonsister chromatid, resulting in repair of the break during the process of recombination.

  (3) Non-homologous end-joining (NHEJ): A double-strand break in DNA can be repaired by proteins that simply bring the broken ends together so that DNA ligase can reform the phosphodiester bonds.

Complementation tests in diploid organisms

Complementation testing is a key technique in genetics whose purpose is to establish whether independently obtained mutations producing abnormal phenotypes are in the same gene or in different genes. A complementation test is quite simple: You simply cross the two mutant strains and then look at the phenotype of the F_1 progeny. If the F_1 display a mutant phenotype (no complementation), the parents have mutations in the same gene; if the F_1 are wild type, the parents harbor mutations in different genes and the two mutations complement each other.

It is critical to note that a complementation test can be interpreted in this way only if the two mutations result in a loss-of-function and are both recessive to the corresponding wild-type allele. The reason is that the test is asking whether the gamete from one parent can supply a function that the other gamete is lacking, and *vice versa*. For example, if the two mutations are in different genes, one parent could be written as *aa BB* and the other parent as *AA bb*. The F_1 progeny would be *Aa Bb*, and the F_1 phenotype would be wild type because one gamete supplies the function of gene *A* and the other the function of gene *B*. The two mutations complement each other. If instead the two mutations were in the same gene, the parents would have both been *aa* and the progeny would also have been *aa*; here the mutations fail to complement each other. When two different mutations fail to complement, the F_1 will have a mutant phenotype whose nature depends on the dominance relationship between the two alleles.

Scientists studying biological pathways often collect mutants for a particular trait (for example, flies with unusual eye colors) and then perform all pairwise complementation tests to determine which mutants are allelic. A **complementation group** is simply a collection of mutants that *fail to complement* one another in all pairwise crosses; all the members of a complementation group are simply different mutant alleles of the same gene.

Benzer's experiments with bacteriophage T4

Seymour Benzer performed two kinds of experiments to test the hypothesis that a gene is composed of subunits (base pairs) that could be separately mutated and can recombine with each other. To follow his landmark investigations, it is important for you to understand the differences in the purposes and methodologies of his complementation and recombination experiments.

Benzer's complementation experiment: The purpose of this experiment was to find many *rII-* mutations that he could demonstrate were in the same gene. Benzer thus performed **complementation tests** with his different *rII-* strains. He introduced pairs of *rII-* mutants into *E. coli* K – the bacterial strain on which no single *rII-* mutant phage could propagate by itself – and tested for bacterial lysis. If lysis failed to occur, the two *rII-* phage had mutations in the same gene. But because the ability to lyse *E. coli* K is the *rII+* (wild-type) phenotype, lysis would mean that two mutants complemented – that each had a mutation in a different gene required for propagation in *E. coli* K – and thus that each *rII-* phage genome provided the wild-type function that the other lacked. Benzer found that his *rII-* point mutants fell into one of two complementation groups (two genes), which he called *rIIA* and *rIIB*. Thus, the *rII* locus consists of two adjacent genes that work together in a pathway essential for lysis of *E. coli* K.

Benzer's recombination experiment: Benzer predicted that *recombination* in the region of DNA in between the base pairs responsible for two mutant alleles of the same gene defined by the complementation test just described (for example, *rIIA₁* and *rIIA₂*) should yield a wild-type gene as one of the reciprocal recombinants. Such recombinants would be extremely rare because the distance between any two mutant base pairs in the same gene would be small.

The key to Benzer's recombination experiment was the system he devised for detecting extremely rare recombinants through a *selection*. Unlike wild-type (*rIIA+*) phage, *rIIA-* mutants cannot form plaques (cannot perform their lytic cycle and propagate) on a lawn of a particular bacterial strain called *E. coli* K. By collecting the progeny phages resulting from coinfection of two different *rIIA-* mutants in *E. coli* B, and plating large numbers of them on *E. coli* K, Benzer was able to detect rare *rIIA+* recombinants produced by crossing-over between the two mutation sites of different *rIIA-* alleles. Detection of the recombinants supported both parts of his hypothesis: that genes were composed of separately mutable subunits, and that crossing-over can occur inside a gene between any two subunits.

Complementation vs. Recombination: Be careful not to confuse Benzer's complementation and recombination experiments. In the complementation experiment, he coinfected *E. coli* K with two different *rII-* mutant phage and looked at the phenotype: Did the cells lyse (complementation) or not lyse (no complementation). The purpose was to show that two mutations were in the same gene. In the recombination experiment, he coinfected *E. coli* B with two phage, each carrying a mutation he already knew to be in the same gene. *rII-* mutants can grow in this host, where their genomes can recombine. The ultimate test was to look at the *progeny* phages resulting from this cross, and to see whether he could find rare *rII+* recombinants that could now grow on *E. coli* K.

Supplementation and the dissection of biochemical pathways

Beadle and Tatum asked whether **supplementation** of minimal media with any of the chemical intermediates in the arginine biosynthesis pathway could support growth of specific Arg- auxotrophs that had previously been separated into complementation groups. It is important not to confuse supplementation with complementation. Supplementation asks at which point in a biochemical pathway a gene's protein product functions, while complementation defines which mutations are allelic. The results of these analyses are not necessarily synonymous. As just one example, two different mutations may affect a single step in a biochemical pathway, yet be in different complementation groups if the enzyme catalyzing that step is a multimer of different subunits.

Problem Solving

Many of the problems in this chapter ask you to interpret data from complementation tests (*complementation matrices*), from supplementation experiments, or from recombination experiments in which pairs of mutants in the same gene are tested for their ability to produce wild-type genes by crossing-over.

Interpreting complementation matrices. A complementation matrix is simply a table that shows the results of all pairwise complementation tests. Usually, a "–" symbolizes no complementation, and a "+" means that the mutant pair complement. Analysis of complementatation matrices enables you to assign all of the mutants to *complementation groups*, and thereby figure out how many different genes are mutant among your collection of mutants. The simplest way to do this is to first look at all the boxes with a "–" in them; list all the mutants that *don't complement* each other. This process should sort the mutants into complementation groups (groups of alleles of different genes). Next, verify your list by looking at the boxes with "+"s in them. Make sure that all the mutants within each complementation group *do complement* those in other complementation groups. This should be the case if the data given are internally consistent. Make sure to notice any mutants that complement all of the other mutants; these represent complementation groups containing only one allele, and count as genes represented in your collection.

Interpreting supplementation experiments. If an auxotrophic mutant grows in the presence of an intermediate, the gene must encode an enzyme that functions at some step prior to the

chapter 7

formation of that intermediate. If no growth is observed, the gene product functions subsequent to the formation of that intermediate.

Not all biochemical pathways display this simple linear pattern. For example, some pathways may branch, leading to the formation of more than one essential end products. Mutants affecting a step prior to the branch point will grow only if supplemented with all the end products, while those affecting a step after a branch point require supplementation only with the molecule at the end of that branch.

Interpreting recombination experiments. Recombination problems emphasize that crossing-over can occur between different mutant base pairs (point mutants) in the same gene, and when that happens, one recombinant product is a wild-type allele of the gene. When two different mutant alleles of the same gene cannot generate wild-type recombinants in an experiment with sufficient genetic resolution, this means that the same base pairs of the gene are mutant in both alleles: either the same base pair has a point mutation; or one allele is a point mutation and the other is a deletion that includes (deletes) the base pair(s) that are mutant in the other allele; or both mutations are deletions that overlap each other at least partially.

Vocabulary

1.

1. an A-T base pair in a wild-type gene is changed to a G-C pair
   - a. transition
   - b. base substitution
   - f. deamination

2. an A-T base pair is changed to a T-A base pair
   - b. base substitution
   - c. transversion

3. the sequence AAGCTTATCG is changed to AAGCTATCG
   - d. deletion
   - h. intercalator

4. the sequence AAGCTTATCG is changed to AAGCTTTATCG
   - e. insertion
   - h. intercalator

5. the sequence AACGTTATCG is changed to AATGTTATCG
   - a. transition
   - b. base substitution
   - f. deamination

6. the sequence AACGTCACACACACATCG is changed to AACGTCACATCG
   - d. deletion
   - g. X-ray irradiation

chapter 7

Section 7.1

2. We are told the trait in the pedigree is very rare. The trait is first seen in individual III-1 and shows a vertical pattern of inheritance between generations III and IV, suggesting a mutation that is dominant to the wild type. The mutant gene cannot be X-linked because male IV-1 inherits it from his father. Thus **the trait is autosomal dominant**, yet it is not seen in generations I and II.

   It is possible that the mutant allele is incompletely penetrant. In this case one individual in generation I and one individual in generation II (either II-2 or II-3) have the mutant allele but do not display the trait.

   A second possibility is that a mutation occurred in the germ line of either II-2 or II-3, producing a gamete with the dominant autosomal mutation. This gamete was then used to form III-1 (the *propositus* – the person in a pedigree who first shows the phenotype in question).

   There are two potential ways to discriminate between these hypotheses. First, more information about this trait from a more extended family pedigree or from pedigrees of other families might indicate whether the trait is or is not completely penetrant. A pedigree where the trait appears suddenly in a generation, disappears in the next generation, and then reappears in a following generation would argue for incomplete penetrance. Second, if you could obtain DNA sequence information for the gene in question from II-2, II-3 and III-1, you could see if one of the alleles of the gene in the propositus has a mutation not found in his parents.

3. Each independently derived mutation will be caused by a different single base change. When you find a base that differs in only one of the sequences, that base is the mutation. Determine the wild-type sequence by finding the base that is present at that position in the other two sequences. The wild-type sequence is therefore:

   5' ACCGTA G TCGACT G GTA AAA CTTTGCGCG 3'

4. **Dominant mutations can be detected immediately in the heterozygous progeny who receive the mutant gamete** (see Problem 7-5). **Recessive mutations can be detected only when they are homozygous. To detect the appearance of new recessive alleles, you must testcross with a recessive homozygote.** Testcrosses can be done in mice, where the researcher can control the mating, but it cannot be done in humans.

5. Of the achondroplasia births observed, 23/27 are due to new mutations because no family history of dwarfism exists. Achondroplasia is an autosomal dominant trait, so it will be expressed in the child that receives the mutant gamete. There were 120,000 births registered, so there were 240,000 parents in which the mutation could have occurred during meiosis, leading to a mutant gamete. **The mutation rate = 23 mutant gametes/240,000 gametes = 9.5×10^{-5}. This rate is somewhat higher than 2 to 12×10^{-6} mutations per gene per generation, which is the average mutation rate for humans.**

6. **To ensure that the mutants you isolate are independent you should follow procedure #2.** If you follow procedure #1, a mutation causing resistance to the phage could have arisen several generations before the time when you spread the culture; thus, several of the colonies you isolate could be descendants of the same cell and have the identical mutation.

chapter 7

Different mutations in the same gene often give different information about the role of the gene product. Geneticists therefore generally strive to find many independent, different mutations in the same gene in order to understand as much about the gene as possible.

7. Kim's hypothesis is that the bacteriophages are able to induce resistance in ~1 in every 10^4 bacteria. If she is right, then several of the colonies on each of the replica plates (~10 if there are 10^5 bacteria on each plate) should contain resistant cells that will continue to grow after exposure to phage. These resistant cells would be generated only after the replica plating. **If Kim is correct, resistant colonies will be distributed in random locations on the three replica plates**. Maria's hypothesis is that the resistant cells are already present in the population of cells they plated on the original (master) plate. If she is right then some of the colonies on the original plate (about 10 of them) should have already contained multiple resistant bacteria that will allow colonies to grow on the phage-treated replica plates. **If Maria is right, resistant colonies will appear at the same locations on all three of the replica plates** (Fig. 7.6b, p.212).

8. **Several equal aliquots of each of the 11 culture tubes could be spread on different petri plates. If differences in the petri plates are not a factor, every plate corresponding to a single one of the 11 culture tubes should contain approximately the same number of colonies.**

9. As the mutation is not found in the somatic tissue (blood) from any of the parents in generation II, **the mutation must have occurred in the germ line of one of the parents. This parent must be II-2** because the two affected children had the same father but different mothers. But mutations are rare, so how can III-1 and III-2 have exactly the same mutation? **The mutation must have occurred in the germ line of II-2 at a point before the formation of separate sperm**. In the male germ line, cells called spermatogonia divide mitotically to make many cells that will subsequently undergo meiosis (primary spermatocytes; see Fig. 4.19, p. 108). **It is likely that a mutation occurred in an early spermatogonial cell, leading to a large proportion of spermatocytes and therefore a large proportion of sperm in the testes that carry the same mutation. This pedigree suggests that the testes of II-2 exhibit germ-line mosaicism** - some of the spermatocytes carry the mutation while others do not.

    Note that this inference is similar to the pattern seen in Hypothesis 2 of the Luria-Delbrück fluctuation test (Fig. 7.5, p. 211): **If a mutation occurs early in the growth of a bacterial colony (or in the mitotic division of the germ line), then many of the cells in the colony (or in the germ line) can have the mutation**. Germ-line mosaicism in which a large proportion of the germ line carries a new mutation is likely to be rare because the mutation must occur early in the multiplication of the germ line. At this early stage there are fewer cells than later, so fewer chances exist for mutations to occur.

10. Because the germ cells that produce human sperm undergo mitosis continually, the **older the father, the more mutations, on average, each of his sperm contains**. Recall that the fixation of mutation requires DNA replication (Fig. 7.7, p. 212). Human females, on the other hand, are born with all (or most) of the eggs that they are ever going to have, so maternal age is not a factor in the accumulation of gene mutations. In addition, female germ cells underwent a far smaller number of mitoses than the germ cells of a even a

young, 20-year-old male. Because mutant alleles of a variety of genes are likely to promote autism, the correlation between autism and older fathers makes sense.

Section 7.2

11. Without the Balancer X chromosome, X-ray-induced mutations on the irradiated X chromosome and the *Bar* mutation on the nonirradiated X chromosome could recombine in the germ line of the F_1 females. The experiment would be uninterpretable because Bar eyes would no longer serve as a marker for males that inherited no alleles from the irradiated X chromosome. Therefore, **the Balancer was crucial because, by preventing the appearance of recombinant X chromosomes in the gametes of the F_1 females, it enabled the absence of non-Bar-eyed F_2 males to signify that an irradiated X chromosome contained a lethal mutation** (that is, a mutation in an essential gene).

12. a. *Essential* genes are those that produce a recessive lethal phenotype when mutant - if one such gene doesn't function because of a mutation, the animal does not live. H.J. Muller's control results showed that 0.3% of the females produced only wild type (non-Bar) sons. These females had a recessive lethal mutation on the (non-irradiated) X chromosome with the *Bar* mutation. In other words, mutations in essential genes on the X chromosome occur at a rate of 3×10^{-3} mutations per gamete. The average spontaneous mutation rate for *Drosophila* genes is 3.5×10^{-6} mutations per gene per gamete. Therefore, 3×10^{-3} X-linked lethal mutations per gamete / 3.5×10^{-6} mutations per gene per gamete = **857 essential (lethal) X-linked genes**. This estimate depends on the accuracy of the estimate for the average mutation rate per gene in *Drosophila*.

    b. The estimated 857 essential genes on the X chromosome / 2279 total genes on the X chromosome $\times$ 100 = **37.6% of the genes on the X chromosome are essential**. Although this is a rough figure, it matches fairly well to other estimates made with other techniques suggesting that only 1/4 - 1/3 of the genes in the *Drosophila* genome are essential. In other words, 66-75% of the genes in the genome could be disrupted by mutation and the flies could still survive.

    c. *Note:* Your copy of the text might say incorrectly that 12% of females produced male progeny that were all "wild type". Instead of "wild type," the problem should say "Bar-eyed."

    Clearly the mutation rate has increased with exposure to X-rays, demonstrating that X-rays are mutagenic. Genetics became much easier to do because scientists could increase the chances they would find rare mutations. (H.J. Muller received a Nobel Prize for this finding.) You can calculate the average mutation rate per gene upon exposure to this dosage of X-rays if you use the answer to part (a) indicating the number of essential genes. There are 857 essential genes on the X chromosome $\times$ (X-ray induced mutation rate/gamete) = 0.12 X-linked lethal mutations/gamete. Thus **the X-ray induced mutation rate = 0.12/857 = 1.4×10^{-4}**. In other words, this high dosage of X-rays has **increased the mutation rate about 40-fold** over the spontaneous rate (1.4×10^{-4} / 3.5×10^{-6}).

chapter 7

13. In all of these cases, a purine is replaced by the other purine, and a pyrimidine is replaced by the other pyrimidine; **all of these mutagens cause transitions**.

14. A perfect reversion is a very rare event, even when induced by two-way mutagens. Such a reversion demands one particular change in the exact same base pair that was originally mutated (Fig. 7.4b on p. 209). These kinds of reversion events normally can be seen only in experimental organisms like bacteria or bacteriophage where it is possible to examine so many individuals that rare events can be detected.

    a. Figure 7.14 (pp. 219-220) shows how the base analog 5-bromouracil (5-BU) can cause a T:A to C:G substitution. 5-BU can sometimes behave like thymine and sometimes like cytosine. In the figure, 5-BU behaves like T (because it base pairs with A), but then shifts to a state where it behaves like C, so that during replication, DNA polymerase will incorporate a G in the newly forming strand. **5-BU is a two-way mutagen**. A reversion can occur if 5-BU is incorporated into DNA in the C-like state the DNA will have a 5-BU:G base pair. If the 5-BU shifts into its alternative form during replication, it will act like a T, so the result will be a 5-BU:A base pair; following the next round of DNA replication, the result will be T:A.

    b. **Hydroxylamine is a one-way mutagen** because it changes C to hydroxylated C (C* in the figure), and C* can only pair with A. This will yield a T:A pair after the next round of replication. Hydoxylamine can not mutagenize a T:A pair.

    c. **Ethylmethane sulfonate (EMS) is a two-way mutagen** because it can modify either G or T. Figure 7.14 shows that ethylated G (G*) pairs with T, resulting in a G:C to A:T substitution. If EMS now ethylates T forming T* (this part is not shown in the figure), then the modified T will pair with G. This will result in an A:T to G:C substitution, which is an exact reversion.

    d. Nitrous acid changes C to U and also A to hypoxanthine. Figure 7.14 shows how this mutagen can cause both a C:G to T:A substitution as well as a reverse T:A to C:G substitution, making it a **two-way mutagen**.

    e. **Proflavin is a two-way mutagen** because it can add a single base pair or delete a single base pair; the same base pair that was added can be deleted.

15. a. EMS induced new mutations in the *dumpy* gene in the sperm of the treated males. However, as shown in Fig. 7.14 (pp. 219-220), EMS-induced mutations are fixed in both strands of the DNA only after two rounds of DNA replication. **The DNA in the *dumpy* gene of a sperm just treated with EMS would have one DNA strand with the normal G and the other DNA strand with an ethylated G (G*)**. This sperm now fertilized a *dumpy* egg. **After several rounds of DNA replication and mitosis, some cells will have the normal G-C base pair, while other cells will have a *dumpy* mutant A-T base pair**. An animal in which different cells have different genotypes is called a *mosaic* and will be described in more detail in other chapters. **The phenotype of the fly that grows from this fertilized egg would depend upon which particular cells in the body had the wild-type or mutant sequence of the *dumpy* DNA**. This fact explains why one wing might be short and the other long. Whether or not the new dumpy mutation in the F_1 animals can be transmitted to future progeny depends on whether or not some of the cells in the germ line carried the mutant A-T base pair.

b. If the progeny of the second cross have short wings, then the germ-line cells of their F_1 male parent carried the mutant A-T base pair. Many rounds of DNA replication are involved in producing the male germ line from the zygote. Therefore **these progeny are not mosaics because they have received an A-T base pair from their F_1 parent in which both DNA strands are mutant; these progeny will have two short wings.**

16. a. DNA in most organisms is not exposed to high levels of aflatoxin B_1. It is therefore unlikely that most cells would have evolved genes for enzymes that could directly remove the aflatoxin B_1 from guanine, or a glycosylase that could specifically remove the adduced guanine-aflatoxin B_1 base (as in base excision repair). **The nucleotide excision repair system is most likely to repair damage due to aflatoxin B_1.** This system would treat the adduct in a similar fashion to a thymine-thymine dimer, which is also a bulky group that distorts the double helix (review Fig. 7.16 on p. 221). **It is also possible that the SOS-type error-prone repair system will repair aflatoxin B_1 damage.** Most other repair systems make less sense. Methyl-directed mismatch repair is unlikely because the mutagen works independently of DNA replication. Double-strand break repair is not involved because this mutagen does not produce double-strand breaks.

    b. This new information changes the picture somewhat. Apurinic (AP) sites are intermediaries in base excision repair, so **AP endonuclease and other enzymes in the base excision repair system could remove the damage** (review Fig. 7.15 on p. 221). A glycosylase would not be required because the adduct removes itself from the sugar. It is also known that **SOS repair systems can work at AP sites, adding any of the 4 bases at random** across from the AP site during DNA replication.

17. **The mutagen induces mutations in somatic cells, not in gamete-producing cells in the germ line.** These somatic cell mutations give rise to the tumor cells. When the tumor cells are injected into a new mouse, they will divide in an uncontrolled manner and cause a tumor to develop. The somatic mutation that caused the original cell to become cancerous is not present in the germ-line cells of the mouse. Thus, the cancer phenotype cannot inherited in a Mendelian fashion.

18. Yes. **The rat liver supernatant contains enzymes that convert substance X to a mutagen, and *his*+ revertants occur. Our livers contain similar enzymes** that process various substances, converting them into other forms that cause mutation and can lead to cancer.

19. a. Yes; particular mutagens can revert only particular types of mutations. Proflavin, for example, can revert only single base insertions or deletions. The Ames test deals with this issue by testing potentially mutagenic compounds for their ability to revert *his*- strains that have different types of mutations at the molecular level.

    b. **A wild-type (*his*+) strain could be grown in the presence of a potentially mutagenic compound (plus rat liver enzymes) and then plated for single colonies on minimal medium + histidine. Replica-plating on minimal medium (without a histidine supplement) would identify colonies that are *his*-; they would fail to grow.**

chapter 7

c. **Identification of forward mutations as described in part (b) is through a *screen*, not a *selection*.** In the technique used to find forward mutations after exposure to the compound, both His+ and His- colonies were allowed to grow, and they were "screened" by replica-plating for the His- mutant phenotype. **Even though the forward mutation rate is higher than the reversion rate, this screening process is much more labor-intensive than the *selection* that is employed when testing for revertants.** In the selection for *his+* revertants, millions of bacteria can be spread on a single petri plate containing minimal medium and only the rare revertants will form colonies, making them simple to identify and quantify even though the reversion rate is low.

20. a. As shown in Figure 7.8b on p. 213, deamination of (nonmethylated) cytosine converts it to uracil. A base excision repair pathway, in which uracil glycosylase enzyme removes the uracil and a cytosine is replaced opposite the G on the other DNA strand, usually corrects the problem before DNA replication fixes it permanently in the DNA. **A reasonable hypothesis to explain the fact that methylated CpG dinucleotides are hotspots for point mutation would be that a thymine glycosylase enzyme does not exist in human cells, because unlike uracil, thymine is a normal component of DNA.** In reality however, thymine glycosylase enzymes *do* exist in humans, and they remove Ts from T-G base pairs caused by deamination of methylated cytosines. However, thymine glycoyslases are much less efficient enzymes than uracil glycosylases, and so errant Ts formed by deamination of 5–methylcytosines are less frequently repaired than Us, resulting in a high frequency of C-G to T-A transitions.

    b. The frequency of each base is 1/4, and thus the expected frequency of CpG at a given location in the genome is the same as that of any particular dinucleotide: $1/4 \times 1/4 = 1/16 \approx 6.25\%$.

    c. **Any CpG dinucleotides in the genome that are not important functionally would mutate to TpG over time.** Another interesting idea is that evolution may select against some CpG dinucleotides whose methylation could have deleterious consequences.

Section 7.3

21. The formal difference between albinism and leucism is that the former affects only melanin formation or deposition, while the latter affects all skin pigments, not just melanin.

    a. The progeny of two albinos will be either be **white (albino)** because the albino parents are both homozygous for recessive alleles of the same gene **or colored** due to *complementation* (if the albino birds are homozygous for recessive alleles of different genes). (*Note:* Your copy of the text may have wording suggesting that mutations in only one gene can cause albinism. This is actually incorrect; mutations in any one of several genes whose products participate in melanin synthesis or deposition can cause albinism.)

    b. The progeny of two leucistic birds can be **either white** (if the leucistic birds are homozygous for recessive alleles of the same gene) **or colored** (if the leucistic birds are homozygous for recessive alleles of different genes) due to *complementation*.

c. The progeny of a mating between an albino bird and a leucistic bird will be **colored** because the parents are necessarily homozygous for recessive alleles of different genes.

22. **Do a complementation test by mating the two mice.** If the mutations in each mouse are in the same gene all the progeny will be mutant (albino). If the mutations causing albininsm in each mouse are in different genes, all the progeny will be wild type (the genes will complement).

23. a. Each complementation test involves taking pollen from a plant homozygous for one mutation and using it to fertilize ovules from a plant homozygous for another mutation. This creates plants heterozygous for the two mutations. The diagonal (mutant 1 crossed with mutant 1, mutant 2 crossed with mutant 2, etc.) represents self-fertilization; a recessive mutation cannot complement itself. A '-' indicates a lack of complementation and the resulting plants have serrated leaves. A '+' means that the two mutations complement each other, and so the resulting plant has wild-type leaves with smooth edges. The colored boxes in the table represent crosses that were not done because the information is redundant (that is, the same crosses had already been done in the opposite direction with the parents switched). For example a cross with mutant 1 pollen and mutant 2 ovules will produce the same kinds of progeny as a cross with mutant 2 pollen and mutant 1 ovules.

    b. If a box has a '-' then the two mutants must be in the same complementation group or gene. If the box has a '+' the two mutations are in different complementation groups or genes. From the results given, mutation 1 is in the same complementation group as mutation 3 (-). Mutation 1 is in a different complementation group than mutation 2 (+). Therefore mutation 2 must be in a different complementation group than mutation 3 (+). **The completed boxes are circled in the following diagram.**

	1	2	3	4	5	6
1	–	+	–	⊖	+	⊕
2		–	⊕	⊕	⊕	–
3			–	–	⊕	⊕
4				–	⊕	⊕
5					–	+
6						–

    c. There are **3 complementation groups (genes). One includes mutations 1, 3, and 4. A second group has mutations 2 and 6. The third consists only of mutation 5.**

24. a. Assuming that the man and woman represented in the pedigree in Fig. 3.24c each have a recessively inherited form of albinism then **this is a complementation test**. In this example, **the two albinism-causing mutations are in different genes** (complement each other) and none of the children is albino. If the two mutations were in the same gene, then all the children should be albinos.

    b. Complementation tests upon demand are not possible in humans. One way to tell if a trait can be caused by mutations in more than one gene (that is, the trait exhibits *locus heterogeneity*) is to **map the mutations in a number of independent families.** Mapping involves recombination analysis of the gene causing the mutant trait with other markers. If the mutations in different families map to different chromosomes or are far apart on the same chromosome, then they must be in different genes. If the mutations map close to each other on the same chromosome they may or may not be in the same gene. For example, the two mutations could be in adjacent, closely related genes such as the red and green photoreceptor genes shown in Fig. 7-32c on p. 242. In such a case, researchers can determine the DNA sequence of these closely related genes in various mutant individuals to see if the mutations affect the same gene. All of these techniques will be described in detail in Chapters 9 and 10.

c. Again, as explained in part (b) above, you can try to **map the dominant mutations or analyze the DNA of candidate genes** you think might be involved in the conditions.

25. a. Deletion mutations in the *rII* region can be identified in a couple of ways. **First, deletions are mutations that never revert to wild type. Second, phage strains with *rII-* deletion mutations fail to recombine so as to produce wild-type progeny in separate coinfections (recombinations) with *rII-* point mutations that affect different base pairs in the *rII* region.** This second method is based on the idea that if a point mutation is included in (deleted by) the deletion, it is impossible to recover *rII+* phage by recombination because neither *rII-* mutant gene has the wild-type base pair to contribute to the recombinant.

    As an example of this second method, consider a putative deletion *rIIΔ* and two point mutations *rIIA₁* and *rIIA₂*. You know the latter are point mutations because they can revert. You also know that *rIIA₁* and *rIIA₂* affect different nucleotides because they can recombine with each other during a coinfection to produce wild-type progeny. But if *rIIΔ* fails to recombine with *rIIA₁* and it also fails to recombine with *rIIA₂*, **then *rIIΔ* is lacking more than one nucleotide pair and thus must be a deletion**.

    b. *rII-* mutants in the same nucleotide pair cannot recombine with each other to produce *rII+* phage.

26. a. The starting tube (call it tube A) contains 5 ml of bacteriophage with 1.5×10^{10} phages. Take a 1 μl (0.001 ml) sample of tube A, corresponding to 3×10^6 phages, and add it to 999 μl (in practice, 1 ml) of diluent in tube B. This step is a 2×10^{-4} dilution (1 μl / 5000 μl). Repeat this step with 1 μl of tube B (3×10^3 phages) and mix it with 999 μl of diluent in tube C (2×10^{-7} dilution). Next take 10 μl of tube C (30 phages) and mix it with bacteria [about 100 × more cells than phage = a low multiplicity of infection (MOI)]. Allow the phages to infect the cells, then add cells to a top agar and pour on an agar plate. Repeat the infection/top agar step with 100 μl of tube C (300 phages) and plate. **There should be 30 plaques on the first plate and about 300 plaques on the second plate**. The above describes only one of many possible protocols; other dilution steps, such as 10^{-2} dilutions or 10^{-1} dilutions, could also be employed.

    b. To figure out the total number of phage, you need to look at a particular dilution in the electron microscope and count all the phage particles. The ratio of plaques to total phage is the *plating efficiency*. In part (a), it is fair to assume that only one phage initiated each plaque because of the very low MOI. Because there were many more bacterial cells than phage, the chances are very high that any individual bacterial cell would have been infected only by a single phage.

27. a. The data shown is a complementation matrix. Analysis of the matrix (see suggestions at the beginning of this chapter's Solutions Manual) indicates that **there are two complementation groups and therefore two genes**.

   b. **The complementation groups are (1, 4) and (2, 3, 5).**

28. To understand this problem, it is important to realize the parallels between the cross described here and Benzer's phage coinfection experiments which showed that the *rIIA* gene is composed of subunits (base pairs) that can mutate independently and also recombine with each other. In Benzer's experiments, rare recombination events in the region between distinct nucleotides causing different mutations in the *rIIA* gene generated wild-type (*rIIA+*) progeny phage genomes. Here, rare recombination events in the region between different mutations in the *rosy* (*ry*) gene (the ry^{41} and ry^{564} alleles are mutations in distinct nucleotides) during meiosis in *Drosophila* females produce wild-type (*ry+*) chromosomes in the progeny.

   a. Diagram the cross:

   $Ly^+\ ry^{41}\ Sb\ /\ Ly\ ry^{564}\ Sb^+$ ♀ × $ry^{41}\ /\ ry^{41}$ ♂ → 8 $Ly\ ry^+\ Sb$ and lots of *ry* progeny

   Recombination events within the *rosy* gene, between the two *ry* mutations in the heterozygous female, generate the eight offspring with wild-type eyes. As shown in the following diagram, such a recombination event will produce one recombinant gamete with a wild-type *rosy* gene, thus giving a ry^+ phenotype in the progeny of this cross. The reciprocal recombinant gamete will be a double mutant, $ry^{41}\ ry^{564}$, which will yield *ry* progeny indistinguishable from the parental type *ry* progeny. The wing and bristle phenotypes of those eight recombinant offspring are a consequence of the order of the ry^{41} and ry^{564} with respect to the flanking markers, the *Ly* and *Sb* genes. Try both orientations of the two *ry* mutations with respect to the flanking markers to see which order produces $Ly\ ry^+\ Sb$ as the result of a crossover between the *rosy* mutations.

   Orientation II (see diagram that follows) produces the $Ly\ ry^+\ Sb$ recombinants obtained, so the order must be: $Ly\ [ry^{41}\ ry^{564}]\ Sb$. (The brackets around the two *ry* mutations shows that these are alleles of the same gene.)

   b. We assume the $ry^{41}\ ry^{564}$ progeny are present in equal numbers to the ry^+ recombinants (even though we cannot discriminate them from the other *ry* progeny), so the **recombination frequency = (8 + 8) / 100,000 = 0.00016 = 0.016%. The distance between ry^{41} and ry^{564} = .016 mu.**

chapter 7

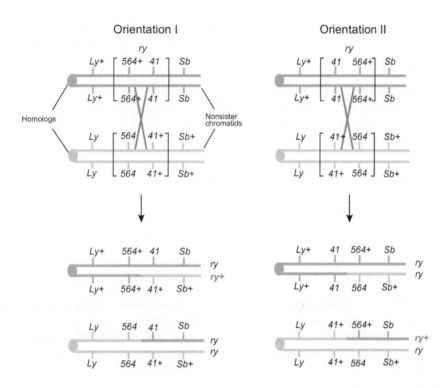

29. The first table is a complementation matrix, because in each entry you are coinfecting *E. coli* K with two kinds of phage and then simply asking for the phenotype (whether [+] or not [-] the cells lyse to produce large numbers of progeny). The second table shows the results of recombination experiments in which each entry represents a coinfection of *E. coli* B with two kinds of phage, and then asking whether any progeny are *rII*+ recombinants. In the latter table, "+" indicates recombination, while both "-" and "0" mean no recombination. The few *rII*+ phage seen in the "-" cases are due to reversion, whereas the "0" cases cannot revert, presumably because the two phages are deletions.

    a. The simplest way to identify the deletions in these particular experiments is to focus on the second table and see which phage cannot revert to wild type. Remember that point mutations can revert to wild type, while deletion mutations cannot. **Mutants 3, 6, and 7 did not form any plaques in single infections (as designated by '0' along the diagonal), so these three must be deletions. Thus, mutants 3, 6, and 7 are deletion mutants (non-reverting).**

        Another criterion that identifies mutation 6 as a deletion is its failure to complement any of the other mutations in the first table (because, as you will see, this deletion removes nucleotides from both the *rIIA* and the *rIIB* complementation groups).

    b. The first table (the complementation matrix) lets us place the point mutations in the two *rII* complementation groups. The two complementation groups are (1, 2, 5) and (4, 8, 9). The second complementation group is the *rIIB* gene as defined by the problem; the first must therefore be the *rIIA* gene.

Notice that deletion 6 does not complement any of the mutants so it must delete at least part of each of the two *rII* genes, whereas deletion 3 removes nucleotides only from *rIIB* and deletion 7 deletes nucleotides only from *rIIA*.

Next use the recombination data to order the mutations with respect to the deletions. Deletion 6 does not recombine with mutation 1 (*rIIA*) nor with mutations 8 and 4 (*rIIB*) to produce wild-type phage, so these mutations must be near each other because deletion 6 spans both genes. Deletion 3 allows you to order 8 (outside deletion 3), then mutation 4 (overlaps both deletions 6 and 3), then mutation 9 (only overlaps deletion 3). Mutations 2 and 5 cannot be ordered relative to each other, except to say that they do not overlap deletion 6, and so they are surrounded by {}. **The map is:**

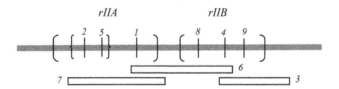

c. **The order of mutations 2 and 5 in *rIIA* cannot be determined from this data. To determine the order, you could cross each of them to mutant 1 in *E. coli* B, and quantify the number of *rII*+ phage progeny produced by recombination in each case by plating on *E. coli* K.** (In other words, these numbers show the map distances.)

30. Mutations 5 and 6 do not revert, so they are deletions. Deletions are not included in the complementation groups shown in part (a) below. All of the other mutations do revert, so they are point mutations.

    a. Remember that when analyzing complementation data, you should group the mutants that do *not* complement as these are mutations in the same gene. Mutation 1 does not complement mutation 8, but these 2 mutations *do* complement all of the other point mutations. Therefore, mutations 1 and 8 make up one complementation group. Mutation 2 complements every other mutation; therefore it is the sole mutation in another complementation group. Mutation 3 does not complement 4 or 7, so these form a third complementation group. **In total there are three complementation groups: (1, 8); (2); and (3, 4, 7)**. Note that mutation 5 fails to complement mutations (3, 4, 7) and mutation 6; this means that the deletion in mutation 5 includes at least part of the gene that is represented by the (3, 4, 7) complementation group, and it overlaps deletion 6. Mutation 6 complements only mutation 2, meaning that its deletion includes both the (1, 8) and (3, 4, 7) genes.

    b. The diploid cells from part (a) are allowed to undergo meiosis. The fact that the prototrophs are rare indicates that all of the point mutations are on the same chromosome and linked. We already know that deletions 5 and 6 and genes (1, 8) and (3, 4, 7) are all on the same chromosome because the deletions overlap each other and also fail to complement one or both of these genes. If mutation 2 were on a different chromosome from all of the other mutations (*m*), 2^+ m^+ prototrophs would be produced frequently by independent assortment.

chapter 7

If the two mutations in the heterozygous diploid are in the *same gene* (mutations 1 and 8 or mutations 4, 3 and 7), then recombination can occur between two mutations producing prototrophic (Lys+) spores. The example below shows the result for mutations 1 and 8, which are in the same complementation group. After the recombination event shown, the 4 spores will be 1 1^+ 8 (Lys-) : 1 1^+ 8^+ (Lys+) : 1 1 8 (Lys-) : 1 1 8^+ (Lys-) = **3 Lys- : 1 Lys+**. The numbers of tetrads showing the 3 Lys- : 1 Lys+ ratio will depend on the distance between the mutations. This example is similar to Problem 28(a) above.

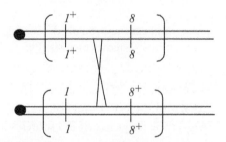

Any two point mutations in *different linked genes* on the same chromosome also give relatively rare prototrophic spores as a result of crossing-over between the two genes; as the mutations in different genes are further apart than two mutations in the same gene, the number of 3 Lys- : 1 Lys+ tetrads will be greater than when the mutations are in the same gene.

c. A '-' result in the table in part (b) means that the two mutations in the diploid cannot recombine onto the same chromosome. Because recombination occurs between adjacent nucleotides, the two mutations must affect the same nucleotide. Diploids that were generated by mating the same mutation (e.g., 1 × 1) have the mutation in the same position on both homologs. All point mutations recombine with all other point mutations, so no two point mutations affect the same nucleotide.

Note that mutations 5 and 6, which are known deletions (do not revert) do not recombine with various point mutants. In the complementation data, deletion 5 acts like any of the point mutations in the complementation group (4, 3, 7). However, it acts differently in the recombination data. It does not recombine with (overlaps) point mutations 4 and 3, but it does recombine with 7. Thus it is possible to genetically define mutation 5 as <u>a deletion = a mutation that does not recombine with 2 other mutations that *do* recombine with each other</u>. If two point mutations recombine, they must affect different nucleotides, and because the deletion fails to recombine with the two mutations, it must remove more than one nucleotide.

Deletion 6 does not recombine with mutations 1 or 4. Because 1 and 4 are in different genes (complementation groups), this deletion must span the distance between these 2 genes as well as *uncover* the point mutations themselves. Deletion 6 <u>does</u> recombine with mutation 8, which places 8 on the far side of its gene from the gene containing mutation 4.

Combining this information with the complementation data from part (a) allows you to **draw the map below:**

7-19

chapter 7

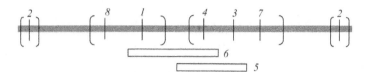

The location of gene 2 (to the right or left side) cannot be determined from these data.

Section 7.4

31. a. Diagram the cross, assuming the genes are unlinked. [See Fig. 5.21a-d on p. 152; also remember that in *Neurospora*, a mitosis follows meiosis (Fig. 5.25, p. 155)]:

ARG-E⁻ ARG-H⁺ × ARG-E⁺ ARG-H⁻ → ARG-E⁻ /ARG-E⁺ ; ARG-H⁻ /ARG-H⁺

When this diploid is sporulated, the PD asci are: 4 *ARG-E⁻ ARG-H⁺* (Arg-) : 4 *ARG-E⁺ ARG-H⁻* (Arg-), and the NPD asci are: 4 *ARG-E⁺ ARG-H⁺* (Arg+) : 4 *ARG-E⁻ ARG-H⁻* (Arg-). The frequency of PD spores = frequency of NPD.

Next, diagram the cross assuming the genes are linked. [See Fig. 5.23a-f on p. 153 and again, remember that in *Neurospora*, a mitosis follows meiosis (Fig. 5.25, p. 155)]:

ARG-E⁻ ARG-H⁺ × ARG-E⁺ ARG-H⁻ → ARG-E⁻ ARG-H⁺ / ARG-E⁺ ARG-H⁻

In this case, the PD ascus is the same as above: 4 *ARG-E⁻ ARG-H⁺* (Arg-) : 4 *ARG-E⁺ ARG-H⁻* (Arg-). The NPD asci are the same as above also: 4 *ARG-E⁺ ARG-H⁺* (Arg+) : 4 *ARG-E⁻ ARG-H⁻* (Arg-). However, because PD asci come from both NCO and 2-strand DCO meioses (Fig. 5.23a and c, respectively), **but NPD asci are produced only by 4-strand DCOs** (Fig. 5.23f), **there would be many more PD asci than NPD asci.**

PD and NPD spore types can show either MI or MII segregation for both genes, whether they are linked or not. When the genes are unlinked, most PDs and NPDs come from meioses where no crossing-over occurred between either gene and the centromere (Fig. 5.21b and c); a single crossover between the centromere and one of the two genes results in a T tetrad (Fig. 5.21d). However, if such a crossover were to take place between each of the two genes and the corresponding centromeres, then a PD or an NPD tetrad with an MII segregation pattern would result. When the genes are linked, crossing-over between the centromere and the gene (*ARG-E* or *ARG-H*) closest to it (not shown in Fig. 5.23) would not alter the fact that PD asci are produced by NCO (Fig. 5.23a) or 2-strand DCO (Fig. 5.23c) meioses, nor that NPD asci are produced by 4-strand DCOs (Fig. 5.23f). However, a single such crossover [between the centromere and the gene closest to it (*ARG-E* or *ARG-H*)] would result in an ascus with an MII segregation pattern.

  b. The *Arg-E⁻ Arg-H⁺* spores will grow when you supplement the media with ornithine, citrulline, argininosuccinate or arginine. For the *ARG-E⁺ ARG-H⁻* or *ARG-E⁻ ARG-H⁻* spores, only arginine itself in the media allows growth. The *ARG-E⁺ ARG-H⁺* spores are prototrophs that grow on minimal medium without supplementation.

chapter 7

32. a. Diagram the cross:

orange × black → F₁ brown

The problem says that orange is caused by one autosomal mutation and black is caused by another. This information implies that the two mutations are in different genes, and therefore the F₁ is doubly heterozygous. Thus, the cross is: *BB oo* (orange) × *bb OO* (black) → F₁ *Bb Oo* (brown) → F₂

And the underlying genotypic ratio of the F₂ is:

9 *B– O–* : 3 *B– oo* : 3 *bb O–* : 1 *oo bb*.

The phenotypic ratio depends on the order of orange and black in the pathway to brown. If the pathway is orange → black → brown, with the O^+ gene product carrying out the conversion from orange to black and the B^+ gene product catalyzing the conversion from black to brown, **the F₂ would have 9 brown (*B– O–*): 3 black (*bb O–*): 4 orange (*B– oo* or *bb oo*).** In other words, *oo* (orange) would be epistatic to both alleles of gene *B*. **If the order were black → orange → brown,** with the O^+ gene product carrying out the conversion from black to orange and the B^+ gene product catalyzing the conversion from orange to brown, **the F₂ would have a 9 brown (*B– O–*) : 3 orange (*bb O–*) : 4 black (*B– oo* or *bb oo*) ratio.** In this case, black (*oo*) is epistatic to both alleles of gene *B*.

b. If there are two pathways, one producing orange (requiring the O^+ gene product) and the other black (requiring the B^+ gene product), then there would be four different phenotypes in the F₂ generation: **9 brown (*B– O–*) : 3 black (*B– oo*) : 3 orange (*bb O–*) : 1 nonpigmented (*bb oo*)**.

33. Designate the genes and alleles. The *W* gene product converts a colorless (white) pigment to green. The *G* gene product converts green to blue flowers; the mutant (nonfunctional) allele is *g*. Either of two gene products *B* or *L* can convert blue to purple flowers; *b* and *l* are the mutant alleles. Diagram the cross. Note that both parents are *WW*. All progeny will be *WW*, and it will not affect the array of phenotypes in the progeny. For this reason, it is not considered in this cross:

gg BB LL (green) x *GG bb ll* (blue) → F₁ *Gg Bb Ll* (purple) → F₂:

3/4 *G–* × 3/4 *B–* × 3/4 *L–* = 27/64 *G– B– L–* (purple);
3/4 *G–* × 1/4 *bb* × 3/4 *L–* = 9/64 *G– bb L–* (purple);
3/4 *G–* × 3/4 *B–* × 1/4 *ll* = 9/64 *G– B– ll* (purple);
3/4 *G–* × 1/4 *bb* × 1/4 *ll* = 3/64 *G– bb ll* (blue);
1/4 *gg* × 3/4 *B–* × 3/4 *L–* = 9/64 *gg B– L–* (green);
1/4 *gg* × 1/4 *bb* × 3/4 *L–* = 3/64 *gg bb L–* (green);
1/4 *gg* × 3/4 *B–* × 1/4 *ll* = 3/64 *gg B– ll* (green);
1/4 *gg* × 1/4 *bb* × 1/4 *ll* = 1/64 *gg bb ll* (green).

The ratio is 45 purple : 16 green : 3 blue. You can see why the problem specified that the green parent was mutant in only a single gene, as *gg bb LL* or *gg BB ll* plants would still be green yet would yield a very different ratio of phenotypes in the F₂.

chapter 7

34. First, order the compounds from final product to first one in the pathway. The final compound is the one on which all of the mutants in the pathway will grow; you were told in the question that the final product is G, and you can see in the table that the data supports that idea. The compound before that (E in this example) is the one that allows all the mutants except one class to grow. Continue working toward the beginning of the pathway in this manner.

    Next, order the mutants. Again, you can do this by working backwards from the final product through the intermediates. Start by looking for the mutant that grows only when supplied with G; in this problem it is mutant 2. The mutation must be in the gene encoding the enzyme catalyzing the last step in the synthesis of compound G. Then look for the mutant that grows only when supplied with G or one other intermediate. Mutant 7 can grow only when supplied with intermediate E or with G. This verifies our earlier assignment of E as the intermediate that precedes G, and it also tells us that the gene in which mutation 7 is located encodes the enzyme that allows the synthesis of E. In this way, continue working back through the pathway to get the answer.

    $$\underset{X}{\overset{6}{}} \to \underset{F}{\overset{1}{}} \to \underset{D}{\overset{5}{}} \to \underset{A}{\overset{3}{}} \to \underset{C}{\overset{4}{}} \to \underset{B}{\overset{7}{}} \to \underset{E}{\overset{2}{}} \to G$$

35. a. Diagram the crosses:

    1. blue × white → F_1 purple → F_2 9 purple : 4 white : 3 blue
    2. white × white → F_1 purple → F_2 9 purple : 7 white
    3. red × blue → F_1 purple → F_2 9 purple : 3 red : 3 blue : 1 white
    4. purple × purple → F_1 purple → F_2 15 purple : 1 white

    Two observations apply to the four crosses:

    - All four crosses show modifications of the 9:3:3:1 ratios in the F_2, implying that 2 genes are controlling the colors.
    - In all 4 crosses, the purple phenotype corresponds to the "*A– B–*" class.
    - Remember that the parents are pure-breeding

    Cross 1. *AA bb* (blue) x *aa BB* (white) → *Aa Bb* (purple) →
    9 *A– B–* (purple) : 4 *aa – –* (white) : 3 *A– bb* (blue)

    Cross 2. *AA bb* (white) x *aa BB* (white) → *Aa Bb* (purple) →
    9 *A– B–* (purple) : 7 *aa – –* + *– – bb* (white)

    Cross 3. *AA bb* (red) x *aa BB* (blue) → *Aa Bb* (purple) →
    9 *A– B–* (purple) : 3 *A– bb* (red) : 3 *aa B–* (blue) : 1 *aa bb* (white)

    Cross 4. *AA bb* (purple) x *aa BB* (purple) → *Aa Bb* (purple) →
    15 *A– – –* + *– – B–* (purple) : 1 *aa bb* (white)

    b. **The simplest biochemical pathways that could explain the data are diagrammed** as follows. In each case, the dominant alleles of genes *A* and *B* encode proteins (A and B, respectively) that perform the conversions indicated, and the recessive alleles are nonfunctional. Recall from Chapter 3 that in Cross 1, *bb* is epistatic to *A;* in Cross 2,

chapter 7

gene *A* and gene *B* are complementary; in Cross 3, genes *A* and *B* work in independent pathways and so do not interact; in Cross 4, gene *A* and gene *B* are redundant.

1. colorless $\xrightarrow{A}$ blue $\xrightarrow{B}$ purple

2. colorless1 $\xrightarrow{A}$ colorless2 $\xrightarrow{B}$ purple
   or
   colorless $\xrightarrow{A,B}$ purple

3. colorless $\xrightarrow{A}$ red
   colorless $\xrightarrow{B}$ blue
   ———————
   purple

4. colorless $\xrightarrow{A}$ purple
   colorless $\xrightarrow{B}$ purple

c. **Cross 2:** The results of this cross series are compatible with two possible pathways. In one scenario, proteins A and B are separate enzymes that perform successive conversions. In a second pathway, proteins A and B are subunits of a single enzyme, and both subunits are required for the enzyme to function.

d. To simplify the answer, let's assume initially that "tightly linked" means that the distance between the *A* and *B* genes is 0 mu (no crossing-over occurs between the two genes).

We can rewrite all four of the crosses as:

$A\,b\,/\,A\,b\ \times\ a\,B\,/\,a\,B\ \to\ A\,b\,/\,a\,B$ (selfed) $\to$

$1\ A\,b\,/\,A\,b\ :\ 2\ A\,b\,/\,a\,B\ :\ 1\ a\,B\,/\,a\,B$.

The phenotypes for each cross are:

Cross 1. 2 purple (*A b* / *a B*) : 1 blue (*A b* / *A b*) : 1 white (*a B* / *a B*)
Cross 2. 2 purple [(*A b* / *a B*)] : 2 white [(*A b* / *A b*) + (*a B* / *a B*)]
Cross 3. 2 purple (*A b* / *a B*) : 1 red (*A b* / *A b*) : 1 blue (*a B* / *a B*)
Cross 4. All purple [(*A b* / *a B*) + (*A b* / *A b*) + (*a B* / *a B*)]

Note that if any recombination occurs between the two genes, then recombinant *a b* gametes will be produced, allowing the emergence of white F$_2$ plants (*a b* / *a b*) in Crosses 3 and 4. The smaller the distance between genes *A* and *B*, the more the ratios will resemble those shown above. The greater the distance between the two genes, the more the ratios will resemble those in part (a).

36. To solve this problem involving a branched biosynthetic pathway, consider first only those mutants that are defective solely in biosynthesis of one amino acid. The mutants defective in only the proline pathway are those that grow when given proline in the media but not when given glutamine. Mutants 2, 6, and 1 are of this type. No intermediate exists that allows the growth of mutant 2, so the defect must be in the final enzyme that produces proline. Working backward from this point in the pathway, mutant 6 grows when supplied

with intermediate A, so A is the final intermediate and mutant 6 is blocked in the step that leads to A. Mutant 1 grows when supplied with intermediates E or A, indicating that E is prior to A in the proline pathway.

Now conduct the same analysis for glutamine. Mutants 7 and 4 are defective only in glutamine biosynthesis. Mutant 4 grows only on glutamine, while mutant 7 grows when supplied with B or glutamine, indicating that mutant 7 is blocked in the production of B, and mutant 4 cannot convert intermediate B to glutamine.

Finally, look at the mutants that are defective in both glutamine <u>and</u> proline biosynthesis. Mutants 5 and 3 are of this type. Mutant 3 grows only if given intermediate C, so it must be blocked just prior to this step. Mutant 5 grows if given C or D, so it is blocked prior to the D intermediate. This represents the first part of the pathway that is used both in proline and glutamine biosynthesis.

Putting all of this information together, we have the following branching pathway:

$$X \xrightarrow{5} D \xrightarrow{3} C \begin{array}{c} \xrightarrow{7} B \xrightarrow{4} \text{Glutamine} \\ \xrightarrow{1} E \xrightarrow{6} A \xrightarrow{2} \text{Proline} \end{array}$$

37. a. This problem [is similar to the previous one. The] mutants all complement each other, meaning that each inactivates a different gene. Briefly, one mutant can grow only when supplemented with C, so C must be the final intermediate before thymine, etc. Mutant 21 cannot grow with any supplement other than thymine itself, so it must block the final step in the pathway (the conversion of C to thymine), etc. **The pathway is:**

$$X \xrightarrow{18} D \xrightarrow{14} B \xrightarrow{9} A \xrightarrow{10} C \xrightarrow{21} \text{Thymidine}$$

b. **Double mutant 9 and 10 accumulates B**; single mutant 9 would also accumulate B. In both cases, conversion of B to A is blocked by mutant 9, causing accumulation of B. The additional block of the conversion of A to C in the double mutant is inconsequential when no A is being made. By similar logic, **double mutant 10 and 14 accumulates D.** You can see the general rule here is that the mutant that acts first in the synthesis pathway dictates which compound accumulates.

38. The data represent complementation experiments done at the biochemical level in the test tube. The results suggest that **two different X-linked genes cause hemophila when mutant. Individuals 1 and 2 are mutant in one gene (call it gene *A*) and individuals 3 and 4 are mutant in the other gene (gene *B*).** Individuals 1 and 2 thus lack the function of one factor needed for clotting, while 3 and 4 lack a different factor. When the two kinds of blood complement (as in the mixture of blood from individuals 1 and 3), clotting occurs because the blood of each patient supplies the factor lacking in the other patient.

Several types of biochemical pathways involving possible functions of the products of these two genes are consistent with the data. The two genes could encode enzymes that sequentially convert a compound in the blood serum into an end product that is required for clotting. Another of several possibilities is that one of the proteins (say A) is the substrate for a reaction catalyzed by enzyme B, and this reaction produces a needed clotting factor. The results exclude a pathway in which the product of one of these genes (say the *A* gene) is required for the synthesis of the protein encoded by other gene (*B*). In such a case, the blood plasma of a patient mutant for gene *A* would have neither protein A nor protein B, so the mixture of mutant plasmas would have no source of compound B. (The plasmas are cell-free so new proteins could not be synthesized.)

As an interesting historical sidelight, the cited article in the *British Medical Journal* was published in the December 27th (Christmas) issue of 1952. The first patient whose blood could complement that of most other hemophiliacs in the test tube (thus indicating the existence of two different kinds of X-linked hemophilia) was a 5 year old boy whose family name was Christmas. Because of these facts, the rarer form of the disease, usually called hemophilia B, is still often called Christmas disease.

39. a. This should be a **successful treatment** because the normal plasma contains both vWF and factor VIII. The **effect should be immediate** because both factors are present **and prolonged** because vWF stabilizes factor VIII.

   b. This **treatment should be unsuccessful** because vWD plasma has neither vWF nor factor VIII.

   c. This should be a **successful treatment** because the hemophilia A plasma contains vWF even though it does not have factor VIII. The **effect should be delayed** because the patient has no factor VIII and the added vWF will only stabilize factor VIII newly synthesized by blood cells in the patient, and this takes time. The **effect should also be prolonged** because vWF stabilizes factor VIII.

   d. This should be a **successful treatment** because the normal blood contains both vWF and factor VIII. The **effect should be immediate** because both factors are present **and prolonged** because factor VIII in the transfused plasma is already stabilized.

   e. This **treatment will be unsuccessful**. The vWD plasma has neither vWF nor factor VIII, and the patient cannot synthesize any factor VIII.

   f. This **treatment should be unsuccessful**. Neither the patient's blood nor the transfused plasma has any factor VIII.

   g. This treatment **should be successful, but only after a delay** to allow the patient's blood cells to synthesize enough factor VIII that can be stabilized by the injected vWF. The **effects should be prolonged** because of the stabilization.

   h. This treatment should be **unsuccessful** because no factor VIII can be made.

   i. This treatment **should be successful immediately** because you are injecting factor VIII, but the **effects will be only very short-term** because the injected factor VIII will be degraded in the absence of vWF.

   j. This treatment **should be successful immediately, and the effects will be prolonged** because the patient's blood has vWF, which can stabilize the injected factor VIII.

40. This problem is similar to an ordered enzymatic pathway except the gene products are a series of nonenzymatic proteins that form a structure assembled in a particular order. The loss of one protein due to mutation will prevent all the subsequent proteins from being

added. The loss of the first protein at the surface would prevent all others from being at the cell surface. The mutant that fits this description is E. Mutants A and C have a similar pattern in which only E and C, or E and A, respectively, are at the surface, so genes A and C encode the two proteins that form the dimer structure shown second from the embryo surface. The logic is continued to place the remaining three gene products in their order, as **shown in the following diagram.**

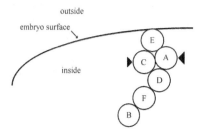

41. a. **Two loci are needed**, one for the α globin polypeptides and one for the β globin polypeptides.

    b. Assuming that both alleles of both genes are expressed into polypeptides at the same levels, then you would see 1 α1 : 1 α2 for the two forms of the hemoglobin α subunit and 1 β1 : 1 β2 for the forms of the β subunits. The α subunits will form the following sorts of dimers: 1 α1α1 : 2 α1α2 : 1 α2α2. The β subunits will assemble into dimers in the same way, giving a genotypic monohybrid ratio of: 1 β1β1 : 2 β1β2 : 1 β2β2. In order to figure out the types of heterotetramers and their frequencies, apply the product rule to the two monohybrid ratios: **1 α1α1 β1β1 : 2 α1α2 β1β1 : 1 α2α2 β1β1 : 2 α1α1 β1β2 : 4 α1α2 β1β2 : 2 α1α2 β2β2 : 1 α1α1 β2β2 : 2 α1α2 β2β2 : 1 α2α2 β2β2.**

42. *Important Note:* The Table in your copy of the text may have an error in the *ARG-K⁻* row. The '+' should be in the "Ornithine" column, not in the "Argininosuccinate" column.

    The particular subset of supplements that will support growth of any specific Arg⁻ mutant is dictated by the order of the chemical reactions in the arginine biosynthesis pathway (Fig. 7.27c on p. 235). Only four different complementation groups are possible because the pathway consists of only four steps, each catalyzed by a unique enzyme encoded by a single gene. A particular Arg⁻ mutant will grow only on supplements that are pathway intermediates downstream of the blocked step in the pathway. The reason is that a mutant lacking the enzyme for that (blocked) step nevertheless contains the enzymes for all the upstream and downstream steps. The upstream enzymes and chemical intermediates are rendered irrelevant by the downstream block. However, the same is not true for the downstream enzymes and intermediates; if supplied an intermediate downstream of the blocked step, the downstream enyzmes present can use it to make arginine. Therefore, it is logical that the subsets of intermediates that are able to support arginine biosynthesis in any one mutant represent all of the intermediates downstream of the step catalyzed by the mutation.

    The theoretical Arg⁻ mutants shown would not be found because in every case, the subsets of intermediates that support growth are not contiguous in the pathway. In other words, in each case, there is a pathway intermediate that fails to support growth that is downstream in the arginine biosynthesis pathway of one that does support growth.

chapter 7

43. In unequal crossing-over, the homologous chromosomes align out of register according to DNA sequence similarity between related, but different genes (see the following diagram). **The result of crossing-over is a homolog with a duplication, β (δ/β) δ, and a homolog with a deletion, (β/δ).** The genes with a slash and enclosed in parentheses indicate hybrid genes.

As a point of interest not asked in the problem, these globin genes have differences even though they are very similar in DNA sequence. One of the more important differences between the δ and β genes is the time of expression. In these hybrid genes the regulatory region of the β gene may have been replaced with the regulatory information of the δ gene, or *vice versa*, so the hybrid gene may be expressed inappropriately (that is, during the wrong time in development). In addition, the polypeptides formed from the hybrid genes may affect the affinity of hemoglobin for oxygen in unpredictable ways.

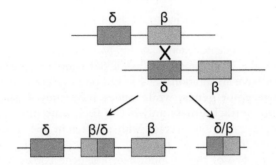

44. a. The information in the problem is that most mammals, including New World primates, are dichromats (two kinds of photoreceptors), while Old World primates are trichromats (three kinds of photoreceptors). Primates diverged from mammals 65 million years ago (65 Myr), while Old World primates diverged from each other about 35 Myr ago. This information means that the final gene duplication event that led to different red and green photoreceptors most likely occurred in the lineage leading to the trichromatic Old World primates (like humans), but not in the lineage leading to the dichromatic New World primates (like marmosets). This interpretation dates the final gene duplication event to some time subsequent to 35 Myr. Because all mammals have rhodopsin plus at least two color photoreceptors, you can also conclude that the two previous gene duplication events that led first to the blue gene and then to the green gene occurred some time prior to 65 Myr. **These conclusions can be summarized by annotating Fig. 7.32d as shown in the following figure:**

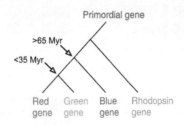

7-27

b. We assume here that only one allele for the autosomal receptor gene exists. For the X-linked receptor gene, any given female monkey will have two alleles while any male will only have one. **Males will be dichromats**, as they have only two kinds of photoreceptors: They are homozygous for the autosomal receptor gene but are hemizygous for the X-linked gene. However, **there will be 3 classes of males that see colors in different ways. Females can be dichromats** if they are homozygous for an allele of the X-linked gene, again there will be 3 classes. **Females can also be trichromats** if they are heterozygous for this gene, and there are 3 different combinations: 1-2, 1-3 and 2-3.

c. The fact that 95% of our receptors work best in low light conditions suggests that **early mammals were most active at night.** Natural selection would have favored ancestors with the best night vision.

45. a. Tetrachromats must have four different kinds of photoreceptor proteins. As shown in the following diagrams (next page), men and women both have two copies of the autosomal blue photoreceptor gene (*darker blue rectangles;* one on each homolog). However, because the X chromosome has a red gene and green gene (*red* or *green rectangles*), **females have** an additional 2 X-linked photoreceptor gene copies and thus **a total of 6 photoreceptor genes, while males have only 4 photoreceptor genes. Males can become tetrachromats only if 1 of their blue photoreceptor genes is a mutant** that produces a photoreceptor with a different spectral sensitivity than the normal blue receptor (*light blue rectangle*). Females, on the other hand, can be tetrachromats if any one of their 6 photoreceptor gene copies make a receptor with altered spectral sensitivity (*light blue* or *yellow rectangles*). (This answer assumes that all of the retinal cells do not inactivate the same X chromosome.) **The chance that one of 6 genes will acquire such a mutation is obviously much higher than the chance that a single gene will do so.**

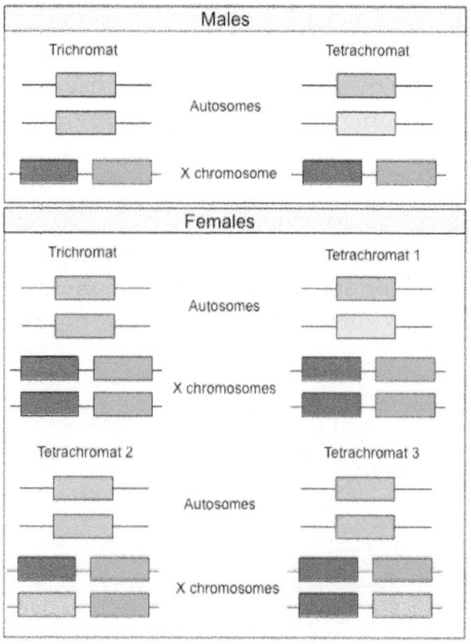

b. A non-colorblind woman who has red/green colorblind sons will have X chromosomes like those of Tetrachromat 2 or 3 in the figure for part (a) above. The sons who inherit her X chromosome that lacks either a normal red or green photoreceptor gene will be colorblind.

c. A rare XY male could be a tetrachromat if he has 3 different photoreceptor genes on an X chromosome that was a product of unequal crossing-over between red and green genes (Fig. 7.33d, p. 243): a red gene, a green gene, and a red/green hybrid that encodes a photoreceptor with a different spectral sensitivity than either the red protein or the green protein.

chapter 8

Gene Expression: The Flow of Information from DNA to RNA to Protein

Synopsis

This chapter describes how genes are *transcribed* into RNAs, and how some of those RNAs are *translated* to make proteins using the *genetic code*. Mutations in genes were an extremely important tool in the experiments that elucidated the structure of the gene and the nature of the genetic code. Thus, two key elements of this chapter are systems for describing mutations according to: (1) how the mutation affects the function of the gene (gene expression) at the molecular level, and (2) how the product of the mutant gene functions as compared to wild-type gene product. These descriptive systems will help you understand how mutations cause aberrant phenotypes including disease syndromes in humans.

Several concepts in this chapter will reappear many times in the remainder of the book and thus must be understood in detail:

- **The "central dogma" of molecular genetics:** Genes are templates for transcription of RNAs by RNA polymerase; many RNAs are processed into mRNAs; mRNAs are translated into proteins using the genetic code. (See Fig. 8.1 on p. 255.)

- **DNA, RNA transcripts, and protein molecules each have polarities, and they are related to one another in specific linear manner.** (See Fig. 8.7 on p. 261.)

- **The structures within the base pair sequences of all genes – promoters, exons/introns (in eukaryotes), 5' UTRs, 3' UTRs, ORFs – are tied intimately to gene function.** (Fig. 8.14 on p. 271 is a key illustration.)

- **The genetic code is related intimately to the base sequences of tRNAs, and the activities of aminoacyl-tRNA synthetases.** It is essential to understand how codon sequences relate to anticodon sequences, and how the tRNA synthetases ensure that the genetic code is obeyed. (See Figs. 8.18 and 8.19 on p. 274.)

- **The most generally useful system for classifying mutant alleles is that distinguishing between loss-of-function and gain-of-function alleles.** (See Table 8.2 on p. 288.) Other systems, such as describing the effects of mutations on gene expression (Fig. 8.27 on p. 284) also provide important details about the consequences of particular mutations.

Key terms

transcription – use of one strand of DNA (the **template strand**) by **RNA polymerase** to synthesize, from ribonucleotide subunits, a complementary strand of RNA (a **transcript**) whose sequence corresponds to the other DNA strand (the **RNA-like strand**)

promoter – the DNA sequence in a gene to which RNA polymerase binds in order to initiate transcription

upstream and **downstream** – terms describing directions or positions with respect to a gene. *Upstream* signifies the direction opposite to that traveled by RNA polymerase as it transcribes the gene, while the *downstream* direction is the direction in which RNA polymerase moves during transcription.

messenger RNA (mRNA) – transcripts destined to be used as blueprints for the synthesis of proteins by translation

RNA processing – events that occur in eukaryotic cells to change a **primary transcript** synthesized by RNA polymerase into a mature mRNA. These processing steps include: addition of a run of As (a **poly-A** tail) near the 3' end of the primary RNA; addition of a modified G nucleotide called a **5' cap** to the 5' end of the RNA; and **RNA splicing** – the process by which long stretches of RNA called **introns** are removed and degraded, and the remaining RNA fragments, called **exons,** are connected together.

alternative splicing – a phenomenon where the primary transcripts of a single gene may be spliced differently by combining different exons into mature mRNAs. Alternatively spliced mRNAs may encode related but different protein products.

translation – the synthesis of polypeptides (proteins) from amino acids using an mRNA as a blueprint; the ribozyme **peptidyl transferase** forms peptide bonds between amino acids

ribosome – a multisubunit ribonucleoprotein complex in the cytoplasm that performs translation

genetic code – the rules for translating mRNA base sequences into amino acid sequences. Groups of 3 bases (triplet sequences) are code words **(codons)** for amino acids; a particular codon (5' AUG) indicates the beginning of the code for the protein **(start codon)**, and 3 different codons (5' UAA, 5' UGA, and 5' UAG) are stop signals **(stop codons).**

reading frame – the pattern of triplet codons read as determined by the first nucleotide where reading of the genetic code begins. Every RNA molecule has three potential reading frames, while a double-stranded DNA molecule has six potential reading frames (three on each DNA strand).

open reading frame (ORF) – a base sequence, which when read as triplets as if it were an mRNA, contains no stop codons. The parts of a gene that contain its ORF are called the gene's **coding regions.**

chapter 8

polycistronic and **monocistronic** – Many prokaryotic mRNAs are *polycistronic* – a single mRNA contains more than one ORF, each encoding a separate protein; eukaryotic mRNAs are *monocistronic* – each has one ORF.

transfer RNA – RNA adapters that relate amino acids to the codons in the genetic code. At the ribosome, the three nucleotides that constitute the tRNA's **anticodon** base-pair with the corresponding three nucleotides in the mRNA's codon. The 3' ends of tRNAs become covalently linked to a specific amino acid by **aminoacyl-tRNA synthetases** (one synthetase enzyme exists for each of the 20 common amino acids); such tRNAs are **charged tRNAs.** The ribosome positions adjacent tRNAs at the ribosome so that peptide bonds can be formed between the amino acids the tRNAs carry.

degeneracy – Because there are 61 codons that specify 20 amino acids, several different codons must specify the same amino acid. In addition, a single tRNA species can often decode more than 1 codon (see *wobble* just below). Thus, a 1 amino acid: 1 codon: 1 tRNA relationship does not exist.

wobble – describes how the anticodon of a single tRNA species can base pair with several different codons for the same amino acid. The reason is that the 5' base of the anticodon can form certain unusual, noncomplementary base pairs with the 3' base of the codon.

posttranslational modifications – a variety of chemical changes (including phosphorylation and cleavage among others) that enzymes make to proteins after the proteins have been translated at the ribosome. These modifications can alter protein function significantly.

missense, nonsense, frameshift, and silent mutations – point mutations within the coding regions of a gene. *Missense* mutations are base substitutions in a codon sequence that alter the amino acid specificity of the codon; *nonsense* mutations are base substitutions that alter an amino-specifying codon (a **sense codon**) so that it becomes a stop codon; *frameshift* mutations are deletions or additions of one or a few base pairs (not multiples of 3) that change the reading frame downstream of the mutation; *silent* mutations are base substitutions, usually in the third base of a codon, that do not alter the amino acid specificity of the codon.

nonsense suppressor mutations – base substitutions within the anticodon regions of genes encoding tRNAs that enable the tRNA to recognize a stop codon (a **nonsense codon**); the mutant tRNAs are called **nonsense suppressor tRNAs.**

frameshift suppressor mutations – base addition or deletion in the coding region of a gene containing a frameshift mutation that compensates for the original mutation by restoring the reading frame

loss-of-function allele – mutant allele that retains no wild-type gene activity (**amorphic** or **null allele**) or that retains only partial gene activity (**hypomorphic** or **leaky allele**)

gain-of-function allele – mutant allele that has more function or a different function from the wild-type gene. Gain-of-function alleles are divided into three classes: (1) **hypermorphic alleles** perform the same function as wild-type alleles, but are overactive; (2) **neomorphic alleles** have acquired a new function – they either

chapter 8

make the normal gene product in a new context (time or cell type), or they make a new gene product; **antimorphic alleles (dominant negatives)** make a gene product that antagonizes the function of the wild-type gene product.

haploinsufficiency – a property of certain genes where one wild-type allele of the gene in a diploid does not produce enough gene product to prevent the organism from displaying a mutant phenotype

Problem Solving

Keep a copy of the genetic code table and the wobble rules handy. You do not need to memorize the meaning of all 64 codons, but it will increase your problem-solving efficiency to remember the code's punctuation marks (5′ AUG is the start codon; 5′ UAA, 5′ UAG, and 5′ UGA are the stop codons.) The most difficult part about solving problems relating to the genetic code is to keep the strands straight. Remember that codons and anticodons have polarity – and that like all base pairing interactions – codons and anticodons base pair in *antiparallel* configuration. Realize also that there's nothing at all wrong with an *anticodon* having the base sequence 5′ UAG (it's not a stop codon in the anticodon – only in the codon); a tRNA with such an anticodon would pair with the codon 5′ CUG and other codons also, depending on how the 5′U was modified enzymatically.

Vocabulary

1.

a.	codon	5.	a group of three mRNA bases signifying one amino acid
b.	colinearity	10.	the linear sequence of amino acids in the polypeptide corresponds to the linear sequence of nucleotide pairs in the gene
c.	reading frame	8.	the grouping of mRNA bases in threes to be read as codons
d.	frameshift mutation	12.	addition or deletion of a number of base pairs other than three into the coding sequence
e.	degeneracy of the genetic code	6.	most amino acids are not specified by a single codon
f.	nonsense codon	2.	UAA, UGA, or UAG
g.	initiation codon	9.	AUG in a particular context
h.	template strand	14.	the strand of DNA having the base sequence complementary to that of the primary transcript
i.	RNA-like strand	3.	the strand of DNA that has the same base sequence as the primary transcript
j.	intron	13.	a sequence of base pairs within a gene that is not represented by any bases in the mature mRNA

chapter 8

k.	RNA splicing	1.	removing base sequences corresponding to introns from the primary transcript
l.	transcription	7.	using the information in the nucleotide sequence of a strand of DNA to specify the nucleotide sequence of a strand of RNA
m.	translation	15.	using the information encoded in the nucleotide sequence of an mRNA molecule to specify the amino acid sequence of a polypeptide molecule
n.	alternative splicing	11.	produces different mature mRNAs from the same primary transcript
o.	charged tRNA	4.	a transfer RNA molecule to which the appropriate amino acid has been attached
p.	reverse transcription	16.	copying RNA into DNA

Section 8.1

2.

a.	existence of an intermediate messenger between DNA and protein	4.	protein synthesis occurs in the cytoplasm, while DNA resides in the nucleus.
b.	the genetic code is nonoverlapping	6.	single base substitutions affect only one amino acid in the protein chain
c.	the codon is more than one nucleotide	1.	two mutations affecting the same amino acid can recombine to give wild type
d.	the genetic code is based on triplets of bases	2.	one or two base deletions (or insertions) in a gene disrupt its function three base deletions (or insertions) are often compatible with function
e.	stop codons exist and terminate translation	3.	artificial messages containing certain codons produced shorter proteins than messages not containing those codons
f.	the amino acid sequence of a protein depends on the base sequence of an mRNA	5.	artificial messages with different base sequences gave rise to different proteins in an *in vitro* translation system

3. The key to this problem is to realize that when using *in vitro* systems with artificial mRNAs, translation does not always start at the same place on all mRNAs. (See Problem 9 below.)

   a. ... GU GU GU GU GU ... or ... UG UG UG UG UG ...

   b. ... GU UG GU UG GU UG GU UG GU ...

c. ... GUG UGU GUG UGU GUG ...

d. ... GUG UGU GUG UGU GUG UGU GUG UGU... This is the result of reading the first 3 nucleotides then going back to the second nucleotide and reading a codon, etc. Overlapping codes will always give more coding information for the same number of bases compared to the non-overlapping code.

e. ... GUGU GUGU... or ...UGUG UGUG...

4. a. By comparing the two sequences, you can see that the 4th A from the 5' end of the wild-type sequence is deleted (-) in the mutant, and a G is inserted (+) just upstream of the "GCC" at the 3' end of the wild-type sequence. [See part (b) below.]

b. The amino acids in the wild-type and mutant proteins are shown below:

```
                    Lys  Ser  Pro  Ser  Leu  Asn  Ala
wild-type:    5'    AAA  AGT  CCA  TCA  CTT  AAT  GCC  3'
                         (-)                 (+)
mutant:       5'    AAA  GTC  CAT  CAC  TTA  ATG  GCC  3'
                    Lys  Val  His  His  Leu  Met  Ala
```

The four bold amino acids between the – and + mutations other than Leu are different from wild-type.

c. Those four amino acids must not be in a crucial function domain of the protein (for example, they do not carry out catalysis), nor do they alter the overall structure of the protein significantly enough to affect its function.

5. a. FC0 was a single-base insertion and FC7 was a single-base deletion nearby that restored the reading frame. **A reversion of FC0 back to the wild-type sequence (deletion of the inserted base) could also have restored *rIIB+* function.**

b. The *rIIB+* phage obtained by intragenic suppression is a double mutant (FC0 FC7) that has both a single-base insertion and a single-base deletion. In a coinfection, the double mutant *rIIB+* can undergo recombination with the original wild-type *rIIB+* strain – in the region between the FC0 and FC7 mutations – to generate *rIIB-* phage that are either FC0 or FC7. **A true revertant cannot produce *rIIB-* progeny through recombination with wild-type *rIIB+* phage** because the true revertant and the wild-type have the same base pair sequence.

c. FC7 single mutants (*rIIB-*) produced by recombination between true wild types and FC0 FC7 double mutants have a property that distinguishes them from FC0 single mutants (*rIIB-*): **Only FC7 mutants can recombine with FC0 to produce *rIIB+* phage.**

d. To obtain many deletion mutations, several different intragenic suppressors of FC0 could have been obtained, including FC7; all of them are single-base deletions like FC7. Each would have been separated from FC0 by recombination with a true wild-type (*rIIB+*) strain, and the resulting *rIIB-* recombinants that are deletions distinguished from FC0 by their ability to recombine with FC0 to produce *rIIB+* phage. The resulting strains would be *rIIB-* phage with single deletion mutations.

chapter 8

Multiple addition mutations (like FC0) could have been obtained by isolating suppressors of deletion mutants like FC7, recombining them away from FC7, etc. This process of isolating suppressors of suppressors of suppressors to obtain more and more mutations of opposite sign could go on *ad infinitum*.

6. Glutamic acid (in the wild-type protein) can be encoded by either GAA or GAG. In sickle cell anemia this amino acid is changed to valine by a single base change. Valine is encoded by GUN with N representing any of the four bases. Therefore, the second base of the triplet was altered from A to U in the $Hb\beta^S$ allele. In $Hb\beta^C$ the glutamic acid codon (GAA or GAG) is changed to a lysine (AAA or AAG). The change here is in the first base of the codon. **The mutation causing the $Hb\beta^C$ allele therefore precedes the $Hb\beta^S$ mutation in the sequence of the β-globin gene when reading the RNA-like strand in the 5'-to-3' direction.** (Stated another way, the $Hb\beta^C$ mutation is upstream of the $Hb\beta^S$ mutation.)

7. a. If the Asn (5' AAC) is changed to a Tyr residue, the nucleotide change is to a UAC. In protein B this means that the Gln (5' CAA) at position 3 becomes a Leu (5' CUA).

   b. Leu (5' CUA) at position 12 is changed to Pro (5' CCA). In protein B the Thr (5' ACU) at position 5 is still a Thr residue even though the codon is different (5' ACC).

   c. **When Gln (5' CAA) at position 8 in protein B is changed to a Leu (5' CUA), the Lys codon (5' AAG) at position 11 in protein A is changed to a stop codon (5' UAG).** This would cause the production of a truncated (shortened) form of protein A only 10 amino acids long.

   d. This is a thought question that involves some speculation; the following are two reasonable possibilities. (1) As seen in parts (a)-(c) above, **a mutation in the region of overlap has a high probability of causing alterations in both of the proteins simultaneously. Any change in this DNA sequence has the potential to affect two proteins instead of just one, and an organism would be less likely to tolerate mutations affecting the production of two proteins** than mutations affecting the production of a single protein. **There would thus be strong evolutionary selection against overlapping reading frames. (2) If a region of DNA evolved so as to encode a protein with a stable three-dimensional conformation, it is very unlikely that a stable protein could be produced by the sequence shifted by one nucleotide.** The reason is that the alternate reading frame is likely to have a stop codon every 3 codons out of 64. This means that in reality, among the very few examples of overlapping reading frames that exist, either one or both of the proteins is very small, being composed of only a few amino acids.

8. Note that the nucleotide sequences that follow are written as mRNA sequences that can be converted easily to the DNA sequence of the gene. All mutagens work at the level of the DNA, even though the changes are often written at the level of the mRNA. From the amino acid sequences given, we can deduce that the mRNA sequences of the wild-type and mutant genes are:

8-7

	Gly	Ala	Pro	Arg	Lys
wild-type mRNA:	5' GGN	GCN	CCN	AGA/G CGN	AAA/G 3'
mutant mRNA:	5' GGN	CAU/C	CAA/G	GGN	AAA/G 3'
	Gly	His	Gln	Gly	Lys

Remember that proflavin causes frameshift mutations (single base insertions and deletions) in the DNA. Notice that there are several ambiguous bases in the wild-type sequence. Some codons can end with either of 2 bases, indicated with a slash, and other codons can end with any one of the four bases (indicated by an N); in addition, Arg can be encoded by six different codons.

From inspection of the invariant bases in the two sequences, it is clear that a single base deletion occurred in the mutant: Either the first nucleotide of the second codon (in *red* below) was deleted to make the mutant sequence, or the third nucleotide of the first codon (in *blue* below) was deleted. In either case, the deduced DNA sequence (14 nt) of the wild-type is:

5' GGN GCA CCA AGG AAA 3'

9. a. As there are no AUGs in the RNA, **polypeptides will be produced only at high Mg^{2+} concentrations.** The codons in the RNA will be read as …UGU GUG UGU GUG… and so **polypeptides with alternating Cys and Val amino acids will be made.**

   b. At low Mg^{2+} concentrations, translation will start with AUG, and the RNA will be read as AUG CAU GCA UGC AUG …; **polypeptides with repeats of Met-His-Ala-Cys will be synthesized.** At high Mg^{2+} concentrations, because translation can initiate anywhere on the RNA, **the same polypeptides will be produced except they will not all begin with Met.** (RNAs not read starting with AUG are read as UGC AUG CAU GCA… or GCA UGC AUG CUA…)

   c. At low Mg^{2+} concentrations, translation will start with AUG: AUG UAU GUA UGU… and thus **polypeptides with the repeat Met-Tyr-Val-Cys will be made.** At high Mg^{2+} concentrations, translation can start anywhere on the RNA. This results only in **polypeptides containing Met, Tyr, Val, and Cys in that order, but they will not all begin with Met.** (RNAs that are not read starting with AUG are read as UGU AUG UAU GUA… or GUA UGU AUG UAU…)

10. A nonsense mutation is a single nucleotide change that turns a sense codon (one coding for an amino acid) into one of the three stop codons. The change can occur at any of the three positions in the codon. The easiest way to identify such sense codons is to begin with each nonsense codon and systematically change each position to all other possible nucleotides. This procedure gives nine different possible codons that could be altered by a single base change to that particular stop codon. Use the genetic code table (Fig. 8.2 on page 256) to translate the resulting codons. Not all of these changes lead to sense codons – some of them will result in nonsense (STP) codons.

chapter 8

Stop Codon Change		UAA		UAG		UGA	
1st position		AAA	Lys	AAG	Lys	AGA	Arg
		CAA	Gln	CAG	Gln	CGA	Arg
		GAA	Glu	GAG	Glu	GGA	Gly
2nd position		UUA	Leu	UUG	Leu	UUA	Leu
		UCA	Ser	UCG	Ser	UCA	Ser
		UGA	STP	UGG	Trp	UAA	STP
3rd position		UAU	Tyr	UAA	STP	UGU	Cys
		UAC	Tyr	UAC	Tyr	UGC	Cys
		UAG	STP	UAU	Tyr	UGG	Trp

11. a. **Eight** codons exist in RNAs made up of Us and Gs in random order: UUU, UUG, UGU, GUU UGG, GUG GGU, GGG.

   b. The eight codons in part (a) correspond to **six** different amino acids: Phe, Leu, Val, Cys, Trp, and Gly.

   c. **The answers to part (a) and (b) are different because the genetic code is degenerate in that the same amino acid is specified by several different codons:** UUU = Phe, UUG = Leu, UGU = Cys, GUU or GUG = Val, UGG = Trp, GGU and GGG = Gly.

   d. The frequency of U is 3/4, and the frequency of G is 1/4. Therefore: f(UUU) = $(3/4)^3$ = 27/64; f(UUG) = f(UGU) = f(GUU) = 3/4 × 3/4 × 1/4 = 9/64; f(UGG) = f(GUG) = f(GGU) = 3/4 × 1/4 × 1/4 = 3/64; f(GGG) = $(1/4)^3$ = 1/64.

   e. Phe (UUU) = 27/64; Leu (UUG) = 9/64; Cys (UGU) = 9/64; Val (GUU or GUG) = 9/64 + 3/64 = 12/64; Trp (UGG) = 3/64; Gly (GGU or GGG) = 3/64 + 1/64 = 4/64.

   f. By doing this experiment, **you would have been able to determine that UUU specifies Phe; that Val, Leu, and Cys are specified by codons composed of G+2U; and that Trp and Gly are specified by codons composed of 2G+U.**

12. a. The protein would terminate after the His codon due to a nonsense mutation. **The Trp codon (UGG) could have been changed to either a UGA or a UAG codon.** These stop codons result from changing the second or third base of the Trp codon to an A.

b. "Reverse translate" the amino acid sequence - that is, deduce the possible sequence of the mRNA from the identity of the amino acids in the protein (N =any base):

Protein	N	Ala	Pro	His	Trp	Arg	Lys	Gly	Val	Thr	C
mRNA	5'	GCN	CCN	CAU CAC	UGG	CGN AGA AGG	AAA AAG	GGN	GUN	ACN	

If you restrict the possibilities to mutations that substitute one base pair for another, seven possible ways exist to generate a nonsense mutation from this sequence. (i) A UAG stop codon will result from a change of the second base of the Trp codon to A; (ii) If the third base of the Trp codon UGG changes to A, a UGA stop codon will result; (iii) If the Lys codon is AAA, and there is an A to T substitution at the first position, a UAA stop codon would be produced; (iv) If the Gly codon is GGA, a mutation of G to T at the first position will generate a UGA nonsense codon; (v) If the Arg codon is CGA, then a change of the first base to U would result in a UGA stop codon; (vi) If the Arg codon is AGA, then a change of the first base to U would generate a UGA stop codon also; (vii) If the Lys codon is AAG, then a change of the first base to U would result in a UAG stop codon.

Many other changes could also cause the premature termination of the protein encoded by this sequence; for instance, the insertion or deletion of a single base pair causing a frameshift mutation. As just one example, if the Arg codon is CGU, a single base insertion in the DNA before or within this codon would lead to a UAA codon in the mRNA, with the U would come from the former Arg codon and AA from the Lys codon.

13. The extra amino acids could come from an intron that is not spliced out due to a mutation in a splice site. The genomic DNA sequence in normal cells should contain this sequence. An alternative is that the extra amino acids could come from the insertion of DNA from some other part of the genome, such as the insertion of a small transposable element. In this case, the normal allele of the gene will not have this sequence. Note that the defect in this case could not be caused by a mutation that changes the stop codon at the end of the open reading frame to an amino-acid-specifying codon. If this were the case, the extra amino acids would be found at the C terminus of the longer protein, not in the middle.

14. Both strands of a double-helical DNA molecule are possible template strands. Three possible reading frames exist on both the top and bottom strands. If a stop codon is found in a particular reading frame, then it is <u>not</u> an *open reading frame* (that is, the entire sequence cannot not code for a protein). Assume first that the bottom strand is the template strand, so that the top strand is the RNA-like strand. Treat the top strand as the equivalent of an mRNA, which means that it would be translated in the 5'-to-3' direction and that you would substitute T for U to allow translation from the genetic code. Thus, as shown in the following diagram, the first potential reading frame begins with the first nucleotide, the second reading frame with the second nucleotide, etc. Scan

chapter 8

the sequence looking for stop codons (or their direct DNA equivalents: TAA, TAG and TGA, shown in *red* in the diagram). Repeat this process assuming that the top strand of the DNA is the template strand and the bottom strand is the RNA-like strand, with three possible reading frames beginning at the 5' end of the sequence. **When you analyze this DNA sequence you find that the top strand has two open reading frames and the bottom strand has one open reading frame (ORF).**

Top strand is RNA-like strand

1   5' ... CTT ACA GTT TAT TGA TAC GGA GAA GG ... 3'

2   5' ... C TTA CAG TTT ATT GAT ACG GAG AAG G ... 3'   ORF

3   5' ... CT TAC AGT TTA TTG ATA CGG AGA AGG ... 3'   ORF

Bottom strand is RNA-like strand

4   3' ... GA ATG TCA AAT AAC TAT GCC TCT TCC ... 5'

5   3' ... GAA TGT CAA ATA ACT ATG CCT CTT CC ... 5'   ORF

6   3' ... G AAT GTC AAA TAA CTA TGC CTC TTC C ... 5'

15. a. **The physical map is based on the number of base pairs.** In Fig. 8.3a, the *red bar* represents the physical map. On this map of the *trpA* gene in *E. coli* the distance between amino acids 1 and 10 will be the same as the distance between amino acids 51 and 60. **The genetic map, shown in purple in this figure, is based on numbers of crossovers. Different regions in the gene have different rates of crossing over, and so the genetic map varies proportionally.** (It is not the case that crossing over is equally likely to occur between any two base pairs in a genome.) Notice that amino acids 15 and 22 (21 nucleotides apart) are about the same genetic distance apart as amino acids 22 and 49 (81 nucleotides apart).

    b. Regions with a high frequency of crossing-over ("hotspots") will have a proportionally larger genetic map relative to the physical map. Regions with a low crossover frequency ("coldspots") will appear smaller relative to the physical map. A comparison of the physical and genetic maps in Fig. 8.3a reveals that **the N-terminal third and the C-terminal half of the gene have relatively high recombination frequencies, while the central portion of the gene has a lower incidence of recombination.**

16. a. Yanofsky could have screened for Trp- auxotrophs by growing wild-type bacteria in the presence of a mutagen, and then plating the culture to obtain single colonies on a plate containing minimal medium supplemented with tryptophan. Each colony on the plate would then be replica-plated onto a minimal medium plate; colonies that fail to grow on the replica plate are Trp- auxotrophs.

    b. The Trp- auxotrophs include mutants in the gene for each enzyme in the tryptophan biosynthesis pathway. As the TrpA enzyme catalyzes the first step in tryptophan biosynthesis, the *trpA-* colonies produce all of the other enzymes in the pathway, and thus they are distinct from all the other Trp- auxotrophs in that they are able to grow when supplemented with any downstream intermediate in the pathway.

17. One aspect of the degeneracy of the genetic code is *wobble:* Due to unusual base pairing between the 5' base of the anticodon and the 3' base of the codon, a single tRNA species can bind to more than one codon. A further complication is that when the wobble base of the tRNA anticodon is either A or U, that base is modified enzymatically after transcription of the tRNA. In the case of wobble U, three different modifications are possible, and the particular modification dictates the base pairing ability of the anticodon. It is important to realize that the wobble base modifying enzymes recognize not just the anticodon sequence of a tRNA, but other aspects of tRNA structure (dictated by the tRNA base sequence) as well. This fact means that two different tRNAs transcribed from two different genes, both of which have the same anticodon and correspond to the same amino acid, can have their wobble bases modified differently.

   a. As inosine (I) in the wobble position at the 5' end of the anticodon can pair with U, A, or C at the 3' end of the codon, anticodons with wobble base inosines must correspond to codons where 5' XYU, 5' XYA, or 5' XYC all specify the same amino acid (X and Y are any two specific bases). By examining the genetic code table (Fig. 8.2 on page 256), you can see that those codons are: 5' CGN (Arg), 5' GGN (Gly), 5' CCN (Pro), 5' UCN (Ser), 5' ACN (Thr), 5' GCN (Ala), 5' CUN (Leu), 5' GUN (Val), and 5' AUN (Ile). Therefore, **the anticodons that can have I as the wobble base are: 5' ICG, 5' ICC, 5' IGG, 5' IGA, 5' IGU, 5' IGC, 5' IAG, 5' IAC, and 5' IAU.**

   b. Remember that when U is in the Wobble position at the 5' end of the anticodon, it is modified in one of the three ways shown in Fig. 8.21 (p. 275). Any one type of tRNA (that is, a tRNA encoded by a single tRNA gene) with U in the wobble position is always modified in only one of these three ways (it is not the case that some molecules of one type of tRNA are modified in one way and other molecules of this type are modified in another way). Therefore, tRNAs with wobble U modified to xm^5U are specific for codons whose 3' ends are G; tRNAs with wobble U modified to xm^5s^2U respond to codons whose 3' ends are A/G; and tRNAs with wobble U modified to xo^5U will recognize codons with A/G/U (and rarely C) at the 3' ends.

   No matter how the wobble U at the 5' end of the anticodon is modified, it can always base pair with G at the 3' end of the codon. This fact means that **the only anticodon sequence that could never be allowed to have a 5' U is 5' UUA**, because whatever way this anticodon is modified, it would always base pair with the stop codon 5' UAG. Any other 5' UNN anticodon sequence could potentially be modified to xm^5U so as to be specific for a single codon with a 3' base of G (5' NNG).

   Certain 5' UNN codons can be modified to xm^5U but not to xm^5s^2U and/or to xo^5U. For example, 5' UAU anticodons in which the wobble U is modified either to xm^5s^2U or to xo^5U would improperly recognize codons for both Met and Ile amino acids. However, tRNAs with 5' UAU anticodons modified to xm^5U would be

chapter 8

specific for Met as required by the genetic code. In a second example, 5' UAA codons could recognize only Leu codons if the wobble U was modified to xm^5U or to xm^5s^2U, but if the wobble U was altered to xo^5U, such a tRNA would mistakenly recognize both Phe and Leu codons.

c. As explained in part (b), **xm^5U can be the wobble base for all anticodons except for 5' UUA.**
    A tRNA with a U converted to xm^5s^2U could recognize codons where the 3' nucleotide is A or G and the two specify the same amino acid. This includes all sense codons except 5' AUA, 5' AUG, and 5' UGG (and also excepts the stop codons 5' UAA, 5' UAG, and 5' UGA). Therefore, **xm^5s^2U can be the wobble base for all anticodons except for: 5'UAU, 5'UCA, and 5' UUA.**
    A tRNA with a U converted to xo^5U could recognize only codons for which the 3' base could be any nucleotide and the codon would specify the same amino acid. This would include only the codons 5' CUN (Leu), 5' GUN (Val), 5' UCN (Ser), 5' CCN (Pro), 5' ACN (Thr), 5' GCN (Ala), 5' CGN (Arg), and 5' GGN (Gly). Therefore, the **anticodons that would be able to have a xo^5U as the wobble base are: 5' UAG, 5' UAC, 5' UGA, 5' UGG, 5' UGU, 5' UGC, 5' UCG, and 5' UCC.**

d. All but one box of 4 codons in the genetic code table in Fig. 8.2 on p. 256 requires a minimum of 2 different anticodons to recognize all 4 codons in the box. Two possible scenarios for these two different anticodons include one anticodon with a wobble G for codons with 3' base U or C, and a second anticodon with either a wobble xm^5s^2U for codons with 3' base A or G or a wobble xo^5U for codons with A, G, or U as the 3' base. Other scenarios involve one anticodon with a wobble I for codons with the 3' base U, C, or A, and a second anticodon with either a wobble C for codons with G as the 3' base or a wobble U modified in any of the 3 possible ways shown in Fig. 8.21 on p. 275. The exceptional box is the one with 5' UAN codons, as 5' UAA and 5' UAG are stop codons that do not base pair with anticodons. This box would require only one anticodon with a wobble G. Therefore, as there are 16 boxes, the number of anticodons needed are (16 × 2) − 1 = 31. Add one special tRNA for selenocysteine, and **the minimal number of tRNAs required in humans is 32.**
    Any of the scenarios above are possible for boxes in which all 4 codons are for the same amino acid. But only some of the scenarios would apply to boxes that contain codons for two amino acids. As an exercise, you should work out for yourself which combinations of two anticodons are possible for each such box.

18. a. The mutant nucleotides are marked in *red*, and *arrows* indicate the sites of deletions or inversion breakpoints. (Inversions are 180^0 rotations of a stretch of DNA sequence; inversions and other sorts of chromosomal rearrangements will be discussed in detail in Chapter 12. For the time being, you should note that in the inverted region, the 5'-to-3' polarity of the DNA can only be maintained if the strands are flipped simultaneously with the sequence rotation.) The corresponding change to the amino acid sequence is shown below the RNA sequence.

wild type	5' AUG	ACA	CAU	CGA	GGG	GUG	GUA	AAC	CCU	AAG
	Met	Thr	His	Arg	Gly	Val	Val	Asn	Pro	Lys
mutant 1 transversion	5' AUG	ACA	CAU	CCA Pro	GGG	GUG	GUA	AAC	CCU	AAG
mutant 2 deletion	5' AUG	ACA	CAU	CGA	GGG↓(G)	UGG Trp	UAA STOP	ACC	CUA	AG
mutant 3 transition	5' AUG	ACG Thr (no change)	CAU	CGA	GGG	GUG	GUA	AAC	CCU	AAG
mutant 4 insertion	5' AUG	ACA	CAU	CGA	GGG	GUU Val	GGU Gly	AAA Lys	CCC Pro	UAA G STOP
mutant 5 transition	5' AUG	ACA	CAU	UGA STP	GGG	GUG	GUA	AAC	CCU	AAG
mutant 6 inversion	5' AUG	ACA	UUU Phe	ACC Thr	ACC Thr	CCU Pro	CGA Arg	UGC Cys	CCU Pro	AAG Lys

b. **The frameshift mutations 2 and 4 (single base insertions and deletions) can be reverted with proflavin.** Mutations 1, 3 and 5 are single nucleotide substitutions, either transitions (changes from one purine to the other or from one pyrimidine to the other) or transversions (changes a C-G base pair to a G-C base pair). **EMS causes transitions, so it can revert mutations 3 and 5 (transition mutations) back to the original DNA sequences.** Mutation 1 is a transversion, and EMS cannot change the G-C back to C-G.

Often the term "reversion" is used in a more general sense, meaning simply a restoration of wild-type protein function rather than a change back to exactly the original DNA sequence. In this case, it is possible that another, non-wild-type amino acid at the mutant position will restore function. An EMS-induced transition in the mutant codon could give a functional protein. With this second meaning of "reversion", **all three point mutations (1, 3, and 5) could potentially be reverted by EMS.**

19. Proflavin creates frameshift mutations. Thus the mutant protein is expected to have the normal amino acid sequence up until the point of the frameshift mutation (+1 insertion or -1 deletion). At the site of the frameshift mutation, the amino acid sequence of the mutant protein will change because the one of the other two reading frames in the mRNA is begin translated. A second equal (but opposite) frameshift in the gene would be expected to restore the correct frame and suppress the original mutation. This is how the intragenic suppressors located between the N terminus and the original mutation restore a functional polypeptide.

Notice that the original mutant protein is shorter (110 amino acids) than the normal protein (157 amino acids). Thus **the initial frameshift mutation changes the reading frame and there is a stop codon in this alternate reading frame.** This stop codon

chapter 8

terminates translation early and generates a shorter polypeptide. The second frameshift mutation of the opposite sign cannot restore function if it occurs after the premature stop codon. Because no suppressors can be identified after the original mutation (in other words, between the original mutation and the premature stop codon), **we can predict that the stop codon is likely to be very close to the site of the original frameshift mutation**. For example:

5'...GUG GCA AUA GAC...	Nonmutated sequence
5'...GUG GCA **UAG** AC....	Original frameshift mutation (-A)
5'...GUG GCA **UAG** AC....	

In this example only an insertion in the *blue* region (just upstream of the U, between the U/A, or just downstream of the A) within three base pairs of the original deletion, could potentially restore the original, normal open reading frame.

20. a. First, the mechanisms by which tRNA^{Pyl} and tRNA^{Sec} are charged with amino acids differ. A special Pyl tRNA synthetase exists to charge tRNA^{Pyl} with Pyl, but tRNA^{Sec} is recognized by a Ser tRNA synthetase that charges the tRNA with Ser; the Ser on the Ser-charged tRNA^{Sec} is subsequently modified to Sec enzymatically.

    Second, although both Sec and Pyl are encoded by nonsense (stop) codons, the way in which UGA is recognized as a Sec codon differs from the way in which UAG is used as a Pyl codon. For Sec incorporation, a special mRNA structure called the SECIS element, downstream of the UGA, prevents the ribosome from terminating translation. The tRNA^{Pyl} instead likely works like a nonsense suppressor mutant tRNA. In bacteria that incorporate Pyl into polypeptides, you might expect that for genes whose open reading frame terminates with UAG, some longer-than-normal protein would be made. Perhaps the cells can deal with these longer polypeptides if the amounts made are small, or perhaps evolution selects against the appearance of UAG codons at the end of the ORFs of genes where a longer-than-normal polypeptide product would be deleterious.

    Third, something about the structure of the tRNA^{Sec} and/or the SECIS element allows for unusual base-pairing of xm^{5}U in the 5' position of the anticodon with A in the codon. The codon-anticodon interaction involving tRNA^{Pyl} does not require any non-canonical wobble.

    b. The Pyl-charged tRNA^{Pyl} must get to the UAG at the ribosome before the ribosome stops translation and releases the mRNA. Also, as just described, both the Pyl incorporation mechanism and nonsense suppression could lead to the production of some longer-than-normal polypeptides.

Section 8.2

21. **In transcription, complementary base pairing is required to add the appropriate ribonucleotide to a growing RNA chain.**

22. DNA replication is a permanent event – the new daughter cells will each get only one copy of the DNA molecule. That one copy must serve as the template for all transcription and the template for further cell divisions. This is not true for

transcription, and consequently there are several reasons why the higher level of error during transcription is tolerated:

(i) **Transcription produces a transient product**. mRNA molecules persist only for relatively short periods (the half life of a bacterial mRNA is typically only a few minutes, whereas those for eukaryotic mRNAs can vary from ~15 min to ~1 day), while DNA molecules must be maintained throughout and even beyond the life of the organism.

(ii) **Cells contain multiple copies of each kind of mRNA.** Even if a particular mRNA molecule contains critical errors that produce nonfunctional proteins, the cell has many other mRNA molecules transcribed from that same gene and the majority of them do not contain errors in critical locations.

(iii) Further, **the degeneracy of the genetic code limits the effects of errors in transcription** because many single errors will produce silent changes, and others will lead to conservative substitutions.

(iv) In most cases, portions of the primary RNA transcript produced by transcription will be spliced out of the molecule. For instance, only 0.6% of the dystrophin primary transcript will be present in the mature mRNA; the range for this value in different genes varies from 100% to 0.6%. **Mistakes that are made in transcription of the untranslated regions (5' and 3' UTRs and introns) will have no effect on the amino acid sequence of the polypeptide.**

(v) Lastly, **the body produces much more protein than is actually required for most genes**. Many disorders caused by lack of protein function (e.g. cystic fibrosis, muscular dystrophy, tyrosinase-negative albinism, hemophilia and other clotting disorders) are recessive. This means an individual with one nonfunctional, mutant allele and one normal, functional allele of these genes will produce only about one half the normal levels of functional protein. Yet these individuals have a completely normal phenotype. Therefore a small percentage of abnormal mRNA molecules that give rise to no protein (or perhaps even an abnormal protein) will not affect the phenotype of the individual.

23. DNA sequences are generally written with the 5'-to-3' strand (read left-to-right) on top and the 3'-to-5' strand on bottom. If the protein-coding sequence for *gene F* is read from left (N terminus) to right (C terminus), then the top strand is the RNA-like strand which is read from 5'-to-3', and the **template strand for gene F must be the bottom strand of DNA.** The template for *gene G* is the opposite because the coding sequence is read in the opposite direction (right-to-left). **The template strand for *gene G* is the top strand.** Note that this means that the enzyme RNA polymerase moves from left-to-right along the DNA in transcribing gene F, and from right-to-left in transcribing *gene G*.

24. The *green* and *orange* lines in the figure below represent mRNAs from two different genes; the *blue* lines are single strands of a double-stranded DNA molecule. Two genes are portrayed to show that the results would be slightly different depending upon whether or not the gene has introns. In general the DNA strands form a double-stranded DNA structure, except where their pairing is interrupted by the presence of the mRNA/DNA heteroduplex. The *orange* gene mRNA pairs with the template strand of the DNA, forcing the other strand to loop out. The DNA (*blue*) corresponding to the *green* gene has an intron which has been processed out of the mature mRNA (*green*).

chapter 8

When the mRNA pairs with the template strand of DNA there is a loop-out of the DNA in the region corresponding to the intron.

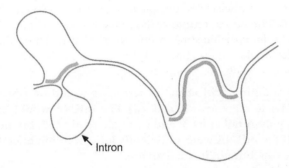

Section 8.3

25. In translation, complementary base pairing between the codon in the mRNA and the anticodon in the tRNA is responsible for aligning the tRNA that carries the appropriate amino acid to be added to the polypeptide chain.

26. The diagram that follows indicates the different items on the figure. Note that in many cases, only one of the many examples of the item are indicated. Also, **the entire mRNA ( ) is composed of exons (v). Items (a), (d), (g), (i), (j), (n), and (t) are not present in the diagram.**

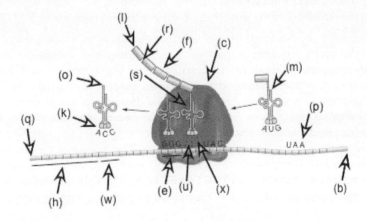

27. a. The figure accompanying Problem 26 represents **translation**.

   b. **The next anticodon is 5′ GUA; it is complementary to the codon 5′ UAC which codes for tyrosine. The tyrosine is added to the C-terminal end of the growing polypeptide chain. The protein will be 9 amino acids long when completed.**

chapter 8

c. Four additional amino acids have known identities. **The first amino acid in the polypeptide chain must be fMet. The third amino acid from the N terminus of the growing polypeptide chain is tryptophan** (the anticodon of the departing tRNA is 5' CCA, which is complementary to the codon 5' UGG specifying tryptophan). **The fourth amino acid is glycine** (the codon is 5' GGG). **The sixth amino acid is tyrosine** (the codon is 5' UAC). Thus, the sequence of the polypeptide would be N f(Met)-X-Trp-Gly-X-Tyr-X-X-X C (where X is an unknown amino acid).

d. (i) **The first amino acid at the N terminus would be f-Met in a prokaryotic cell and Met in a eukaryotic cell.** (ii) **The mRNA would have a cap at its 5' end and a poly-A tail at its 3' end in a eukaryotic cell, but not in a prokaryotic cell.** (iii) **If the mRNA were sufficiently long, it might encode several proteins in a prokaryote but not in a eukaryote.**

Section 8.4

28. **Eukaryotic genes contain introns that interrupt the open reading frame.** Because introns are spliced out of the primary transcript and thus are not included in the mature mRNA, they do not have to contain open reading frames. In fact, by chance, introns almost always contain stop codons that would halt all possible reading frames. Thus, the reading frames of almost all eukaryotic genes are interrupted by introns that contain stop codons in that frame.

29. a. **The minimum length of the coding region is 477 amino acids × 3 bases/codon = 1431 base pairs** (not counting the stop codon). The gene could be much longer if it contained introns. In addition, the coding region does not include the sequences in the gene that are transcribed into the 5' and 3' UTRs.

    b. Look at both strands of this sequence for the open reading frame (remember that the given sequence is part of a exon composed exclusively of protein-coding sequences, so there must be an open reading frame). This ORF occurs on the bottom strand, starting with the second base from the right (X = closed reading frames, O = open reading frames). The direction of the protein is N terminal-to-C terminal. going from right to left in the coding sequence on the RNA-like strand (that is, the bottom strand of the DNA).

    ```
                          XXX
    template strand   5'  GTAAGTTAACTTTCGACTAGTCCAGGGT  3'
    RNA-like strand   3'  CATTCAATTGAAAGCTGATCAGGTCCCA  5'
                                                    XOX

    mRNA              5'  ACCCUGGACUAGUCGAAAGUUAACUUAC  3'
    ```

    c. The amino acid sequence of this part of the mitotic spindle protein is: **N...Pro Trp Thr Ser Arg Lys Leu Thr Tyr...C.** Notice that the open reading frame begins with the second nucleotide. You cannot determine the amino acid corresponding to the

chapter 8

A nucleotide at the 5' end of the mRNA because you don't know the first two nucleotides of the codon.

30. First look at the sequence to determine where the Met Tyr Arg Gly Ala amino acids are encoded. The top strand clearly does not encode these amino acids. On the bottom strand, there are Met Tyr codons on the far right (reading 5'-to-3') and Arg Gly Ala much farther down the same strand. Why aren't these codons adjacent? An intron in the DNA sequence may be spliced out of the primary transcript. Thus, the mature mRNA would encode this short protein.

    a. **The bottom strand is the RNA-like strand, so the top strand is the template. The RNA polymerase moves 3'-to-5' along the template; that is, in the right-to-left direction with respect to the written sequence.**

    b. The sequence of the processed nucleotides (after splicing) in the mature mRNA is shown below. The junction between the exons is marked by a vertical line:

        5' GCC AUG UAC AG|G GGG GCA UAG GGG 3'

    The sequence of this mRNA contains nucleotides prior to the AUG initiation codon (in *green*) and subsequent to the UAG stop codon (in *red*), emphasizing that the mRNA does not begin and end with these codons: Remember that mRNAs contain both 5' and 3' untranslated regions (UTRs).

    In fact, this problem has a simplified DNA sequence to allow the sequence to fit on the page and to facilitate your analysis of the sequence in a reasonable amount of time. If you look carefully at the sequence, you can see that there are canonical splice donor and splice acceptor sequences at the borders between the intron and the two exons that flank it. However, the intron does not contain a canonical branch site (see Fig. 8.15 on p. 272 to review the nature of the three sequences needed for splicing, and then try to verify these statements yourself). In reality, the shortest introns are about 50 nt long, instead of the 24 nt in the problem, and must contain branch sites.

    c. **A Thr residue at this position could occur if the G base on the bottom strand that just precedes the junction between the intron and the first exon (underlined in the answer to part (b) above) was mutated to a C.** This base change would also alter the splice donor site, so splicing does not occur. The next codon after the ACG for Thr is a UAA stop codon, so the polypeptide encoded by the unspliced RNA is only three amino acids long. The sequence of the mutant mRNA is:

5' GCC AUG UAC ACG UAA GUG UUU UGA UCC UCC CCC AGG GGG GCA UAG GGG 3'

31. **Mitochondria do not use the same genetic code as do the chromosomes in the nucleus.** In yeast mitochondria, the codon 5' CUA 3' codes for Thr, not Leu as it does in yeast or human nuclear genes. If you want to ensure that the correct protein will be made by the yeast cell, **you should mutate all the 5' CUA 3' codons in the mitochondrial gene to 5' ACN 3' before putting the gene into a chromosome in the yeast nucleus.** This step ensures that the cellular translation machinery will put a Thr at all positions it is required in the protein. (Mitochondrial genetics will be discussed in detail in Chapter 13.)

chapter 8

32. a. The differences in transcription and translation prokaryotes (bacteria) and eukaryotes (humans) mean that the prokaryotic transcription and translation systems cannot produce a functional insulin protein from the human gene. (i) **The promoters for the human insulin gene may not work in** *E. coli*, thus blocking transcription of the gene. (ii) One of the main problems is that the **human insulin gene has introns** which are transcribed into the primary transcript and then spliced out by the spliceosome. **Bacterial cells do not have introns and thus have not evolved the machinery necessary to remove them from the RNA**. (iii) Other eukaryotic post-transcriptional modifications such as **addition of the 5' cap are found in eukaryotes but are absent in prokaryotes**. . (iv) **The human insulin mRNA may not have sequences like the Shine-Delgarno box** needed for efficient initiation of translation at the AUG. (v) Finally, it is also possible that **correct folding of the polypeptide and other post translational modifications may not occur in the same ways in prokaryotic cells.**

    b. To make these bacterial insulin factories, you would have to transform *E. coli* cells with a composite (or *fusion*) gene in which some parts would be from the human insulin gene and other parts from a highly expressed *E. coli* gene. **The only parts of the human insulin gene that would be needed are the protein-coding sequences in the exons. These must be properly spliced together before they are transformed into bacteria**. The easiest way to get these sequences is to make a DNA copy of an insulin mRNA molecule (known as a cDNA), a method that will be explained in Chapter 9. **All of the DNA sequences controlling gene expression (promoter, Shine-Dalgarno sequence, transcription terminator) should come from an** *E. coli* **gene that is transcribed and translated at very high levels**. In essence, you would replace the protein-coding part of the bacterial gene with the (properly spliced) protein-coding part of the human insulin gene.

33. The order of these elements in the gene is: **c; e; i; f; a; k; h; d; b; j; g**. This order represents the upstream → downstream direction in which RNA polymerase would move as it transcribes the gene.

34. a. **All of these elements are abbreviated (that is, they are in the RNA), with the exceptions of the promoter (c) and the transcription terminator (g)**. These 2 are the only structures in this list that the RNA polymerase enzyme recognizes on the DNA. RNA processing and translation occur posttranscriptionally.

    b. The **splice-donor site (a), the nucleotide to which the methylated cap is added (e), the initiation codon (f), and the 5' UTR (i) are found partly or completely in the first exon.**

    The **splice-donor site (a), the splice-acceptor site (h), and the splice branch site (k) are found partly or completely in the intron.**

    The **3' UTR (b), the stop codon (d), the poly-A addition site (j), and the transcription terminator (g) are found partly or completely in the second exon.**

    Note that the splice-donor site shown in Fig. 8.15 on p. 272 includes nucleotides in both the upstream exon and the intron, while the splice-acceptor site is completely within the intron. You should also be clear that these assignments are

chapter 8

specific to the instructions in the problem. For example, in many genes the 5' UTR is not restricted to the first exon, nor is the initiation codon always in the first exon.

35. The diagram following the answer to part (q) summarizes how to approach all parts of this problem.

   a. To calculate gene size, you simply add up all the exons and introns and ignore the 5'UTR and 3'UTR (because these are parts of exons): 119 + 532 + 337 + 1431 + 208 + 380 + 444 + 99 + 546 = **4096 bp**.

   b. To calculate the size of the primary transcript, you take the gene size (exons plus introns) = **4096 nucleotides = 4096 nt**. This answer would be correct based on Fig. 8.14 on p. 271. However, in actuality the 5' cap and the poly-A tail at the 3' end are added very rapidly to the primary transcript, usually as it is being made prior to splicing, so some researchers consider that the primary transcript contains these modifications, as was just shown in the preceding figure. With this definition of primary transcript, you could thus add on 150 bases of poly-A tail at the 3' end: 4096 + 150 = **4246 bp**. (You could also add **plus** one nucleotide **to account for the 5' cap**: 4096 + 150 + 1 = **4247 nt**.

   c. To calculate the size of the mature mRNA, you add up all the exons alone and add on 150 bases of poly-A tail (and possibly one nucleotide for the 5' cap): 119 + 337 + 208 + 444 + 546 + 150 (+1) = **1804 nt (1805 nt with the 5' cap)**.

   d. To determine where the ATG initiation codon is located, compare the size of the 5' UTR to that of the first exon. Part or all of the first exon must be the 5' UTR. In this case, the 5' UTR (174 bp) is larger than the first exon (119 bp); the 5' UTR must therefore include the entire first exon plus an additional 55 bp (174-119 = 55). The second exon (337 bp) is larger than 55 bp, so the initiation codon must be in **Exon 2**. The first 55 bp of the second exon are thus the remainder of the 5' UTR, while (337-55 =) 282 bp of the second exon are coding sequence (that is, they correspond to the triplet codons in the mRNA). 282/3 = exactly 94 amino acids. Intron 2 is thus located in between codons.

   e. You should now determine the extent of the 3' UTR in exactly the same way. In this problem, the 3' UTR is 715 bp. Exon 5 (the last exon) is 546 bp, so all of Exon 5 is 3' UTR, and the 3' UTR extends (715-546 =) 169 bp further into Exon 4. Exon 4 is more than 169 bp (it is 444 bp), so the stop codon must be in **Exon 4**. There is some semantic confusion about whether the 3' UTR contains the stop codon or not, but it is easiest to calculate things if you consider the 3' UTR to include the stop codon. Then the coding part of Exon 4 is (444-169 =) 275 bp.

   f. **No base pairs encode the 5' cap.**

   g. **The promoter is upstream of the 5' UTR and so is not included in any of the DNA sequences listed in the problem.**

   h. **Intron 4** [see part (e)].

   i. **Exon 4** contains the stop codon [see part (e)], and the codon preceding it encodes the final (C-terminal) amino acid.

   j. **The sequence encoding the poly-A tail does not exist in the gene** (poly-A is added posttranscriptionally by poly-A polymerase).

k. We've already established that there are 282 bp of coding region in Exon 2 [see part (d)], and that this corresponds to 94 codons. This means that Exon 3 starts with the first nucleotide of a codon. There are 208 bp in Exon 3, which corresponds to 208/3 = 69.33 codons = 69 codons + one extra nucleotide. Therefore Intron 3 is placed between the 1st and 2nd nucleotide of the next codon. The coding region of Exon 4 is 275 bp [see part (e)], which corresponds to (275/3 =) 91.67 codons or 91 codons + 2 extra nucleotides. Therefore, the coding region of the gene is: 282 + 208 + 275 = **765 bp**. (*Note:* This assumes that the 3' UTR includes the stop codon and that the coding region does not include the stop codon.)

l. As explained in part (k): 94 (Exon 2) + 69 1/3 (Exon 3) + 91 2/3 (Exon 4) = **255 amino acids.**

m. **Intron 3** interrupts a codon [see part(k)].

n. **Intron 2** is located between codons [see part(k)].

o. The sequence specifying poly-A addition is the sequence AAUAAA (or near variants of this), and this sequence occurs in the final exon of the gene (**Exon 5**) **in the 3' UTR**, and also in the primary transcript and mature mRNA. A ribonuclease cleaves the primary transcript about 20 nucleotides downstream (towards the 3' end) of this sequence, and poly-A polymerase then adds 150-200 As.

p. As this is a normal (nonmutant) process, it could be explained by alternative **splicing in which Exon 2 is connected either to Exon 4 or Exon 5, or Intron 2 is not removed at all. The 94 amino acids are those encoded by Exon 2.**

q. We are assuming that, although this alternative splicing is a natural process, that this small protein has no function. Sometimes primary transcripts are spliced at low levels into alternative mRNAs with no functions; the unusual splice junctions are called "cryptic splice junctions". Assuming that this is a cryptic splice product, Intron 2 does not have codons, Exon 4 is out of frame with respect to Exon 2, and Exon 5 doesn't encode any ß2 lens crystallin protein. In any of these cases, the processed material would encode "junk" when spliced to the 2nd exon. **The reason you would expect 20 amino acids of junk is because ~1/20th of the codons in the genetic code are stop codons, so you would run into a stop codon within 20 amino acids when junk is being translated if the sequence of DNA is random.**

chapter 8

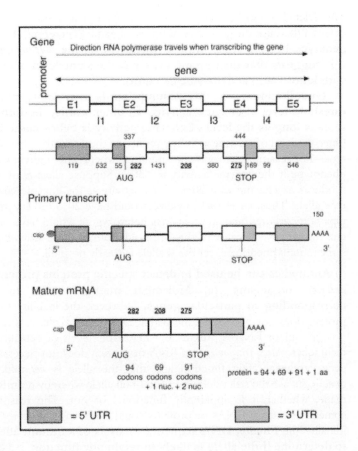

Section 8.5

36. Nonsense or frameshift mutations that affect codons for amino acids near the C terminus are usually less severe than nonsense or frameshift mutations in codons for amino acids near the N terminus because less of the protein will be affected.

   a. **very severe effect** as there will be no functional protein

   b. **probably mild effect** if none of the last few amino acids are important for function

   c. **very severe effect** as most of the amino acids in the protein will be incorrect

   d. **probably mild effect** as only the last few amino acids will be affected

   e. **no effect**, by definition a silent mutation maintains the same amino acid

   f. **mild to no effect** as the replacement is with an amino acid with similar chemical properties

   g. **severe effect** as the mutation is likely to destroy the protein's activity

   h. **could be severe** if it affects the protein structure enough to hinder the action of the protein, **or could be mild** if the protein's function can tolerate the substitution

chapter 8

37. a. The deleted homolog has a known null activity allele for all the genes within the deletion (because the genes are not present at all). Therefore, **if a *mutant / deletion* genotype has the same mutant phenotype as the *mutant / mutant* genotype, this suggests that the mutant allele has the same level of activity as an allele with known zero activity (the deletion).**

    One limitation of this assumption is that **some phenotypes have a threshold level of enzyme activity. In other words, the mutant phenotype is seen as long as the level of enzyme activity is below some critical threshold.** Once the level of enzyme activity rises above this level, the phenotype becomes wild type. For example, imagine a situation where any individual has the mutant phenotype if the enzyme activity is <30%. Suppose allele *m* of an autosomal gene produces an enzyme with 20% of the activity as the enzyme encoded by the wild-type allele. Thus, an *m / m* individual would have 20% the enzyme activity of wild-type homozygotes, and an *m / deletion* heterozygote would have only 10% of normal enzyme activity. Because of the threshold, both of these individuals would have the same mutant phenotype, yet the *m* allele is clearly not null.

    b. (i) **Antibodies can be used to detect specific proteins present in *m / m* or *m / deletion* organisms.** (ii) **Molecular methods exist to detect mRNAs corresponding to particular genes.** However, the inability to detect mRNA or protein does not necessarily mean that *m* is a null allele; it is always possible that protein and/or RNA is present, but below the levels of detection of these various techniques. Also, the *presence* of RNA or protein does not necessarily mean that the gene products retain function and that the allele is *not* null. (However, some geneticists use the strict definition for a null allele whereby no mRNA nor protein is made, whether it is apparently functional or not. The reason for this is that sometimes mutant RNAs or proteins could function in ways that are not apparent initially.) (iii) **Finally, DNA sequence analysis of a mutant allele can often help to determine if the allele is likely to retain any function.** All of these techniques will be discussed in detail in later Chapters. The bottom line is that it can be quite difficult to determine for certain whether or not an allele is null, especially if using the strict definition above.

38. The size of the protein is 2532 amino acids. Thus the mRNA must be 2532 × 3 = 7.6 kb. To be detectable, any changes must be more than 1% of normal size or amount. **The answers below are presented in the order of (i) mRNA size; (ii) mRNA amount; (iii) protein size; (iv) protein amount. A '+' means there will be a >1% change, while a '-' means there will be no change.** We assume in the answers below [excepting part (j)] that mutant mRNAs or proteins have normal stability in the cell, although this is not always true in practice.

    a. – – – –. This is a nonconservative amino acid substitution changing the identity of only a single nucleotide in the mRNA and a single amino acid in the protein, so it will probably affect the ability of the protein to function normally, but it will not affect the size or amount of mRNA or protein.

    b. – – – –. This is a conservative amino acid change, so it probably won't affect the function of the protein.

    c. – – – –. This is a silent change, so it won't affect any detectable parameters.

chapter 8

d. − − + −. This is a nonsense mutation, so it won't affect mRNA size or amount, nor amount of protein. It will affect the size of the protein.

e. − − + − or +. Met1Arg could mean that no protein is made, as the ribosome can't initiate translation at an Arg codon. However, it is possible that a protein could be made if there is another downstream, in frame 5' AUG codon that can be used to initiate translation. This second possibility would result in a smaller protein.

f. − + − +. A mutation in the promoter would most likely make it a weaker promoter, so RNA polymerase would bind less often. With less mRNA, less protein would be produced.

g. − − + −. This change would not be detectable in the mRNA, but it will cause a frameshift mutation in the protein which will obviously affect the size of the protein.

h. − − − −. A deletion of 3 bases (a codon) would remove only 1 amino acid / 2532 amino acids; this is a <1% change in size of the mRNA or protein.

i. + − + −. This mutation causes alternate splicing to occur, removing exon 19. Depending on the size of exon 19, this could easily have a >1% affect on the size of the mRNA and protein.

j. + + − +. Lack of a poly-A tail could make the mRNA noticeably smaller. It could also decrease the stability of the mRNA. If the mRNA is degraded faster then less protein will be translated.

k. − − + or − −. This substitution in the 5' UTR could potentially affect mRNA localization or stability. It could have an affect on the amount of protein made if it affects ribosome binding, for example. Conversely, it may have absolutely no effect on the mRNA itself or how well the protein is translated.

l. − − − −. If this insertion is into an intron, then it should be spliced out of the primary transcript, producing a wild-type mature mRNA.

39. a. Null mutants are those with no protein activity. (We are using the looser definition of null allele here, where nonfunctional gene product could be made.) **The following changes could lead to loss of enzyme activity in the worst case scenarios: a** (if this amino acid substitution blocks protein activity); **b** (if this amino acid substitution blocks protein activity); **d** (would destroy most of the protein activity); **e** (if there is no translation initiation); **f** (if promoter does not function); **g** (frameshift would alter most of the amino acid sequence of the protein); **h** (it is possible that the deletion of 1 amino acid blocks protein function or alters the tertiary or quaternary structure of the protein); **i** (will produce an altered protein that might be unable to function); **j** (could make the mRNA completely unstable); and **k** (losing exon 19 could inactivate the protein). In practice, mutations like j and k seem to cause less severe reductions in gene expression.

b. Any mutation that makes an altered protein with impaired function or lower levels of otherwise functional protein is likely to be recessive to wild type. Most null or hypomorphic alleles are in fact recessive, loss-of-function mutations. They are recessive because for most genes, homozygosity for the wild-type allele results in more than twice the amount of protein required for a wild-type phenotype. Loss of one of the two copies of the gene will thus not alter the wild-type phenotype. For this reason, **any of the mutations in the list (except c and l) could be recessive. Those listed in the answer to part (a) are also the most likely to be recessive.**

c. **Any of the mutations, other than c and l, could potentially be dominant to wild type.** For most of these, if the mutation was null or strongly hypomorphic there could be a **dominant effect from haploinsufficiency**, where one wild-type allele of the gene does not express enough gene product for a wild-type phenotype. This is true for any mutation that alters the protein product in size or amount. Any of the mutations that change the size of the protein, as well as a, b and h (which alter the protein to a lesser degree) could potentially have **dominant negative or neomorphic dominant effects** that would depend on the protein (for example, whether it is multimeric) and the exact consequences of the mutation. Finally, it is possible that a promoter mutation, as in f, could turn on a gene in the wrong tissue or at the wrong time, leading to a **neomorphic allele due to ectopic expression**. However, you should remember that most of the scenarios leading to dominant phenotypes are generally much more rare than recessive effects due to loss of function.

40. Cross each of the reverted colonies to a wild-type haploid to generate a heterozygous (wild-type phenotype) diploid. Then sporulate the diploid and examine the phenotype of the spores. **If the Met+ phenotype is due to a true reversion, then the cross was:** *met-* × *met+* → *met+ / met-* → 2 *met+* : 2 *met-*. **If there is an unlinked suppressor mutation in another gene, the suppressor ($su-$) and original *met-* mutation should assort from each other during meiosis:** *met- su-* (phenotypically Met+) × *met+ su+* (wild type) → *met / met+* ; *su- / su+* → 1 *met- su-* (Met+) : 1 *met- su+* (Met-) : 1 *met+ su-* (Met+) : 1 *met+ su+* (Met+) = 3 Met+ : 1 Met-. If the two genes are linked, you will still get the four classes but the proportions of the recombinant gametes (*met- su+* being the one you can detect as it is the only one with a Met- phenotype) will increase proportionate to the distance between the two genes.

It is now (in 2014) possible to determine the DNA sequence of the entire yeast genome relatively inexpensively, so you could also sequence the genomes of the original *met-* strain and of the five apparent reversions to see whether the base pair responsible for the mutation is still present.

41. tRNA synthetases bind specifically to particular tRNAs. The enzymes recognize many aspects of the shape of a particular type of tRNA dictated by the base sequence, and the anticodon is only a minor aspect of the tRNA's three-dimensional shape. A nonsense suppressor mutant tRNA^{Tyr}, because its anticodon has an unusual base sequence, may not be recognized quite as efficiently as wild-type tRNA^{Tyr} by Tyr tRNA synthetase – but it must be charged with Tyr well enough to function as a nonsense suppressor.

42. a. The anticodon is complementary to the codon. **The anticodon in this nonsense suppressor tRNA is 3' AUC 5'.** Because the 5'-most nucleotide in the anticodon is C, the wobble rules (Fig. 8.21b, p. 275) predict that it can pair only with a 5' UAG 3' stop codon and not with any other triplet.

b. The wild-type tRNA^{Gln} recognizes either a 5' CAA 3' or a 5' CAG 3' codon. Only a single nucleotide was changed to turn the wild-type tRNA into the nonsense suppressor. The suppressor tRNA recognizes a 5' UAG 3' codon, so the wild-type

chapter 8

tRNA must have recognized the 5' CAG 3' codon if only one base change occurred. The anticodon sequence in the wild-type tRNA is therefore 3' GUC 5'. The template strand employed to produce the tRNA is complementary to the tRNA sequence itself, so **the sequence of the template strand of DNA for the wild-type tRNA^{Gln} gene is 5' CAG 3'**.

c. In any wild-type cell of any species there is a minimum of one tRNA^{Gln} gene. The one tRNA it encodes has to be able to recognize both of the normal Gln codons, 5' CAA 3' and 5' CAG 3'. The wobble rules say that a tRNA with a 3' GUU 5' anticodon can recognize both Gln codons if the wobble U in the anticodon is modified to xm^{5}s^{2}U, so this would be the minimum tRNA^{Gln} gene number in any organism. However, *B. adonis* is a species that can harbor the Gln nonsense suppressing tRNA discussed in part (a). **Therefore two tRNA^{Gln} genes must exist in a wild-type *B. adonis* cell. One would code for the tRNA^{Gln} (anticodon 3' GUC 5') that was changed into a nonsense suppressor, and the other gene would code for the tRNA^{Gln} (anticodon of 3' GUU 5') that recognizes both of the normal Gln codons.**

43. a. There is a progression of mutations from Pro (5'CCN, wild type) to Ser (5'UCN) to Trp (5'UGG, strain B) codons. **The original wild-type proline codon must have been 5' CCG, so the sequence of the DNA in this region must have been:**
    5' CCG 3'
    3' GGC 5'

    b. The wild-type amino acid is Pro (proline). The amino acid at the same position in strain B is Trp (tryptophan). Although these are not the same amino acids, the phenotype of strain B is wild type. This means that **Trp at position 5 is compatible with the function of the enzyme encoded by the gene.** This fact implies that the change from Pro to Trp is a conservative substitution. In fact, if you look at Fig 7.28b on page 237, you can see that Pro and Trp are both amino acids with nonpolar R groups. The original mutation changed Pro to Ser (serine). This mutant is nonfunctional, so the **Ser at position 5 is not compatible with enzyme function - it is a nonconservative substitution.** Figure 7.28b shows that Ser has an uncharged polar R group, so its chemical properties are likely to be very different than those of Pro.

    c. **Strain C does not have any detectable protein, so it is likely to be a nonsense mutation that stops translation after only a few amino acids have been added.** One way in which thus could have happened is that the codon for Ser in the mutant (5' UCG) could have been changed by mutation to a 5' UAG. However, the nonsense mutation could also have occurred at other locations in the gene's coding region.

    d. Strain C-1 is either a same-site revertant (5' UAG to sense codon) or a second site revertant (nonsense suppressing tRNA mutation). **Because the reversion mutation does not map to the enzyme locus, it must be a nonsense suppressing tRNA.**

44. A missense suppressing tRNA has a mutation in its anticodon so that it recognizes a different codon and inserts an inappropriate amino acid. Problem 8-42 dealt with the

change of the tRNA^Gly anticodon from 3' GUC 5' to 3' AUC 5', making a nonsense suppressor tRNA. Imagine instead that the tRNA^Gln anticodon had mutated to 3' GCC 5'. This mutant tRNA will respond to 5' CGG 3' codons, thus putting Gln into a protein in place of Arg. This is a missense suppressor.

a. Consider here the effect of the presence of a missense or nonsense tRNA on the normal proteins in a cell that are encoded by wild-type genes without missense or nonsense mutations. **A missense suppressing tRNA has the potential to change the identity of a particular amino acid found in many places in many normal proteins.** Using the example above of a missense suppressor that replaces Arg with Gln, most proteins have many Arg amino acids, so a missense suppressor could potentially substitute Gln for Arg at many sites in any single protein. **In contrast, a nonsense suppressing mutation can only affect a single location in the expression of any wild-type gene (that is, the stop codon that terminates translation), making a normal protein longer.** The longer protein will still have all of the amino acids comprising its normal counterpart. Together, these considerations mean that the presence of a missense suppressing tRNA has more likelihood of altering proteins synthesized in the cell than the presence of a nonsense suppressing tRNA.

b. In addition to the situation described above in which a mutation in a tRNA gene would change the anticodon to recognize a different codon, there are other possible ways to generate missense suppression, including: **(i) a mutation in a tRNA gene in a region other than that encoding the anticodon itself, so that the wrong aminoacyl-tRNA synthetase would sometimes recognize the tRNA and charge it with the wrong amino acid; (ii) a mutation in an aminoacyl-tRNA synthetase gene, making an enzyme that would sometimes put the wrong amino acid on a tRNA; (iii) a mutation in a gene encoding either a ribosomal protein, a ribosomal RNA or a translation factor that would make the ribosome more error-prone, inserting the wrong amino acid in the polypeptide; (iv) a mutation in a gene encoding a subunit of RNA polymerase that would sometimes cause the enzyme to transcribe the sequence incorrectly.**

45. If a tRNA was suppressing +1 frameshift mutations, then **it must have an anticodon that is complementary to 4 bases, instead of 3**.

46. Use the wobble rules (Fig. 8.21b on p. 275) to help solve this problem.

a. The nonsense codons that differ only at the 3' end are 5' UAG and 5' UAA. **A tRNA with the anticodon 3' AUU 5' (U modified to xm^{5}s^{2}U) could recognize both of these nonsense codons** because the U at the 5' end of the anticodon could pair with G or A at the 3' end of the nonsense codons.

b. **This nonsense suppressing tRNA would suppress 5' UAA 3' and 5' UAG 3'.**

c. **No.** The nonsense suppressor tRNA described in part (a) would recognize only the two stop codons.

d. This question asks which wild-type tRNAs could have their anticodons changed to 3' AUU 5' with a single nucleotide change. It is easier to answer this question from the perspective of the mRNA sequences. In other words, which sense codons could

chapter 8

have mutated to become 5' UAA/G? **These are: 5' CAA/G = Gln, 5' GAA/G = Glu, 5' AAA/G = Lys, 5' UCA/G = Ser, 5' UUA/G = Leu, 5' UAU/C = Tyr, and 5' UGG = Trp**. Note that in the anticodons reacting with all of the codons above except the Trp codon, a wobble U modified to xm^5s^2U would allow a single tRNA to recognize the codons that end with either A or G as the 3' base. A normal Trp tRNA could have a wobble U (an anticodon of 3' ACU 5'), but the wobble U would have to modified as xm^5U so that the tRNA does not recognize a 5' UGA 3' stop codon; in other words, the anticodon of a normal Trp tRNA has to be specific for the codon 5' UGG.

47. **In the second bacterial species where the isolation of nonsense suppressors was not possible, there must be only a single tRNA^{Tyr} gene and a single tRNA^{Gln} gene**. Thus, if either gene mutated to a nonsense suppressor, it would be lethal to the cell, as there would not be any tRNA that could put Tyr or Gln where they belong. In this scenario, the single tRNA^{Tyr} would have to have an anticodon of 3' AUG 5' to recognize the two Tyr codons of 5' UAU 3' and 5' UAC 3' based on the wobble rules (Fig. 8.21b on p. 275). The single tRNA^{Gln} would have to have an anticodon of 3' GUU 5' (wobble U modified to xm^5s^2U) to recognize both 5' CAG 3' and 5' CAA 3' Gln codons.

48. a. **The mutant gene was a tRNA whose anticodon could recognize the stop codon 5' UAG 3'**. Assuming that the nonsense suppressor tRNA was specific for that stop codon, the mutant anticodon sequence was 5' CUA 3' or 5' xm^5UUA 3'.

   b. **Each M protein had a single amino acid change in its polypeptide sequence; the amino acid that the nonsense suppressor tRNA was charged with was inserted in place of the amino acid specified by the normal codon that was changed to UAG by the nonsense mutation.**
   (*Notes:* (i) Even if the amino acid sequence is not identical to wild type, an M protein made from the mutant phages in the *su-* bacteria could still have wild-type function as long as the amino acid inserted by the nonsense suppressor tRNA into that specific location in the protein is compatible with the protein's function. (ii) Although the problem states that the particular mutants shown in Fig. 8.8 produce proteins with amino acid sequences different from wild type, some other M protein made from a phage with a different nonsense mutation could have a wild-type amino acid sequence if, by chance, the nonsense suppressor tRNA was charged with the amino acid specified by the original codon.)

chapter 9

Digital Analysis of Genomes

Synopsis

The method of genome analysis described in Chapter 9 is at its heart very simple: Scientists chop up the DNA comprising a large genome into bite-sized pieces, using restriction enzymes or by mechanical shearing. The researchers then purify large amounts of individual bite-sized pieces by making and amplifying recombinant DNA molecules through the process of molecular cloning. The next step is to determine the DNA sequence of many bite-sized pieces, and then to use computer programs that recognize overlapping DNA sequences in order to reconstruct the sequence of the original whole genome.

The job is still far from done even after the genome sequence data is available. The sequence by itself is not very useful until it is *annotated*; in particular, geneticists want to identify the various genes and other elements (like centromeres and telomeres) that allow the genome to function. The chapter concludes with a description of some of the lessons learned about genome structure through the Human Genome Project.

Key terms

genome – the DNA contained within all the chromosomes of an organism. **Genomics** is the study of genomes.

Human Genome Project – the collaborative effort to determine the sequence of all the nucleotides in the human genome and to **annotate** these sequences according to their functions

restriction enzyme – an enzyme used by bacteria to attack the genomes of bacteriophage by cutting foreign DNA at specific sequences, producing smaller, linear **restriction fragments.** Bacteria protect their own genomes from restriction enzymes by making **modification enzymes** that add methyl groups to specific nucleotides so that **restriction sites** are no longer recognized by the restriction enzymes.

sticky ends versus blunt ends – If a restriction enzyme cuts the two strands of DNA at phosphodiester bonds between the same base pairs, the ends of the restriction fragments are *blunt*. If the cuts are offset from each other, the restriction fragments have either 5' or 3' overhangs (depending on the enzyme). If the overhangs of any two fragments produced by an enzyme can base-pair with each other, the ends are *sticky* and **compatible**.

mechanical shearing – breaking DNA at random locations by passing a DNA solution forcefully through a thin needle or by applying ultrasound energy

gel electrophoresis – separating molecules according to their size by using an electric field to drive their migration through pores in a gel made of *agarose* or *acrylamide*. Smaller molecules move faster through the gel than large molecules because they can pass through more of the pores, so their path is less circuitous and shorter than that of larger molecules.

molecular cloning – isolation of large amounts of a given fragment of DNA that is joined in the test tube with a **vector** DNA, allowing propagation of the resultant **recombinant DNA** molecule. A preparation containing many identical copies of the same one recombinant DNA molecule is a **DNA clone.**

plasmid – a small circular DNA molecule that can replicate within bacterial cells independently of the bacterial chromosome. Plasmid DNAs are often used as vectors to construct recombinant DNA molecules with relatively small inserts (<20 kb) of foreign DNA.

selectable marker – gene carried on a vector that allows the selection of cells that contain recombinant DNA molecules

BACs and **YACs** – Bacterial artificial chromosome (*BACs*) and yeast artificial chromosomes (*YACs*) are vectors that allow the construction of recombinant DNA molecules with very large inserts of foreign DNA (hundreds to thousands of kb).

hybridization/annealing – the process by which single strands of DNA or RNA (such as a template and a primer) can form double-stranded molecules by base complementarity

Sanger sequencing – a method invented by Frederick Sanger for determining the nucleotide sequence of a DNA molecule by the incorporation of fluorescently-labeled **dideoxyribonuclotide triphosphates** (ddNTPs) into a newly polymerized chain.

read – a digital file containing the nucleotide sequence determined from one molecule of DNA (or the many identical copies of this molecule found in a DNA clone)

whole-genome shotgun strategy – a method to sequence large genomes by determining a very large number of small, random sequences and then using the computer to assemble these back into the sequence of the genome

hierarchical strategy – a method to sequence large genomes by first separating them into intermediate sized fragments cloned into BAC or YAC vectors, and then determining the *mimimal tiling path* that could represent the entire genome from the smallest number of overlapping fragments

ORF (open reading frame) – a sequence of adjacent nucleotides in a given reading frame that is uninterrupted by a stop codon. The presence of a long ORF often signals the presence of a gene.

DNA homology – two sequences of DNA that have similar nucleotide sequences because they were derived from a common ancestor. Homologous sequences of DNA are found in the genomes of different species are said to be **conserved.**

cDNA (complementary DNA) – a DNA copy of an RNA molecule made by the enzyme **reverse transcriptase** (RNA-dependent DNA polymerase). cDNAs can be either single-stranded or double-stranded.

exome – the collection of all the exons of all the genes in a genome

chapter 9

gene desert – a large region of a genome relatively devoid of genes

protein domain – a contiguous sequence of amino acids within a protein that folds into a particular shape and supplies a particular unit of function

multigene family – a group of genes that are closely related in sequence and function because they evolved from a process of gene duplication and divergence.

orthologous genes – genes in two different species that arose from the same gene in a common ancestor of the two species (e.g., the ε-globin genes in humans and chimps). Orthologous genes are likely to provide the same function in both species.

paralogous genes – genes that arose from a duplication event. Usually, this term refers to different members of a gene family in a single organism (e.g., the β-globin and ε-globin genes in the human genome).

homologous genes – genes that have similar DNA sequences because they derive from a common ancestor; includes both orthologous and paralogous genes

pseudogene – a DNA sequence that originated from a copy of a gene, but that has since diverged sufficiently to become nonfunctional

syntenic block/conserved synteny – a region in which the identity and order of genes is very similar in the genomes of two different species

GenBank – a database maintained by the U.S. National Institutes of Health that is a repository for DNA sequences and their annotations

bioinformatics – using computer software to analyze DNA sequences and other biological information

RefSeq – a single, annotated version of a species' genome stored in GenBank to which the sequences of other genomes may be compared

genome browser – computerized software, usually accessible through the Internet, by which researchers can look at genomes in graphical fashion

BLAST – a software tool that allows scientists to find DNA or protein sequences that are similar to each other

LCR (locus control region) – a region adjacent to a locus containing several genes of a multigene family that controls the expression of all the genes in the locus

anemia – a decrease in the amount of red blood cells or hemoglobin in the blood

thalassemia – an inherited form of anemia due to decreased amounts of one of the two kinds of chains (α or β) in adult hemoglobin

Genomic versus cDNA libraries

Figure 9.14 on p. 320 discriminates between genomic and cDNA libraries. Genomic libraries are collections of recombinant DNA molecules in which the inserts are random pieces of genomic DNA of a particular size class. Any region of the genome is equally likely to be represented in any clone of a genomic library. Whether a region is in an exon, and intron, or an intergenic region has no bearing on its appearance in the library;

all DNA is equivalent to the restriction enzymes or mechanical shearing forces used to fragment the DNA to make the genomic library. You should understand that for the reconstruction of the sequence of the whole genome (whether by the shotgun or the hierarchical approach), it is necessary that overlapping fragments of the genome are analyzed so that the relationships between the component sequences can be inferred.

For a cDNA library, the starting material is the collection of all the mature mRNA molecules that are found in a particular tissue. Because mRNAs have already been spliced, the inserts in the cDNA library will include only the exons of genes; introns or intergenic regions are not found. The frequency of finding cDNA clones corresponding to a particular mRNA type provides an estimate of the proportion of mRNA molecules in the tissue that are of this type. Thus, if a gene is not transcribed in this tissue, you will not find any corresponding cDNA clones in the library.

Problem Solving

Restriction enzyme recognition sites: When examining restriction enzyme recognition sites in DNA being digested to make recombinant DNA molecules, you should draw out the site on both strands with careful attention to the 5'-to-3' polarity, and then draw out the ends of the resultant molecule(s) after digestion with the restriction enzyme. If the cut DNAs represent the vector and insert being used to make a recombinant DNA molecule, draw out how the ends would go together in both strands, and look for complementarity and also the nature of the resultant sequence after ligation.

The likelihood of finding a particular restriction site in DNA of random sequence depends on the length of the recognition site. A 4-bp recognition site is found once in every $(4)^4 = 256$ bp, while a 6-bp site is found once in every $(4)^6 = 4096$ bp. Remember that an N (any base pair) has a probability of 1 at any location, while a Y (pyrimidine) or R (purine) has a probability of 1/2 at any location.

To determine if a restriction enzyme cleavage produces blunt ends or ends with overhangs, ask whether the cleavage sites are at the same position on the two strands (blunt) or whether they are staggered with respect to each other (5' or 3' overhangs). Single-stranded overhangs will be sticky if the original site and the positions of the two cleavages exhibit rotational symmetry.

Vectors and the formation of recombinant DNA molecules: Remember that vectors must have (i) an *origin of replication* allowing their DNA to be copied independently of the host cell genome, and (ii) a *selectable marker* so that researchers can identify cells containing the recombinant DNAs. Most vectors also have one or more restriction enzyme sites into which the foreign DNA insert can be added to the vector; these restriction enzyme sites cannot interrupt either the origin of replication or the selectable marker.

Sanger DNA sequencing: Sequence traces such as that shown in Fig. 9.7f on p. 311 depict the newly synthesized molecules in which nucleotides are added successively onto the 3' end of the primer due to complementarity with nucleotides in the template. The result of the sequencing reaction is a *nested set* of new DNA molecules. In all molecules in the nested set, the 5' end is the 5' end of the primer, while the 3' end is a fluorescently labeled dideoxyribonucleotide.

chapter 9

Vocabulary

1.

a.	oligonucleotide	10.	a short DNA fragment that can be synthesized by a machine
b.	vector	12.	a DNA molecule used for transporting, replicating, and purifying a DNA fragment
c.	sticky ends	9.	short single-stranded sequences found at the ends of many restriction fragments
d.	recombinant DNA	7.	contains genetic material from two different organisms
e.	syntenic block	6.	chromosomal region with the same genes in the same order in two different species
f.	genomic library	2.	a collection of the DNA fragments of a given species, inserted into a vector
g.	genomic equivalent	8.	the number of DNA fragments that are sufficient in aggregate length to contain the entire genome of a specified organism
h.	cDNA	11.	DNA copied from RNA by reverse transcriptase
i.	gene family	5.	set of genes related by processes of duplication and divergence
j.	hybridization	4.	stable binding of single-stranded DNA molecules to each other
k.	alternative RNA	3.	the joining together of exons in a gene in different combinations
l.	protein domain	1.	a discrete part of a protein that provides a unit of function

Section 9.1

2. a. *Sau*3A (^GATC) (i) *Sau*3A recognition sites are 4 base pairs (bp) long and are expected to occur randomly every 4^4 or 256 bp. The human genome contains about 3×10^9 bp, one would expect $3 \times 10^9 / 256 = 1.2 \times 10^7$ **~12,000,000 fragments.** (ii) The enzyme would cut on average once every 4^4 or **256 base pairs,** so this should be the size of the average piece of the human genome cut with this restriction enzyme. (iii) Ends produced by cleavage with this enzyme have **sticky ends with a 5' overhang.** Note that the cut site (^) is not in the middle of the recognition sequence, meaning that the enzyme cuts the two strands of the DNA target at different base pairs, so this is not a blunt end. The cut site is to the left of the recognition sequence's middle, so this is a 5' overhang. (iv) All the ends of human

DNA fragments produced by digestion with Sau3A would be **identical** (they are all 5' GATC overhangs).

b. *Bam*H1 (G^GATCC) (i) The enzyme would cut random DNA once every 4^6 or 4096 bp. $3 \times 10^9/4096 =$ **~732,400 fragments.** (ii) **4096 base pairs = ~4.1 kilobase pairs (kb).** (iii) **Sticky ends with a 5' overhang.** (iv) **All ends are identical** (they are all 5' GATC overhangs).

c. *Hpa*II (C^CGG) (i) $3 \times 10^9/256 =$ **~12,000,000 fragments.** (ii) **256 base pairs.** (iii) **Sticky ends with a 5' overhang.** (iv) **All ends are identical** (these are 5' CG overhangs).

d. *Sph*I (GCATG^C) (i) $3 \times 10^9/4096 =$ **~732,400 fragments.** (ii) **4096 base pairs.** (iii) **Sticky ends with a 3' overhang.** The cut site is not in the middle of the recognition sequence, so this is not a blunt end, but in this case the cut site is to the right of the middle of the recognition site, so this is a 3' overhang. (iv) **All ends are identical** (these are CATG 3' overhangs).

e. *Nae*I (GCC^GGC) (i) $3 \times 10^9/4096 =$ **~732,400 fragments.** (ii) **4096 base pairs.** (iii) **Blunt ends.** Note that the cut site is in the middle of the recognition sequence, and that the restriction enzyme cuts both strands at the same place. (iv) **All ends are identical.**

f. *Ban*I (G^GYRCC) (i) In this case, the restriction fragment would be expected to cut DNA once every (4)(4)(2)(2)(4)(4) = 1024 bp. (This is because Y bases could be either C or T, while R bases could be A or G). Thus, $3 \times 10^9/1024 =$ **~2,930,000 fragments.** (ii) **1024 base pairs.** (iii) **Sticky ends with a 5' overhang.** (iv) **All ends are NOT identical.** After digestion, some resultant sticky ends will be 5' GCAC while others will be 5' GCGC, 5' GTAC, or 5' GTGC overhangs.

g. *Bst*YI (R^GATCY) (i) $3 \times 10^9/1024 =$ **~2,930,000 fragments.** The 1024 value is calculated as in part (f). (ii) **1024 base pairs.** (iii) **Sticky ends with a 5' overhang.** (iv) **All ends are identical.** In contrast with part (f), the actual overhangs produced with this restriction enzyme will all be 5' GATC.

h. *Bsl*I (CCNNNNN^NNGG) (i) $3 \times 10^9/256 =$ **~12,000,000 fragments.** Even though this recognition sequence is long, only 4 base pairs are defined, while all the others can be any base pair. So the chance of any random 11 base pair sequence being recognized by this enzyme is (1/4)(1/4)(1)(1)(1)(1)(1)(1)(1)(1/4)(1/4) = 1/256. (ii) **256 base pairs.** (iii) **Sticky ends with a 3' overhang.** (iv) **All ends are NOT identical**, the 3' overhangs would be 3 bases long but they could be any 3 bases.

i. *Sbf*I (CCTGCA^GG) (i) This recognition site is 8 bp long, so the enzyme would cut on average every 4^8 or 65,536 bp. $3 \times 10^9/65,536 =$ **~45,800 fragments.** (ii) **65,536 base pairs = 65 kb.** (iii) **Sticky ends with a 3' overhang.** (iv) **All ends are identical** and would be TGCA 3'.

3. Note that the DNA molecule depicted has two recognition sites for the *Eco*RI enzyme. When you cut a linear piece of DNA twice, you end up with **three fragments.** These fragments are shown in different colors as follows:

chapter 9

```
5' AGATG  3'           5' AATTCGCTGAAGAACCAAG 3'        5' AATTCGATT 3'
3' TCTACTTAA 5'        3' GCGACTTCTTGGTTCTTAA 5'        3' GCTAA    5'
```

Because the *Eco*RI restriction site is 6 bp long (G^AATTC), the enzyme would on average digest DNA of random sequence only once in every 4^6 or 4096 bp. **The fragment shown in this problem is much smaller than 4096 bp, so it would be unusual to find two recognition sites for *Eco*RI so close together.** However, if the DNA that was digested is long (like the 3 billion bp in the human genome), it is likely that somewhere in the genome two such sites would be as close together as this.

4. The agarose gel through which the DNA is being electrophoresed has holes that are roughly the size of the DNA fragments to be analyzed. **Smaller DNA fragments will find more holes through which they can travel, so they will move faster than larger DNA fragments, whose movements will be retarded when they bump into holes that are not large enough for them to go through.** You should note that the motive force on the DNA fragments exerted by the electric field is the same for all DNA fragments regardless of their sizes, because all nucleotides have about the same mass and same charge. Thus, the ability of a fragment to find holes through which is can travel is the main determinant of the mobility of a DNA fragment in the gel.

5. **Lane A: bacteriophage genome digested with *Eco*RI.** Only a few bands are seen in the gel, so the genome must have been small. There are fewer bands of higher size than is seen in lane C (the other sample in which the bacteriophage genome was digested, but in that case with an enzyme that recognizes a 4-bp sequence rather than the 6-bp sequence for *Eco*RI).

   Lane B: human genomic DNA digested with *Eco*RI. A smear consisting of many bands is produced, so the genome must have been very large. These fragments are of larger average size than those in lane D (where the human genome was digested with the *Hpa*II enzyme).

   Lane C: bacteriophage genome digested with *Hpa*II. Only a few bands are seen (so this must be bacteriophage DNA), but these are smaller than those seen in Lane A where the same DNA was digested with *Eco*RI, which has a larger recognition site.

   Lane D: human genomic DNA digested with *Hpa*II. A smear of many bands is produced, so this must be a digest of human genomic DNA. These are smaller on average than those seen in Lane B, so the recognition site for the enzyme used in Lane D must be smaller than the recognition site for the enzyme used to digest the same DNA in Lane B.

6. [*Note:* In your copy of the text, the lanes on the gel diagram may be mislabeled. For both Sample A and Sample B, the order of the lanes should be *Eco*RI, *Bam*HI, *Eco*RI + *Bam*HI.]

   a. **Sample A represents the circular form.** Compare the *Eco*RI digests of Sample A and Sample B. Sample A shows 3 bands while Sample B shows 4 bands. If you digest a circular molecule at 3 places, you will end up with 3 bands, while if you digest a linear molecule at the same 3 places, you will end up with 4 bands. The same pattern is seen for all of the other digests: Sample A always has 1 band fewer than Sample B when the two samples are cut with the same enzyme(s).

b. You could add up the sizes of the restriction fragments seen in any one digest of Sample B (the linear form of the bacteriophage DNA) to find the total length of this molecule. For example, in the *Eco*RI digest, this would be 4 + 3 + 2 + 1 = **10 kb.** You would find the same value for any of the other digests of Sample B. (*Note*: In actuality, the bacteriophage λ genome is 48.5 kb.)

c. Using the same addition procedure as in part (b), the total length of the circular form of the bacteriophage DNA is 5 + 3 + 2 = **10 kb.** The size of the bacteriophage DNA does not change when it is circularized.

The following figure diagrams the locations of all the restriction enzyme recognition sites (E = *Eco*RI; B = *Bam*HI) and the sizes of the fragments that would be produced by digestion of either the circular (*left*) or linearized (*right*) form of the chromosome. The *red triangles* and *red diamond* indicate the location of the sticky ends of the bacteriophage λ chromosome.

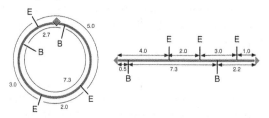

7. a. With partial restriction digestion, the enzyme does not cut the genomic DNA at every recognition site that is present in the genome. Thus, **the partial digest would produce fewer fragments of longer average sizes than would be produced by digesting the same DNA sample to completion with the same enzyme.**

b. **The fragments of DNA produced from the genomes of different cells by partial restriction digestion would be different from each other.** In contrast with a complete digestion, in which every *Eco*RI recognition site would be cut in the DNA molecules from all the cells, in a partial digestion you would get a range of different fragments produced from different molecules of genomic DNA. The figure that follows illustrates that 3 genomic DNA molecules from three different cells would in fact yield different fragments upon a partial digestion. (In this particular case, conditions were adjusted so that on average only one site out of five was cut; by digesting the DNA for longer or shorter times with the enzyme, researchers can produce fragments with different average sizes.) Because different genomic DNA molecules produce different fragments, scientists can use partial restriction digestion to make genomic libraries with overlapping fragments that allow reconstruction of the genomic DNA sequence.

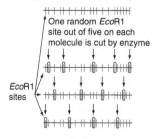

chapter 9

Section 9.2

8. a. **The purpose of molecular cloning is to isolate DNA corresponding to a small region of a complex genome,** so that scientists can purify enough of this small region of DNA to study in detail (for example, to determine the nucleotide sequence of this region).

    b. Selectable markers in vectors provide a **means of determining which cells in the transformation mix take up the vector.** These markers are often drug resistance genes, so a drug can be added to the media and only those cells that have received and maintained the vector will grow into colonies.

    c. **The origin of replication in a plasmid vector allows the vector (and any recombinant DNA molecule that incorporates the vector) to grow in a bacterial host cell independently of the bacterial chromosome.** This property allows the plasmid to be transferred to daughter cells when the cells divide, and also allows researchers to purify the plasmid away from the bacterial host's chromosome.

9. To make a recombinant DNA molecule, you would need (in addition to vector and insert DNA) the following enzymes: **(c) a restriction enzyme to cut the vector and the genomic DNA so that the resultant pieces have compatible sticky ends; and (d) DNA ligase to stitch together the vector and insert.**

10. First, work through the digestion and ligation of the DNA fragments and the vector. The vector (*red*) is cut with *Bam*HI, leaving the following ends:

    ```
    5' -G    3'         5' GATCC- 3'
    3' -CCTAG 5'        3'     G- 5'
    ```

    The insert DNA (*blue*) is cut with *Mbo*I, leaving the following sticky ends. The N means any base; of course, the bases in the two strands should be complementary.

    ```
    5' -N    3'         5' GATCN- 3'
    3' -NCTAG 5'        3'     N- 5'
    ```

    The ligation of an *Mbo*I fragment to a *Bam*HI sticky end will only occasionally create a sequence that can be digested by *Bam*HI. It depends on the exact base sequence at the ends of the *Mbo*I fragment. The N in the sequence below indicates this ambiguity. In all cases the following sequence will be found: The sequences from the inserted *Mbo*I fragment are in bold.

    ```
    5' -GGATCN----------NGATCC- 3'
    3' -CCTAGN----------NCTAGG- 5'
    ```

    a. **100%** of the junctions can be digested with *Mbo*I because they all contain the recognition sequence 5' GATC 3'.

    b. A junction that can be digested with *Bam*HI must have a C at the 3' end of the *Mbo*I recognition sequence; in other words, N would need to be C. This would occur 1/4 or **25% of the time**.

    c. **None** of the junctions will be cleavable by *Xor*II because none of the possible junctions has the sequence 5' CGATCG 3'.

d. A junction that can be digested *Eco*RII must have a pyrimidine (C or T) in the position of the N at the 3' end. This will occur by chance **50%** of the time.

e. For the restriction site to be a *Bam*HI site in the human genome it must have had a G at the 5' end. This G was in the vector sequence in the clones created. The chance that the 5' end was NOT a G = **3/4**.

11. After digestion, the vector (*red*) will look as follows. (It is now a linear molecule; the dots at the two ends are actually connected to each other.)

    5' ...CGGATCCCCTAAGATG 3'   5' AATTCCGCGCGCATCGGC... 3'
    3' ...GCCTAGGGGATTCTACTTAA 5'   3' GGCGCGCGTAGCCG... 5'

    The insert fragment, which is the *blue* fragment from Problem 3 that has *Eco*RI sticky ends at both sides, is the following:

    5' AATTCGCTGAAGAACCAAG 3'
    3' GCGACTTCTTGGTTCTTAA 5'

    a. **The two possible recombinant molecules are:**

       5' ...CGGATCCCCTAAGATGAATTCGCTGAAGAACCAAGAATTCCGCGCGCATCGGC... 3'
       3' ...GCCTAGGGGATTCTACTTAAGCGACTTCTTGGTTCTTAAGGCGCGCGTAGCCG... 5'

    or 5' ...CGGATCCCCTAAGATGAATTCTTGGTTCTTCAGCGAATTCCGCGCGCATCGGC... 3'
       3' ...GCCTAGGGGATTCTACTTAAGAACCAAGAAGTCGCTTAAGGCGCGCGTAGCCG... 5'

    b. **The fragment could be inserted into the vector in either of two orientations that still preserve the 5'-to-3' polarity of the resultant recombinant molecules.** The two orientations are flipped relative to each other (in terms of the strands) and turned 180°.

    c. The recombinant DNA molecule made (regardless of the orientation of the insert) will have **only a single site for *Bam*HI.** This comes from the vector and is underlined in the answers to part (a). The *blue* insert has no recognition sites for this enzyme.

    d. After the recombinant DNA is digested with *Bam*H1, there will be only one fragment (in essence, the molecule will be linearized). The size of this fragment would be the length of the pMBG36 vector (4271 bp) plus the size of the insert (19 bp), or **4290 bp.**

    e. The recombinant DNA molecule would have **two sites for *Eco*RI**; one at the left junction of the vector and insert, and the other at the right junction of the vector and insert. This is an important point: After ligation, the insert will be flanked by *Eco*RI sites at either side because both the vector and insert are contributing sequences that will make up these sites in the recombinant DNA.

    f. After the recombinant DNA is digested with *Eco*RI, there will be **two fragments**; one is the vector and the other is the insert. **One fragment will thus be 4271 bp and the other will be 19 bp.** (The single-stranded sticky ends are counted as if they were double-stranded.)

12. a. You need **4-5 genome equivalents** to reach a 95% confidence level that you will find a particular unique DNA sequence.

chapter 9

b. The number of clones needed depends on the total size of the genome of your research organism and the average insert size in the vector. Plasmid vectors normally have inserts smaller than 15 kb, while inserts into bacterial artificial chromosome vectors (BACs) can be 300 kb and those into yeast artificial chromosomes can have inserts as large as 2,000 kb (= 3 Mb). **Divide the number of base pairs in the genome by the average insert size then multiply by five** to get the number of clones in five genome equivalents.

13. One of the issues encountered in molecular cloning (the construction of recombinant DNA molecules) is that after the vector is cut by a restriction enzyme, the two ends of the vector can come back together and be resealed by DNA ligase without any insert being incorporated at all. One way to deal with this problem is to use vectors like that shown in the figure. **If the vector resealed without any insert, the resultant colonies growing on ampicillin plates supplemented with X-gal would be blue in color, because these cells would make the β-galactosidase enzyme, and this would turn the colorless X-gal into a blue compound. If the vector contained an insert, the resultant colonies growing on the same plate would be white, because no enzyme and thus no blue compound would be made.** This property of the vector allows researchers to identify white colonies that have recombinant DNA molecules and to ignore blue colonies that have resealed vectors without inserts.

14. a. The goal of the ligation is to generate recombinant DNA molecules that have attached one piece of frog DNA to one vector molecule. A ligation mixture consists of linear double-stranded vector DNA with complementary *Eco*RI sticky ends (**Fig. 9.2b** on p. 302 and **Fig. 9.6** on p. 308) at both ends; and linear double-stranded frog DNA, also with complementary *Eco*RI sticky ends at both ends.

    DNA ligase simply connects a 3'-OH (hydroxyl) group to a 5'-P (phosphate). The most abundant products that occur in a ligation mixture are those produced by *intramolecular* ligation of the 5' and 3' ends of a vector molecule or an insert molecule to form a circle. The reason is that these 5' and 3' ends are on the same DNA molecule, so they will bump up against each other the most often in the ligation solution. As the frog insert DNA alone has no origin of replication and no selectable marker, only the recircularized (reconstituted) vector molecules will be recovered in transformants (see Problem 12). The desired ligation product is produced by more rare *intermolecular* ligation of a single vector molecule with a single frog insert molecule. (Other intermolecular ligations that produce chains with multiple vectors and/or inserts can also occur.)

    To decrease the amount of reconstituted vector produced, you treat the linear, digested vector with alkaline phosphatase. The following figures show why you would perform this procedure. The figures redraw the vector (after cleavage) previously shown in Problem 11. The dots at the ends mean that the strands of the digested vector are contiguous except for where the cleavage took place. The figure also illustrates in *black* that after cleavage by *Eco*RI, the 5' carbon of the nucleotide at the 5' end of the sticky overhang is attached to a phosphate group, while the 3' carbon of the nucleotide at the 3'-end is has only an hydroxyl group and no phosphate. The vector is in *red* and the insert in *blue*.

    5' ...CGGATCCCCTAAGATG_{OH} 3'   5' pAATTCCGCGCGCATCGGC... 3'
    3' ...GCCTAGGGGATTCTACTTAAp 5'   3' _{HO}GGCGCGCGTAGCCG... 5'

Alkaline phosphatase removes the 5'-phosphate groups on the linear DNA molecule:

5' ...CGGATCCCCTAAGATG$_{OH}$ 3'   5' $_{HO}$AATTCCGCGCGCATCGGC... 3'
3' ...GCCTAGGGGATTCTACTTAA$_{OH}$ 5'   3' $_{HO}$GGCGCGCGTAGCCG... 5'

After the treatment with alkaline phosphatase, **ligase cannot join a hydroxyl group to the dephosphorylated 5' ends**. Therefore the two ends of the vector cannot be ligated to each other and this treated molecule will remain linear. If insert DNA is added then the ligase will join the 3'-OHs on the vector with the 5'Ps on the insert:

5'...CGGATCCCCTAAGATG$_{OH}$ pAATTCGCTGAAGAACCAAG$_{OH}$ $_{HO}$AATTCCGCGCGCATCGGC... 3'
3'...GCCTAGGGGATTCTACTTAA$_{OH}$ $_{HO}$GCGACTTCTTGGTTCTTAAp $_{HO}$GGCGCGCGTAGCCG... 5'

You should note that after this ligation, each strand will still have one nick (a position without a phosphodiester bond) where an –OH from the vector abuts and –OH from the insert. But after the ligation mix is transformed into *Escherichia coli* cells, enzymes in the bacterial cells repair these nicks in the phosphate backbone.

As discussed in Problem 12, certain vectors are constructed so that they contain the *lacZ* gene with a restriction site right in the middle of the gene. If the vector reanneals to itself without inclusion of an insert, the *lacZ* gene will remain uninterrupted; if an insert has been cloned into the vector the *lacZ* gene will be interrupted. The ligation mix is transformed into *E. coli* cells such that about one cell out of 1,000 cells takes up a plasmid. The transformed cells are plated on media containing ampicillin. Only the cells with a plasmid will grow, thus removing the intramolecular ligation products that consist of inserts. The media also contains X-gal. This is a substrate for the β-galactosidase protein that is coded for by the *lacZ* gene. The β-galactosidase enzyme cleaves X-gal and produces a molecule that turns the cell blue. Those cells that took up an intact, recircularized vector with no insert will produce β-galactosidase and form blue colonies. The bacterial cells that took up a vector + insert (clone) will not be able to produce functional β-galactosidase and will form white colonies.

The non-dephosphorylated vector (that is, with no prior alkaline phosphatase treatment) reanneals to itself at a high frequency, leading to 99/100 blue colonies after transformation with the ligation mix. The dephosphorylated vector (that had been treated with alkaline phosphatase) formed 99/100 white colonies after transformation of the ligation mix; almost all of the vectors had an insert.

b. **Yes**, the suggestion was a good one. **Dephosphorylation of the vector increased the fraction of colonies with recombinant DNA molecules (vector + insert) 100-fold.**

c. The choice of whether to dephosphorylate the vector versus the insert DNA is based on an understanding of the mechanics of the bacterial transformation that is carried out after the ligation. If the vector is dephosphorylated it cannot self-ligate. The insert can self-ligate. As mentioned in part (a), the self-ligated inserts do not have any vector DNA, so they do not have a bacterial origin of replication (ORI) nor do they have a gene encoding antibiotic resistance. Therefore, these recircularized DNAs will not allow the transformed bacteria to grow on the selective media. **If the insert were dephosphorylated, it will not self-ligate, but the vector WILL self-ligate. The vector has the antibiotic resistance gene and ORI, so the "empty" vector will be propagated in** *E. coli*, **generating a high level of "background" – colonies that survive drug selection but are transformed only with empty vector.**

Section 9.3

15. Assuming the DNA molecule to be sequenced is already in hand, **the only enzyme you would need to sequence DNA is DNA polymerase (a).**

16. a. Reading the peaks from left-to-right (corresponding to the 5'-to-3' direction because DNA polymerase successively adds nucleotides to the 3' end) the sequence would be **5' TTTGCTTTGTGAGCGGATAACAA 3'**.

    b. **The sequence written in part (a) is of the strand that is being newly synthesized in the sequencing reaction.**

    c. You know how to design the primer because: (1) You added the insert into the vector at a known location defined by a specific restriction enzyme; and (2) you know the sequence of the vector that flanks the position of the insert.

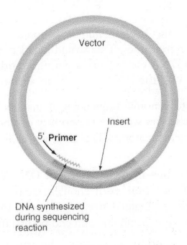

    d. The smallest molecule synthesized in the sequencing reaction that contains dideoxy G would be:

    5' GCCTCGAATCGGGTACCTTTG* 3'

    The asterisk indicates the dideoxy G, which must be at the 3' end because the chain terminates when this dideoxy nucleotide has been incorporated.

Section 9.4

17. a. There are **two contigs.** They are:
    (sequences 1 3 5)
    5' AGCAAATTACAGCAATATGAAGAGATCATACAGT 3'
    3' TCGTTTAATGTCGTTATACTTCTCTAGTATGTCA 5'
    (sequences 2 4 6)
    5' TCCTTTTAAAAATCTCATTTCCTTTAGGGCATTT 3'
    3' AGGAAAATTTTTAGAGTAAAGGAAATCCCGTAAAA 5'

    The orientation in which these fragments are written does not matter.

    b. You are sequencing different molecules of DNA in clones that are overlapping. **Some of the sequences read the same strand but start in different places. Some of the sequences read the complementary strand.**

    c. If you only had 6 short sequences from the entire 3 billion bp human genome, the **chances would be vanishingly small that these few sequences all come at random from the same small region of the genome.** Thus, 6 random sequences of the human genome should not overlap. Clearly, these sequences were not derived at random but were instead selected. (For example, these could be sequences all derived from subfragments of a particular human DNA insert that had already been obtained from a particular clone of a human genomic library.

    d. If you had enough sequence reads to cover all the human genome, these should in theory resolve into a number of contigs equal to the number of separate DNA molecules in the haploid human genome, which in turn equals the number of different chromosomes. This would be **23 contigs (22 autosomes and the X) if the human DNA came from a woman, and 24 contigs (22 autosomes plus the X plus the Y) if the DNA came from a man.**

18. The key point to genomic sequencing is that the fragments to be sequenced need to overlap; otherwise you would have no way of understanding their relationship to each other in the original genome. This problem explores several methods to achieve such overlap.

    a. **Mechanical shearing will fragment DNA at random locations, ensuring that overlap will be obtained** if enough fragments are analyzed. Because the genomic DNA sample is obtained from many cells, the fragments from different cells will not have the same beginning and end points; thus, they will overlap.

    b. In this case, **overlap is obtained by making different libraries of the same genome, each library constructed with a different restriction enzyme (for example, one library made with the *Eco*RI enzyme and a second library with the *Bam*HI enzyme). The recognition sites for different enzymes are at different positions in the genome to be analyzed, so that the fragments in any region obtained from the libraries will overlap.**

    One problem with constructing a library by complete digestion of a genome with a given restriction enzyme is that by chance, some fragments may be too large to be cloned or analyzed, while others are so small. Thus, another advantage of using multiple libraries made with different restriction enzymes is to minimize the chance that a region will not be covered. For example, a too-long fragment with *Eco*RI sites at its ends may have internal sites for *Bam*HI or some other enzyme.

chapter 9

c. **You could perform a partial digestion with the single restriction enzyme.** As shown in the figure accompanying the answer to Problem 7 above, partial digestion will ensure overlap because different genomic DNA molecules will be cut at different sets of recognition sites.

If the restriction enzyme recognizes sites that are relatively far apart in the genome (for example, a 6 bp recognition sequence will be found on average every 4 kb, but an 8 bp sequence will be found on average every 64 kb) any approach using a single enzyme will suffer because some fragments are too large to clone. **It will be preferable to use an enzyme that recognizes a 4 bp sequence because the sites will be spaced closer together. The more potential locations of digestion, the more the fragments would resemble a library made by mechanical shearing with cuts made at random locations.** If you adjust the partial digestion conditions properly, you could construct a library that would have inserts with an average size of any desired length.

19. The primers must be designed on the basis of DNA sequences you already know. In the case of the recombinant DNA construct, you already know the sequence of the pMBG36 plasmid vector surrounding the site into which foreign DNA will be inserted. These primers must "point into" the insert, in the sense that at each vector/insert junction, the 3' end of the primer should be closer to the insert than the 5' end of the primer. In this way, the new nucleotides added during the sequencing reaction will be complementary to nucleotides in the insert. (After all, the insert is ultimately what you want to sequence.) One primer would be:

    5' CGGATCCCCTAAGATGAATTC 3'

    The other primer would be:

    3' CTTAAGGCGCGCGTAGCCG 5' = 5' GCCGATGCGCGCGGAATTC 3'

    Note that to be as long as possible, both primers will include the *Eco*RI recognition site (underlined) at their 3' ends, because you know that all the inserts must have this site at both junctions with the vector.

Section 9.5

20. Some possible methods for identifying gene sequences in the human genome are: (1) **Compare complete genome sequences from a related mammalian species such as mouse.** Non-functional regions of the genomes have significantly diverged while functional regions are relatively conserved. (2) **Use computational analysis to search the DNA sequence for open reading frames and potential intron/exon boundaries.** (Although not discussed in the text, a refinement of this method would be to look at the potential open reading frames and see if the codons used in the cases of amino acids with degenerate codons match the codon usage for the species; certain codons are used much more often than other codons for the same amino acid in a species-specific way.) (3) **Locate transcribed regions by comparing the genomic sequence with the sequences of cDNA clones copied from mRNAs in various tissues.**

21. a. **Humans and chimps last shared a common ancestor 6-7 Mya (million years ago). The last common ancestor of humans and mice was 75 Mya; of humans and dogs 92 Mya; of humans and chickens, 310 Mya; of humans and frogs, 360 Mya.**

    b. **The slowest-clicking clock would be missense mutations (i).** The reason is that missense mutations would affect the amino acids in proteins and this could have deleterious consequences on phenotype that would be disfavored by natural selection. Many missense mutations that occur over history would thus be lost, and so the missense mutations would accumulate only slowly in surviving genomes, meaning that the clock ticks slowly. Silent mutations would not affect phenotypes because no amino acids would be changed; thus, silent mutations would have the fastest-ticking clock of the three possibilities. Most mutations in introns would not affect gene function, although some could affect the efficiency or accuracy of splicing, so a minority could affect gene function in ways that would be acted upon by natural selection.

    c. Note that 400 Mya is a very long time - further back than the last common ancestor of humans and frogs. **For species that last shared a common ancestor so long ago, you would want to use a slow-ticking clock like missense mutations (i).** If you used a faster clock (ii or iii), so many mutations may have accumulated that you couldn't make a very accurate comparison as you could often not be able to even see that the sequences in the genomes of the two species had any relationship at all. If you refer to Fig. 9.12 on p. 318, you will see that the only sequences that have clear homologies between humans and zebrafish (last common ancestor was ~416 Mya) are the coding sequences in the most highly conserved genes.

    Interestingly, if species are more closely related in evolutionary terms, like humans and chimps, you would want to use a faster clock like comparing changes in introns. If you used missense mutations, so few such mutations may have accumulated between these genomes that you could not get a very accurate estimate of the time of divergence.

    d. The clock *least* likely to vary in the rate of accumulation of mutations in different genes is clearly silent (synonymous) mutations. **The clock *most* likely to vary in the rate of accumulation of mutations in different genes is missense mutations (i) because some genes would be more sensitive than others to mutations that would alter amino acids.** (Some proteins are very highly conserved because most amino acid substitutions would disrupt their function, while other proteins are poorly conserved because many changes in amino acid composition are still consistent with protein function.) An answer of (ii) would also have been reasonable, because some changes in introns could affect gene function; however, the majority of intron changes have no effects, because you saw in Fig. 9.12 on p. 318 that the sequences of introns are generally poorly conserved.

22. The first enzyme you would need is **reverse transcriptase (g) to copy mRNA into the first strand of cDNA.** You would probably also need **DNA polymerase (a) to copy the first strand of cDNA in order to make double-stranded cDNA.** Although it was not shown in Fig. 9.13 on p. 320, the DNA polymerase is not strictly required because reverse transcriptase is not only an RNA-dependent DNA polymerase, but it

also can serve as a DNA-dependent DNA polymerase (that is, it has the potential to copy the first strand of cDNA to make double-stranded cDNA). You next need to insert the double-stranded cDNAs into the vector. This was also not shown in Fig. 9.13, but the most straightforward way is to use **a restriction enzyme (c).** Here, you would add oligonucleotides containing the recognition site for the restriction enzyme to the ends of the double-stranded cDNA, and then make recombinant DNA molecules as in Fig. 9.5 on p. 307. (Other more complicated methods exist that would allow you to insert the double-stranded cDNA into the vector without a restriction enzyme.) After mixing vector cut with the restriction enzyme together with double-stranded cDNA fragments that have the same sticky ends, you would need to add **DNA ligase (d) to seal up the recombinant DNA molecule.**

23. a. **Sequence 1 is the genomic fragment; Sequence 2 is the cDNA sequence.** On the genomic fragment written below, the exons are in *purple* and the intron is in *blue:*

    5' TAGGTGAAAGAGTAGCCTAGAATCAGTTA 3'
    3' ATCCACTTTCTCATCGGATCTTAGTCAAT 5'

    You can make these assignments because the reverse complement of Sequence 2 is the same as Sequence 1 (the top strand above), except that Sequence 2 lacks the intron sequence (*blue*).

    b. Because you don't know the direction of transcription, the primary transcript could correspond to either strand written in the answer to part (a) above. To decide between these two possibilities, you need to search for the splicing consensus sequences at 5' and 3' ends of the intron (exon|GU.......AG|exon). These sequences are found in the proper positions only if the top strand of DNA is the RNA-like strand, producing the following primary transcript:

    5' UAGGUGAAAGAGUAGCCUAGAAUCAGUUA 3'

    After splicing, the mRNA made from this primary transcript would be as follows, with the stop codons indicated in *red:*

    5' UAGGUGAAAGAAAUCAGUUA 3'

    As no start codon (AUG) is present in any reading frame, we can assume that this part of the mRNA contains sequences that are in the middle of the open reading frame. Only one of the three possible reading frames on this mRNA is "open":

    5' UA GGU GAA AGA AAU CAG UUA 3'

    The translation of this mRNA would be:
    N ... Gly Glu Arg Asn Gln Leu... C.

24. a. **Because cDNAs are made from mRNAs, they lack introns and they also do not have regulatory region information such as promoters and enhancers.**

    b. **Clones in a cDNA most definitely can include 5' UTR and 3' UTR sequences.** The reason is that the 5' UTR and 3' UTR sequences are found in exons, and the cDNA clones contain these exons (review Fig. 8.14 on p. 271).

    c. **You would be more likely to find longer ORFs on average in cDNA clones that in genomic clones.** In DNA of random sequence, you would expect to find that the average open reading frame only encodes about 20 amino acids (because there are 64 possible codons of which 3 signify stop). But most proteins have many more than 20 amino acids. If you obtained gene sequences from a genomic DNA clone, the exons would be interrupted by introns; if the introns were of random sequence, a reading frame begun in an exon would stop within about 60 nucleotides (20 amino acids) into the intron. cDNA clones don't have introns; they contain the entire coding sequence. So if the protein was say 1000 amino acids long, a cDNA clone has the potential to have an open reading frame that would encode all of these amino acids.

25. **The main reason geneticists studying eukaryotic cells make cDNA libraries is to determine how a particular gene's primary transcript is spliced in a given cell type. Almost all eukaryotic genes are interrupted by introns. In fact, only a small proportion (~1-2% in humans) of most eukaryotic genomes are made of protein-coding exons. In contrast, the genomes of bacteria do not have introns (with very few rare exceptions) and the large majority of the base pairs in bacterial genomes are protein-coding regions.** cDNA libraries corresponding to eukaryotic mRNAs are thus an enriched source of protein-coding exons. In addition, lacking introns, cDNA clones have the complete open reading frames for the protein products of the genes, while genomic clones will not have such open reading frames (see the preceding Problem 24).

    Although bacterial geneticists do not usually need to make cDNA libraries, such libraries are valuable for certain specific uses, such as determining which genes are transcribed in different environmental conditions. The main difficulty for bacterial geneticists in constructing cDNA libraries is that bacterial mRNAs do not have poly-A tails. As you saw in Fig. 9.13 on p. 320, the first step in making eukaryotic cDNA libraries is to anneal oligo-dT to the poly-A tails so that the oligo-dT can serve as a primer for reverse transcriptase. There are ways in which bacterial geneticists can get around this problem, though these methods are outside the scope of this book.

26. a. **The genomic library would contain the greatest number of different clones.** The library would have to represent all of the sequences in the genome. In contrast, cDNA libraries at most would have to represent just that fraction of the genome found in exons, which is only about 1-2% of the human genome. Because not all genes are transcribed in any one particular type of cell, brain or liver cDNA libraries would represent even smaller fractions of the total human genome.

    b. **All three of these libraries should include some of the same sequences.** The genomic library will include the exons found in the cDNA libraries. Some genes are expressed (transcribed) in all cell types, so cDNAs for these genes would be found in both brain and liver cDNA libraries. However, some genes are expressed in brain but not in liver or *vice versa*, so brain and liver cDNA libraries will also have tissue-specific cDNA clones.

    c. **The genomic library starts from human genomic DNA; the liver cDNA library begins with the total mRNA pool extracted from liver; the starting material for the brain cDNA library is the total mRNA pool extracted from brains.**

    d. **To annotate a genome, you would have to sequence many clones from many cDNA libraries because different genes are transcribed in different tissues and because some genes are transcribed only rarely in any tissue.**

chapter 9

27. **Several types of genes are difficult to find, so current counts of gene number are almost certainly incomplete.** These difficult-to-find genes include: (1) genes that are transcribed only rarely (so cDNAs corresponding to the mRNA would account for only a very small proportion of all cDNAs sequenced); (2) genes that encode very small proteins (so the open reading frame would be too short to identify by computer analysis; (3) genes that encode proteins that are poorly conserved through evolution (so between-species comparisons would not be informative); and (4) genes that are transcribed into *non-coding RNAs (ncRNAs)* that are not translated into proteins. Several classes of ncRNAs have been found only within the last few years prior to 2014, so it is highly possible that additional ncRNAs (and the genes that encode them) remain to be found.

28. a. **Yes, if you sequenced many clones from a cDNA library, you would sequence many independent copies of the same mRNA molecules.** For example, if a particular mRNA was very abundant among all the mRNAs in a tissue, then many clones in the library would have in effect the same cDNA inserts corresponding to this mRNA. However, this repetitive information still has value. Most obviously, the frequency at which a particular type of cDNA clone is found in the library reflects the abundance of the corresponding mRNA, so you could estimate relative levels of gene expression from sequencing many clones in the library. Other uses of this information might be the identification of alternatively spliced mRNAs made from the same gene, or more speculatively, finding unexpected post-transcriptional alterations made to some of the mRNAs

    b. **A diagram of the primer walking procedure is shown below.** You would use the sequence information you obtain in each step of the procedure to design the primers needed for the next step. These primers would be located near the 3' end of the sequences obtained, and the 5'-to-3' orientation of the primers would "point" in the direction of the insert DNA sequences you would like to obtain in the next step. In the following figure, the *black arrows* represent the primers you would use. Primers 1A and 1B represent the two primers you would use in the first step (in separate sequencing reactions coming in from opposite vector/insert junctions), 2A and 2B the two primers in the next step, etc. The *dotted green lines* indicate the new DNA sequences you would obtain with each primer. (The primers are drawn larger in scale than they would be in reality: the primers are only about 20-25 bases, while you can sequence ~1000 bases of DNA from each primer.) The *blue* dots at the ends of the vector signify that the vector is actually a circle.

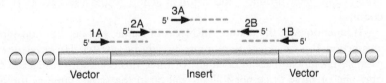

Section 9.6

29. (1) Genes in the human genome are frequently interrupted by introns, whereas bacterial genomes lack introns. (2) The intergenic regions (regions between genes) can be quite long in the human genome, but are almost all very short in

bacterial genomes. (3) The human genome has long stretches of repetitive DNA sequences, for example at centromeres and telomeres, which are lacking in bacterial genomes.

30. **The two different cDNA clones represent alternatively spliced products of the primary transcript of the same gene.** These alternatively spliced mRNAs would have different sets of exons brought together. For example, suppose the gene had 5 exons. One spliced product could have exons 1, 2, 3, and 5, whereas the other could have exons 1, 2, 4 and 5. The ends of the two mRNAs (exon 1 and exon 5) are the same, but the middle exons are different.

31. With regard to the scenario depicted in Fig. 9.17 on p. 323 and reprinted as follows, suppose that all the exons are protein-coding so that a unit consisting of an exon from Gene 1 plus part of its flanking introns (*red*) is shuffled into an intron of Gene 2 (*blue*). The new gene can, after splicing, produce a new mature mRNA with all the exons of Gene 2 plus the new exon of Gene 1. We will label the two junctions between the *blue* and *red* parts of this mRNA junctions A and B (see figure).

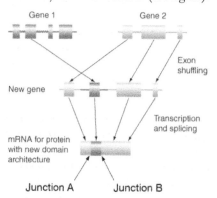

At junction A, the *blue* exon could be joined to the *red* exon in any of 6 ways, only one of which is in frame. [First, the shuffled part of Gene 1 would have to be inserted into Gene 2 with the same (versus the opposite) polarity (probability = 1/2). Second, even if the relative polarities are correct, the introns need to occur in the same location with respect to the codons in both genes (probability = 1/3). For example, if the intron in Gene 2 was located between codons (rather than in between two nucleotides that are part of the same codon), then to make a new protein that consists of domains from both original gene products, the intron in gene 1 would also have to be located between codons.]

If junction A in the mature mRNA is in frame, you need to consider what happens at junction B. The orientations must already be correct. However, the end of the *red* exon at this junction could be at any of three positions with respect to the codons, and only one of these three possibilities will be in the same frame as the *blue* exon at this same border.

Thus, the probability that an exon shuffling event like that shown in Fig. 9.17 on p. 323 would produce a new protein with the domains that were present in the original Gene 2 product plus a new domain from the original Gene 1 is **(1/6) x (1/3) = 1/18.** If the new mRNA at the bottom of the figure had to be in frame only at one junction rather than two (as would be the case if for example the initiation codon was in the *red* exon, then the probability would be 1/6.

Even though only a fraction of the possible exon reshuffling events that could occur would produce a new protein with a novel domain architecture, exon reshuffling is still much more efficient than simply joining random pieces of the genome together to make new genes. In exon shuffling, the chromosome breakages can occur pretty much anywhere within the introns. If random pieces were joined together, new domain architectures would result only when the breakages were exactly at the proper codons and the joined products remained in frame.

32. The human genome contains two types of pseudogenes: "duplicated pseudogenes" produced as the result of duplication of genes and the subsequent divergence of their sequences (see Fig. 9.19 on p. 324), and "processed pseudogenes" produced through reverse transcription of mRNAs (in which double-stranded cDNA copies of the mRNAs are inserted into the genome. Pseudogenes of both types will decay over evolutionary time, changing until they are no longer recognizable.

   a. The processed pseudogenes should retain remnants of mRNA structure. **Such pseudogenes should not contain introns (which were spliced out of the primary transcript to make the mRNA), and may also have a poly-A tail at the region corresponding to the 3' end of the transcript.** Note that these processed pseudogenes are themselves unlikely to be transcribed, because the mRNA does not include the promoter.

   b. **The mechanisms giving rise to processed pseudogenes (those copied from mRNAs) are most likely to result in pseudogenes scattered around the genome.** The mRNAs do not remain associated with the genes from which they are transcribed; instead, they eventually leave the nucleus so that they can be translated in the cytoplasm. As a result, there would be little possibility of a double-stranded cDNA being incorporated into the genome at a location adjacent to that of the original gene. In contrast, certain mechanisms of gene duplication such as unequal crossing-over would favor the creation of gene families whose members are clustered together in the same area of the genome.

33. a. **Because zinc-finger domains allow DNA binding, you would be most likely to suggest the hypothesis that the protein is involved in some kind of DNA-dependent metabolism.** In terms of gene expression, such a protein might be a *transcription factor* that modulates the transcription of genes near the DNA binding sites recognized by the protein. DNA-binding proteins might also be involved in other processes such as DNA replication, DNA packaging into chromosomes, or DNA segregation upon cell division.

   b. **If two genes in the same organism share significant base similarity, it is highly likely the two genes are homologous - related to each other through a common ancestral gene that underwent duplication and whose descendant genes then diverged from each other.** If the similarity is very high, either the duplication event was in the recent evolutionary past, or both genes fulfill essential functions that are sensitive to mutation (that is, natural selection would remove mutant alleles of either gene). With respect to the terminology introduced in Fig.

9.20 on p. 325, two genes in the same organism that are closely related to each other would be *paralogous genes*.

34. a. The best consensus would be:

    ```
    5' XXATATAAAAXXXXX 3'
        G       T
    ```

    Eight of the positions in the group of four sequences can have any nucleotide. This is denoted with an X. Position 3 can have an A or a G, position 8 can have an A or a T. The remaining nucleotides are invariant – they always have T or A as shown.

    b. **A consensus sequence shows the nucleotides that are likely to be the most important to gene or chromosome function.** These nucleotides have been conserved during the evolutionary time since the four organisms last shared a common evolutionary ancestor. Because you can compare any short stretches of random sequences and just by chance see what appear to be conserved nucleotides you must first determine that these sequences might have some functional relationship. **You must choose DNA sequences that are in the same evolutionarily conserved orthologous gene and in the same relative position (like the same exon) in the four species.**

    c. If you **compare the amino acids of the proteins encoded by homologous genes, you might be able to find particular conserved amino acids that you can use to define a consensus**. Even if the identical amino acid is not conserved, it is possible that the amino acid might be replaced by another amino acid with similar chemical properties in other species. For example, replacing the acidic amino acid aspartic acid with glutamic acid would be defined as a conservative substitution. This sort of comparison can only be done properly with amino acid sequences because the genetic code is degenerate.

35. To examine what combinatorial events happen at the DNA level, **you would make genomic DNA libraries of non-immune system cells (in which no combinatorial events would have occurred) and from immune system cells in various stages of development into B cells, including the mature B cells themselves that are producing the antibodies. You would then isolate several copies of the antibody-encoding genes from each of these libraries and determine their DNA sequences.** Any differences you detected would show the type of events that occur to rearrange or alter the DNA of the antibody genes so that they can produce so many kinds of antibodies.

    To check for the possibility of alternative splicing, you would first make a cDNA library from the antibody-producing B cells. (You don't have to make cDNA libraries from other types of cells because they are not making antibody mRNAs.) **You would then sequence many of the cDNAs corresponding to antibody-encoding mRNAs.** If alternative splicing was occurring, you could detect this by comparing the sequences of different clones. You could also characterize the splicing involved by comparing these cDNA sequences to the genomic sequences of the antibody genes in the B cells.

36. a. **All of these genes, in both humans and chimpanzees would be considered homologous since they were all derived from a common ancestor.** However,

chapter 9

the type of homology would differ for different pairs of genes [see part (c) below]. The $α_1$, $α_2$, β, Gγ, Aγ, δ, ε, and ζ genes in humans would comprise one set of **paralogous genes**; the $α_1$, $α_2$, β, Gγ, Aγ, δ, ε, and ζ genes in chimpanzees would also comprise a set of paralogous genes. **The $α_1$ gene in humans would be orthologous to the $α_1$ gene in chimpanzees**, the $α_2$ gene in humans would be orthologous to the $α_2$ gene in chimpanzees, etc.

b. Orthologous genes are more likely to have closely related functions, as they have been more recently derived from a common ancestor than paralogous genes.

c. The **orthologous genes (the human β gene and the chimpanzee β gene) should have a greater degree of nucleotide similarity, as they were more recently derived from a common ancestor** than the paralogous human α and β genes.

d. Most of the gene duplication events depicted in <u>Fig. 9.19</u> that gave rise to the **different kinds of hemoglobin genes ($α_1$, $α_2$, β, Gγ, Aγ, δ, ε, and ζ) must have occurred before the human and chimpanzee lineages diverged**. That is, the last common ancestor of these species must have already had all of these genes. In fact, most of these genes must have been present in an ancient species ancestral to all mammals since most of these genes are found in all present-day mammals.

37. **The human proteome is much more complex than that of the nematode**, even though the human genome has only about 3,000 more protein-coding genes. There are several underlying reasons for this fact, although not all of these are completely understood. A small part of the explanation involves the fact that the human genome tends to have more members of multigene families (more paralogs), although in some cases a particular multigene family may have more members in worms than in humans. Another possible reason is that human proteins might be subject to more different kinds of chemical modifications than proteins in nematodes. There are more than 400 different kinds of chemical reactions that can affect proteins in human cells, so our proteome is much larger than that of our genome. Most of these modification reactions also occur in nematodes, but it is possible that some do not. A large part of the explanation is that combinatorial amplification at both the DNA (e.g., V-D-J joining of antibody genes) and RNA levels (alternative splicing) occurs more frequently in humans than in worms.

Section 9.7

38. a. **The *CFTR* gene in humans has 27 exons** (and 26 introns). (In the browser, the exons are thicker than the introns. The introns are indicated by a thin line with arrowheads showing the direction of transcription. Don't worry if you missed counting one of the small exons.)

b. **The 5' UTR is completely contained within Exon 1. The 3' UTR is completely contained in Exon 27.** (In the browser, the UTRs are indicated by regions that are relatively thinner, while the protein-coding parts of the exons are relatively thicker.)

c. On the idiogram of the top of the browser window, you can see that *CFTR* is located roughly in the middle of the long (*q*) arm of chromosome 7 (band 7q31.3). **The direction of transcription** (indicated by the arrowheads along the introns) **is from the centromere to the telomere.** That is, RNA polymerase moves towards the telomere as it transcribes the *CFTR* gene.

d. As just noted, **the *CFTR* gene is located on the long (*q*) arm of chromosome 7.**

e. **The gene to the left of *CFTR*** (that is, the next gene closer to the centromere) **is called *ASZ1*. It is transcribed in the opposite direction from that for *CFTR*, so it is transcribed from the other strand. The gene just to the right of *CFTR*** (closer to the telomere) **is called *CTTNBP2*. It is also transcribed from the other strand** (its template for transcription is the opposite strand from the template strand for *CFTR*).

f. Chromosome 7 is approximately **160 Mb** (million base pairs) long. You can see this number at the top of the browser window once you have zoomed out far enough that the entire chromosome is highlighted in *red* on the idiogram. This number does not include certain sequences like those around the centromere (see below).

g. The centromere is located **at ~60 million base pairs**, with numbering starting at the left end [that is, the telomere of the short (*p*) arm].

h. The genes appear to pile up because **there is not room to display all of the genes on the chromosome one after the other on the same line when you are viewing the entire chromosome in a single window.**

39. When you do your BLAST search, the response window should have three main components. The first is a graphical representation (the beginning of which looks as follows). The *green lines* indicate that two sequences in the database have strong homologies with the Query sequence from Fig. 9.7f. The **black lines** are other sequences with lower degrees of similarity; we will ignore those.

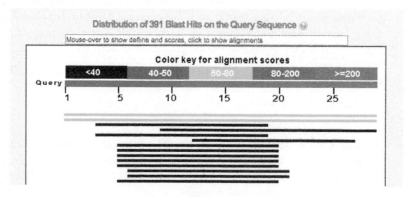

The next component of the response window is a list of the similar "hits", corresponding to the lines in the graphical representation. The top two of these are shown in the following reproduction; these are the sequences that correspond to the two *green lines* in the graphical window. Note that the E-values of these hits are both 3e-

chapter 9

07, roughly meaning that the chance that a random sequence of DNA would have the same degree of similarity to the Query is only about 3×10^{-7}, or less than one in a million.

The third component of the response window shows the actual matches between the Query sequence and the genes corresponding to the *green lines*. The output should be as follows for the top hit:

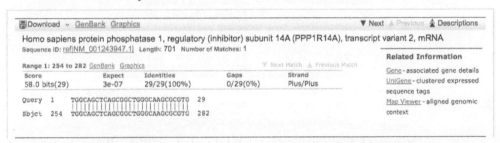

a. As seen in the preceding window, **the human gene (the Subject) most similar to the Query sequence encodes protein phosphatase 1, regulatory (inhibitor) subunit 14A (also known as PPP1R14A)**; this protein inhibits a specific kind of enzyme that removes phosphate groups from other proteins.

b. **The match is exact.** All the nucleotides of the Query match perfectly with the Subject sequence, so this is almost certainly the gene that was sequenced in Fig. 9.7f on p. 311. The Subject sequence is actually a cDNA sequence corresponding to one alternatively spliced variant of the mRNA for this gene. (The other green hit is a different splice variant of the same gene that still has the query sequence.)

c. **Most of this part of the *PPP1R14A* gene is fairly well-conserved in the mouse genome;** the sequence alignment that you should be seeing is shown in the following screenshot. Note however that the sequence identity only starts at nucleotide 9 of the query, so **the first 8 nucleotides at the 5' end of the Query are not well-conserved in the mouse copy of this gene.**

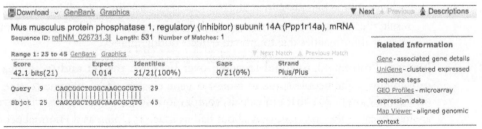

Source of Figures for Problem 9-39: National Institutes of Health

Section 9.8

40. a. **Unequal crossing-over between the two homologous copies of the β gene cluster between the δ gene on one chromosome and the β gene on the homologous chromosomes** would produce novel genes that encode polypeptide chains that could be incorporated into either the Lepore $\alpha_2(\delta-\beta)_2$ hemoglobin or the anti-Lepore hemoglobin $\alpha_2(\beta-\delta)_2$. See the following figure. Unequal crossing-over can take place because the δ and β genes are homologous, with similar (though not identical) DNA sequences.

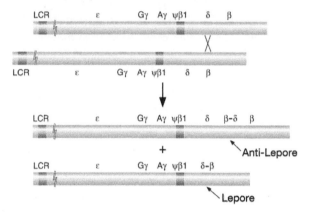

b. **The mildly thalassemic individuals are heterozygotes for the unusual Lepore allele,** meaning that they have somewhat less hemoglobin in total than would be found in a completely normal individual (less than 100% but more than 50%).

c. **The reason these heterozygotes would have less Lepore hemoglobin than normal adult hemoglobin has to do with the fact that the δ gene is transcribed much less frequently than is the β gene, so normal adults have much more $\alpha_2\beta_2$ hemoglobin than $\alpha_2\delta_2$ (see Fig. 9.25c on p. 329).** The promoter of each gene is located at the left end (because the genes are transcribed from left-to-right as seen in Fig. 9.26 on p. 331). Thus, the transcription of the Lepore hemoglobin depends on the very weak promoter for the δ gene, so relatively little Lepore hemoglobin is made.

For this reason, rare individuals who are homozygous for the Lepore allele make very little hemoglobin at all (and this hemoglobin would all be $\alpha_2(\delta-\beta)_2$. As a result, these homozygotes have strong β-thalassemia; they exhibit severe anemia and skeletal abnormalities due to growth defects during infancy.

41. Just as in Problem 40, **unequal crossing-over between the α1 and α2 genes** could explain how some people come to inherit α gene clusters with only one α gene. Note on Fig. 9.27 on p. 331 that individuals homozygous for such a deletion will have mild anemia, because they produce only about half as much α globin as do normal people.

chapter 9

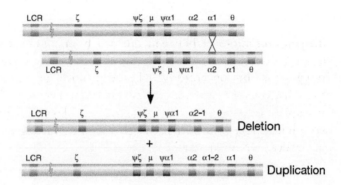

Chapter 10

Analyzing Genomic Variation

Synopsis

The genomes of any two humans differ from each other at more than 3 million nucleotides on average; it is these differences that make us individually distinct. This chapter describes the kinds of sequence variations that differentiate individual genomes, how we can detect this variation, and how we can use and interpret this information about genome variation. The analysis of genome variation has many practical purposes: from genotyping prospective parents, fetuses, or just-fertilized embryos; to the identification of tissues from individuals (DNA fingerprinting); to pinpointing the genes responsible for diseases; to understanding human history.

As you read the chapter, you should be aware of the level of resolution afforded by different methods of analyzing individual genomes. Some techniques, particularly those based on the polymerase chain reaction (PCR), allow researchers to look at sequence differences (*DNA polymorphisms*) at just one or a handful of locations in the genome. Other methods, particularly those involving hybridization of probes to DNA microarrays, are suitable for looking at several million locations in the genome at one fell swoop, but these microarrays still sample less than 0.1% of all the 3 billion nucleotide pairs in the human haploid genome. Finally, due to spectacular advances in DNA sequencing technology, we can now begin to look at all the polymorphisms that distinguish individual genomes.

Key terms

DNA polymorphisms – sequence differences between individual genomes within a species

anonymous DNA polymorphisms/DNA markers – DNA polymorphisms that do not affect phenotype but can be used to track specific regions of the genome

nonanonymous DNA polymorphisms – DNA polymorphisms that do affect phenotype by altering gene function, such as frameshift, nonsense, or missense mutations

SNP (single nucleotide polymorphism) – a polymorphism that substitutes one base pair for another

DIP/Indel – alternative terms for describing polymorphisms associated with the insertion or deletion of a few base pairs of DNA at a given locus

SSR (simple sequence repeat)/microsatellite – a polymorphism caused by different numbers of repeating units of a short, tandemly repeated sequence (< 10 bp long) such as ACACACAC

CNV (copy number variant) – a polymorphism caused by different numbers of copies of a region of DNA >10 bp long; for example, different numbers of copies of a gene

ancestral allele/derived allele – The *ancestral* allele of a polymorphism was inherited by two individuals from a common ancestor; the *derived* allele of the polymorphism is that due to a mutation in the lineage leading to one of the two individuals since the time the lineages diverged from the common ancestor.

unequal crossing-over – the result of recombination between mispaired copies of a tandemly repeated sequence. Unequal crossing-over can create new alleles of a CNV, with an increase or a decrease in the number of copies of the tandemly repeated unit.

PCR (polymerase chain reaction) – a method for amplifying a particular region of the genome (usually <10 kb) located between the 5' ends of two *PCR primers*. See Fig. 10.9 on p. 348 and Fig. 10.10 on p. 350 for details.

prenatal genetic diagnosis – genotyping fetal cells, which are usually isolated by **amniocentesis** (recovering cells from the amniotic fluid in the mother's womb). In the near future, prenatal genetic diagnosis is likely to be done by obtaining fetal DNA from the mother's blood.

preimplantation genetic diagnosis – genotyping a single cell obtained from an embryo recently generated by *in vitro* fertilization. Desired embryos are then implanted into the uterus of the prospective mother or surrogate mother.

DNA fingerprinting – analyzing a sufficient number of DNA polymorphisms in a tissue sample to make a match with a single individual in the human population. DNA fingerprint results using a defined set of polymorphisms are maintained in a database called **CODIS**.

nucleic acid hybridization – the ability of single-stranded DNA or RNA molecules to come together to form double-stranded molecules on the basis of nucleotide complementarity

probe – In nucleic acid hybridization, a probe is a short fragment of single-stranded DNA or RNA that is labeled by radioactivity or a fluorescent tag to test for its complementarity with a sample of nucleic acid.

ASO (allele-specific oligonucleotide) – a short (usually < 50 nt) molecule of single-stranded DNA used to genotype a SNP locus within a DNA sample

DNA microarray – a wafer made of silicon or some other material to which ASOs are attached for purposes of genotyping. Typically more than 1 million SNP loci can be genotyped simultaneously on a single DNA microarray.

disease gene – shorthand designation for a gene, mutations in which are responsible for a disease syndrome

positional cloning – a strategy to identify a disease gene on the basis of its linkage with anonymous DNA polymorphisms at known positions in the genome. See Fig. 10.18 on p. 357 for an overview.

phase/haplotype – when considering the two loci in a double heterozygote, the phase or haplotype is a description of which alleles of the two loci are present on each of the homologous chromosomes

chapter 10

> **LOD score** – a statistical test used to determine from pedigree data the likelihood that two loci are genetically linked. See the Tools of Genetics Box on p. 361 for methodological details.
>
> **allelic heterogeneity** – when a genetic condition can be due to different alleles (with different nucleic acid changes) of a disease gene. When the condition is recessive and displays allelic heterogeneity, affected individuals may be *trans*-heterozygotes (also called **compound heterozygotes**) where one chromosome bears one mutant allele and the homologous chromosome bears a different mutant allele.
>
> **exome** – all of the exons in a genome. Because it is less expensive to sequence whole exomes than whole genomes, and because many (though by no means all) disease-causing mutations are found in the exome, scientists often choose to sequence whole exomes rather than whole genomes. However, the cost of whole-genome sequencing is plummeting so rapidly that whole-exome sequencing may become obsolete in the near future.

Types of polymorphisms

Table 10.1 on p. 345 is reproduced here to remind you of the different classes of DNA polymorphisms that can differentiate genomes.

TABLE 10.1 Categories of Genetic Variants

The right column shows how frequently on average you would find a polymorphism of the indicated class when comparing any two haploid human genomes.

	Size	Frequency (1 per....)
SNP Single nucleotide polymorphism	1 bp	1 kb
DIP or **Indel** Insertion/deletion	1–100 bp	10 kb
SSR or **Microsatellite** Simple sequence repeat	1–10 bp repeat unit	30 kb
CNV Copy number variant	10 bp–1 Mb	3 Mb

Problem Solving

Genotyping by PCR: The key point about the polymerase chain reaction is that the technique amplifies a specific region of the genome (usually less than 10 kb long) located between the 5' ends of two primers. These primers must be oriented so that their 5'-to-3' polarities point toward each other. This means that each primer uses a different DNA strand (Watson or Crick) as a template for DNA synthesis. To detect polymorphisms in the amplified region, you can either sequence the PCR product (for any polymorphism) or determine its size by gel electrophoresis (for Indel or SSR polymorphisms that alter the number of nucleotide pairs in the PCR product).

Genotyping by DNA microarrays: This method is generally used for analysis of SNP polymorphisms. For all the end-of-chapter problems based on this technique, each *column* represents an individual locus (that is, a specific nucleotide pair somewhere in the genome), and each row represents one of the four possible alleles of this SNP (that is, A, C, G, or T as read off of one strand of the chromosome).

The other important aspect of microarray data concerns the intensity (color) of the signal for each of the four alleles in a column. *White* means that the allele in question is not found in the genomic DNA, *orange* means that the genome contains one copy of the relevant allele, and *red* indicates that the genome contains two copies of the relevant allele. For an autosomal gene in a diploid organism, you would expect to see that the total number of copies in each column is 2.

Whole-genome sequencing: Because this technology is evolving so rapidly, you should not concern yourself with understanding the nuts-and-bolts of the various methods. What is essential is to understand the concept of a **read:** a small snippet of sequence obtained ultimately from a single molecule of DNA. Reads are distinguished of course first by their actual DNA sequence. But the next critical piece of information concerns the number of reads of a particular sequence that was obtained in the experiment. For example, if an individual is a heterozygote for two alternative SNP alleles of a locus, then roughly equal numbers of reads corresponding to each allele should be observed. Finally, the diagram shown in Fig. 10.25 on p. 365 is critical; it shows you how to interpret whole-genome sequence information so as to find the polymorphisms responsible for a genetic disease.

Positional cloning: This procedure aims to locate the gene responsible for a genetic disease based on the gene's proximity to anonymous DNA polymorphisms (that are usually followed on microarrays). Positional cloning is in essence nothing more than recombination mapping: if the disease gene and the DNA polymorphism are close together, they will exhibit genetic linkage. Remember that gene mapping requires that at least one parent must be a double heterozygote for the two loci in question (in this case, one parent must be simultaneously heterozygous for the disease gene and for a particular DNA polymorphism). In determining recombination distances, you can only consider progeny from a cross in which the *phase* (that is, the haplotypes in the doubly heterozygous parent) is known or can be inferred with considerable certainty.

Finally, two numerical generalizations will be of value:
  (i) on average, 1 map unit corresponds to 1 Mb of human DNA, and
  (ii) on average, there is ~1 gene/100 kb of human DNA

Once the recombination frequency between the disease gene and polymorphism is determined, these numbers will allow you to approximate the physical distance between them and the number of candidate genes likely to be located in this interval. Because the genomic locations of the polymorphisms used for positional cloning are known, you can consult the human RefSeq database website to identify the candidate genes.

Allelic heterogeneity: In evaluating various candidate genes that might be responsible for a genetic disease (as identified either by positional cloning or by whole-genome sequencing), you need to be aware that many genetic conditions can be caused by a variety of different

chapter 10

mutations within the disease gene; this is *allelic heterogeneity*. If the aberrant phenotype is recessive to wild type, you will see allelic heterogeneity as *trans-heterozygosity:* an affected person can have one mutant allele on one chromosome but a different mutant allele on the homologous chromosome. (See Fig. 10.25b on p. 365.)

Vocabulary

1.

a.	DNA polymorphism	5.	a DNA sequence that occurs in two or more variant forms
b.	phase	3.	arrangement of alleles of two linked gene in a diploid
c.	informative cross	8.	allows identification of a gamete as recombinant or nonrecombinant
d.	ASO	6.	a short oligonucleotide that will hybridize to only one allele at a chosen SNP locus
e.	SNP	2.	two different nucleotides appear at the same position in genomic DNA from different individuals
f.	DNA fingerprinting	7.	detection of genotype at a number of unlinked highly polymorphic loci
g.	SSR	1.	DNA element composed of short tandemly repeated sequences
h.	locus	4.	location on a chromosome
i.	compound heterozygote	10.	individual with two different mutations in the same gene
j.	exome	9.	all exons in a genome

Section 10.1

2. When examining anonymous DNA markers you are looking directly at the DNA sequence of an individual. The terms dominant and recessive are usually used only when discussing the phenotype of an organism, so in one sense this question is meaningless. Also, the majority of these anonymous DNA loci are not located in genes and so have no effect on the phenotype of the organism (Problem 11-4). However, geneticists often say that DNA markers are inherited in a **codominant** fashion to denote that both alleles can be seen in the DNA sequence and that the genotype of a heterozygote depends equally on both alleles. This use of the word "codominant" makes sense if you think of the phenotype being described as the DNA sequence of the marker locus (SNP), or the size of PCR product band on a gel (SSR).

3. SNPs in protein coding regions are more likely to have a deleterious effect on the organism as they are more likely to affect protein function, so they are expected to be relatively rare. Also **most SNPs are found in noncoding DNA** simply because

noncoding DNA in humans comprises a much greater proportion of the genome (98%) than coding DNA.

4.   a. The genome in a gamete is ~3 × 10^9 bp long. This number times 1 × 10^{-8} base substitutions per bp per gamete yields ~30 base substitutions in the genome of a gamete. Because individuals are formed from two gametes (an egg and a sperm), **your own genome should contain ~60 new base substitution mutations that were not found in the genomes of either of your parents.**

   b. Because these mutations were passed on to you, **these *de novo* (new) mutations must have occurred in the germ lines of your parents.** They could have happened during the mitotic divisions that increase the number of germ-line cells (that is, in the spermatogonia or oogonia), or during meiosis (in spermatocytes or oocytes). You should note that every time a cell divides provides an opportunity for mutations to accumulate.

   c. The more rounds of cell division that take place before the production of sperm or eggs, the more mutations will be found in the gamete. Because more rounds of cell division are needed for the production of sperm than eggs, sperm should contain more base substitution mutations than eggs. (Because each gamete is the result of a single meiosis, most of these divisions are in the spermatogonial and oogonial cells.) The average rate of base substitutions at SNP loci mentioned in the problem (1 × 10^{-8} base substitutions per bp per gamete) is actually a composite of very different values for sperm and egg: **The average rate of new base substitutions in sperm is actually considerably higher than that for eggs. Thus, most of the *de novo* mutations that are found in your genome but not in the genomes of either of your parents must have been obtained from the sperm.**

   Recent research (as of 2014) suggests that this conclusion may be behind findings that the rate of autism and behavioral problems such as attention deficit disorder (ADD) is higher among the children of older fathers than younger fathers. The older the father, the more rounds of cell division took place before sperm are produced. (In testes, spermatogonial cells are continually dividing throughout adulthood.) Thus, the older the father, the more new mutations will appear in the sperm. (A good arguments for younger males to freeze and store their sperm!) In contrast, human females are born with a full complement of oocytes (or nearly so), so maternal age is not a factor in the number of base substitutions that accumulate in eggs. (Of course, older women are subject to increased rates of chromosome nondisjunction leading to Down syndrome among their children.)

5. **The regions in which Watson and Venter share the same SNPs imply that these two men must have shared a recent common ancestor (within several generations) from whom they derived these SNPs. The interspersion of such regions of shared SNPs with regions in which the two men do not share SNPs is the result of recombination.** That is, the one particular chromosome from their recent common ancestor began to recombine with other homologous chromosomes that were different in the separate lineages leading to Watson or to Venter. These homologous chromosomes had different SNPs. Thus, the present-day chromosomes in Watson and Venter are pastiches of parts obtained from the common recent ancestor and parts

obtained from other ancestors who were not common to the two men; recombination over generations has shuffled and reshuffled these different parts.

You should understand that any comparison of the genomes of any two people alive today will also show patterns in which similar and nonsimilar SNPs will be interspersed. The more closely related the people, the longer will be the regions of SNP similarity. For people more distantly separated, like people from Africa versus those from Asia or Europe, the similarities will be harder to find, but they will still be present because humanity as a species is quite young.

6. a. 3 billion − 50 million = **2.95 billion.** The main point here is that the majority of nucleotides in the human genome are the same in all people.

   b. **The studies to date with thousands of human genomes will have already found most of the SNPs that are common in human populations (~50 million). But there are many more rare SNPs that will be found in just one or a very few people (for example, very recently occurring mutations). To find all the rare mutations in all the human genomes on earth, you would have to sequence the genomes of all present-day humans.**

   It is interesting to consider why the human population has so many rare SNPs and not so many common SNPs. Remember that the accumulation of mutations depends on cell division, so the each new individual can have dozens or more mutations not found in the genomes of his or her parents (see Problem 4). The total population of humans on the earth was historically very small, but starting with the Industrial Revolution in ~1750, there was a tremendous population explosion. Many new mutations would have been introduced into the species during this short time period, and these mutations would be rare because only a few people would share them (the person who first inherited a rare mutation would not have been able to engender many descendants given the few generations involved). In contrast, the mutations giving rise to common SNPs must have occurred many more generations in the past, but there are relatively few of such mutations because the population prior to 1750 was relatively so small.

   c. The chance that any one bp is bi-allelic in the thousands of humans examined to date is $5 \times 10^7 / 3 \times 10^9 = 1.67 \times 10^{-2}$. The chance that another mutation happened at this same base pair would be $(1.67 \times 10^{-2})^2 = 2.79 \times 10^{-4}$. This number times $3 \times 10^9 = 8.37 \times 10^5$, or about **840,000 positions at which 3 different alleles might be found at any one nucleotide position in the human genome. For 4 alleles,** this would be $(1.67 \times 10^{-2})^4 = 7.78 \times 10^{-8} \times 3 \times 10^9 = 2.3 \times 10^2 =$ **230 positions.** You can see that for the large majority of SNP loci, the human population has only two alternative alleles.

   The calculations above are only very rough approximations. They assume that all possible mutations that change one nucleotide pair to another are equally likely, but in fact transitions are about twice as likely as transversions because the mechanisms that generate transitions spontaneously are more prevalent. The calculations for 3 or 4 different alleles also do not account for the fact that if 2 or 3 alleles already exist in the population, then only a subset of possible mutations would change the base pair into a new allele. (For example, if A, C, and G alleles were already present in the population, then a pre-existing allele could only change to a T in order to make the

SNP locus tetra-allelic.) These factors would tend to make tri- and tetra-allelic loci even less common in human genomes than the calculations above would indicate.

7. a. **SSRs are polymorphisms in which the alleles differ in the number of tandem repeats of a simple sequence less than about 10 bp long.**

    b. **SSRs are likely generated by a mechanism of DNA polymerase stuttering (slipped mispairing of DNA strands) at repeated sequences during DNA replication.**

    c. **CNVs are repeats of units longer than 10 bp.** DNA polymerase stuttering is unlikely to generate these kinds of changes. Instead, it is thought that **unequal crossing over between repeats is responsible for most changes in the number of CNV repeats.** In contrast with the short repeat units of SSRs, the repeat units of CNVs are sufficiently long that they can mispair in prophase of meiosis I.

    d. **SNPs could potentially occur at any of the 3 billion bp in the human genome. SSRs could occur only at positions that already contain several repeats of a small DNA sequence.** Thus, there are many fewer SSR loci than SNP loci.

8. The figure that follows illustrates the answers to parts (a), (b), and (c). In all the red designations, the letter to the left of the arrow is the ancestral nucleotide and that to the right of the arrow is derived. The sequence in the common ancestor of all three species is at the top. For position #3, you don't know if the ancestral sequence was A or T. You do know that the mutation causing this polymorphism arose after the gorilla lineage split from the chimp/human lineage.

    Note that the data indicates considerable substructure to the human population. The G -> T mutation at position #4 occurred at one moment in time, and then among the descendants inheriting that mutation, a mutation at position #7 occurred.

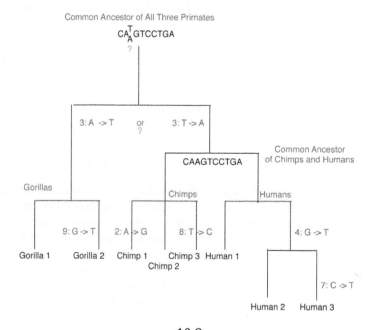

chapter 10

Section 10.2

9. a. The human genome sequence shows the sequence of the normal allele of PKU. **You wish to know whether or not the PKU syndrome in this patient is caused by a mutation in the phenylalanine hydroxylase gene.** You suspect that there might be such a mutation in this particular exon, so you will sequence the PCR product. If there is a mutation in this 1 kb exon, you want to know exactly what it is, how it affects the enzyme, and perhaps something about the history of this mutation in human populations. For example, if you compare the sequence in many patients and track where the patients are from, you might get an idea of where this mutation arose in time and geographical space. If you do not find a mutation in this 1 kb exon that changes the amino acid sequence of the enzyme, there might still be a mutation in a different exon.

    b. One haploid human genome contains 3×10^9 bp. Therefore $(3 \times 10^9$ bp/haploid genome$) \times (6.6 \times 10^2$ g/mole$) \times ($mole$/6.02 \times 10^{23}$ bp$) = 3.3 \times 10^{-12}$ g/haploid genome. In other words, one haploid genome weighs 3.3×10^{-12} g or 3.3 picograms. Each haploid genome will contain only one phenylalanine hydroxylase gene to be used as the template for the PCR reaction. You start the PCR reaction with 1 ng (1×10^{-9} g) of human DNA. Therefore $(1 \times 10^{-9}$ g DNA$) \times (1$ haploid genome$/3.3 \times 10^{-12}$ g$) \times (1$ template molecule/1 haploid genome$) = 0.3 \times 10^3$ template molecules = **300 template molecules in 1 ng of DNA.**

    c. You begin the PCR with 300 template molecules. If the PCR runs for 25 cycles then this number of molecules doubles exponentially 25 times. Therefore you will end up with 300 molecules $\times 2^{25} = 10^{10}$ or about 10 billion molecules. This result explains the power of PCR: you started with only 300 template molecules and end up with 10 billion copies of the region you are amplifying. In practice the yields are not quite as high because not all potential template molecules get amplified each cycle. However the amplification is still substantial. The PCR product is 1 kb long, so $(10^{10}$ molecules of PCR product$) \times (10^3$ bp/molecule of PCR product$) \times ($mole$/6.02 \times 10^{23}$ bp$) \times (6.6 \times 10^2$ g/mole$) = 1.1 \times 10^{-8}$ g = 110 ng. You started with 1 ng of the whole genome and ended up with **110 ng of a 1 kb section of the genome after the PCR!**

10. The two primers should flank the target sequence and should be oriented so that their 5'-to-3' orientations drawn as arrows should point to each other (with the 3' ends towards the center). This reflects the fact that DNA polymerase adds nucleotides sequentially to the 3'-ends of the primers. You want DNA polymerase to extend the primer into the sequence being amplified. Only **set b** satisfies these criteria.

11. a. If human DNA were a random sequence of equal proportions of A, C, G and T (not entirely accurate, but close enough), then **the chance that either primer would anneal to a random region of DNA would be $(1/4)^{18}$, or about 1 chance in 7×10^{10}.** In other words, any particular 18 base sequence would occur at random only once in every 70 billion nucleotides. As the human genome is 3 billion nucleotide pairs long, it is very unlikely that even one of the primers will anneal anywhere else except to the *CFTR* gene. **The probability is even much lower that *both* primers would anneal to other random stretches of DNA that happen to be close enough together to**

allow the formation of a PCR product. (This probability is hard to calculate exactly because of the variation in the possible distance between the two primers.)

b. The lower limit on the size of the primers is governed by several factors. The most important is that the primers must anneal to the genomic DNA. Hydrogen bonding between 15 nucleotide pairs is ordinarily required to allow DNA to be double stranded stably. A second factor is that the primers need to be sufficiently long so that the PCR amplifies specifically only the target DNA. Using the same logic as we did in part (a), the chance that one primer of 16 nucleotides would anneal to random DNA is $(1/4)^{16}$ = one in 4 billion. As there are only 3 billion nucleotides in the human haploid genome, **a 16 bp primer will likely hybridize to only one sequence in the human genome.** A particular 15 bp primer will be present every $(1/4)^{15}$ bases, which is about 1 time every 1 billion bp. That means that **a 15 bp primer sequence is likely to be in the 3-billion bp human genome 3 times.** The chance that two such primers would anneal to random DNA close enough together for PCR amplification is extremely low. However, because a primer that is at least 16 bp primer is likely to hybridize to a unique genomic sequence, most researchers would view 16 bp as the lower limit.

Interestingly (though not asked in the problem), very long PCR primers (more than 40 or so nucleotides) also create some problems. In particular, the longer primers might hybridize with genomic DNA sequences to which they are not perfectly matched. (Internal mismatches are tolerated and hybridization can occur as long as there are enough surrounding base-paired nucleotides, particularly at the 3' end of the primer. Thus, very long primers might amplify regions of the genome other than the target.

c. **You would be more likely to obtain a PCR product if the mismatch were at the 5' end.** The 3' end of the primer is the business end, where DNA polymerase adds additional nucleotides to the chain. But mismatches at the 3' end would prevent DNA polymerase from adding any new nucleotides. (This question is important because it means you can add nucleotides unrelated to the template to the 5' ends of PCR primers. In this way, for example, you could add a restriction enzyme recognition site to the ends of the PCR product to facilitate cloning.)

12. a. If you see a double peak and this represents a mistake, much of the PCR product must have the same error. **This scenario could occur only if the mistake happened early in the PCR amplification and was then copied over many PCR cycles.**

b. **It is more likely that a mistake would happen in a late round of PCR,** because there are more molecules present and thus more chances for mistakes. Fortunately, the later these mistakes happen, the less likely you would be to see them because only a small proportion of the PCR product would have this mistake.

c. Basically, **you would want to determine the person's genotype several times with independent PCR reactions.** These mistakes are so rare that it is almost impossible that the same mistake would occur in multiple independent PCR reactions. You could not do multiple independent PCRs easily in preimplantation genotyping, because you would have to remove multiple single cells from the same ~8 cell embryo.

chapter 10

d. This exonuclease function would prevent the *P. furiosa* DNA polymerase from making as many mistakes in base incorporation as the *T. aquaticus* enzyme, so the PCR reaction would be more accurate and reliable.

13. This problem requires you to remember that people are diploids and therefore have two copies of autosomal loci. They could be homozygotes for any of the three sequences in the table, or they could be heterozygotes for any two of the three sequences. If you were doing Sanger sequencing of PCR products, then for heterozygotes you would see double peaks where the two sequences in a heterozygote are different.

   a. **The three possible answers are thus:**

   CAAGTCCTGA/CAATTCCTGA (1/2) + CAATTCTTGA/CAATTCTTGA (3/3)
   CAAGTCCTGA/CAAGTCCTGA (1/1) + CAATTCCTGA/CAATTCTTGA (2/3)
   CAAGTCCTGA/CAATTCCTGA (1/2) + CAATTCCTGA/CAATTCTTGA (2/3)

   Note that the sequence data from the 1/2 or 2/3 heterozygotes reveal both 10 nt haploid sequences unambiguously because there is only one double peak - only one variable nucleotide. Because there are no double peaks in the sequences from the homozygotes (1/1 or 3/3), the sequence traces for these individuals reveal both haploid sequences (which are of course identical) in their genomes.

   b. If a **person was (1/3),** there are two differences and thus two double peaks. One possible solution is:
   CAAGTCCTGA/CAATTCTTGA (1/3)
   However, the gel trace is also compatible with:
   CAAGTCTTGA/CAATTCCTGA (4/2)
   where 4 is a new allele that is neither 1, 2 or 3. You could not distinguish these two possibilities using this methodology. You could not conclude that a person whose sequence trace is that produced by (1/3) is in fact (1/3), as they could instead be (4/2).
   Note that this issue would not exist if you examined the people's genotypes by single-molecule DNA sequencing (as described in Fig. 10.24 on p. 364) because each sequence would be from a single chromosome, not a mixture of the two homologous chromosomes as is the case with PCR-amplified material.

14. Huntington disease is late-onset dominant lethal disease. The disease is associated with the expansion of a CAG trinucleotide repeat in the *HD* gene. As stated in the text (p. 352), normal individuals have up to 35 copies of the triplet. Individuals with repeat regions containing more than 35 repeats are susceptible to Huntington disease; all disease alleles with 42 or more repeats are completely penetrant. In general, greater numbers of repeats correlate with younger age of onset of the disease (see Fig. 10.13 on p. 352).

   a. **Individuals A, B, C and E** all have one PCR band (one allele) that is much larger (between 270 and 380 nucleotides long) than the second allele (between 200 and 220 nucleotides long). It isn't possible to correlate these band sizes with trinucleotide repeat numbers because we don't know the position (relative to the repeats) of the unique sequences to which the PCR primers anneal. However these much larger bands must have more copies of the repeat than the smaller bands. The repeat region in **individual B** is the longest; so this person is likely to have the earliest onset of the disease.

b. Both PCR bands for **individuals D and F** are smaller, so they appear to have received HD^+ alleles from both of their parents. Thus they should not have the disease.

c. This problem can be solved only if the number of repeats in an allele is known. If we assume that the longest HD^+ allele shown in the figure (the smaller band in individual B) contains the maximum number of repeats for a non-disease-causing allele (35 repeats; see Fig. 10.13 on p. 352), then the 220 bp of this PCR product should include 35 × 3 = 105 bp of CAG repeats, plus 70 bp between the 5' end of one PCR primer to the nearest CAG repeat. Therefore, 220 - 175 = **45 bp** will remain from the end of the CAG repeats to the 5' end of the second PCR primer. Note that the distance between the primers and the beginning and end of the SSR repeats is the same for all alleles.

15. a. In the patient whose graph is shown **on the top**, the CAG repeat number in the HD^+ allele is **approximately 15**; the somatic cells of the patient **on the bottom** has **about 20** CAG repeats in the HD^+ allele.

b. These results say a great deal about the mechanisms that give rise to mutant HD alleles. First, the **repeat number varies among different sperm** from the same individual, so these **processes take place in germ-line cells during spermatogenesis**. Second, it appears that **the larger the original number of repeats in any HD allele, the more likely it is that the number of repeats in the sperm will vary and the greater the degree of potential variation**. In the graph on the top, almost all of the sperm with the mutant HD allele have more than 62 repeats, some sperm have more than 120 repeats and the distribution is quite broad. The distribution of mutant sperm sizes in the graph on the bottom is tighter. In addition, note the fact that there is little if any variation in sperm size of the sperm containing an HD^+ allele. Third, it is interesting that the **sperm mostly seem to accumulate more CAG repeats rather than lose CAG repeats** - few if any sperm have less than the number of CAG repeats in the HD genes in somatic cells.

c. There is some probability that the number of CAG repeats can expand during spermatogenesis, even with lower numbers of CAG repeats. A man with **an HD allele on the high side of normal (~30 CAG repeats) or in the gray area between 36 and 41 repeats** (who may not show any symptoms), **may produce sperm with more than 42 repeats**. Note that these data predict that the unaffected father (but not the unaffected mother) of a Huntington disease patient should have an allele with between 30 and 42 repeats, because this particular repeat expansion occurs during spermatogenesis. Interestingly, in fragile X syndrome, another trinucleotide repeat disease, the expansion usually occurs during meiosis in the mother (see Fig. 7.11 on p. 215).

d. You would expect that **each of the two blood cell samples would show a very tight distribution of CAG repeat numbers around the numbers seen in the normal and mutant HD alleles (15 and 62 for the patient at the top; 20 and 48 for the patient at the bottom)**. There should be very little variation in repeat number

chapter 10

between different blood cells from the same person since the expansion in repeat number seems to be restricted to the germ line.

Section 10.3

16. **The evidence could be used to show that the accused's pickup truck was at a specific location in the Arizona desert where the Palo Verde tree was growing.** Presumably this location would be very close to where the corpse was discovered.

17. a. **In general, SNPs are bi-allelic and have only 2 alleles in human populations (see Problem 6 above). The SSRs used in CODIS have many alleles that vary in the number of repeating units. This means that the chance of any two people sharing the same set of alleles at 14 SNP loci is much higher than the chance they would share the same set of alleles at 14 SSR loci.**

    In Chapter 20, you will read about the *Hardy-Weinberg equilibrium*, which allows you to calculate the probability that any two random people would share the same alleles of any polymorphism, as long as you know the frequency of that allele among all of the alleles in the population being studied. Of course, the lower the frequencies at which the alleles in the test subject are found in the population, the lower the probability that a random individual from the population will have exactly the same alleles. Because the CODIS loci are unlinked to each other (and therefore are inherited independently), you can multiply the low probabilities for a match with the alleles of each locus together, resulting in extremely low probabilities of a match with the alleles of all 14 CODIS loci. This is of course the basis of DNA fingerprinting.

    b. **Dogs and other domesticated animals have been highly inbred.** This means that any two dogs would be much more likely to share the same alleles of any polymorphism than any two humans; moreover, the dogs would be much more likely to be homozygous at any given location. **You would thus have to look through many more DNA sequences to find one that would vary between any two dogs.**

18. a. **C-Tsar; E-Tsarina; ABDH-Daughters, G – Son (Tsarevitch), F- unrelated.** There is no completely systematic way to do analyze these data, so eventually you will need to sort through the most likely alternatives and see which ones fit. The easiest place to start is with the males (C and G), one of whom must be Tsar Nicolas II and the other of whom must be Tsarevitch Alexei. Because G has two alleles for D3S1358, neither of which is found in A and B (who must be sisters because they share all alleles but one of all loci), G does not pass on any alleles to these daughters and so G must be the Tsarevitch rather than the Tsar. Once this determination is made, the rest of the analysis becomes straightforward. An alternative way to begin is to search for the unrelated individual, who must be F because she has at least one allele of each locus that is found in none of the other people.

    b. B (12,12); C (11,12); E (9,11)

    c. **No, none of the daughters has identical alleles at all loci.**

    d. **The sizes of the skeletons might reflect the ages of the daughters.** Perhaps dental records exist might match the teeth of one or more of the skeletons.

e. **All of the daughters are accounted for, so Anastasia must have been one of the skeletons.**

f. **1/8 = 12.5%.** (The Tsarina and Philip's grandmother should share 1/2 of their alleles because they are sisters. There is a 1/2 chance for Prince Philip's mother to inherit a common allele from Philip's grandmother, and a 1/2 chance for Philip to inherit a common allele from his mother. $1/2 \times 1/2 \times 1/2 = 1/8$.)

g. **1/16 = 6.25%.** Another factor of 1/2 for the Tsarevitch to inherit a common allele from the Tsarina.

h. **The Tsarina was Prince Phillip's great aunt.**

19. a. **In this problem, you are not looking at individual sperm, but instead at semen,** which contains millions of different sperm. If the man is heterozygous for two alleles of an SSR locus, about half of his sperm would have one allele and the remaining half the other allele.

b. **The purple locus is on the X chromosome.** Note that there is only one copy in the semen and one copy in individuals 1 and 4. These two people must be males. Individuals 2 and 3 have two copies of the purple locus, and thus 2 copies of the X chromosome; these two people are females.

c. **The green locus is on the Y chromosome.** There is one copy in the semen and in the somatic genomes of males. Individuals 2 and 3 don't have any green bands because they don't have a Y chromosome.

d. **If any individual was the rapist, all the bands of all the colors should match the semen sample.**

e. Note that individual 2 matches the rapist at 5 alleles out of 10 (one **black**, two red, and two orange). **Individual 2 is likely to be the rapist's sister.**

f. The bands are 200 and 212 bp long. Each primer for the PCR is 20 bp long; PCR requires two primers. The repeating unit is 4 bp. **Thus, the 200 bp allele would have (200-40)= 160/4 = 40 repeats. The 212 allele would have 3 more repeats, or 43.**

20. a. **The microarray hybridization technique is not very sensitive,** so you need to generate enough fluorescently labeled probe to see any signals.

b. This question is basically asking you to **redraw Fig. 10.11a on p. 350;** this figure is reproduced below

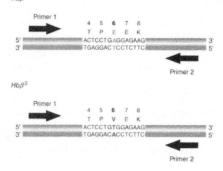

chapter 10

c. At 80°C, the hydrogen bonding between the two strands of DNA (one strand on the silicon chip and the other being the fluorescent probe) is disrupted by the heat, so the probe cannot bind to the DNA on the chip.

d. The incubation temperature affects the accuracy of annealing between complementary sequences. **If the temperature is sufficiently low, then one mismatch will not affect the ability of the two strands of DNA to come together and stay together.**

e. Embryos 2 and 3 are homozygous for the *A* allele (*AA*); embryos 1, 5, and 7 are *AS* heterozygotes, and individuals 4 and 6 are *SS* homozygotes. **To avoid the possibility that the child would have sickle cell anemia, you would choose the *AA* homozygotes.** If only a few embryos were available, you might also implant the *AS* heterozygotes (which are usually phenotypically normal except under exceptional circumstances like long exposure to low amounts of oxygen). You would not choose the *SS* homozygotes, as they would have sickle cell anemia if brought to term.

   In the early days of *in vitro* fertilization, physicians implanted several embryos into the uterus of the expectant mother. As the negative effects of multiple births became increasingly apparent, the number of implanted embryos has been reduced. As of this writing (2014), no more than two or three embryos are chosen in most cases of preimplantation genotyping.

21. The two possible ASOs for the *Hbβ^A* allele are:

    5' CTGACTCCTGAGGAGAAGTCG 3'   or   3' GACTGAGGACTCCTCTTCAGC 5'

    The two possible ASOs for the *Hbβ^S* allele are:

    5' CTGACTCCTGTGGAGAAGTCG 3'   or   3' GACTGAGGACACCTCTTCAGC 5'

22. a. In Fig. 10.17b on p. 356, **you would only need a single primer because the same adapter sequence was already added to both ends of the fragment prior to the PCR.** The primer you would use for the PCR amplification step shown in the figure would correspond to this sequence as it is shown.

    b. **You would want to cut the DNA with a restriction enzyme that made sticky ends; this would make it very easy to add the adapter at the two ends.** (The adapter would be made with one end being blunt and the other having the sticky end, as shown in the figure.) It would not matter if the sticky end had a 5'-overhang or a 3'-overhang. **You would generally want to cut the DNA with an enzyme that recognized only a 4 bp recognition site.** The reason here is mostly technical. PCR can only amplify regions of DNA less than 10 kb or so. So if you used an enzyme that recognized 6 bp or 8 bp, much or even most of the genomic DNA fragments would be too large for PCR amplification.

23. a. Dad: 1-AC 2-GG 3-AG 4-GT 5-CC 6-CG 7-AG 8-GT

    Mom: 1-AA 2-CG 3-GG 4-AT 5-CC 6-CC 7-AG 8-TT

    b. **SNP locus 4 has alleles A, G, and T.**

    c. **You know the whole sequence of the human genome.** You designed the ASOs on the microarray with the benefit of this knowledge.

d. If the 8 loci are each about 10 Mb apart, **SNP1 and SNP8 would be ~70 Mb apart. 70/191 = 36% of chromosome 4.** The answer depends on the average value of 1 cM corresponding to 1 Mb in the human genome that was cited on p. 358 of the text (in the first paragraph on this page). This value masks a great deal of variation in the rate of recombination for a given length of DNA in different parts of the genome; the rate for any given region also varies between males and females.

Section 10.4

24. a. **The disease is dominant:** the condition is rare and you see a vertical pattern of inheritance (the affected children have an affected father). **The disease is also autosomal;** it cannot be X-linked because there is father-to-son transmission. Note also that this cannot be X-linked for another reason: The males and females have two copies of the SNPs (assuming that these SNPs are syntenic with the disease gene).

    b. **Affected son: 1-C   2-G   3-A   4-G   5-C   6-C   7-A   8-T**
    Affected daughter: 1-A   2-G   3-A   4-G   5-C   6-C   7-A   8-G
    Unaffected son: 1-C   2-G   3-A   4-T   5-C   6-G   7-G   8-T
    Unaffected daughter: 1-A   2-G   3-G   4-T   5-C   6-G   7-G   8-T

    c. **SNP2 and SNP5 are uninformative for linkage to the disease gene** because the father is a homozygote for both SNPs. (Remember you can only follow linkage in a double heterozygote.)

    d. **SNP1 is unlinked** because half the affected and half the unaffected have the A allele of SNP1, while the other half of the affected and the other half of the unaffected have the C allele. R = NR, so RF = 50%.

    e and f. **See the figure that follows.** *Dark* blue and *light blue* indicate the two copies of chromosome 4 in the father. The chromosome has the dominant disease-causing allele *D*. The *light blue* chromosome has the normal recessive *d* allele. A recombination event in the father's germ line that gave rise to the affected son occurred somewhere between SNP7 and SNP8. The recombination event in the father giving rise to the unaffected son must be somewhere between SNP3 and SNP4. The disease gene must be between these two crossovers. The disease gene thus must be between SNP3 and SNP8; it must be to the left of SNP 4 and to the right of SNP7. The boxes that are crossed by the Xs signifying crossovers show the regions of uncertainty. The crossovers must occur within the boxes, but we don't know exactly where.

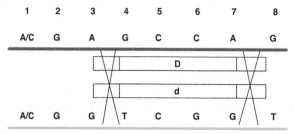

chapter 10

25. a. **Yes, there is evidence of linkage.** There are 2 informative matings I-1 × I-2 and II-7 × II-8 which together yielded 11 progeny, only 1 of which (II-7) is recombinant. **The estimated genetic distance is thus 1/11 × 100 = 9.1 map units.** Note that I-1 produces parental types that are C + dominant disease allele and T + recessive (wild-type) allele. II-7 received a recombinant sperm that is T + dominant disease allele. This combination defines the parental types among the gametes that II-7 produces. Note also that the mating of II-1 × II-2 is uninformative, because neither parent is a double heterozygote.

b. The LOD score is based on the following calculations, following the pattern shown in the Tools of Genetics Box on p. 361:

Likelihood ratio: $[(1 - (1/11))^{10} \times (1/11)^1]/(1/2)^{11} = 71.7$

LOD = $\log_{10}(71.7) = 1.86$

This value means that it is 71.7 times more likely that the disease gene is linked to the SNP than that they are unlinked. This is pretty good evidence for linkage, but it would still not be accepted as definitive because the LOD score is below 3.

26. a. **Matings W, Y and Z are not informative** because neither parent is a double heterozygote. **Mating X is informative** because both parents are double heterozygotes <u>and</u> because all possible permutations of the gametes can be distinguished as parental or recombinant types.

b. **It is likely that both A and B are SSRs rather than SNPs.** Most SNPs are bi-allelic in human populations (see Problem 6c above), but the data show 4 different alleles of A and 4 alleles of B. One A allele is labeled "A5," implying that there might be 5 alleles or more in human populations, so SNP A would be definitely excluded if in fact an A4 allele existed in some person not shown.

27. Note that the disease is recessive. Thus, it for a mating to be informative, it must be the unaffected parent who could potentially be the double heterozygote.

a and b.

Locus 1 - No, neither parent is a double heterozygote.

Locus 2 - No, neither parent is a double heterozygote.

Locus 3 - Yes, but only in some cases. The mother would have to be a heterozygote for the disease gene. If the children were AC, you could not tell which allele they got from which parent, so these cases would be uninformative. Children who are AA or CC will be potentially informative whether they have the disease or not, because you would know what disease allele and what SNP allele they inherited from their mom. For part (b): In terms of phase, such children are almost certainly inheriting the parental combination from their mother if the SNP is close to the disease gene (because it would be unlikely that these children would inherit a rare recombinant combination from both parents).

Locus 4 - Yes, if the mother is a heterozygote for the disease gene (and thus a double heterozygote). All of the children will be potentially informative because you can tell what alleles of the disease gene and the SNP locus they got from their mom. Part (b): As for Locus 3, you could solve the phase problem if you knew that the SNP and disease gene were in fact closely linked. Because this is a consanguineous mating, the carrier mother would be most likely to have one chromosome with both the disease allele and the C allele of the SNP, as was true for both of the chromosomes in her affected partner.

28. a. It is clear that **allele C is common to almost all of the individuals who have Huntington disease, so this allele is highly likely to be part of the same haplotype** (that is, on the same chromosome) as the disease-causing allele.

   b. Surprisingly, if you make no assumptions at all, then **you can't say for certain that any people in the pedigree were formed from parental or recombinant gametes** with respect to the *HD* gene and the G8 marker. The reason is that there are no cases in which the children, parents, and grandparents are all genotyped. You thus don't know the phase without making the assumption of linkage.

   c. Looking first at the affected people in the pedigree who are genotyped at G8, all of them have the C allele. The simplest assumption is thus that they all received the parental configuration *HD* C that was present in I-1. Of the unaffected blood descendants of I-1 who could have inherited the *HD* allele from their affected parent, none have the C allele of H8 except for VI-5. It is possible that this woman could have received C from her unaffected father (in other words, V-3 is definitely AC and V-4 could have been AC as well). Thus, the most straightforward reading of this pedigree is that no one in this pedigree must have received a recombinant gamete from their affected parent (in terms of the *HD* gene and the G8 marker).

   d. With no recombinants, the map distance would be 0 map units. But stated more accurately, you could only say that the map distance would be less than $1/47 \times 100 = 2.1$ cM. (*Note:* there are actually 47 people resulting from informative matings, not 46 as stated in the problem.)

   e. Calculating the LOD score by the method outlined in the Tools of Genetics Box on p. 361 and the existence of 47 (rather than 46) progeny of informative matings:

   Likelihood ratio = $[(1-0)^{47} \times 0^0] / (1/2)^{47} = 1.4 \times 10^{15}$

   LOD = 15.14

   It is 15 orders of magnitude more likely that *HD* and G8 are linked than that they are unlinked. This would be taken as absolutely conclusive evidence for linkage. (If you used the value of 46 informative matings given in the problem, then the likelihood ratio would be 7×10^{14} and the LOD score would be 13.85; still convincing evidence for linkage.)

29. a. **It is possible that the SNP is tightly linked to the disease-causing mutation, but is not the mutation.** You have simply not examined enough children to see any rare recombinants.

   b. You could: **(1) Determine whether or not the SNP locus is rare by asking whether it is found in databases of known SNPs. (2) Determine if the SNP is within a gene, and if so, is it likely to affect gene function (is it nonanonymous)?** Possibly the base pair substitution results in a missense mutation affecting a conserved amino acid or a nonsense mutation. Perhaps the substitution disrupts a splicing junction. **(3) Check other unrelated families, some with the disease and some without.**

Perhaps the same G base is always correlated with the disease. Maybe you would find that in all affected families, the gene in which this substitution occurs consistently has a nonanonymous SNP, even if the polymorphism in the family is not the T/G change noted in the problem. The same gene could harbor different mutations in different affected families in cases of allelic heterogeneity.

30. a. *D* is the dominant, disease-causing allele; *d* is the normal allele; *m1*, *m2* and *m3* are the SSR alleles. In each case, you first need to determine the phase of the disease gene alleles and the SSR alleles; in other words, you need to determine which disease gene alleles (*D* or *d*) and which SSR alleles (*m1* or *m2*) are on the same chromosome as one another in the male parent. Second, given the RF of 10% between the two loci, and the knowledge of the SSR allele that the child inherited from the male parent (the *condition*), you need to use conditional probability to determine the chance that the child also inherited *D*.

    In the pedigree at the left, the male parent's chromosomes are *D m2* / *d m1*. He produces the following gametes at the frequencies shown in parentheses: *D m2* (.45); *d m1* (.45); *D m1* (.05); and *d m2* (.05). Child A inherited *m1* from her father; this means that she inherited either *d m1* or *D m1*. The 45:5 ratio (or 9:1 ratio) of *d m1* : *D m1* gametes means that **the probability is 10% that Child A inherited allele *D* (diseased)**. Child B inherited *m2* from her father, meaning that she inherited either *D m2* or *d m2*. The 9:1 ratio of *D m2* : *d m2* gametes means that **the chance is 90% that Child B is diseased (inherited allele *D*)**.

    In the middle pedigree, Child C inherited the *m2* allele from her father. The probability that Child C is affected is not influenced by the fact that the first child was affected, so the calculation is the same as for child B; that is, **the probability that Child C has the disease is also 90%**.

    In the pedigree at the right, the phase of the alleles in the male parent is not known. However, we can use the information that the sibling of Child D is affected to determine the likelihood of one phase or the other. The brother of Child D inherited *D m1* from his father. This means that the chance is 90% that the genotype of the father is *D m1* / *d m2*, and the chance is 10% that his genotype is *d m1* / *D m2*. Child D inherited *m2* from his father. The probability that Child D inherited *m2 D* from his father is the weighted sum of the chances of inheriting this chromosome given each of the father's possible genotypes above. If the father's genotype is *D m1*/ *d m2*, then using the same logic as we did above for Children A-C, there are 9 *d m2* gametes produced for every 1 *D m2* gamete made, and so the relative chance of *D m2* is 10%. If the father's genotype is *d m1*/ *D m2*, the relative chance of *D m2* is 90%. Therefore, **the chance that Child D has the disease (inherited *D m2*) is: (0.9)(0.1) + (0.1)(0.9) = 0.09 + 0.09 = 0.18, or 18%**.

    b. A human geneticist would not want to use an SSR marker 10 map units (cM) away to diagnose a gene because 10% of the time, the presence in a fetus of the SSR allele linked to the disease allele would lead to an incorrect diagnosis that the fetus is diseased when it was not. 10% of the time when the other SSR allele (usually on the same chromosome as the recessive, non-disease allele) is seen in the fetus, the fetus would actually be affected. Thus, **10% of the negatives would be false negatives, and 10% of the positives false positives**.

31. a. The affected first child shows which marker allele is linked to the disease allele in each parent. The affected child got Dad's large allele and Mom's small allele. Note that the marker is in the *CFTR* gene, but is not the causative mutation. The fetus has Dad's small allele and Mom's small allele. **The fetus is thus a carrier, but the chance the fetus will be affected is very close to 0%.** It is possible that recombination could have occurred between the SSR and the disease mutation in the Mom, but this possibility is very low because the marker and SSR are so close together (both in the same gene). So the likelihood is greater than 0%, but not much greater.

b. The fetus will grow into a woman who is a carrier. Because this is a rare recessive trait, the chances are very low that any of her children would have cystic fibrosis, because they would not only have to get a mutant allele of *CFTR* from this woman (probability = 0.5), but they would also have to get a mutant allele of the *CFTR* gene from their father. The problem states that 3% of the population carries a mutant *CFTR* gene, and we furthermore assume that this man does not himself have cystic fibrosis. Thus, the mating that could produce a child with cystic fibrosis is *CF/CF+* × *CF/CF+*. We already know that the woman is a carrier (probability = 1), and the probability that any given man is *CF/CF+* is 0.03. The chance is 1/4 that any individual child produced from this mating would be *CF/CF* and have the disease. Thus, **the probability that a child of this woman (and a man chosen at random from the population) would have cystic fibrosis is 1 × 0.25 × 0.03 = 0.0075 = 0.75%.**

c. In contrast with the situation described in Problem 30 above, the chance of false positives or false negatives is very low in the scenario described here, so such a test would be quite accurate (though not 100% so). **Because so many different mutations in *CFTR* could cause cystic fibrosis, doctors would not know what mutations to test for, whereas they would be able to follow the SSR marker and make accurate predictions about the fetus's disease status with high (but not perfect) confidence.** You should note that in the future, when whole genome sequencing becomes more routine, you could potentially find all disease-causing mutations in the fetus's *CFTR* gene, so this indirect method would not be needed.

32. a. **The drug should be effective in the G551D heterozygote, although likely less so than in a homozygote as there will be half as much protein in the heterozygote that can be made functional by the drug.**

b. **Younger patients have not built up as much mucus in their lungs as older ones, and it may be easier for the drug to get to the CFTR in the membrane. Furthermore, older patients likely already have some irreversible damage; the drug is not really designed to repair damaged tissue.**

c. **The researchers put compounds into the growth media of the cells homozygous for the G551D mutant and then looked under the microscope for cells that have beating cilia.** This was in fact the approach that led to the development of ivacaftor. You should note that this approach cannot work in the case of patients who have no CFTR protein function (because they are homozygotes or *trans*-heterozygotes for null alleles). This technique depends upon the presence in the membrane of some mutant protein that could be made functional after binding of the drug. If no such protein is present, then this approach would not allow the formation of functional chloride ion channels.

33. a. The poly-A is added for two reasons: (i) To bind the single-stranded fragments of genomic DNA to the surface of the flowcell through the oligo-dT already present on the surface. (ii) Because the oligo-dT can act as a primer for DNA synthesis.

  b. You need to remove the fluorescent tag at the end of each cycle so that you can detect reliably the color of the fluorescent tag of the nucleotide incorporated at the next cycle. Remember that this method ultimately will produce a movie for each template fragment showing successive flashes of color that indicate the sequence of the newly synthesized strand. You could not obtain such a movie if the fluorescent tag added each round remained at the same location.

  c. The blocking group ensures that only one nucleotide will be added each cycle. If you did not remove the blocking group, then you could not add any more nucleotides to the growing chain so you could not sequence anything.

34. a. The researcher should look for possible cases of *trans*-heterozygosity; that is, look at genes to find whether both alleles are mutant but in different ways. In theory, the researcher would want to look at any such genes identified to see if the two mutations are likely to result in a loss of function, such as nonsense or frameshift mutations; and also to see if the two mutations affect highly conserved amino acids (if they are missense).

  b. Unfortunately, it is hard to know what to look for in the whole genome outside of the coding regions. Maybe the patient has a mutation in an intergenic region or in an intron that affects the expression of a target gene, but how would the scientist be able to recognize it among the many polymorphisms that exist in the patient's whole genome? One possible approach is to focus the search on particularly well-conserved nucleotides, or those in the binding motifs known for certain transcription factors.

  Problem 35 below discusses how scientists might be able to determine whether a genetic condition results in quantitative changes in the expression of a particular gene in the tissues from a patient.

35. a. Using this kind of "deep sequencing" of mRNAs, researchers could attempt to find mRNAs whose levels change dramatically in the patient with respect to unaffected controls. This approach might be very useful if no mutations are found in the coding regions of genes, because mutations might affect the amount of expression of a gene. **Scientists might also be able to detect changes in splicing patterns of a gene's transcript.** But there are issues with this technique. How would you know which tissue to examine? If you don't choose the right one, then you might not find any changes in gene expression levels. (Sometimes the disease phenotype would suggest what tissue to look in.) The information from this approach would not be informative unless dramatic changes in the amounts of a specific mRNA were observed.

  b. If major changes were observed in the size or amount of a protein in a patient, then the gene encoding that protein could be responsible for the disease. You might subsequently be able to detect changes in regulatory regions that cause aberrant expression of the gene, or nonsense or frameshift mutations in the coding region. You would not obtain any information from this method if the disease was caused by a missense mutation in the coding region because neither the amount or gross size of the protein would have changed. Note that one problem with this technique is that you can only look at one protein at a time, and you would need a specific antibody to examine the protein. So you would only use this technique if you regarded a particular gene as a very likely suspect for the disease gene.

36. a. **Miller syndrome could be due to a dominant mutation if: (i) The trait was not completely penetrant, so that both of the parents had the mutant allele but did not express the trait; or (ii) There was a *de novo* mutation in the germ line of one of the parents** that occurred in oogonia or spermatogonia, so that many oocytes (and then eggs) or spermatocytes (and then sperm) might have the same mutation. The mutation could not have happened late in germ line development (that is, during meiosis in oocytes or spermatocytes) because more than one child was affected. Part (b) explores the *de novo* mutation scenario in more detail.

b. ***De novo* mutations in the germ line do in fact occur. For a large proportion of the gametes produced by a parent to carry the new mutation, the mutation would have had to occur relatively early in the mitotic proliferation of oogonia or spermatogonia. The earlier in the proliferation of the germ line, the fewer cells would have been present and thus the fewer chances for mutation to occur.** This is why new mutations that occur in the germ line and that cause a large percentage of gametes to be mutant are extremely rare. Human geneticists always need to be aware of such possibilities, even if they might not be very likely.

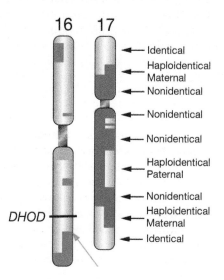

c. Any place there is a color shift from a dark shade to a light shade or *vice versa*, a recombination event must have occurred during the meiosis in one of the parents that produced the gamete making one of the offspring. **The *green arrow* added to the preceding figure shows the recombination event that occurred closest to the**

chapter 10

DHOD gene. This recombination event is seen because *DHOD* is in an Identical region (where the affected son and affected daughter share the same alleles from the mother and the same alleles from the father) to a Haploidentical Maternal region in which they share the same alleles as their mother but have different alleles from their father. Thus, **at the site of the *green arrow*, a recombination event must have occurred during meiosis I in a primary spermatocyte found in the father's testis.** We do not know from the way the data is displayed whether this recombination event happened during the meiosis leading to the sperm that made the son or the daughter.

d. **The data reveal about 2× more crossovers than expected, but the actual numbers are still surprisingly close to expectations.** The figure shows 8 crossovers on chromosome 16 and 10 crossovers on chromosome 17 among the 4 gametes that were used to make the two affected children. The two chromosomes are together about 170 Mb, and thus 170 centiMorgans long. 1 cM means one crossover in 100 gametes, so you would expect about 170 crossovers in 100 gametes on chromosomes 16 and 17 together. In 4 gametes, this would be about 7 expected crossovers, but the data show slightly more than twice this number (18).

e. **You would simply compare the nucleotides in the parents' whole genomes with those in the children's, and look for *de novo* alterations.** Then you divide the number of alterations in the two children by 4× the haploid genome size (because there are 4 haploid genomes in the two children).

Interestingly, note that in many cases (depending on the actual arrangement of alleles), you can actually tell in which parent a particular mutation occurred. The data from this study revealed that the two children in this family each received on average 30 new mutations that occurred in the mother's germ line, and about 50 that occurred in the father's germ line. You would expect that the number of paternal *de novo* mutations would increase in the offspring of older fathers but not mothers, because the male germ line is always being propagated mitotically, and the more divisions, the more chance for mutations to occur (review Problem 4 above).

37. a and b. **The conclusions are summarized in the following Table.** All autosomal loci should have equal numbers of reads, 90% of which should be from the mother's genome and 10% from the fetus's genome. If the mother and the fetus were homozygotes for the same allele, there should be only one type of read representing that allele. If there are two or three types of reads (two or three alleles), the relative proportions depend on whether the alleles are shared by the mother and fetus. For a locus on the X chromosome, you would expect either the same situation as for autosomal loci (the same number of reads; 90% maternal, 10% fetal) if the fetus was an XX female, but there would be 5% fewer reads (of which 95% would be maternal and 5% fetal) if the fetus was an XY male with only 1 X chromosome. For a locus on the Y chromosome, the number of reads would only be 5% of the number of reads for an autosomal locus. (Obviously, these data have been much simplified relative to the actual results researchers would obtain from such a test.)

chapter 10

Locus	Sequences	# of Reads	Mother's Genome		Fetus's Genome		Autosomal X-linked Y-linked
1	α: TCTTTGGTAAACGCAAG	1000	α	α	α	α	autosomal
2	α: GTACCGGAGGCAGCCTC β: GTACCGGCGGCAGCCTC	500 500	α	β	α	β	autosomal
3	α: AGCCATTGCGGATCCGA β: AGCTATTGCGGTTCCGA	950 50	α	α	α	β	autosomal
4	α: GGGGCCTTATGATAAGG	50	–	–	α	–	Y-linked
5	α: CAGTTCCTGGAGTTGTA β: CAGTTCATGGAGTTGTA	550 450	α	β	α	α	autosomal
6	α: GCAGCCCGTGCTGTTAA β: GCAGCCCGTGCTGTCAA	500 450	α	β	α	–	X-linked
7	α: CACTCAGTCCTACGGAC β: CACTCGGTCCTACGGAC γ: CACTCAGTCCTAAGGAC	500 450 50	α	β	α	γ	autosomal

c. The fetus is clearly **male** as one of the loci (4) is Y-linked.

d. The 8th locus with 1500 reads suggest that **there are additional copies of this region**. Most likely, the 8th locus is also autosomal, and both the mother and the fetus have 3 copies of the locus instead of the normal 2. One possibility is that the locus is on chromosome 21, and both the mother and the fetus are trisomic for this chromosome (that is, both have Down syndrome). Alternatively, the locus could be on any autosome and both the mother and the fetus could be heterozygotes for the duplication.

38. a. **The researchers would not have looked only for homozygous variants. Insofar as we know, Nic's parents are not close relatives, so it is entirely possible that Nic is a compound heterozygote** (that is, Nic would have inherited two different mutations in the same gene, one from each unaffected parent).

b. (i) **The mutation causing Nic's disease could have been a *de novo* dominant mutation** that occurred in the germ line of one of his parents (so they would not be affected). (ii) **It is also possible that the condition is dominant but the penetrance is low,** so one of Nic's parents could have had the mutation but not have been affected. Fortunately for Nic, the doctor's original presumptions were correct. It would have been much harder to find the disease mutation if these other hypotheses had to have been considered.

c. You could accomplish this goal very simply using PCR. You would design primers that would amplify the region of the genome containing the causative mutation. You would then **PCR-amplify this region from the genomes of Nic's two parents** (say from a swab from their cheeks). If this mutation was not found in either genome, it must have occurred *de novo*. [This approach would also provide evidence that the causative mutation is X-linked and the condition is recessive if you see heterozygosity for the mutation in the mother's genome.]

chapter 10

39. a. If this is the first exon, it must contain the initiating ATG codon. Thus, the sequence of the protein product that can be deduced from the nucleotide sequence of the first exon is: **N Met-Cys-Gly-Ala.... C.**

b. **Individual 2 is a homozygote** (because only one kind of sequence was found), for the allele 5' CAACGCTTAGGATGTGAGGAGCCT 3'

c. **The disease is caused by a dominant allele,** because heterozygotes with one normal allele (found in RefSeq) and one abnormal allele have the condition.

d. **Yes, the data show the existence of allelic heterogeneity.** Among these four patients, different mutant alleles can cause the disease.

e. **None of these individuals is a compound heterozygote** (also called a *trans*-heterozygote). (A compound heterozygote would have two different mutant alleles, but none of these individuals do.)

f. **The data provide no evidence for locus heterogeneity in Brugada syndrome.** You can explain the disease in all of the affected individuals by the existence of mutations in a single gene. Locus heterogeneity would mean that mutations in different genes could cause the syndrome. The data do not exclude this possibility (because other patients could exist who have the disease because of mutations in some other gene), but this hypothesis is not required to explain the data.

g. **Individual 4 should be normal** (that is, not affected by Brugada syndrome). One allele is the RefSeq allele. The other allele is a silent mutation (changing a TGC codon to a TGT codon, both for Cys).

h and i. **Individual 1's mutation in the bottom allele changes GGC (Ala) to GAC (Asp). This is a missense mutation. This would be classified as a SNP** because only one nucleotide pair is changed. **The mutation causes a loss-of-function,** but you cannot tell just from this mutation alone (see below). **If this is a loss-of-function mutation with dominant effects, the reason must be haploinsufficiency. This is probably a hypomorphic allele with less activity of the protein product** because it is loss-of-function but some protein is likely produced (since all the amino acids are present except one that is changed). **You could perhaps classify this as a null mutation instead of hypomorphic** if the missense mutation resulted in a protein with no function.

Individual 2 is homozygous for a nonsense mutation in the second codon. Again, **this is a SNP** because only one base pair is affected. **This must be a null, loss-of-function allele because no protein is produced.** The fact that individual 2 survives means that the gene cannot be essential. It would be of interest for a scientist to look hard at this person's electrocardiogram. It might be possible that heart function in this individual is affected more than in other patients with Brugada syndrome because he/she is a homozygote without any gene function.

Individual 3 is a heterozygote for a frameshift mutation in which one nucleotide is lost from the reading frame. This mutation would be classified as a DIP, not a SNP. This would be a null mutation because the allele could not produce any normal protein. **The gene is haploinsufficient**, explaining why the heterozygote would have the condition. (This is why the mutation in individual 1 is likely to be loss-of-function as well; heterozygosity for at least one loss-of-function mutation produces the disease.)

In individual 4, the allele that is not the RefSeq allele has a silent mutation (an anonymous SNP) as discussed in part (g) above.

j. **Yes,** the function of the gene is haploinsufficient. You know this to be the case because Individuals 2 and 3, who are heterozygotes for null alleles of the gene (in each person's genome, one allele is wild type, and one allele is null), have aberrant Brugada syndrome symptoms.

chapter 11

The Eukaryotic Chromosome

Synopsis

The DNA in the nucleus of eukaryotic cells is wound up with proteins to form a complex called *chromatin*. Each chromosome is thus a piece of chromatin that includes a single molecule of linear, double-stranded DNA and all the proteins with which this molecule of DNA associates. The organization of the genome into chromatin and chromosomes affects gene expression, the segregation of genes into daughter cells during cell division and gamete formation, DNA replication, and even the lifespan of a cell.

Chromatin proteins compact the DNA, starting with the fundamental unit of chromatin structure called the *nucleosome*, which is made up of a stretch of DNA wound around proteins called *histones*. The presence of nucleosomes at a promoter prevents transcription, and posttranslational modifications of histone proteins can loosen or tighten their association with DNA. Therefore, histones are targets of transcriptional regulatory proteins that bring histone modifying enzymes to genes in order to activate or repress transcription.

The structure of chromosomes is also important for their proper behavior during cell division. Centromere sequences support the formation of *kinetochores* – multisubunit protein structures that attach chromosomes to spindles during mitosis and meiosis, and ensure that sister chromatids stay together or disengage at the appropriate times. In addition, structures called *telomeres* at the chromosome tips allow DNA replication out to the ends of chromosomes and also protect the ends from degradation. Modulation of telomere structure can mean life or death for a cell.

Scientists have been able to exploit their understanding of eukaryotic chromosome structure to create *artificial chromosomes* that can be used as vectors for the construction of recombinant DNAs and may have future significance as agents of gene therapy.

Key terms

chromatin – complex of DNA and protein in the cell nucleus. Some scientists consider certain RNA molecules associated with chromosomes also to be part of chromatin.

histones – small DNA-binding proteins with a preponderance of the positively charged amino acids lysine and arginine; help compact DNA within chromatin

nucleosome – rudimentary DNA packaging unit, composed of DNA wrapped twice around the **core histones:** two copies each of the four different histone proteins called H2A, H2B, H3, and H4

linker DNA – a stretch of ~40 base pairs of DNA that connects one nucleosome with the next

nonhistone chromosomal proteins – chromatin constituents other than histones with a wide variety of functions. Human chromosomes associate with several thousand different kinds of nonhistone proteins.

radial loop-scaffold model – a leading theory to explain the high compaction of DNA in chromosomes. In this model, the looping and gathering of DNA by nonhistone proteins plays an important role.

G bands – alternating dark and light segments (1-10 Mb) of a chromosome as seen under a microscope after staining with Giemsa dye. An artist's diagram of the microscopic image of G-banded chromosomes is called an **idiogram.**

***p* arm** and ***q* arm** – short (*p*) and long (*q*) arms of human chromosomes. (An arm is the interval between a telomere and the centromere.)

fluorescent *in situ* hybridization (FISH) – a technique for physical mapping mapping of chromosomes that uses fluorescent tags to detect hybridization of nucleic acid probes with chromosomes on a microscope slide

spectral karyotyping (SKY) – a type of FISH that results in labeling each of the 24 human chromosomes a different color

DNase hypersensitive site – sites on DNA that contain few or no nucleosomes, and are thus susceptible to cleavage by DNase enzymes

heterochromatin and **euchromatin** – highly condensed chromosomal regions within which genes are usually transcriptionally inactive are called *heterochromatin*; less condensed chromosomal regions within which genes are often transcriptionally active are called *euchromatin*

constitutive heterochromatin and **facultative heterochromatin** – chromosomal regions that are always condensed are *constitutive heterochromatin,* while regions that condensed only under some conditions or in some cell types are *facultative heterochromatin*. Heterochromatin surrounding the centromere is constitutive, while *position-effect variegation* (*PEV;* see below) and X chromosome inactivation are examples of facultative heterochromatin.

position-effect variegation (PEV) – variable expression of a gene in a population of cells, caused by the gene's location near heterochromatin.

histone tails – the N termini of histone proteins whose amino acid residues are enzymatically modified to affect local chromatin structure

histone acetyl transferases and **histone deacetylases** – enzymes that add acetyl groups (*acetyl transferases*) to histone tail lysines, thereby opening chromatin; or that remove acetyl groups (*deacetylases*), thereby closing chromatin

histone methyl transferases and **histone demethylases** – enzymes that influence chromatin structure by adding methyl groups (*methyl transferases*) to histone tail lysines and arginines or that remove methyl groups (*demethylases*) from lysines. (Surprisingly, no arginine demethylases have yet been discovered.)

X inactivation center (XIC) – region of X chromosome (~450 kb) that mediates dosage compensation in mammals

noncoding RNA (ncRNA) – a transcript that lacks an open reading frame and functions as an RNA molecule

chapter 11

Xist (**X** **i**nactive **s**pecific **t**ranscript) – ncRNA transcribed from the XIC of the X chromosome that becomes a Barr body; *Xist* ncRNA binds to the X chromosome and mediates X chromosome inactivation

replicon (replication unit) – the region of DNA synthesized starting from one origin of replication and extending bidirectionally until it merges with DNA from adjacent replication forks

telomeres – specialized terminal structures that ensure the maintenance and accurate replication of the two ends of each linear chromosome in eukaryotes

telomerase – an enzyme critical to the successful replication of telomeres

shelterin – a protein complex that binds to telomeres and protects them from enzymatic activities that would otherwise degrade the chromosomes

centromere – a specialized chromosome region at which chromatids are most tightly connected and which elaborates the *kinetochores* to which spindle fibers attach during cell division (see below)

kinetochore – a specialized chromosomal structure located at the centromere that is composed of DNA and proteins and functions as the site at which chromosomes attach to the spindle fibers

satellite DNAs – blocks of repetitive, simple noncoding sequences (SSRs), usually in the regions around centromeres; these blocks have a different chromatin structure and different higher-order packaging than other chromosomal regions

cohesin – a multisubunit protein complex that holds the sister chromatids of eukaryotic chromosomes together until anaphase; can be found at both the centromere and along the chromosome arms

separase – a enzyme that cleaves the cohesin complexes at anaphase, allowing the sister chromatids to separate

shugoshin – a protein that protects the cohesin complex from separase cleavage during meiosis I

YAC (yeast artificial chromosome) – a vector used to clone DNA fragments up to 400 kb in length; constructed from telomeric, centromeric, and origin-of-replication sequences needed for replication and chromosome segregation in yeast cells

Problem Solving

The problems at the end of Chapter 11 are fairly straightforward, but they require you to integrate information found not only in the chapter but also material that you have encountered previously in the book. You will need to remember how chromosomes move during meiosis and mitosis (Chapter 4), and you must understand the molecular mechanism of DNA replication (Chapter 6). Finally, review what you learned in Chapter 10 about using nucleic acid probes and hybridization to detect particular DNA sequences.

chapter 11

Vocabulary

1.

a.	telomere	4.	specialized structure at the end of a linear chromosome
b.	G bands	10.	regions of a chromosome that are distinguished by staining differences
c.	kinetochore	7.	complex of DNA and proteins where spindle fibers attach to a chromosome
d.	nucleosome	8.	beadlike structure consisting of DNA wound around histone proteins
e.	ARS	2.	origin of replication in yeast
f.	satellite DNA	3.	repetitive DNA found near the centromere in higher eukaryotes
g.	chromatin	5.	complexes of DNA, protein, and RNA in the eukaryotic nucleus
h.	cohesin	1.	protein complex that keeps sister chromatids together until anaphase
i.	histones	6.	small basic proteins that bind to DNA and form the core of the nucleosome
j.	shelterin	9.	protein complex that protects telomeres from degradation and end-to-end fusion

Section 11.1

2. Nonhistone proteins, which make up ~1/2 the mass of proteins associated with DNA, are a heterogeneous group. **Hundreds or even thousands of different kinds of nonhistone proteins exist. Some of these proteins play a purely structural role, (e.g. scaffold proteins) while others are active in replication (DNA polymerase) and the processing of recombination (proteins in the synaptonemal complex). Still others are necessary for chromosome segregation (the motor proteins of the kinetochores) and chromosome integrity (proteins at telomeres). The largest class of nonhistone proteins are those that foster or regulate transcription and RNA processing.** In mammals, 5,000-10,000 of tissue specific transcription factors are found in different tissues at different times in the life cycle. The distribution of the nonhistone proteins along the chromosome is uneven. They are found in different amounts and in varying proportions in different tissues.

Section 11.2

3. **In interphase the chromosomes are compacted 40-fold more than naked DNA, but during metaphase the chromosomes are compacted 10,000-fold more than**

chapter 11

naked DNA. The additional compaction during metaphase is likely needed to allow the chromosomes to behave as independent units during cell division by preventing the DNA of different chromosomes from intertwining. However, the chromosomes must be less compacted during interphase to allow the genes to be transcribed. (Very little gene transcription occurs during M phase.)

4. **The core histones (H2A, H2B, H3 and H4) form the protein portion of the most rudimentary DNA packaging unit, the nucleosome.** The core is an octamer made up of 2 of each core histone. **Roughly 160 bp of DNA wraps twice around the core, leading to a 7-fold compaction over naked DNA.** About 40 bp forms the linker that connects one nucleosome to the next. Histone H1 lies outside the core, apparently associating with the DNA where it enters and leaves the core. Removal of H1 causes some DNA to unwrap from each nucleosome, but 140 bp of DNA stays intact at the core. **H1 is involved in the next level of compaction, formation of a 300 Å fiber.**

5. a. A human haploid genome consists of 3×10^9 bp. If each nucleosome has a spacing of 200 bp, then 3×10^9 bp in a haploid human genome / 2×10^2 bp in a nucleosome = 1.5×10^7 nucleosomes to cover the DNA in a haploid genome. (This estimate is high because not all parts of the genome are arranged uniformly into the most densely packed nucleosomes.) The human genome is diploid, and after S phase each cell would have 2 chromatids for each chromosome, so 4 (1.5×10^7 nucleosomes) = 6×10^7 nucleosomes would be required per cell. Finally, each nucleosome contains two molecules of H2A, so soon after the completion of S phase cells would need roughly **1.2×10^8 molecules of H2A protein.**

b. Histone proteins need to be synthesized **during or just after S phase**, when the chromosomes have just replicated. This time frame is when there would be new "naked" DNA requiring nucleosomes.

c. Each cell needs to make about 6×10^8 molecules of each type of histone during S phase of the cell cycle [see part (a) above]. In human cells, S phase generally lasts between 3 and 6 hours depending on the cell type. This means that a lot of molecules of protein need to be made in a short period of time. **Multiple copies of the histone genes mean more templates that the cells can transcribe simultaneously, allowing for more rapid production of histone mRNAs and thus histone proteins.**

6. Chromatin is the complex of DNA and proteins (histone and nonhistone) that make up eukaryotic chromosomes. The nucleosome protects DNA from being digested with DNase enzymes such as micrococcal nuclease, so such enzymes preferentially attack the DNA in chromatin somewhere in the linker DNA between nucleosomes. Thus, the patterns in lanes A and B reflect the distribution of nucleosomes in chromatin: 200 bp is a single nucleosome, 400 bp is two nucleosomes, etc.

If the nuclease treatment occurs only for a short time, cleavage of double-stranded DNA will occur in some linker regions but not others, producing chromatin fragments with one or more nucleosomes as in lane A.

Longer periods of nuclease treatment result in more cleavage – the enzyme will attack all the linker regions, so that all the chromatin will be reduced to units of single nucleosomes (lane B).

Finally, if the chromatin is treated with the nuclease for a sufficiently long time, then all the linker DNA will be digested, leaving core nucleosomes, each with about 160 bp of DNA as seen in lane C. These core nucleosomes presumably still contain histone H1; if they did not, the DNA remaining at the core would be somewhat shorter at ~140 bp (see Problem 4).

These types of experiments were actually performed in the 1970s, and provided significant support for the emerging picture of the nucleosome.

7. a. **p represents the short arm; q represents the long arm**

    b. In the diagram below, the centromere is gray, and * **indicates the position of a gene at 3p32:**

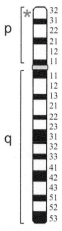

8. Let's assume that 1 G band must be deleted in order to detect a deletion by karyotype analysis. Then the question comes down to: How many genes are in 1 G band? The average number of base pairs per G band is (3×10^9 bp/haploid genome)/(2000 G bands / haploid genome) = 1.5×10^6 = 1,500,000 bp/ G band. The average number of base pairs per gene is (3×10^9 bp/ haploid genome)/(25,000 genes/ haploid genome) = 1.2×10^5 = 120,000 bp/gene. Therefore, the number of genes per G band is (1,5000,000 bp/G band)/(120,000 bp/gene) = **12.5 genes**/G band. This calculation is of course only an extremely rough average estimate.

9. a.

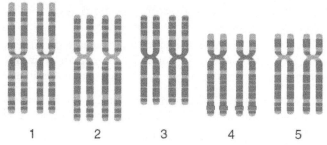

chapter 11

b. Figure 11.9a is a photomicrograph of an actual FISH experiment. **The chromosomes are scattered rather than lined up as in the idiogram above because they were prepared from a cell whose nucleus was burst open.** (To make a karyotype, the pictures of the chromosomes would be cut out and rearranged in pairs and in order of decreasing size.) **The gene probed in the figure is on the q arm of a chromosome, but it looks too small to be chromosome 4** (which is the 4th largest chromosome in the haploid genome).

Section 11.3

10. **The answer is (b).** DNase can cleave DNA that is unprotected because it is not wound up in nucleosomes. Regions of chromatin from which no transcription is occurring [as in answer (a)], including heterochromatic regions, would be extremely resistant to DNase digestion, not hypersensitive.

11. a. 300 Å fiber
    b. **DNA loops attached to a scaffold**
    c. heterochromatin
    d. metaphase chromosomes

12. a. **In *Drosophila*, the centromeric regions of the chromosomes and the Y chromosome are examples of constitutive heterochromatin.** Most of the DNA near the centromeres and on the Y chromosome is bound up as heterochromatin in all cells at all times.

    Facultative heterochromatin in *Drosophila* is seen in cases of Position Effect Variegation (PEV). The example discussed in the text (Fig. 11.12, p. 391) is PEV of the white gene after a chromosomal rearrangement (an inversion that moves the gene next to the X chromosome centromeric heterochromatin). The mosaic of white and red patches seen in the eyes of these animals suggests that the decision about the heterochromatic spreading is the result of a random process which varies from cell to cell during development. Heterochromatization can spread over >1 Mb of previously euchromatic DNA.

    Autosomal genes can also show PEV as the result of a Y:autosome translocation which moves the gene next to Y chromosome DNA. The Y chromosome sequences are constitutively heterochromatic, while the autosomal genes juxtaposed to the Y can become heterochromatic in some cells but not others. The autosomal sequences near the translocation breakpoint would be considered to be facultative heterochromatin in this strain of flies.

    b. **In humans, the centromeric DNA and the great majority of the Y chromosome are constitutive heterochromatin,** as was also the case in *Drosophila*.

    The formation of Barr bodies due to the random inactivation of one of the X chromosomes in each cell early in female human fetal development is an example of **facultative heterochromatin. Any one X chromosome (say the maternal X chromosome) could be a heterochromatic Barr body in some cells but euchromatic in the cells in which it was not inactivated.

13. **To determine if acetylation of lysine 12 in H4 is important for chromatin formation you could alter the 12th codon of the H4 gene so that it specifies a different amino acid and test if this change affects chromatin.** You could assay chromatin structure by performing DNase sensitivity assays such as that outlined in Fig. 11.10 on p. 389 to see if various regions of the genome had an altered distribution of nucleosomes. If nucleosome structure were disrupted, it would be interesting to investigate other potential phenotypic consequences of the amino acid alteration in the H4-encoding gene. The change might affect the survival of the organism (for example, the cells might not grow), or this alteration might have milder consequences like disrupting the expression of certain genes.

    In Chapter 18, you will learn various techniques that would allow you to do alter the H4 gene. One of the methods discussed in that chapter permits investigators to replace the wild-type allele of the gene in the yeast genome with an allele containing the substitution in the 12th codon. A different method permits scientists to add a *transgene* with the altered codon to a yeast cell that also has a wild-type copy of the gene. In this latter case, aberrant phenotypes would be observed only if the mutant allele had dominant effects.

14. a. *Su(var)* mutations decrease the amount of PEV (that is, these mutations increase the number of cells expressing the *white*+ gene). **In the presence of a *Su(var)* mutant allele there will be fewer white patches in the eye and more red patches when the eyes are compared to a homozygous *Su(var)*+ fly. The situation would be reversed with more white patches and fewer red (wild type) patches if the fly were heterozygous for the *E(var)* mutation** (Fig. 11.12a, p. 391).

    b. The *Su(var)* and *E(var)* mutations both have phenotypes that lead you to think the proteins encoded by the genes are involved in chromatin condensation. Assuming the mutations are loss of function (null) alleles, then **the *Su(var)*+ genes encode proteins that establish and assist spreading of heterochromatin. These would include histone deacetylases, certain histone methyl transferases, and certain histone demethylases.** Thus, loss of some of the gene product results in engulfment of neighboring genes by heterochromatin.

    The *E(var)*+ genes seem to encode proteins that restrict the spreading of heterochromatin, as loss of one copy of the gene allows heterochromatin to spread into neighboring genes more often. These would include **histone acetyl transferases, certain histone methyl transferases, and certain histone demethylases.** (Note that the methylation of histones may either promote or inhibit heterochromatization, depending on the particular histone site targeted; in contrast, the acetylation of histones almost always "opens" chromatin structure.)

    The results also suggest that position effect variegation is sensitive to the amounts of either type of protein (suppressors or enhancers) because a reduction of 50% of either type of protein causes the mutant phenotype.

chapter 11

15. Only Woman 2's X chromosome that has the XIC can be inactivated; the X chromosome with the deletion of the XIC must be the X chromosome that is always active. **Woman 2 expresses $A^S B^S C^F C^S D^F D^S$.** (Both alleles of the autosomal genes are expressed because the autosomes are not subject to inactivation.)

    In Woman 3, most likely her intact X chromosome will be inactivated. Otherwise, many autosomal genes on the translocation chromosome containing the XIC will be inactivated and the protein product will be underexpressed, while X chromosome genes on the reciprocal translocation chromosome will not be inactivated and will be therefore overexpressed. Such cells would have gross imbalances in the amount of many gene products and would probably die. For this reason, **Woman 3 will most likely only have cells that express $A^F B^F C^F C^S D^F D^S$.** (If the translocation breakpoints were close to the telomeres of both the X chromosome and the autosome, the cells might tolerate the imbalances in the amount of gene products because only a few proteins would be affected in this way. In such cases, Woman 3 could have some cells that express $A^F B^F C^F C^S D^F D^S$ and other cells that express $A^S B^F B^S C^F C^S D^F$.)

16. a. The chromosome that cannot express *Tsix* is more likely to express *Xist* and become the Barr body.

    b. The *Tsix* transcript could hybridize with the *Xist* transcript (forming a double-stranded RNA) and prevent *Xist* function. Alternatively, *Tsix* transcription could somehow prevent *Xist* transcription.

    c. We are still left with the problem of how the cell "decides" which one of the X chromosomes transcribes *Tsix*.

Section 11.4

17. At 50 nucleotides/second, one molecule of DNA Polymerase (DNAP) could synthesize about 270 kb in 3 hours. Since DNAP can synthesize DNA in both directions from an origin of replication, up to 540 kb of DNA could exist between origins. Thus the minimum number of origins expected for a 3 billion base pair genome would be about 3×10^9 base pairs per genome / 5.4×10^5 base pairs per origin = 0.55×10^4 origins = **5500 origins of replication.**

18. a. In order to replicate the longest chromosome (66Mb) from one bidirectional origin of replication in the middle, 33 Mb would have to be copied along each replication fork during the 8 minute cycle (480 sec) = 33,000,000 bp replicated/480 sec = 68,750 bp/sec = ~69 kb/sec. Therefore, if a single origin of replication was used and replication took the entire 8 minutes of the cycle, **the rate of polymerization would be 0.069 Mb/sec or 69,000 bp (69 kb) per second.**

    b. If bidirectional origins of replication occur every 7 kb, then only 3.5 kb would have to be replicated during the 8 min cell division cycle. **The polymerization rate would have to be a minimum of 3.5 kb/480 sec = 7.3 bp/sec, a <u>much</u> more reasonable rate.**

    Scientists have actually observed progression of the replication forks in early *Drosophila* embryos, and the fastest seen was ~50 bp/sec. If all replication origins "fired" at the same time and were exactly 7 kb apart, and all forks traveled at the same speed, cells would need only slightly more than 1 min to complete S phase. But in actuality, some origins are further apart, and some origins start firing at different times (see Fig. 11.16 on p. 395), explaining why the fastest S phase is instead about 8 min.

19. a. **Repeats of the sequence 5' TTAGGG 3'** are found at the telomeres of human chromosomes.

    b. **Telomerase enables the ends of the chromosomes to be replicated;** the enzyme provides the primer that would be otherwise be lacking for lagging strand replication. In the absence of the telomerase primer, the chromosomes would shorten each cycle of cell division.

    Shelterin protects the free ends of the chromosome from nuclease digestion and end-to-end fusions. When chromosomes are broken (for example by X-rays), the broken ends that form are not bound by shelterin. As a result, these chromosome fragments often become shorter due to nuclease digestion, and the broken fragments can fuse to each other out of order (for example, mistakenly connecting parts of nonhomologous chromosomes).

20. a. **If telomerase is unavailable, telomeres will become shorter at each cell division. Eventually cells will die when essential genes are removed.** The absence of telomerase could have a major effect on the supply of stem cells that proliferate to renew tissues. The animals will age prematurely if the supply of stem cells is interrupted.

    b. **Most somatic cells are meant to have finite lifespans. Somatic cells that are not stem cells, and are not meant to proliferate indefinitely, do not express telomerase. If somatic cells inappropriately express telomerase, they might continue to divide when they should not. You will learn in Chapter 19 that immortality and overproliferation are two phenotypes of cancer cells. Thus, individuals whose somatic cells overexpress telomerase would be likely to develop cancers.**

21. a. The probe 5' CCCTAA 3' will hybridize to **telomeres**, as shown in *red* on the idiogram below. Because the telomeres of all the chromosomes have the same repeating sequence, all of the telomeres will light up with the probe.

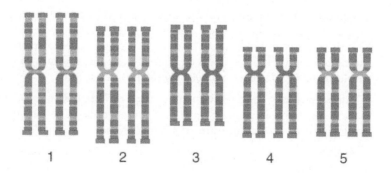

chapter 11

b. To track one end of a particular chromosome, **a probe corresponding to a unique gene sequence** should be used. If that gene is near the telomere of the q arm of chromosome 2, for example, the idiogram would look like the one below (the hybridization signal is *yellow*).

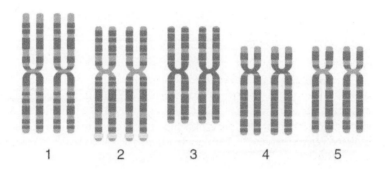

22. **In none of the cases (a-d) would chromosomes be expected to have <u>exactly</u> the same number of telomeric repeat sequences.** The reason is that the telomerase enzyme does not copy the repeats that are already present, instead, it adds repeat units. The enzyme can add different numbers of repeating units to each different telomere depending on the number of translocation steps it undergoes (see Fig. 11.20 on p. 397).

Section 11.5

23. a. The centromeric regions of human chromosomes are made up of **SSRs called** *satellite DNAs.*

    b. **Cohesin holds sister chromatids together until anaphase,** when cohesin complexes are cleaved and the sister chromatids are released. **Kinetochores attach chromosomes to the spindle poles and contain motor proteins that move the separated chromosomes to the poles.**

24. Shugoshin prevents Rec8 cleavage, and in the context of mitosis, this would mean that **sister chromatids would not separate and move to opposite poles. The effect would be** *aneuploidy*; both chromatids of each homolog would end up in one daughter cell or the other at random. You will learn in Chapter 12 that cells with incomplete sets of chromosomes are called *aneuploid* cells.

25. a. The chromosomes shown at the left in the figure in the textbook are in prophase/metaphase of meiosis I; they are undergoing crossing over. At right, they are in anaphase of meiosis I. The light blue lines are sister chromatids of one homolog, and the dark blue lines are sister chromatids of the other homolog.

b. In the diagram below, cohesin complexes are indicated as red ovals. Along the arms, the only cohesin complexes shown are those distal to the recombination site (that is, further away from the centromere than the site of recombination). The cohesin complexes keep sister chromatids together by forming a "basket" of protein that surrounds both double helices. (See Fig. 11.25 on p. 400 for a detailed view of one molecular model for the manner in which cohesin complexes keep sister chromatids together.)

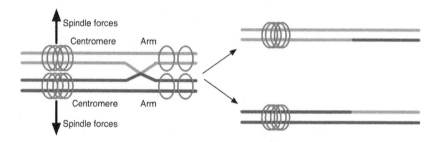

c. The bivalent at left in the diagram above is held together by **cohesin complexes located distal to the recombination site**. To help you visualize this fact, we reproduce here a figure previously shown in the answer to Problem 17 in Chapter 5. Cohesin proximal to the recombination site (that is, the chiasma) cannot hold the homologous chromosomes making up the bivalent together.

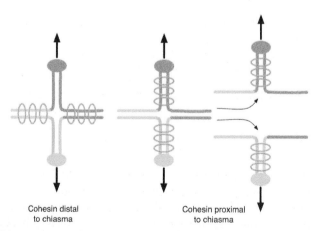

Cohesin distal to chiasma   Cohesin proximal to chiasma

d. **Shugoshin protects the cohesin complexes at the centromere (but not those along the chromosome arms) from cleavage by separase enzyme during meiosis I.** Note that the cohesin complexes at the centromere must remain intact throughout meiosis I because both sister chromatids need to move as a unit during this kind of division; it is only starting with anaphase of meiosis II that centromeric cohesin can be cleaved. In contrast, the cohesin complexes along the arms

chapter 11

(particularly those distal to the chiasmata) must be cleaved at the start of anaphase of meiosis I so that the homologous chromosomes can segregate to opposite spindle poles.

26. a. The ends of chromosomes do not normally fuse together because they are protected by proteins called the **shelterin** complex (Fig. 11.21 on p. 398).

   b.

   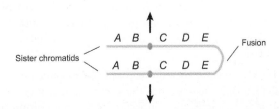

   c. The fusion chromosome shown in part (b) above behaves as a bridge during mitosis because **the centromeres are pulled to opposite poles.** (Remember that these centromeres are those of sister chromatids that must be pulled to opposite poles during mitotic anaphase.)

   d. **The bridge is likely to break during anaphase, which means that each daughter cell will have a broken chromosome.** This cleavage could occur anywhere in the "stretched" region between the two centromeres. During the following mitosis, the broken chromosomes will fuse and form a bridge again.

27. a. **Opening of chromatin near promoters to allow transcription involves histone acetylation; closing of chromatin to silence transcription or form heterochromatin (including Barr body formation) involves histone methylation and deacetylation.**

   b. **Variant histones are required near the centromere for assembly of kinetochores.**

Section 11.6

28. a. Proper mitotic chromosome segregation would be disrupted by **mutations in genes encoding (i) cohesin proteins, (ii) separase, (iii) kinetochore proteins, including the motor proteins that help chromosomes move on the spindle apparatus, and (v) components of the spindle checkpoint** that makes the beginning of anaphase depend on the proper connections of spindle fibers and kinetochores. **(vi) Mutations that alter the DNA comprising a centromere** might also have similar effects in disrupting mitotic chromosome segregation.

   b. In order to look for mutations that affect mitotic chromosome segregation, you need cells containing a YAC with a marker conferring a visible phenotype like colony color. Treat these yeast cells with a mutagen and **look for colonies where**

11-13

some cells lost the YAC because of mitotic chromosome mis-segregation. You could also try to **mutate the centromeric DNA of this YAC using *in vitro* mutagenesis. If the centromere were disrupted the YAC would not segregate properly and would be lost**. Again, you could follow this loss if the YAC carried a genetic marker that resulted in a visible phenotype like colony color.

29. a. A plasmid containing only the *URA+* gene must **integrate** into the chromosome to be replicated and maintained because it has no origin of replication.

    b. A *URA+*, ARS plasmid can be **maintained as a plasmid or it can integrate** into the chromosome. **If it remains as a plasmid, it will not be very stable if the selection for Ura+ is no longer applied;** because it lacks a centromere the plasmid would be lost from many of the daughter cells during subsequent rounds of mitotic division.

    c. A *URA*$^+$, ARS, CEN plasmid could only be **maintained as a separate plasmid in the cell**. If it did integrate into the chromosome, there would be two centromeres on that chromosome and during mitosis the chromosome would break. **The plasmid would be stable from one generation to the next without selection** because the centromere sequence directs its segregation.

    d. In addition to ARS and CEN sequences, a linear artificial chromosome requires **telomeres**. Also, as explained on p. 402 of the text, an artificial yeast chromosome requires ~100 kb of DNA to segregate with great accuracy. Researchers do not yet understand the reason for this latter requirement.

30. The subcloned fragments that contain the centromeric DNA are those that show a high percentage of Trp+ colonies after 20 generations without selection for the plasmid. These subclones include the 5.5 kb *Bam*HI, the 2.0 kb *Bam*HI-*Hind*III, and the 0.6 kb *Sau*3A fragments. **Because the smallest of these fragments has high mitotic stability and its ends are within the boundaries of the other fragments, the centromere sequence must be contained within the 0.6 kb *Sau*3A fragment.**

chapter 12

Chromosomal Rearrangements and Changes in Chromosome Number

Synopsis

Previous chapters mainly addressed point mutations – changes in single base pairs or a small number of base pairs. This chapter is about different sorts of mutations that change the organization of entire chromosomes, or alter the number of chromosomes in each nucleus. These large-scale mutations can affect phenotype at the level of individual genes, but in addition they can affect meiosis, and thereby an organism's fertility and the genetic make-up of gametes it produces. Transposable elements – DNA segments that can move around the genome – are included in this chapter not only because their mobility can be considered to be a type of chromosomal reorganization, but also because transposable elements are a major cause of the other kinds of chromosomal rearrangements discussed in this chapter.

Key terms

deletion – a chromosomal mutation where a segment of DNA – sometimes a large region that contains several genes – is missing from the chromosome

duplication – a chromosomal aberration where a segment of DNA (often large enough to contain many genes) is present twice on the chromosome. The two identical DNA segments are adjacent in **tandem duplications,** while the repeated segments are in different chromosomal locations in **nontandem duplications.**

inversion – a chromosomal aberration where a segment of the DNA is rotated 180^0 with respect to its normal orientation. To maintain the 5'-to-3' orientation of the chromosome strands, the Watson and Crick strands in the inverted region must also be flipped. An inversion is **pericentric** if the inverted region includes the centromere, and **paracentric** if the inverted region does not include the centromere.

reciprocal translocation – a chromosomal rearrangement where parts of two nonhomologous chromosomes switch places

rearrangement breakpoints – the locations on rearranged chromosomes where the DNA was broken and then stitched back together (or recombined aberrantly). At rearrangement breakpoints, DNA sequences that would not normally be adjacent to each other are now juxtaposed.

gene dosage – the number of gene copies in a cell or organism. In a monoploid, the normal "dose" of single-copy genes is 1, while in a diploid, the normal "dose" is 2.

deletion loop – when homologous chromosomes pair during meiosis I in a deletion heterozygote, the structure formed by the segment of the nondeleted homolog that has no counterpart with which to pair on the deleted homolog

inversion loop – When homologous chromosomes pair during meiosis I in an inversion heterozygote, a "loop" must form to allow the inverted and normal homologs to pair in the region of the inversion.

balanced gametes and **unbalanced gametes** – gametes with normal dosages of all genes are *balanced*; gametes with abnormal dosages of some genes – often due to deletions and/or duplications - are *unbalanced*. If the dosages of a sufficient number of genes are unbalanced, the zygotes formed from such gametes will not survive.

unequal crossing-over – aberrant crossovers between two homologs that misalign during meiosis I in a DNA region where two or more similar copies of a nucleotide sequence exist. When unequal crossing-over occurs, one recombinant chromosome, loses copies of the repeated sequence, while and the other recombinant chromosome gains copies.

crossover suppressor – describes a chromosome that contains an inversion and therefore in inversion heterozygotes prevents recovery of recombinant progeny for alleles within the inversion

balancer chromosome – a chromosome made by researchers as an experimental tool that has numerous inversions preventing recovery of any recombinant chromosomes at all from a heterozygote. Commonly used balancer chromosomes also carry a recessive lethal mutant (so that Balancer homozygotes do not survive) and a dominant phenotypic marker to distinguish the Balancer from the other chromosome in heterozygotes.

semisterility – a phenotype shared by heterozygotes for inversions or reciprocal translocations because a large fraction of the gametes they produce are genetically imbalanced

pseudolinkage – the phenomenon where genes on two nonhomologous chromosomes behave as if they are linked because the organism is heterozygous for a reciprocal translocation. Pseudolinkage is observed for genes close to the translocation breakpoints because only one of three possible segregation patterns for the two chromosomes produces balanced gametes.

Robertsonian translocation – reciprocal exchange between two acrocentric chromosomes that results in a large metacentric compound chromosome and a small chromosome that is sometimes lost

transposable elements (TEs) – segments of DNA that move from place to place in the genome by a process called **transposition.** Two classes of TEs exist: **DNA transposons (transposons)** mobilize by a "cut-and-paste" mechanism that involves only DNA; **retrotransposon** mobilization is through transcription of the element into an RNA intermediate.

transposase – *trans*-acting protein encoded by DNA transposons that catalyzes their transposition, usually by acting on the inverted repeats at the transposon ends. **Autonomous elements** are transposons with intact transposase genes;

nonautonomous elements have mutant or deleted transposase genes but can mobilize nevertheless through the activity of transposase from autonomous elements in the same family.

euploid and **aneuploid** – Cells with complete sets of chromosome are *euploid*; if one or more chromosomes is missing from a set, or if extra chromosomes are present, the cells or organisms are *aneuploid*.

ploidy – describes the number of chromosome sets in a cell. **Monoploids** have one set, **diploids** have two sets, **polyploids** have more than two sets: **triploids** have three sets, **tetraploids** have four sets, etc.

monosomy and **trisomy** – forms of aneuploidy in diploids. In *monosomy*, an otherwise diploid cell is missing one chromosome, while in *trisomy*, an extra chromosome is present in addition to the two complete sets.

X chromosome reactivation – reversal of Barr body formation in oogonial cells of the female germ line

Turner syndrome and **Klinefelter syndrome** – abnormal human phenotypes due to X chromosome aneuploidy. Females who are XO have *Turner syndrome*; the phenotype includes unusually short stature and sterility due to lower than normal doses of X-linked genes. Males with more than one X chromosome (typically XXY) have *Klinefelter syndrome*; the phenotype includes unusually tall stature and sterility due to abnormally high X-linked gene dosages.

meiotic nondisjunction – the failure of homologs to separate during meiosis I, or the failure of sister chromatids to separate during meiosis II, resulting in aneuploid meiotic products

Down syndrome – the phenotype in humans caused by chromosome 21 trisomy; sometimes called **trisomy 21.**

basic chromosome number (x) – the number of chromosomes that constitute a complete set. In diploids, $x = n$, where n is the number of chromosomes in gametes.

autopolyploids – polyploids where all of the chromosome sets originated from the same species

allopolyploids – usually sterile hybrids with complete chromosome sets from different, though related species. **Amphidiploids** are fertile allopolyploids made by crossing two diploid species, and then doubling the chromosome number to generate homologous pairs.

gene family – a group of genes with similar base pair sequences and similar functions that derived from a single ancestral gene

syntenic segments – regions of chromosomes in two different species where the genes and gene order are conserved

chapter 12

Problem Solving

As you approach the problems – keep several important general concepts in mind:

- Rearranged chromosomes can affect the functions of particular genes that are located at or near rearrangement breakpoints.
    - The rearrangement breakpoint can interrupt the gene and abolish its function (e.g., Fig. 12.12 on p. 420), or connect parts of two genes together to make a *fusion* gene whose products may have novel functions (e.g., Fig. 12.16 on p. 423).
    - The rearrangement breakpoint may put a gene in an environment that alters its expression. One possible scenario is that the rearrangement now juxtaposes a gene normally found in euchromatin to a position next to heterochromatin. The heterochromatin can spread, leading to *position effect variegation* (*PEV*; review Fig. 11.12 on p. 391 in Chapter 11). Another possibility is that the coding region of the gene is placed nearby an enhancer for a different gene, causing misexpression. (The *Antennapedia* mutation in *Drosophila* that results in the development of a leg coming out of the fly's head [see Fig. 8.32 on p. 287] is one example of this kind of event.

- Chromosomal rearrangements, and also changes in chromosome number such as aneuploidy, alter gene dosage. Decreased or increased dosage of some genes may have little or no obvious effects, but changes in the dosage of specific genes may alter phenotypes (e.g., Fig. 12.9 on p. 419) or even in rare cases result in lethality. Almost always, rearrangements creating imbalances in the dosage of many genes causes lethality.

- Rearranged chromosomes, and also changes in ploidy, can affect the process of meiosis – the way chromosomes pair, crossover, and segregate into gametes. For this reason, when predicting the expected gametes from an individual who is heterozygous or homozygous for a chromosomal rearrangement (or from an individual with an unusual number of chromosomes) it is critical to diagram the pairing of homologous chromosomes at meiosis I. These diagrams will ensure that you understand the chromosome composition of the individual and the kinds of gametes this individual can produce.
    - Remember to trace out the meiotic products beginning from the centromere, because the chromosomes are pulled to the spindle poles by fibers that attach to the kinetochores that form at the centromeres. Keeping track of the centromeres is particularly important when considering heterozygotes for inversions and translocations.
    - Pay attention to whether or not the meiotic products are balanced (with one allele of each gene and one centromere). If the meiotic products are imbalanced either for many genes or for the centromere, then the progeny formed from them are usually inviable.

Deletions. One sign of the presence of a deletion is the "uncovering" of recessive mutant alleles: That is, a heterozygote for the deletion may show a mutant phenotype that is normally recessive, if the non-deleted homolog carries a copy of the gene with the recessive allele and the deleted homolog has no copy of the same gene. Another indication of the

chapter 12

presence of a deletion is the fact that in deletion heterozygotes, no recombination will be seen for genes in the deleted region, and the map distances for genes outside the deleted region will be reduced.

Inversions. Severe reduction of recombination is also a sign that heterozygosity for an inversion may be involved. Single crossovers within the inversion loop lead to recombinant gametes that are imbalanced for genes outside the loop. Interestingly, all gametes have the expected amount of DNA for everything within the inversion loop. If the centromere is in the inversion (*pericentric*) then each meiotic product has a centromere. If the centromere is outside of the inversion loop (*paracentric*) then the meiotic products that result from a single crossover within the loop are imbalanced for the centromere as well as for some genes – one recombinant product will be *dicentric* while the other will be *acentric* (see Fig. 12.14 on p. 422).

Reciprocal translocations. In a translocation heterozygote, only the products of the *alternate* segregation pattern are balanced, while the products of the *adjacent-1* and *adjacent-2* patterns are imbalanced (see Fig. 12.17 on p. 424). This fact underlies two indications that an individual might be a translocation heterozygote. (1) *Semisterility* – less than half the gametes produced by a heterozygote for a reciprocal translocation will be balanced, so many zygotes formed from the gametes produced by this individual will be inviable. (2) *Pseudolinkage* – genes known normally to be on nonhomologous chromosomes will act as if they are genetically linked.

Changes in ploidy. The central ideas to consider in dealing with problems involving non-diploid, non-euploid chromosome numbers are that chromosome pairing is essential for proper chromosome segregation during meiosis I, and that pairing requires exactly two homologous chromosomes. If chromosomes cannot pair, they will missegregate and unbalanced gametes will result. One possible way to overcome this issue in non-diploid but euploid cases (either in nature or in experiments) is to disrupt cell division so as to create cells that now have two copies of each separate chromosome, creating polyploids. These two copies can now pair with each other and segregate properly.

Vocabulary

1.

   a. reciprocal translocation — 4. exact exchange of parts of two nonhomologous chromosomes

   b. gynandromorph — 8. mosaic combination of male and female tissue

   c. pericentric — 6. including the centromere

   d. paracentric — 5. excluding the centromere

   e. euploids — 7. having complete sets of chromosomes

   f. polyploidy — 3. having more than two complete sets of chromosomes

   g. transposition — 2. movement of short DNA elements

   h. aneuploids — 14. lacking one or more chromosomes or having one or more extra chromosomes

chapter 12

Section 12.1

2. a. *Deletion 1* removes the DNA homologous to Probe A and Probe B, while *Deletion 2* removes the DNA homologous to Probe B and Probe C.

   b. Note that the order of the probes along the non-deleted chromosome (+) is telomere-A-B-C-centromere.

   c. *Inv1* is a *pericentric* inversion. This means that one breakpoint of the inversion is the middle of Probe A (as shown in the diagram in the text) while the other breakpoint of the inversion is on the other side of the centromere. This will have three effects: (i) Probe A will hybridize to two separate locations that will be located on different arms of the inverted chromosome. (ii) The centromere will shift location relative to its position in the normal (non-inverted) chromosome. Because the instructions say that the other breakpoint of the inversion is near the centromere, in this case the inversion will change the chromosome from a metacentric to a more acrocentric configuration. (iii) Probe B and Probe C will be found, along with part of Probe A, on the long arm of the chromosome in the order shown in the diagram that follows.

   d. In contrast with *Inv1*, *Inv2* is *paracentric*, meaning that both rearrangement breakpoints are found on the same side of the centromere and that the centromere stays in the same relative position on the chromosome as found in the normal homolog. The inversion moves Probe C away from the other two probes to a position closer to the centromere. (The problem stated that the other inversion breakpoint not shown in the figure is close to the centromere.)

e. See answer to part (d) above for a description of the hybridization of the three probes to the *Inv2* chromosome.

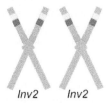

Section 12.2

3. a. (i) **No.** Loops will not be seen in a heterozygote for a reciprocal translocation; instead during prophase of meiosis I a cruciform structure will form (see Fig. 12.17b on p. 424). (ii) **No.** Chromosomal bridges form consistently only during prophase of meiosis I in a heterozygote for a paracentric inversion (Fig. 12.14b on p. 422), not in a heterozygote for a reciprocal translocation. (iii) **No.** Same as answer to part (ii). (iv) **No.** The apparent suppression of recombination occurs only during meiosis in a heterozygote for an inversion, not in a heterozygote for a reciprocal translocation. (v) **Different chromosomes.** A reciprocal translocation involves two nonhomologous chromosomes.

    b. (i) **Yes** (Fig. 12.13 on p. 421). (ii) **Yes, if recombination occurs within the inversion loop** (Fig. 12.14b). (iii) **Yes, if recombination occurs within the inversion loop** (Fig. 12.14b). (iv) **Yes.** (v) **Same side of a single centromere** (this is the definition of a paracentric inversion).

    c. (i) **Yes.** A loop will form during meiosis I in a heterozygote for a duplication because the extra copy of the region cannot pair with any DNA sequences in the normal homolog. (ii) **No.** [See answer to part (a ii).] (iii) **No.** [See answer to part (a iii).] (iv) **No.** The apparent suppression of recombination occurs only during meiosis in a heterozygote for an inversion, not in a heterozygote for a tandem duplication. (v) **Same side of a single centromere.** Note that the duplicated region cannot include the centromere because the result would be a dicentric chromosome that could not be maintained when the cells divide.

    d. (i) **No,** (ii) **No.** (iii) **No.** (iv) **No.** (v) **Different chromosomes.** [Same reasoning as for part (a); a Robertsonian translocation is one subtype of reciprocal translocation.]

    e. (i) **Yes**; (ii) **No**; (iii) **No,** (iv) **Yes**; (v) **Opposite sides of a single centromere.** Heterozygotes for pericentric inversions will behave similarly to heterozygotes for paracentric inversions as was described in part (b) with the following exceptions: Chromosome bridges and acentric chromosomes are formed consistently only during prophase of meiosis I in a heterozygote for a paracentric inversion, not for a pericentric inversion (Fig. 12.14b on p. 422). And by definition, in a pericentric inversion the centromere is within the inverted region, so the rearrangement breakpoints must be on opposite sides of the chromosomes's centromere.

f. (i) **Yes.** See Fig. 12.7 on p. 418 for a diagram of a deletion loop. (ii) **No.** (iii) **No.** (iv) **Yes.** The region of the normal chromosome that is removed in the deletion has nothing with which it can pair or recombine. Thus, no recombination occurs in the interval of the deletion during meiosis in a deletion heterozygote. (v) **Same side of a single centromere.** If the deletion breakpoints were on opposite sides of a chromosome's centromere, the resulting chromosome would be acentric and thus lost during cell division.

4. a. Most deletions remove a large amount of DNA. No mechanism exists for restoring this DNA, so **reversion of this deletion is impossible**. In theory a very small deletion of a single nucleotide, for example, could revert using a mutagen like proflavin.

   b. This rearrangement could revert because all of the original information still exists in the genome. **Reversion within the duplication could occur fairly frequently** because there is a mechanism that generates revertants - they can occur as a result of **unequal crossing over** in an individual homozygous for the duplication (Fig. 12.10, p. 420).

   c. In theory, a pericentric inversion could revert because all of the information still exists in the genome. However there is **no mechanism to ensure that the breaks will occur in the same locations as the original breaks that gave rise to the pericentric inversion, so the rate of reversion should be extremely low.** One exception to this are inversions that result from intrachromosomal crossing-over between some sequence of DNA that is present at two locations on the same chromosome but in reversed order (Fig. 12.3b, p. 414). A similar crossover in the inverted chromosome could regenerate the original gene order, and such crossovers could occur with substantial frequency.

   d. **If the organism with the Robertsonian translocation has already lost the very small reciprocal chromosome generated in the process of translocation then the translocation cannot be restored. If the small reciprocal chromosome still exists, then it is possible that the translocation could revert if it came about through crossing-over between repeated elements** (Fig. 12.3d, p. 414).

   e. **For some types of transposable elements, the mechanism that allows the transposable element to jump into the gene can also allow the element to jump out again, often restoring the original DNA sequence of the gene** (Fig. 12.23b, p. 430). **In these cases the mutation will revert fairly frequently.** If this jumping mechanism is lost (for example, if the inverted repeat sequences at the ends of a DNA transposon were deleted), then these types of mutations revert very rarely.

5. a. During meiosis in an inversion heterozygote, a loop of the inverted region is formed when the homologous genes align. In the following simplified drawing, each line represents both chromatids comprising each homolog. Note that the inversion is paracentric.

chapter 12

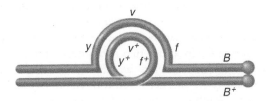

If a single crossover occurs within the inversion loop, a dicentric and an acentric chromosome are formed (see Fig. 12.14b, p. 422). **Cells containing these types of chromosomes are not viable, so the resulting allele combinations from such single crossovers are not recovered.** The four phenotypic classes of missing male offspring would be formed by single crossovers between the *y* and *v* or between the *v* and *f* genes in the female inversion heterozygote and therefore are not recovered.

b. **The *y v f* B^+ and $y^+ f^+ v^+$ *B* offspring are the result of single crossovers outside of the inversion loop, between the end of the inversion (just to the right of *f* on the preceding diagram) and the *B* gene. This region is approximately 16.7 m.u. long (19 recombinants out of 114 total progeny).**

c. **The $y^+ v f^+ B^+$ and $y v^+ f B$ offspring would result from two crossovers within the inversion loop, one between the *y* and *v* genes and the other between the *v* and *f* genes.** You should note that these could be either 2-strand or 3-strand double crossovers, but they could not be 4-strand double crossovers.

6. a. Each of the strains is mutant for one or more of the marker genes, suggesting that in each a **deletion** has removed several adjacent genes, and thus their recessive loss-of-function phenotypes are uncovered. Furthermore, after the diploids undergo meiosis, two of the spores die; deletion mutations are often lethal in haploids.

   b. **Two spores in each ascus die because they receive the deleted homolog. The deletions may remove one or more essential genes from the chromosomes and this is lethal in a haploid** (which would have no copies of these genes). Alternatively, the combined effect of deletion of many individually nonessential genes may be lethal.

   c. There is only one X-ray-induced mutation per strain, so all genes whose recessive loss-of-function phenotypes are uncovered by a single deletion are on the same chromosome. Using this logic, **all four genes: *w*, *x*, *y*, and *z*, are on the same chromosome.**

   d. **The order is *w y z x* = *x z y w*.** Genes *w* and *y* are deleted in strain 1, uncovering the *w* and *y* alleles, so *w* and *y* are adjacent. Genes *x*, *y*, and *z* are deleted in strain 2 so they must be adjacent. Combined from the information from strain 1, this means the order must be *w y* [*x z*], with the brackets indicating that you don't know the relative order of *x* and *z*. Strain 3 is deleted for *w*, *y* and *z*, therefore the gene that follows *y* must be *z*. Note that two answers are given because you cannot determine the left-to-right orientation of this group of four genes.

7. a. **Each fly has a chromosome containing a different deletion;** in a deletion heterozygote, half the number of sequence reads (relative to reads from elsewhere in the genome) are obtained for the deleted region. **The deletion breakpoints in Fly #1 are ~16.4 Mb and ~16.6 Mb, and in Fly #2 are ~16.45 Mb and ~16.55 Mb.** The breakpoints could be defined to the nucleotide by comparing the sequences of Fly #1 and Fly #2 in this region. **Sequence reads from Fly #1 will contain the deletion breakpoints; these reads will be present at half the frequency of reads elsewhere in the genome, and this particular juxtaposition of sequences will not be present in any sequence reads of Fly #2 or of a wild-type fly without any deletion. The same logic applies for sequence reads from Fly #2.**

    b. At least part of the *st* gene must lie in the region of the chromosome deleted in both Fly #1 and Fly #2. That is, **part of the *st* gene must be in the region from 16.45 Mb – 16.55 Mb.**

    c. **Yes,** some of the *st* gene could easily lie outside the region where the two deletions overlap; **deletion of even a small part of the gene could result in a loss-of-function *st* allele.**

    d. Mapping using deletions is a **complementation** experiment. The deletion chromosome, if the deletion has removed the *st* gene, fails to complement the chromosome containing the *st*⁻ point mutation.

8. a. The diagram at the top of the next page is a summary interpretation of the data given; Del = Deletion, and Inv = Inversion. Two key logical assumptions are used to map genes with deletions or inversions: (1) Deletions that uncover a recessive loss-of-function mutant must delete at least part of that gene, while deletions that do *not* uncover a mutant delete *none* of that gene. (2) Inversions that uncover a recessive loss-of-function mutant must have at least one breakpoint inside that gene (see Fig. 12.12 on p. 420), while the failure of an inversion to uncover a mutant indicates that the gene is completely contained on one side or the other of the inversion breakpoints.

    Deletions A and B together tell us that *javelin* is located between 6,000,587 and 6,703,444, and that *henna* is located between 6,000,587 and 7,220,113. Deletion C refines the location of *henna* to between 6,000,587 and 6,880,255. (Deletion D adds no new information.) Because the deletions tell us that *javelin* and *henna* are both located on 3L (that is, the left arm of chromosome 3), the breakpoints of the inversions on 3R (the right arm of chromosome 3) cannot possibly break in either gene, and so the breakpoints in 3L are the meaningful ones. The *javelin* gene must be on both sides of Inversion A, and thus completely to the left of Inversion B. Thus, the inversions refine **the location of *javelin* to the region between 6,000,587 and 6,520,488, and parts of the gene must be on either side of 6,100,792.** As part of *henna* is within Deletion B, *henna* must be to the right of Inversion B. Thus, Inversion B refines **the location of *henna* to the region between 6,520,488 and 6,880,255.**

chapter 12

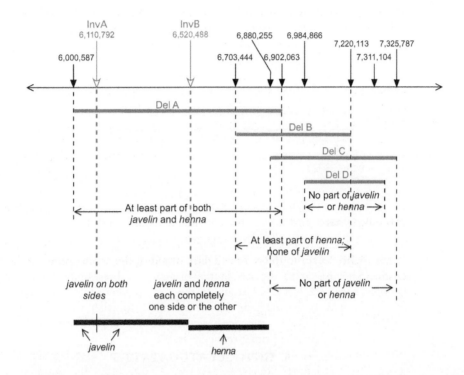

b. The following diagram shows PCR primer pairs (*arrows*, color-coded as indicated) that would amplify the breakpoints of each rearrangement using the genomic DNA of heterozygotes (*rearrangement*/+) as the template. The *arrowheads* are the 3' ends of the primers. Primer pairs for the deletions will amplify a PCR product only from the deleted template because the distance between the primers on the wild-type genomic DNA is too great for PCR amplification (usually PCR products cannot be more than ~20 kb long). Note that the diagram is not drawn to scale.

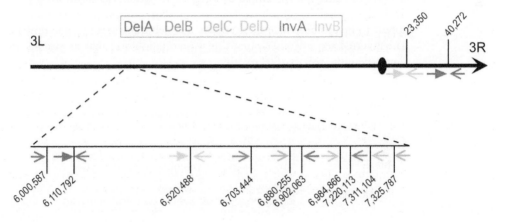

Two different primer pairs, denoted by different kinds of arrowheads, will amplify each of the two breakpoints of each inversion. As an example, amplification of the breakpoints of InvA is shown in the following diagram.

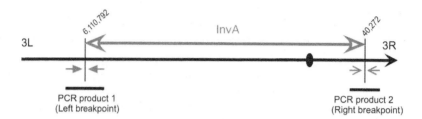

9. a. If **breakpoints 1 and 2** were broken simultaneously, and the DNA between them lost when they were stitched back together, a deletion would result.

    b. Primer A hybridizes to the bottom DNA strand of the chromosome at the top just upstream (5' to, or to the left in the diagram) of breakpoint 1. A primer that hybridizes with the top DNA strand of the same chromosome just upstream (5' to, or to the right in the diagram) of breakpoint 2, together with primer A, would amplify a 32 bp DNA fragment uniquely from a chromosome containing the deletion in part (a). That primer sequence is:

    5' GGTGCCCATGGATATT 3'

    c. **Breakpoints 1 and 2** could yield an inversion if broken simultaneously, and the DNA between them inverted 180⁰, the strands flipped, and then included in the product when both breakpoints are stitched back together.

    d. As shown in the following diagram, primer* will hybridize with DNA in the inverted region, and when paired with primer A, will generate a PCR amplification product only from a template containing the inversion. The sequence of primer* is :
    5' TAGTTTCCCGGGACGT 3'. Note that the primer drawn as a dotted line would work also, but the amplification product would be the shortest possible, not the longest possible. (Don't let it worry you that the base pairs within the inversion appear to be upside-down; there's really no such thing as an upside-down base pair. The inverted sequence is written this way to emphasize that its strands are flipped simultaneously with its 180⁰ rotation.)

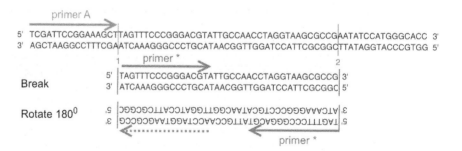

chapter 12

e. **Breakpoints 1 and 3, or breakpoints 2 and 3** could yield a reciprocal translocation. You should note on the diagram that follows in part (f) that the translocated pieces must be put back together in the same orientation as shown in the figure. A translocated chromosome that includes primer A must also have the centromere of the chromosome at the top. For example, if the part of the bottom chromosome to the left of breakpoint 3 was rotated and then connected to the fragment to the left of breakpoint 1 (or to the left of breakpoint 2), the resulting chromosome would be dicentric.

f. The diagram that follows shows that a single primer (primer**) will, together with primer A, PCR-amplify a DNA fragment from translocation 1/3 or translocation 2/3. The sequence of primer** is: **5' TTAATTGGCCTAAAAG 3'**. As shown in the diagram, because each primer hybridizes to a different chromosome, the primer pair cannot amplify a product from normal (nontranslocated) chromosomes.

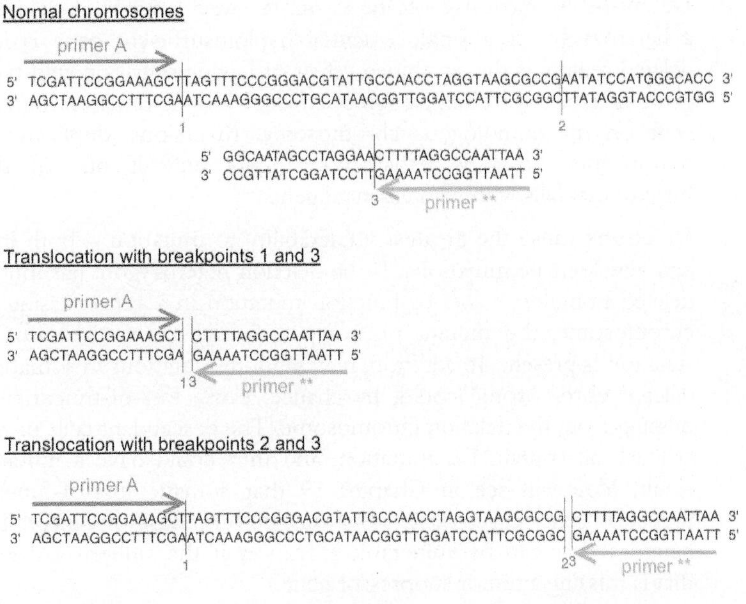

Normal chromosomes

```
                primer A
5' TCGATTCCGGAAAGCT|TAGTTTCCCGGGACGTATTGCCAACCTAGGTAAGCGCC|GAATATCCATGGCACC 3'
3' AGCTAAGGCCTTTCGA|ATCAAAGGGCCCTGCATAACGGTTGGATCCATTCGCGG|CTTATAGGTACCCGTGG 5'
                 1                                        2

             5' GGCAATAGCCTAGGAA|CTTTTAGGCCAATTAA 3'
             3' CCGTTATCGGATCCTT|GAAAATCCGGTTAATT 5'
                                3    primer **
```

Translocation with breakpoints 1 and 3

```
        primer A
5' TCGATTCCGGAAAGCT|CTTTTAGGCCAATTAA 3'
3' AGCTAAGGCCTTTCGA|GAAAATCCGGTTAATT 5'
                 13    primer **
```

Translocation with breakpoints 2 and 3

```
                primer A
5' TCGATTCCGGAAAGCT|TAGTTTCCCGGGACGTATTGCCAACCTAGGTAAGCGCCG|CTTTTAGGCCAATTAA 3'
3' AGCTAAGGCCTTTCGA|ATCAAAGGGCCCTGCATAACGGTTGGATCCATTCGCGGC|GAAAATCCGGTTAATT 5'
                 1                                         23    primer **
```

g. **The translocation is not Robertsonian.** In Robertsonian translocations, one acrocentric chromosome breaks in the short arm, and the other in the long arm. The result of the translocation is a long, compound, metacentric chromosome made up of one long arm and most of a second, and a tiny reciprocal translocation chromosome that sometimes can be lost (see Fig. 12.18 on p. 425). Here, both acrocentric chromosomes were broken in their long arms. The pieces that result that contain a centromere would also necessarily have a short arm, so it would be impossible to generate a Robertsonian translocation in this way.

10. In all of the answers below, the assumptions are: (i) the organism is diploid; and (ii) because rearranged chromosomes are rare, the organism is heterozygous for the rearrangement in question. The 4 classes considered are deletions, duplications, inversions, and reciprocal translocations.

a. **Inversions and reciprocal translocations** are most likely to cause semisterility. Crossovers within inversion loops formed during meiosis in inversion heterozygotes can result in chromosomes with large deletions and sometimes also duplications (Fig. 12.14, p. 422). This result holds whether the inversion is paracentric or pericentric.

   Gametes formed by reciprocal translocation heterozygotes can be balanced only if translocated chromosomes segregate by chance during meiosis I according to the *alternate segregation pattern*. (That is, when the reciprocal translocation pair segregate together into one daughter cell, and the normal chromosome pair segregate together into the other daughter cell.) Because all four chromosomes participate together in homologous chromosome pairing during meiosis, alternate segregation occurs even less than half the time. The gametes formed by the *adjacent-1* and *adjacent-2* segregation patterns, which account for somewhat more than 50% of the gametes, are unbalanced and inviable. (See Fig. 12.17b and c on p. 424.)

b. **Deletions** are most likely to be associated with lethality. A deletion can be lethal in a heterozygote if a single essential haploinsufficient gene is located within the deleted region; if the combined effect of having only one wild-type allele of several genes is lethal; or if the deletion uncovers a loss-of-function mutation in an essential gene on the homologous chromosome. Inversions, duplications, and reciprocal translocations can be lethal more rarely only if one of the rearrangement breakpoints falls within an essential gene.

c. **Deletions** cause the greatest vulnerability to mutation – both inherited mutations and new somatic mutations. If the deletion heterozygote has inherited, on the non-deleted homolog, a loss-of-function mutation in a gene missing from the deletion chromosome, the mutant phenotype (if any) associated with loss of that gene function is present. In addition, new random mutations in somatic cells on the non-deleted chromosome could, by chance, cause loss-of-function of a gene that is missing from the deletion chromosome. The descendant cells of that newly mutated cell will all contain the mutation, and they could have a mutant phenotype as a result. You will see in Chapter 19 that somatic loss-of-function mutations in particular genes called *tumor suppressors* are one cause of cancer. Therefore, deletion heterozygotes can be vulnerable to cancer if they inherit a deletion chromosome that is missing a tumor suppressor gene.

d. **Deletions, inversions, and reciprocal translocations** most obviously alter genetic maps.

   In deletion heterozygotes, the DNA region on the normal homolog that is missing in the deletion homolog cannot crossover because there is nothing for it to crossover with. This means that the genes on the normal homolog that are within the deletion cannot recombine, so the RF between any genes in that region will be zero (when measured in a deletion heterozygote). In addition, the genes flanking the deletion will have a shorter than normal genetic distance between them, because no crossing-over can take place in much of the DNA between them.

   A different phenomenon takes place in inversion heterozygotes. Crossovers between the two homologs within the inverted region do take place, but those recombinant chromosomes are not recovered because they are imbalanced, containing deletions or duplications, and they can be acentric or dicentric (Fig.

12.14, p. 422). Thus, similar to what happens in deletion heterozygotes, the RFs between genes within the inversion are much reduced, and the RFs between genes that flank the inversion are reduced. (The genetic distances between genes within the inversion are not zero because certain kinds of DCOs can take place without producing unbalanced gametes.)

In reciprocal translocation heterozygotes, genes near the translocation breakpoints on nonhomologous chromosomes display a phenomenon called *pseudolinkage*. As only gametes resulting from the alternate segregation pattern produce viable progeny, the alleles of genes on the two chromosomes involved in the translocation appear to be linked (Fig. 12.17c, p.424). Because of the way that the homologs pair during meiosis I (in a cruciform structure – see Fig. 12.17b, p. 424), the genes closet to the translocation breakpoint are the least likely to crossover, and so the pseudolinkage effect is strongest for these genes.

e. **Deletions** are most associated with haploinsufficiency. If a haploinsufficient gene is missing from the deleted chromosome, the haploinsufficient mutant phenotype will be apparent in a deletion heterozygote. (Other rearrangements could break in a haploinsufficient gene, but this is less likely as usually fewer genes are disrupted in inversions, translocations, or duplications than in deletions.)

f. **Deletions, duplications, reciprocal translocations, and inversions** are as an approximation equally likely to generate neomorphic mutations. The breakpoints of these arrangements could fuse the ORFs of two different genes, or fuse the promoter of one gene to the coding region of another gene.

g. **Duplications** are the most likely to generate hypermorphic mutations, as extra copies of a duplicated gene result in gene overexpression. However, deletions, inversions, and reciprocal translocations could more rarely result in a hypermorphic mutation by fusing a more highly active promoter to the coding region of a gene.

h. Crossover suppression is caused mainly by **deletions and inversions.** [See part (d).]

i. **Reciprocal translocations and inversions** are most likely to cause aneuploidy in heterozygotes. Robertsonian translocations (Fig. 12.19, p. 426) can result in aneuploid gametes and progeny. Paracentric inversions can generate tiny chromosomes that could be lost (the acentric fragments in Fig. 12.14b, p. 422), and thereby aneuploid gametes and progeny. (Cells missing an acentric fragment would have one fewer chromosome than a normal cell.)

11. a. The remainder of the male progeny (76,671) are **the parental types, so they will be** y^+ z^1 w^{+R} spl^+ / Y (zeste) and y z^1 w^{+R} spl / Y (yellow zeste split).

b. Classes A and B are a reciprocal pair of products. **They are the result of crossing over anywhere between the *y* and *spl* genes** resulting in the reciprocal classes: y^+ z^1 w^{+R} spl / Y (zeste split) and y z^1 w^{+R} spl^+ / Y (yellow zeste).

c. Remember that the w^{+R} allele is really a tandem duplication of the w^+ gene and that the zeste eye color depends on having a mutant z^1 allele in a genome that also contains two or more copies of w^+. **Classes C and D, are the result of mispairing and unequal crossing over between the two copies of the w^+ gene.** The misalignment can occur in two different ways, as shown in the following figure. Mispairing I gives y^+ z^1 [3 copies w^+] spl (zeste split = same phenotype as class B) as one product and y z¹ [1 copy w⁺] spl⁺ (yellow wild-type eye = class C) as the reciprocal product. In mispairing II, the recombinant products are: y⁺ z¹ [1 copy of w⁺] spl (wild-type eyes split = class D) and y z¹ [3 copies of w⁺] spl⁺ (yellow zeste = same phenotype as class A). Each of these misalignments thus produces one class of recombinant products that results in a wild-type eye color.

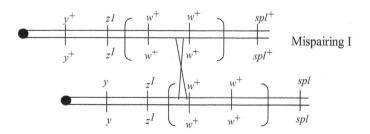

Mispairing I

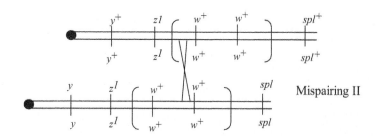

Mispairing II

d. The **genetic distance between y and spl** = # recombinants between y and spl / total progeny = 2430 (class A) + 2394 (class B) + 23 (class C) + 22 (class D) / 81,540 = **5.9 m.u.** This recombination frequency includes all the recombinants, because classes A and B contain the reciprocal events to classes D and C, respectively.

12. Each of the original strains is pure breeding and shows the same recombination frequency of 21 m.u. between genes *a* and *b*. However, in the F_1 heterozygote genes *a* and *b* are only 1.5 m.u. apart. The only rearrangements that affect recombination frequency in this way in the heterozygote are deletions and inversions. There are two reasons why this cross cannot involve a deletion: (1) Both parental strains are homozygous (true breeding) and homozygous deletions are usually lethal; (2) In both parents, genes *a* and *b* are 21 m.u. apart, but if one parent were a deletion homozygote, the genes would be less than 21 m.u. apart.

This reduction of recombination between the 2 genes in the F_1 is therefore caused by an inversion. One parental strain has normal chromosomes and the other parental strain is homozygous for an inversion. There are 2 possibilities for the inverted region: (i) It includes almost all the region between genes *a* and *b*, but does not include the genes themselves, or (ii) the inversion includes both genes and the DNA in between them.

Two other possibilities for the location of the inversion with respect to the genes can be ruled out. (iii) the inversion includes gene *a* and almost all of the DNA between the genes; and (iv) the inversion includes gene *b* and almost all of the DNA between the genes. These possibilities can be ruled out because in both cases the recombination frequency between *a* and *b* in the inversion homozygote would be much less than 21 m.u., as the inversion would bring the genes closer together.

In any case, the F_1 progeny are inversion heterozygotes. The inversion loop occupies either (i) almost all the region between genes *a* and *b*, although the genes are not included in the loop, or (ii) the loop includes both genes and the DNA in between them. Any single crossovers within the inversion loop (the huge majority of crossovers between the genes) will result in inviable gametes. The few recombination events that occur between gene *a* and the inversion loop or between the inversion loop and gene *b* will result in balanced, viable, recombinant gametes in scenario (i), as will some of the double crossover events within the inversion loop in both scenarios (i) and (ii).

13. a. **2, 4**. Inversion loops are seen during meiosis I only if the cells are heterozygous for an inversion.

    b. **2, 4**. Single crossovers within the inversion loop in inversion heterozygotes generate genetically unbalanced chromosomes. The genetic imbalance involves deletions and duplications of regions outside the inversion loop. If the inversion is paracentric, then the recombinant products are also dicentric and acentric as well as imbalanced with respect to genes. Crossovers in inversion homozygotes do not cause genetic imbalance.

    c. **2**. An acentric (and the reciprocal dicentric) fragment is produced from a single crossover within a paracentric inversion in an inversion heterozygote.

    d. **1, 3**. In an inversion homozygote, crossovers within the inversion yield 4 viable, balanced, spores all of which have the inverted gene order.

14. The data shows unexpectedly reduced recombination frequencies between certain pairs of genes in Bravo/X-ray and Bravo/Zorro heterozygotes. This reduction in recombination will be seen in both deletion heterozygotes and in inversion heterozygotes. You are told that the 3 strains have variant forms of the same chromosome, and that karyotype analysis has shown that none of the strains is missing any part of the chromosome. Thus, none of the strains is a deletion homozygote. The chromosomal rearrangements here must be inversions.

    a. Recombination frequencies between genes in the Bravo/X-ray heterozygote are normal for *a-b* and *b-c* and *g-h* but are reduced for all other gene pairs. Thus, the inversion in the X-ray strain breaks between genes *c-d* and between genes *f-g* and inverts genes *d, e,* and *f*. The *c-d* and *f-g* intervals must still include some non-inverted DNA to allow recombination that produces viable gametes. Similarly, the genetic distance is reduced in the *b-c* and *f-g* intervals for Zorro, where the inversion end points are found and minimal in those intervals completely within the inversion. **The order of the genes in X-ray is:** *a b c f e d g h*. **The order of the genes in Zorro is:** *a b f e d c g h*.

    b. **The physical distance in the X-ray homozygotes between *c* and *d* is greater** than that found in the original Bravo homozygotes. The inversion occurred in this

12-17

portion of the chromosome, so *c* and *d* are now separated by many more genes (all the inverted DNA).

c. **The physical distance between *d* and *e* in the X-ray homozygotes is the same** as that found in the Bravo homozygotes because this interval is completely within the inverted segment. The relationship of *d* to *e* has not changed.

15. The diploid cell contains a pericentric inversion on one homolog. The pairing of the homologous chromosomes during metaphase I of meiosis is shown in the following diagram, where each line represents a single chromatid in a bivalent. Use this drawing to trace the consequences of crossovers in different regions.

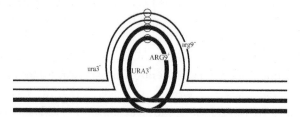

a. A single crossover outside the inversion produces **four viable spores: 2 *URA3 ARG9* (prototrophic) : 2 *ura3 arg9* (auxotrophic for uracil and arginine).**

b. A single crossover within the inversion loop, in this case between *URA3* and the centromere, results in unbalanced recombinant gametes. Both recombinant gametes have a duplication and a deletion of the material outside of the inversion loop. One recombinant is duplicated for the region outside the loop on the left and deleted for the information outside the loop on the right. The other recombinant is the reciprocal - deleted for the information on the DNA to the left of the loop and duplicated for the information outside the loop and to the right. This genetic imbalance is usually lethal, so the two spores containing the products of the recombination will die. The single crossover inside the loop gives rise to **2 parental spores (viable) and 2 recombinant, lethal spores as in the following ascus: 1 *URA3 ARG9* (viable) : 1 *ura3 arg9* (viable) : 1 *URA3 arg9* (lethal) : 1 *ura3 ARG9* (lethal).**

c. This 2-strand DCO within the inversion loop produces **four viable parental spores: 2 *URA3 ARG9* (1 NCO, 1 DCO) : 2 *ura3 arg9* (1 NCO, 1 DCO).**

16. In this problem, the diploid cell contains a paracentric inversion on one homolog. The pairing of the chromosomes in the inversion heterozygote is shown in the following figure, where each line represents a single chromatid in a bivalent. Use this drawing to trace the consequences of crossovers in different regions.

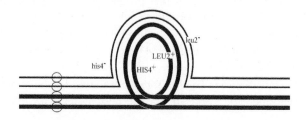

12-18

chapter 12

a. A single crossover within the inversion (between *HIS4* and *LEU2*) leads to 2 parental spores (one of each type) and 2 recombinant spores. The recombinants are duplicated and deleted for the regions outside the inversion loop. In this case, one of the recombinant spores will be duplicated for the region containing the centromere and deleted for the DNA on the other side of the inversion loop (dicentric). The reciprocal recombinant spore will be deleted for the region containing the centromere and duplicated for the region to the right of the inversion loop (acentric). Spores that receive these chromosomes will definitely die. Assuming that the acentric and dicentric chromosomes segregate properly, **the ascus will contain: 1 *HIS4 LEU2* (prototrophic) : 1 *his4 leu2* (auxotrophic for histidine and leucine) : 1 dicentric (lethal) : 1 acentric (lethal)**. (Even if these acentric and dicentric chromosomes do not segregate properly so that they are lost or broken, two lethal spores would result.)

b. This is a 2-strand double crossover within the inversion loop. All four spores are viable: **2 *HIS4 LEU2* (1 NCO and 1 DCO) : 2 *his4 leu2* (1 NCO and 1 DCO)**.

c. A single crossover between the centromere and the inversion loop will give 4 parental spores: **2 *HIS4 LEU2* (1 NCO and 1 SCO) : 2 *his4 leu2* (1 NCO and 1 SCO)**.

17. One way to approach this problem is to realize that any one chromatid that participates in just one crossover event that occurs in a paracentric inversion loop will end up with one part being in a dicentric chromosome and the other part in an acentric fragment. In an SCO meiosis, two of the spores will be inviable for this reason (Fig. 12.14b on p. 422). In a four-strand DCO with both crossovers taking place in the inversion loop, no spores will be viable because two dicentric chromosomes and two acentric fragments would be formed. In a three-strand DCO, one of the strands participates in only one crossover, so again part of this chromatid will be part of a dicentric chromosome and the other part of an acentric chromosome.

    The only possibilities that could create a tetratype ascus with 4 viable spores involve a 2-strand double crossover (DCO). This could happen in two ways. In one scenario, one recombination occurs between *LEU* and *HIS* and the second recombination occurs between *HIS* and the end of the inversion loop. In the second scenario, one crossover occurs between *LEU* and *HIS* and the second one takes place between *LEU* and the end of the inversion loop. In both cases, the second recombination event must occur within the inversion loop. Either such kind of 2 strand double cross over will give the following tetratype ascus: 1 *HIS4 LEU2* : 1 *his4 leu2* : 1 *HIS4 leu2* : 1 *his4 LEU2*.

18. Any haploid spores with a deletion are dead (white). An octad has 8 spores.

    a. **0 white spores.** The inversion has no effect if recombination does not occur. (That is, none of the spores has a chromosomal deletion.)

    b. **4 white spores.** Only two chromatids are involved in the crossover. The recombination gives 2 unbalanced gametes (which have a duplication for some genes but a deletion for others; see Fig. 12.14a on p. 422). The remaining two chromatids survive as haploid products and divide mitotically to form 4 viable spores in the ascus.

c. **0 white spores.** If a crossover occurs outside the inversion loop all the products are viable.

d. **8 white spores.** All of the gametes resulting from adjacent-1 segregation in a translocation heterozygote would be genetically imbalanced and would die (Fig. 12.17c on p. 424).

e. **0 white spores.** Alternate segregation produces balanced gametes (Fig. 12.17c).

f. **0 white spores.** The crossover in the translocated region would simply cause the reciprocal exchange of DNA between homologous portions of the chromosome. No genetic imbalance would occur, so all spores live.

19. a. **1, 3, 5 and 6.** Translocation heterozygotes can produce gametes with any pairwise combination of N1, N2, T1, and T2.

    b. **2 and 4.** Translocation heterozygotes cannot produce gametes with two copies of the same chromatid through any segregation pattern occurring during meiosis I (see Fig. 12.17c on p. 424).

    c. **1 and 3.** These arise from alternate segregation, so they are balanced. You can also establish this point by noticing that the gametes in 1 and 3 have one copy of each region of both chromosomes.

    d. **5 and 6;** these arise from adjacent-1 segregations. **2 and 4;** these arise from adjacent-2 segregations. Note that the adjacent-1 segregations are much more frequent than adjacent-2 segregations. The reason is that the adjacent-1 pattern represents a normal disjunction that occurs when homologous centromeres go to opposite spindle poles, while the adjacent-2 pattern is a form of nondisjunction that takes place as a result of the unusual cruciform structure that forms when the normal and translocated chromosomes pair during meiosis I.

20. a. Diagram the cross. Because the two genes are on different autosomes, they should assort independently. In the cross scheme below, the alleles written in *blue* come from one parent and those written in *black* come from the other parent.

    $cn\ cn^+$; $st\ st^+$ × $cn\ cn$; $st\ st$ → 1/4 $cn\ st$ / $cn\ st$ (white) : 1/4 $cn\ st^+$ / $cn\ st$ (cinnabar) : 1/4 $cn^+\ st$ / $cn\ st$ (scarlet) : 1/4 $cn^+\ st^+$ / $cn\ st$ (wild type).

    b. The genes show pseudolinkage in this male. The $cn\ st$ and $cn^+\ st^+$ allele combinations seem to be linked. **This result suggests that the unusual male has a translocation between chromosome 2 and 3 with the mutant *cn* and *st* alleles either on the translocated chromosomes or on the normal chromosomes.** The following figures show two of the four possible genotypes for the unusual male fly. Both diagrams show the mutant alleles on the normal order chromosomes. Instead, the mutant alleles of the genes could both be on the translocated chromosomes. Although both the *cn* and *st* genes may actually be on the same chromosome after the translocation (right hand panel), this is not necessary. The pseudolinkage will be seen even if the 2 genes are still on separate chromosomes (left hand panel). Also remember that the in the germ line of male *Drosophila*, chromosomes do not recombine, so only parental type progeny will be produced.

chapter 12

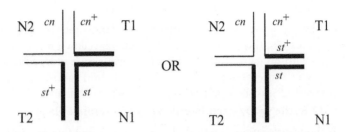

c. The wild-type F_1 females are translocation heterozygotes containing the cn^+ and st^+ alleles on the translocated chromosomes (she obtained the cn and st alleles from her non-translocated mother). The pairing in meiosis would be the same as shown for the male in part (b), except that recombination can occur in the female. **A crossover either between cn and the translocation breakpoint or between st and the translocation breakpoint, followed by alternate segregation produces gametes with the genotypes $cn\ st^+$ (cinnabar) and $cn^+\ st$ (scarlet).** These classes allow you to calculate the map distance between the st and cn genes in the translocation: RF= 10 m.u. (See the following diagram.)

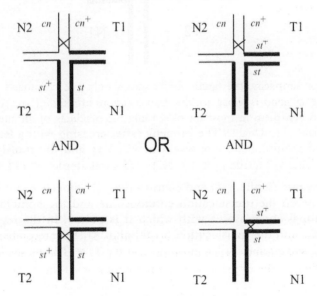

21. The semisterile F_1 is a translocation heterozygote and will produce 1/2 fertile : 1/2 semisterile progeny from alternate segregation. Products of adjacent-1 or adjacent-2 segregation are imbalanced and therefore inviable - this is the basis of the semisterility. Because the only viable gametes are the result of alternate segregation, genes that are on the chromosomes involved in the translocation will not assort independently. Instead, if the genes are located very close to the translocation breakpoints, only the parental classes will be viable, so the genes will display pseudolinkage. Genes that are on any

12-21

other chromosome will assort independently of the translocation (i.e., such genes will assort independently from fertility/semisterility).

a. If the *yg* gene is on a different chromosome than those involved in the translocation, the traits [leaf color (green vs. yellow-green) and fertility/semisterility (ears with no gaps vs. ears with gaps)] will assort independently. The product rule says you can cross-multiply the 2 monohybrid ratios: 1/2 *yg*⁺ (normal green leaf color); 1/2 *yg* (yellow green) and 1/2 fertile : 1/2 semisterile to give: **1/4 fertile *yg*⁺ : 1/4 fertile *yg* : 1/4 semisterile *yg*⁺ : 1/4 semisterile *yg*.**

b. If the translocation involved chromosome 9, the fate of the fertility and leaf color phenotypes are connected – these phenotypes will show pseudolinkage. The original cross was: semisterile *yg*⁺ × fertile *yg* → F₁ semisterile × fertile *yg*. This means the normal, non-translocated chromosome 9 has the *yg* allele, while the translocated chromosome 9 has the *yg*⁺ allele. The chromosomes of the heterozygous F₁ at meiosis I would look like:

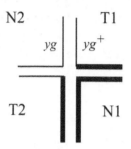

(For simplicity, the figure above shows only one chromatid per chromosome.) If the *yg* gene is close to the translocation breakpoint, no crossovers will occur between them in most meiosis. Only the products of an alternate segregation are balanced and viable. **The progeny (after crossing with a fertile *yg* homozygote, and assuming no crossovers between *yg* and the translocation breakpoint) will be: 1/2 fertile *yg* (N1 + N2) : 1/2 semi-sterile *yg*⁺ (T1 + T2).**

c. The rare (fertile, *yg*⁺) and (semisterile, *yg*) gametes result from recombination between the translocation chromosome and the homologous region on the normal chromosome with which it is paired, in the region between the *yg* gene and the translocation breakpoint. After the recombination event, the N1 + N2 fertile gamete will contain *yg*⁺ and the T1 + T2 semisterile gamete will contain *yg*.

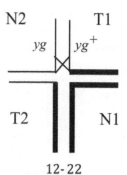

chapter 12

The frequency of crossing-over, as represented by the rare fertile green and semisterile yellow-green progeny, will give you the genetic distance between the translocation breakpoint and the *yg* gene.

22. Individuals homozygous for a translocation are fertile. However, **when insects that are translocation homozygotes mate with insects with normal chromosomes, the F_1 progeny will be translocation heterozygotes. The fertility of these F_1s should be reduced by about 50% and half of their progeny will also have reduced fertility.** If the released insects are homozygous for several different translocations, then the fertility of the F_1 individuals should be reduced by a further 50% for each translocation for which they are heterozygous. For example, in an F_1 insect that was heterozygous for three different translocations (T1 - T3), only 1/8 of its gametes (1/2 balanced for T1 x 1/2 balanced for T2 x 1/2 balanced for T3 = 1/8 balanced gametes) would be balanced and give rise to progeny.

23. Remember that the Y chromosome pairs with the X chromosome during meiosis, so the N1 + N2 chromosomes in the male translocation heterozygote will be the autosome on which *Lyra* is normally found and the X chromosome, as shown on the following figure. (The chromosomes are drawn as bars, with a single bar representing both sister chromatids. The autosome is **black**, the Y chromosome is *blue*, and the X chromosome is *red*.)

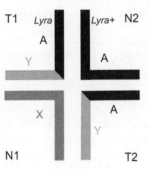

This male can make only two kinds of genetically balanced gametes, which are the products of alternate segregation. (Remember as well that no recombination occurs in *Drosophila* males.) The T1 + T2 segregant has the Y chromosome and the autosome with the mutant *Lyra* allele. This gamete would fertilize an X-bearing egg, producing Lyra males (the *Lyra* allele is dominant). N1 + N2 would yield a gamete with the X chromosome and the autosome with *Lyra*⁺. This gamete would produce wild-type females when fertilized with *Lyra*⁺ gametes from the wild-type mother. **The progeny will be 1/2 Lyra males : 1/2 Lyra⁺ females.**

24. a. A translocation heterozygote will make two types of gametes as a result of alternate segregation and two types of gametes as a result of adjacent-1 segregation in a 1:1:1:1 ratio (these two segregation patterns represent normal disjunction and are equally likely). The 2 products of alternate segregation are N1 + N2 and T1 + T2, both of which are balanced gametes. The two products of adjacent-1 segregation are N1 + T2 and N2 + T1, both of which are imbalanced gametes. When a

translocation heterozygote is crossed to a homozygous normal individual, the imbalanced gametes from the translocation parent never give rise to viable progeny. However, if <u>both</u> parents of a cross are translocation heterozygotes (as is the case in self-fertilization), then it is possible for an imbalanced gamete from one parent to be fertilized by the reciprocally imbalanced gamete from the other parent. For example, a N1 + T2 gamete from one parent can fertilize an N2 + T1 gamete from the other parent, creating a zygote that is a balanced translocation heterozygote.

The following table summarizes all of the possible fertilizations in the cross of two translocation heterozygotes. The gametes produced by alternate segregation are shown in *blue* while the gametes from adjacent-1 segregation are in *red*. Viable progeny that are fertile (homozygotes either for the normal chromosomes or the translocated chromosomes) are in *green*, while semisterile progeny (translocation heterozygotes) are in *purple*.

	N1 + N2	T1 + T2	N1 + T2	N2 + T1
N1 + N2	homozygous normal	translocation heterozygote	imbalanced, lethal	imbalanced, lethal
T1 + T2	translocation heterozygote	translocation homozygote	imbalanced, lethal	imbalanced, lethal
N1 + T2	imbalanced, lethal	imbalanced, lethal	imbalanced, lethal	translocation heterozygote
N2 + T1	imbalanced, lethal	imbalanced, lethal	translocation heterozygote	imbalanced, lethal

Thus, among the viable progeny you would expect a ratio of 2/6 fertile (homozygous normal + translocation homozygote) : 4/6 semisterile (translocation heterozygote) = **1/3 fertile : 2/3 semisterile**.

b. This problem involves the self-fertilization of a particular translocation heterozygote. Instead of producing 2/6 fertile : 4/6 semisterile progeny as in part (a), this plant produced a ratio of 1/5 fertile : 4/5 semisterile. These numbers suggest that one out of the two fertile and viable classes in part (a) did not survive in this cross. Thus, one possible explanation for these results is that **the translocation homozygotes die because the translocation breakpoint interrupts an essential gene**. Alternatively, one of the four chromosomes involved in the translocation has a recessive lethal allele, so that homozygotes for this chromosome could not survive.

25. The X and A (autosomal) chromosomes of a female heterozygous for the X-autosome translocations that break within the X-linked *dystrophin* gene are diagrammed in the following figure:

chapter 12

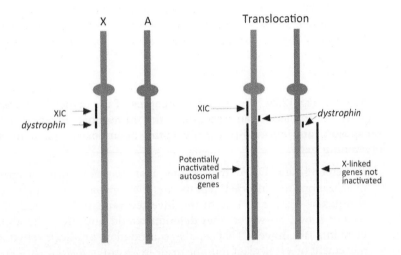

a. Even though both X chromosomes contain a functional XIC, cells in which the normal (intact) X chromosome is inactivated are more likely to be found than cells in which the X-A translocation is inactivated. On the translocation chromosome, the XIC would inactivate many autosomal genes, and this is likely to cause cell lethality because the effective dosage of these genes is then one rather than the normal two. (The genes are present in two copies but only one copy is expressed.) Another consequence of inactivation of the translocation chromosome rather than the intact X is that many X-linked genes on the other translocation chromosome will not get inactivated; overexpression of these X-linked genes could also be lethal. For both of these reasons, most surviving cells would have the normal X as the Barr body, and the X involved in the translocation, which inactivates the *dystrophin* gene, expressed. Therefore, **most cells of the woman will not express *dystrophin*.** This reasoning explains why the woman would have disease symptoms almost as severe as that seen in males hemizygous for a recessive loss-of-function allele of *dystrophin*.

b. **The autosomal genes indicated on the XIC-containing translocation chromosome in the diagram preceding part (a) would be inactivated if that chromosome became the Barr body, and the X-linked genes indicated on the other translocation chromosome would be expressed from both Xs.**

c. The observation that none of the woman's cells express *dystrophin* means that **the only cells that survive are those that inactivated the intact X.** Cells that inactivated the translocation chromosome could have died because of inactivated autosomal genes, failure to inactivate X-linked genes, or a combination of the two effects.

d. In cells with an X-A translocation, X chromosome genes no longer connected to the XIC could escape X inactivation, while autosomal genes now connected to the XIC could become inactivated. **If you know the locations of the translocation breakpoints, by analyzing the expression of genes on either side of the breakpoints, you can infer on which side of the X chromosome breakpoint the XIC is located.** Each different X-A translocation you analyze would refine the map of the XIC. However, interpretation of such experiments would be

12-25

complicated because some cells bearing the translocation die, if too many X chromosome genes are no longer subject to inactivation, and/or if too many autosomal genes are inactivated.

Section 12.3

26. As shown in Fig. 12.26 on p. 432, **if two copies of the same DNA transposon flank a gene, transposase can recognize the outermost inverted repeat of each transposon, and cut-and-paste the entire "composite" element, including the intervening gene.**

27. a. *P* elements are DNA transposons that rely on their inverted repeats for mobilization; the inverted repeats are the binding sites and substrates for the transposase enzyme. **If one of the inverted repeats is deleted, that *P* element cannot move. However, that deletion would not affect the ability of other *P* elements to move.** (That is, the contribution of the inverted repeats to the movement of a *P* element must be made in *cis*, not in *trans*.)

    b. The *yellow* intron (Fig. 12.27, p. 433) must be spliced out of the *P* element primary transcript in order to generate transposase mRNA; with the *yellow* intron remaining, the mRNA instead encodes a repressor of transposition. **A *P* element with a mutation that prevents splicing out of the *yellow* intron can move if somewhere else in the genome resides another *P* element that encodes transposase. A *P* element with such a mutation could in some cases affect the ability of other *P* elements to mobilize.** If the mutant transposase transcript simply retains the *yellow* intron, and thus produces repressor, the added repressor protein could inhibit – at least to some extent – the frequency of mobilization of other *P* elements in the genome. In addition, recall that the previously autonomous *P* element was a source of transposase. Depending on how many other autonomous *P* elements are present in the genome, the loss of the transposase from this one element could have a small or a large effect on the frequency of mobilization of nonautonomous *P* elements.

    c. **If the *P* element in part (b) had been the only autonomous *P* element in the genome, it would be immobile, as would be all other *P* elements in the genome.** Cells would no longer have any gene encoding transposase.

    d. *P* elements are normally mobile only in germ-line cells because only in the germ line is the *yellow* intron Fig. 12.27 removed from the primary transcript; thus, transposase is made only in the germ line. **A mutant *P* element in which the *yellow* intron was deleted exactly could mobilize in somatic cells, because the primary transcript of this mutant *P* element would in essence have the *yellow* intron spliced out already. The mutation would enable the mutant *P* element to make transposase in somatic cells, so other *P* elements would also mobilize in the soma.** Obtaining a deletion that exactly removes the *yellow* intron would be an extremely rare event, but is by no means impossible.

28. The original *ct* mutant allele was caused by the insertion of the *gypsy* transposable element. The key to this problem is to understand that transposable element mobilization can sometimes be a "messy" process. For example, when DNA

chapter 12

transposons like *P* elements are cut out of one location, normally the genomic DNA at that position is stitched back together without deleting any base pairs, a process called *precise excision*. However, occasionally some DNA at the site of the original *P* element insertion is deleted during the process of *P* element mobilization, a process called *imprecise excision*. This imprecise excision may delete nucleotide pairs in the gene, nucleotide pairs in the transposable element, or both. Sometimes imprecise excision even leaves parts of the transposable element at the original insertion site.

The stable ct^+ revertants are likely to be precise excisions of DNA transposons in which the *gypsy* element has moved out of the gene completely, restoring the normal ct^+ sequence. These revertants are stable because no transposable element DNA remains at the site of the original insertion. The unstable ct^+ revertants are likely to be cases in which the transposition process removed part but not all of the *gypsy* element while retaining the critical sequences in the *ct* gene. In such revertants, the *ct* gene could still function normally. However, these revertants would be unstable because the *gypsy* sequence remaining would still be subject to the transposition mechanism. Imprecise excision of these remaining *gypsy* sequences could result in deletions or rearrangements of the *ct* gene, thus creating new, stronger *ct* mutant alleles.

The answer above assumes for simplicity that *gypsy* is a DNA transposon, but in fact, it is a retrotransposon. It turns out that the events resulting in stable and unstable revertants of the ct^{MR2} mutation are generally more complex than those just outlined. For example, several revertants are actually associated with the insertion of other transposable elements into the gene or even into the *gypsy* retrotransposon.

Section 12.4

29. a. **Bob is likely to have Down syndrome.** Each of his genomic ASO probes to loci on chromosome 21 hybridize with 150% the intensity as compared with the probes from his parents' genomes. (We are assuming here that, as suggested by the patterns in the parents Fred and Mary, *yellow* indicates one copy of the region, *orange* two copies, and *red* three copies.) The pattern of hybridization suggests that Bob has 3 alleles rather than the normal 2 alleles of each of these loci.

    b. Bob inherited one chromosome 21 from one parent, and two chromosome 21s from the other parent due to nondisjunction (NDJ). Comparison of the alleles at each locus in the parents and in Bob reveals that Bob's TT alleles at loci 5, 7, and 9 could only have come from Mary, because Fred has no T alleles at any of these loci. This means that **the extra chromosome 21 inherited by Bob came from Mary's gamete.** Because Mary has only one T allele at each of loci 5, 7, and 9, the extra chromosome 21s are most likely sister chromatids and that came **from NDJ at meiosis II** in one of Mary's secondary oocytes.

    c. At locus 10, Bob inherited both G and C alleles from Mary; in order for the sister chromatids to have different alleles at locus 10, a crossover must have occurred between loci 9 and 10 as shown in the following diagram.

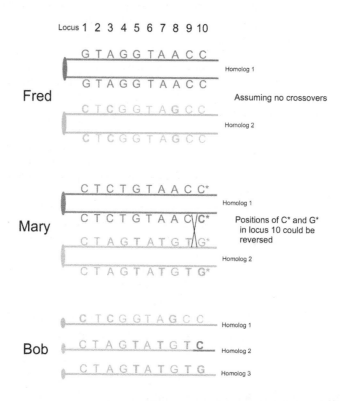

d. Without crossing-over, if the chromosome 21s in the n + 1 egg from Mary were due to NDJ at meiosis I, the expectation is that the two chromosome 21s would have different alleles at the loci that were heterozygous in Mary's genome, because they would be nonsister chromatids. On the other hand, if the two 21s were produced by NDJ at meiosis II, they would be expected to have identical alleles at heterozygous loci, because they are sister chromatids.

But now consider the fact that the further a locus is from the centromere, the more likely a crossover between the centromere and that locus will occur in any given meiosis. Because these crossover events move alleles between nonsister chromatids, recombination can obscure whether NDJ occurred at meiosis I or meiosis II. One solution is to analyze loci close to the centromere. The shorter the distance of the centromere to the locus, the less probability exists that a crossover event could have taken place between them. Also, as can be seen in the examples of loci 5,7, and 9, analysis of multiple loci can also be helpful. The TTs inherited by Bob at those three loci had to come from Mary, but they are unlikely to have resulted from nonsister chromatids and NDJ at meiosis I, as multiple crossovers within a small region of the chromosome would have been required to produce this pattern.

30. a. **The key to this problem is to realize that uniparental disomy can occur only if two rare events take place, explaining why this phenomenon is so rare.** One

chapter 12

homologous chromosome must somehow be lost, and the remaining one has to be duplicated. The following figure diagrams some of the several scenarios that can give rise to uniparental disomy. Paternal and maternal copies of the chromosome are in different shades of *blue*.

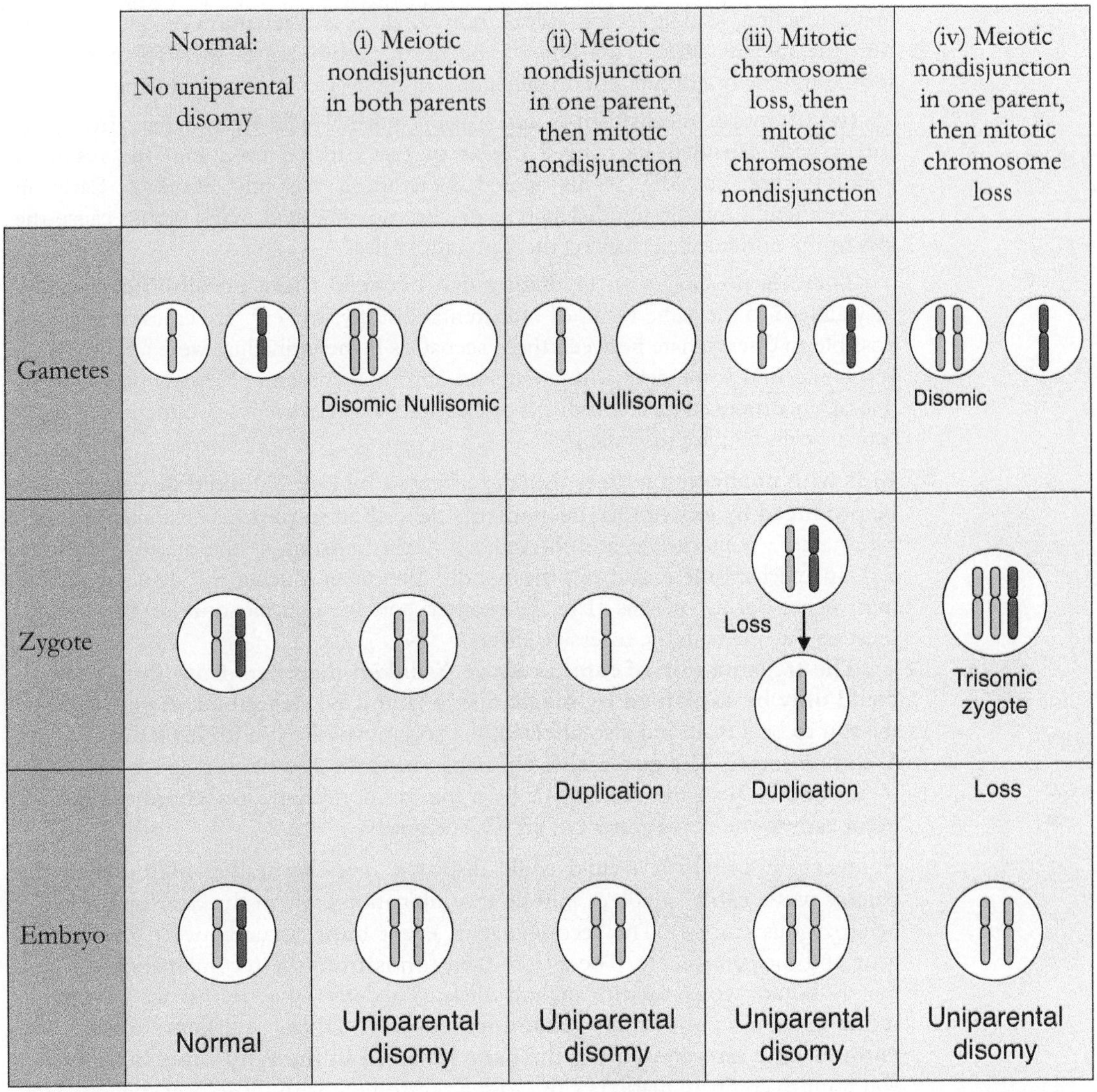

	Normal: No uniparental disomy	(i) Meiotic nondisjunction in both parents	(ii) Meiotic nondisjunction in one parent, then mitotic nondisjunction	(iii) Mitotic chromosome loss, then mitotic chromosome nondisjunction	(iv) Meiotic nondisjunction in one parent, then mitotic chromosome loss
Gametes		Disomic Nullisomic	Nullisomic		Disomic
Zygote				Loss ↓	Trisomic zygote
Embryo	Normal	Duplication Uniparental disomy	Duplication Uniparental disomy	Duplication Uniparental disomy	Loss Uniparental disomy

    (i) Perhaps the most obvious mechanism for uniparental disomy is the fusion of a nullo gamete from one parent (the result of nondisjunction in either MI or MII in that parent) with an MII nondisjunction gamete from the other parent (the mutant homolog must be the one that undergoes MII nondisjunction).

    (ii) A second mechanism begins with the fusion of a nullo gamete from one parent (due to MI or MII nondisjunction) with a normal gamete carrying the mutant allele from the other parent. This generates a monosomic embryo, which then undergoes mitotic nondisjunction of the monosomic chromosome early in development to create an individual homozygous for that chromosome.

(iii) A third possible mechanism begins with the formation of a normal heterozygous embryo. Very early in development this embryo undergoes mitotic nondisjunction, leading to loss of the normal allele and retention of one homolog with the mutant allele. This loss of the normal homolog is then followed by a second mitotic nondisjunction to generate a homozygous mutant genotype.

(iv) A fourth mechanism is fusion of a normal wild-type gamete from one parent with a gamete carrying 2 copies of the affected homolog (the result of nondisjunction in MII). This would generate a trisomic embryo. Early in development a mitotic nondisjunction or chromosome loss event would cause the loss of the normal allele, leaving the 2 mutant alleles.

There is no easy way to distinguish between these possibilities, because they all lead to the same result of uniparental disomy. In very rare cases, it might be possible to discriminate between these scenarios if the individual were a mosaic and you could find some cells with aneuploid chromosome complements predicted by one of the proposed mechanisms. (For example, in scenario (iv) you might see some somatic cells that are trisomic.)

b. **Girls with unaffected fathers that are affected by rare X-linked diseases could be produced by any of the mechanisms described in part (a).** The mother must be a heterozygous carrier, and the father's X chromosome is the one that must be lost. In mechanisms (i and iv), the nondisjunction producing the disomic gamete must occur during meiosis II in the mother, and the chromosome that is disomic must be the one with the recessive allele.

The transmission of rare recessive X-linked disorders from father to son could only be explained by mechanisms (i and iv) described in part (a). For the son to be a male and also affected, the son <u>must</u> inherit <u>both</u> his father's X and Y chromosomes. The maternal X chromosome must also be lost; this could occur by meiotic NDJ in the mother, or by a mitotic nondisjunction/chromosome loss event early in the development of an XXY zygote.

c. Another way in which a child could display a recessive trait if only one of the parents was a carrier involves mitotic recombination early in the development of a heterozygous embryo. The recombination event must occur between the mutant gene and the centromere. The zygote then forms from the recombination product that is homozygous for the mutant allele. To detect the occurrence of mitotic recombination, **you must examine several DNA markers along the chromosome arm containing the gene involved in the syndrome**. In particular, you want to compare markers close to the centromere with those near the telomere. **If the presence of the disease were due to mitotic recombination, then the person would be heterozygous for markers near the centromere but homozygous for those near the telomere. If uniparental disomy caused by any of the mechanisms described in part (a) was instead involved, the individual would be homozygous for all of the DNA markers** (assuming that the individual is not a mosaic).

chapter 12

31. **Both types of Turner's mosaics could arise from chromosome loss or from mitotic nondisjunction early in zygotic development.** Chromosome loss would involve the loss of one of the X chromosomes early in development in an XX embryo (producing a mosaic with both 46, XX and 45, XO cells) or the loss of the Y chromosome in a cell of the XY embryo (yielding a mosaic with 46, XY and 45, XO cells). Mitotic nondisjunction in a normal XX embryo should produce an XXX daughter cell in addition to an XO, while mitotic nondisjunction in an XY embryo yields an XO and an XYY daughter cell. If the XXX or XYY daughter cells did not expand into large clones of cells during development, karyotype analysis would not be able to detect their presence. Note that for mitotic nondisjunction to have given rise to the described mosaic individuals, the nondisjunction event must have occurred after the first mitotic division so that there would be some XX or XY cells.

32. The genotype of the diploid *Neurospora* cell is shown below; the figure shows prophase of meiosis I after mating. In the answers to parts (a) –(f), asci are written as though they have 4 (rather than 8) spores, and all spores are written in order relative the equator of the ascus.

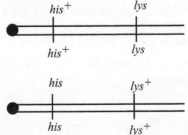

a. A single crossover between the centromere and *his* will give a PD ascus that shows MII segregation for both *his* and *lys*: **1 his+ lys : 1 his lys+ : 1 his+ lys : 1 his lys+**.

b. A single crossover between *his* and *lys* will give a T ascus showing MI segregation for *his* and MII segregation for *lys*: **1 his+ lys : 1 his+ lys+ : 1 his lys : 1 his lys+**.

c. Nondisjunction during meiosis I causes one daughter cell to have both homologs (4 chromatids) and the other daughter cell to have none (nullo). At metaphase II of meiosis, the homologous chromosomes align independently of each other on the metaphase plate, and 1 chromatid from each segregates into the spores. This leads to 2 spores that are disomic for the chromosome that underwent nondisjunction. (The nullo cell will have two nullo daughters.) The ascospores after meiosis II are: **2 his+ lys / his lys+ (disomic, his+ lys+) : 2 nullo (aborted, white)**.

d. Nondisjunction during meiosis II affects one of the 2 daughter cells formed after meiosis I. In the affected daughter cell, both chromatids go to one spore and the other spore is nullo. Two different results are possible, depending on which daughter cell undergoes meiosis II nondisjunction: **2 his+ lys : 1 his lys+ / his lys+ (disomic, his lys+) : 1 nullo (aborted) OR 1 his+ lys / his+ lys (disomic, his+ lys) : 1 nullo (aborted) : 2 his lys+**.

e. The single crossover between the centromere and *his* involving chromatids 2 and 3 (the two "inside" chromatids) gives the following meiotic structure:

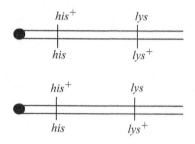

Nondisjunction in meiosis I gives one daughter cell with both homologous centromeres and a second nullo daughter cell. At metaphase of meiosis II, the homologous centromeres line up on the metaphase plate independently of each other. MII segregation then causes one of each sister chromatid to segregate into the 2 resultant spores. After the crossover, the bivalents are no longer homozygous. As a result, there are 2 different segregation patterns that can happen (depending on the orientation of the chromatids of the two chromosomes during meiosis II), leading to the following types of asci: **1 *his+ lys* / *his+ lys* (disomic his+ lys) : 1 *his lys+* / *his lys+* (disomic his lys+) : 2 nullo (aborted) OR 2 *his+ lys* / *his lys+* (disomic his+ lys+) : 2 nullo (aborted)**.

f. The single crossover between *his* and *lys* gives the following meiotic structure at metaphase I of meiosis:

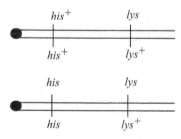

Nondisjunction at meiosis I causes both centromeres to segregate to one daughter cell while the other daughter cell is nullo. Again, the homologous chromosomes line up independently of each other at metaphase II of meiosis, leading to 2 different possible segregation patterns: **1 *his+ lys* / *his lys* (disomic his+ lys) : 1 *his+ lys+* / *his lys+* (disomic his+ lys+) : 2 nullo (aborted) OR 1 *his+ lys* / *his lys+* (disomic his+ lys+) : 1 *his+ lys+* / *his lys* (disomic his+ lys+) : 2 nullo (aborted)**.

33. Meiotic nondisjunction should give roughly equal numbers of autosomal monosomies and autosomal trisomies. In fact, the total number of monosomies would be expected to be greater, because chromosome loss produces only monosomies. The actual results are the opposite of these expectations. The much higher frequency of trisomies seen in the karyotypes of spontaneous abortions suggests that human embryos tolerate the genetic imbalance for 3 copies of a gene much better than one copy. Also, one copy of a chromosome will be lethal if that copy carries any lethal mutations. Monosomies usually arrest

zygotic development so early that a pregnancy is not recognized, and thus they are not seen in karyotypic analysis of spontaneous abortions.

34. The following diagram (in which the aneuploid cells are *yellow*) shows that half the gametes of a Down syndrome individual will be aneuploid ($n+1$). (Only two chromosomes pair at a time, and those that pair would go to opposite spindle poles at meiosis I, leaving the other copy of chromosome 21 to go to one or the other pole at random.) If an $n + 1$ gamete unites with a normal gamete (n) from the other parent, the result would be a Down syndrome child. Assuming that the other parent is normal, the expectation is that 50% of the children with a Down syndrome parent would have Down syndrome.

The actual incidence of Down syndrome among the progeny of Down syndrome parents is actually lower than 50%, likely because there is some lethality associated with trisomy 21; a higher fraction of embryos or fetuses with trisomy 21 die before birth than do euploid ($2n$) embryos or fetuses.

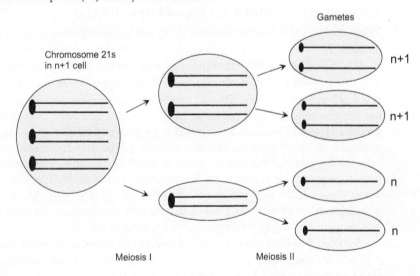

35. You have three marked 4th chromosomes: (i) ci^+ *ey*, (ii) *ci* ey^+, and (iii) *ci ey*. *Drosophila* can survive with 2 or 3 copies of the 4th chromosome, but not with 1 copy or 4 copies. You are looking for mutations that are defective in meiosis and cause an elevated level of nondisjunction.

   a. **Mate potential meiotic mutants that are ci^+ *ey* / *ci* ey^+ with *ey ci* / *ey ci* homozygotes. The normal segregants should be ci^+ *ey* / *ey ci* (ey) and *ci* ey^+ / *ey ci* (ci). Nondisjunction in meiosis I will be seen as the rare ci^+ *ey* / *ci* ey^+ / *ey ci* (trisomic but phenotypically wild type) progeny.** Nullo-4 gametes without any copy of chromosome 4 would produce zygotes with only 1 copy of this chromosome, and such zygotes would not develop into viable progeny.

   b. **The cross in part (a) will detect nondisjunction in meiosis I, but it will not distinguish meiosis II nondisjunction.**

   c. Diagram the testcross: ci^+ *ey* / *ci* ey^+ / *ey ci* × *ey ci* / *ey ci* → ?

Remember that in a trisomic individual, two of the three copies of the chromosome pair normally at metaphase I of meiosis, while the third copy assorts randomly to one pole or the other. There are 3 different ways to pair the 4th chromosomes in the trisomic individual. The first option is: 1/3 (ci^+ ey segregating from ci ey^+ with ey ci assorting independently) = 1/6 probability of (1/2 ci^+ ey / ey ci : 1/2 ci ey^+) and 1/6 probability of (1/2 ci^+ ey : 1/2 ci ey^+ / ey ci). The second option is: 1/3 (ci^+ ey segregating from ey ci with ci ey^+ assorting independently) = 1/6 probability of (1/2 ci^+ ey / ci ey^+ : 1/2 ey ci) and 1/6 probability of (1/2 ci^+ ey : 1/2 ey ci / ci ey^+). The third option is: 1/3 (ci ey^+ segregating from ey ci with ci^+ ey assorting independently) = 1/6 probability of (1/2 ci ey^+ / ci^+ ey : 1/2 ey ci) and 1/6 probability of (1/2 ci ey^+ : 1/2 ey ci / ci^+ ey). Each option gives the following progeny phenotypes (which is the same as the gamete genotypes, as this is a testcross) in the following frequencies: option 1 = 2/12 + ey : 2/12 ci + ; option 2 = 1/12 wild type : 1/12 ci ey : 1/12 ci + : 1/12 + ey; option 3 = same as option 2. The 3 options can be summed to give the final result: **1/3 ci + : 1/3 + ey : 1/6 wild type : 1/6 ci ey.**

d. These compound 4th chromosomes (att4) can be used in crosses to assay potential mutants. For instance, cross a potential mutant that is ci^+ ey / ci ey^+ [as in part (a)] to a fly with attached 4th chromosomes that are not marked (that is, both are ci^+ ey^+); that is, **potential meiotic mutants of genotype ci^+ ey / ci ey^+ × att4 ci^+ ey^+ / ci^+ ey^+ → ?** In this cross all of the normal progeny would have 3 copies of the 4th chromosome and would be phenotypically wild type. **Nondisjunction in meiosis II would be seen as unusual + ey progeny or ci + progeny.** This result would be obtained because half the gametes in the att4-containing parent would have no copy of the 4th chromosome (nullo-4) as the att4 chromosome does not have a partner. The products of nondisjunction in meiosis II in the other parent would thus be ci^+ ey / ci^+ ey (+ ey) or ci ey^+ / ci ey^+ (ci +). **In this case, nondisjunction in meiosis I is not distinguishable** because the resulting progeny would be ci^+ ey / ci ey^+ (wild type, like the normal progeny).

Another possible cross would be: **potential meiotic mutants that are ci^+ ey / ci ey^+ × att4 ci ey / ci ey →** . (Here, the att4 chromosome would have mutant alleles of both genes.) In this cross, the normal progeny will be + ey and ci +. **Nondisjunction in meiosis I would give unusual wild-type progeny** (ci^+ ey / ci ey^+ gametes from the potential mutant fertilizing nullo-4 gametes from the att4 parent). Nondisjunction in meiosis I or meiosis II would yield nullo-4 gametes from the potential mutant, so the progeny (which could only be formed with att4 ci ey / ci ey) would be phenotypically ci ey. **With this second cross you can screen for nondisjunction events both in meiosis I and meiosis II.**

36. a. **The data shown refute the hypothesis that one "Down syndrome critical region" of chromosome 21 exists.** Individuals 1, 2, 3, and 6, for example, each have completely distinct chromosome 21 duplications, yet they all have one or more of the phenotypic abnormalities associated with Down syndrome.

b. **The data also do not support the idea that extra copies of different regions of chromosome 21 cause different aspects of the Down syndrome phenotype.** For example, individuals 1, 2, and 3 all have low IQ and short stature, yet their chromosome 21 duplications have no overlap. Similarly, individuals 5 and 6 share 5

chapter 12

different aspects of the Down syndrome phenotype, and yet their duplicated regions are completely distinct. Several additional observations of this nature that have recently been made similarly refute the hypothesis.

Prior to these recent findings, most researchers expected to find that extra doses of specific diseases would be more closely correlated with specific phenotypes. Scientists are still struggling to explain these new results. One possibility is that each Down syndrome phenotype is caused by overexpression of several different genes on chromosome 21. Another possibility is that the genetic backgrounds of individuals with trisomy 21 could affect phenotypes. That is, genes on chromosomes other than 21 could be required for expression of Down syndrome phenotypes, which would thus depend on the particular alleles of these genes present in the genome. Yet another possibility is that specific genes on chromosome 21 may not even be the main story. Recently, scientists have found that the extra chromosome 21 in people with Down syndrome affects the levels of expression of genes on every chromosome in the genome. It may be that the extra dosage of certain genes on chromosome 21 causes the global gene expression response, and these responses in turn are responsible for the Down syndrome phenotypes.

Section 12.5

37. a. **15 (= 2n + 1)**. Note that $n = 7$ in this species.

    b. **13 (= 2n − 1)**

    c. **21 (= 3n = 3x)**. Note that $n = x$ because the original species was diploid.

    d. **28 (= 4n = 4x)**

38. a. **The x number in *Avena* is 7. This represents the number of different chromosomes that make up one complete set.**

    b. **Sand oats are diploid ($2x = 14$); Slender wild oats are tetraploid ($4x = 28$); Cultivated wild oats are hexaploid ($6x = 42$).**

    c. The number of the chromosomes in the gametes must be half of the number of chromosomes in the somatic cells of that species. **Sand oats: 7; slender wild oats: 14; cultivated wild oats: 21.**

    d. **The n number for each species is the number of chromosomes in the gametes and therefore is the same as the answer in part (c).**

39. a. (i) 5x = 45 chromosomes, (ii) allopentaploid, (iii) should be **sterile** - there are an odd number of chromosomes so there is no way to get an even distribution of chromosomes to the gametes during meiosis.

    b. (i) 4x = 36 chromosomes, (ii) autotetraploid, (iii) should be **fertile if the chromosomes could pair as bivalents or as quadrivalents.** (See Problem 43 for a description of quadrivalents.) The gametes should have half the number of chromosomes, so $n = 18$.

    c. (i) 3x = 27 chromosomes, (ii) autotriploid, (iii) should be **sterile** because there is an odd number of chromosomes.

d. (i) 4x = 36 chromosomes, (ii) allotetraploid, specifically an amphidiploid, (iii) should be **fertile** as the chromosomes in the two B genomes can pair with each other as bivalents and the chromosomes in the two D genomes can do the same, *n* = 18.

e. (i) 3x = 27 chromosomes, (ii) allotriploid, (iii) **infertile**; the chromosomes cannot pair at all.

f. (i) 6x = 54 chromosomes, (ii) allohexaploid, (iii) should be **fertile** as each chromosome has a pairing partner of its own type, *n* = 27.

40. a. (i) aneuploid, (ii) monosomic for chromosome 5, (iii) this will **probably be an embryonic lethal** because of genetic imbalance – this plant has probably evolved to require two alleles of a least some of the genes on chromosome 5.

b. (i) aneuploid, (ii) trisomic for chromosomes 1 and 5, (iii) this will **probably be an embryonic lethal** because of genetic imbalance – there are many genes where too much gene product is detrimental to the organism.

c. (i) euploid, (ii) autotriploid, (iii) adults should be **viable but they will essentially be infertile**.

d. (i) euploid, (ii) autotetraploid, (iii) adults should be **viable and fertile**, particularly if there is a mechanism that makes the chromosomes pair as bivalents or as quadrivalents. (See Problem 43 for a discussion of quadrivalents.)

41. Only plants that are F^A- F^B- will be resistant to all 3 races of pathogen. Because the chromosomes of same ancestral origin still pair, i.e. F^a pairs with F^A and F^b pairs with F^B, the cross can be represented as a dihybrid cross between heterozygotes. This treats the resistance genes from the two ancestral species as independently assorting genes. You are doing a cross between parents that are heterozygous for 2 different genes. Thus, **9/16 of the progeny will have the F^A- F^B- genotypes, and these plants will be resistant to all three pathogens**.

42. Haploid plant cells in culture can be **treated with colchicine (a drug that interferes with formation of the spindle apparatus) to block mitosis (segregation of chromosomes) during cell division and create a daughter cell having the diploid content of chromosomes. Once the diploid resistant cell is obtained, it is grown into an embryoid. Proper hormonal treatments of the embryoid will yield a diploid plant.**

43. a. In an autotetraploid species, all four sets of chromosomes are from the same species. Normally these chromosomes pair two by two and form two bivalents. However all four chromosomes are homologous, so this is not the only pairing option. Any one of the four chromosomes could pair with a second chromosome over part of its length, and with a third chromosome over the rest of its length. The remainder of the third chromosome would be available to pair with the fourth chromosome, making a quadrivalent. The following figure shows one of the possible quadrivalent pairing configurations of the four chromosomes during meiosis I. In the figure, each solid circle is a centromere. Each chromosome is drawn as a single line for simplicity, though of course in meiosis I each chromosome is actually composed of two sister chromatids.

chapter 12

b. **To make euploid gametes, two of the chromosomes in the quadrivalent would have to go to one spindle pole during anaphase I, and the other two chromosomes to the other pole. The arrows in the diagram above show one way this can be done.** Paired centromeres (the two centromeres that are synapsed) will be connected to the spindle so that one centromere attaches to fibers from one spindle pole and the other centromere attaches to fibers from the other spindle pole. The pairing is usually ensured by molecular mechanisms at the centromeres that stabilize centromere/spindle fiber connections only when there is mechanical tension that pulls paired centromeres in opposite directions. In the two sets of paired centromeres shown above, two chromosomes will indeed go to one spindle pole and the other two chromosomes to the other spindle pole, thus giving euploid gametes.

c. An allopolyploid contains chromosome sets from two different, though related, species. Allopolyploids are usually sterile because the chromosomes from the two species cannot pair with each other. Effectively the allopolyploid is haploid for each of two chromosome sets. Occasionally a fertile allopolyploid, or *amphidiploid,* arises when rare chromosomal doubling generates a homolog for each chromosome. (Or researchers can induce chromosome doubling purposely with the drug colchicine.) **As long as the two genomes in the amphidiploid are sufficiently different from each other, none of the chromosomes of one genome should be able to pair and synapse with chromosomes of the other genome. Thus, quadrivalents should not normally form in amphidiploids.** However, if the two species that formed the hybrid amphidiploid had diverged from each other only in the very recent evolutionary past, it is possible that some of these chromosomes would have retained sufficient sequence similarity to form quadrivalents.

44. Allopolyploids contain chromosome sets from two different species; autopolyploids contain multiple chromosome sets from the same species. **In an allopolyploid, you might find some DNA sequences similar to those of one species that has already been characterized, and other DNA sequences that match more closely to those of a different species. This pattern of relationship to two other species would not be seen in the genome of an autopolyploid.**

45. This question is asking: What is the frequency of *AA* – – progeny when an *AAaa* tetraploid self-fertilizes? As shown in Fig. 12.33 on p. 441, an *AAaa* tetraploid produces 3 different gamete types in the following ratio: 1 *AA*: 4 *Aa*: 1 *aa*. The Punnett square that follows shows the types of fertilizations that could result and their relative frequencies. Note that 6 different fertilizations that give *AA* – – progeny with two or more *A* alleles that would show the dominant phenotype (indicated here with *purple* shading).

	AA 1/6	Aa 4/6	aa 1/6
AA 1/6	AAAA 1/36	AAAa 4/36	AAaa 1/36
Aa 4/6	AAAa 4/36	AAaa 16/36	Aaaa
aa 1/6	AAaa 1/36	Aaaa	aaaa

Therefore, the total frequency of $AA--$ progeny is $1/36 + 4/36 + 1/36 + 4/36 + 16/36 + 1/36 =$ **27/36**.

46. **It was of no importance whatsoever** to either the viability or fertility of the hybrid *Raphnobrassica* that $x = 9$ in both cabbages and radishes. *Raphnobrassica* is an $n = 18$ amphidiploid with two complete chromosome sets – one from radish and one from cabbage. The 9 radish chromosomes are not homologs of the 9 cabbage chromosomes. **The reason that *Raphanobrassica* is fertile is the fact that plant breeders doubled the number of chromosomes with the drug colchicine.** The original F$_1$ hybrid formed from the fertiliziation between gametes from the two species was sterile because the non-homologous chromosomes from radish and cabbage had no pairing partners. However, colchicine treatment doubled the number of chromosomes, producing 9 pairs of radish chromosomes and 9 pairs of cabbage chromosomes (total = 36 chromosomes). Because each radish chromosome has a radish homolog, and each cabbage chromosome has a cabbage homolog, the original x numbers of the two parental species made no difference to the fertility of *Raphnobrassica* after the chromosome doubling.

    A hybrid plant will be viable if the gene products of both species can work together to produce a plant. However, as you have seen, the phenotype of such a viable plant is unpredictable.

47. a. Triploids have 3 of each homolog, and during meiosis I, 2 of them migrate to one pole, and 1 of them goes to the opposite pole. Balanced gametes will be generated by triploid watermelons only if, in every case by chance, 2 of the 3 copies of each chromosome migrate to same pole during meiosis, and 1 copy of each migrates to the opposite pole. The balanced gametes that would be produced by this pattern would be two $2n$ gametes and two $1n$ gametes.

    To obtain such gametes that are balanced for all 11 chromosomes, consider the fact that after the first set of homologs moves, the other 10 sets have to move in the same way. As there are two options every time – 2 up and 1 down, or 2 down and 1 up – the chance of any pair of homologs or the single homolog migrating the same way that the first one did is $1/2$. Thus, when $x = 10$, meioses that produce balanced gametes (and the balanced gametes themselves) occur at a frequency of $1/2^{10}$.

chapter 12

b. The most obvious way to produce a viable seed would by a fertilization between two balanced gametes; part (c) below shows that in terms of viability, it does not matter whether these balanced gametes are n or $2n$. The frequency of such an event would be $1/2^{10} \times 1/2^{10} = 1/2^{20}$.

The above calculation is actually an underestimate of the probability of producing a viable seed. The reason is that viable seeds can also be produced not only by the union of balanced gametes, but also by fertilization events where both gametes are unbalanced, but in complementary ways. For example, one gamete might be $2n + 1 + 1 + 1$ with 3 copies of chromosomes 5, 10, and 11, and the other gamete is $2n - 1 - 1 - 1$, and lacks chromosomes 5, 10, and 11. Although it is true that one meiosis would produce both such gametes, and it is also true that watermelons can self-fertilize, fertilization with two such complementary gametes is extremely rare. The reason is interesting: Watermelons have separate male and female flowers, so pollen and eggs must be the products of different meioses.

c. **Viable seeds produced either by two balanced gametes could be 4n (tetraploid; $2n \times 2n$), 3n (triploid; $2n \times n$), or 2n (diploid; $n \times n$).** Two unbalanced gametes could produce seed types listed above, and if each unbalanced gamete had 2 complete chromosome sets plus a complementary set of additional chromosomes, a pentaploid seed could be produced (at least in theory).

48. a. In order for a hybrid to be viable, the gene products of both species have to work together in the context of single cells and the whole organism to build a functional animal. **The closer the two species are in evolutionary terms, the more similar their genes are, and the more likely it is that hybrids would be viable.**

b. **Viable hybrids will be fertile if** the two species crossed are so closely related that they have the same monoploid number (x) of chromosomes, and that **their chromosomes can behave as homologs during meiosis.**

You should note that the definition of the word "species" is inexact and controversial. Under some definitions, organisms whose matings can produce viable and fertile offspring of both sexes cannot be considered to be separate species and are instead regarded as sub-species. But many cases have been described in which hybrid offspring of one sex are fertile while those of these other sex are not, so the two parents would then be regarded as separate species. Other definitions of "species" are broad enough to categorize as separate species certain populations that can produce viable and fertile offspring of both sexes.

Section 12.6

49. **Heterozygotes for one reciprocal translocation are semisterile,** while in homozygotes, fertility is unaffected. Once several translocations accumulate in a genome, that organism could no longer produce fertile progeny in mating with an organism that lacked the translocations. Thus, reciprocal translocations (T), once generated, would tend to accumulate in homozygotes. Populations of T/T and +/+ homozygotes would become reproductively isolated because the progeny within each population could reproduce freely while the progeny resulting from interbreeding could not.

50. a. **Many blocks of genes in mostly conserved order are found on two different chromosomes.** For example, one such block is seen both on chromosomes Os8 and Os9. The majority of the genes in the *Oriza sativa* genome are duplicated in this way; the simplest explanation is that the whole genome duplicated in one event.

   b. **The genome duplication may have originally resulted in tetraploid rice, but over time, the extra copy of the genome may have diverged, so that the four homologs were no longer similar enough to pair.**

   c. **One ancestral chromosome's sequences are now found on rice chromosomes Os8 and Os9 [see part (a)]. A second ancestral chromosome's sequences are now found on chromosomes Os1 and Os5. The third remaining ancestral chromosomes sequences are now found on rice chromosomes 11 and 12.**

   d. **Although most of chromosome Os12 consists of a duplication of sequences also found on Os11 [and therefore were derived from the same ancestral chromosome as explained in the answer to part (c)], part of Os12 consists of sequences that are duplicated on chromosome Os3 and were derived from a different ancestral chromosome.**

51. a. Note that almost all of the *K. waltii* genes are homologous to a *S. cerevisiae* gene, which makes sense because the two species shared a common ancestor. The *K. waltii* genes that are shown in *dark purple* are homologous to two different genes in *S. cerevisiae*. For example the leftmost *K. waltii* gene in *dark purple* is homologous to both *S. cerevisiae* gene 206 on chromosome 4 and gene 233 on chromosome 12. Likewise the other darkly shaded *K. waltii* gene is homologous to both *S. cerevisiae* gene 201 on chromosome 4 and gene 238 on chromosome 12. Therefore the Scer 4 gene 206 and Scer 12 gene 233 are the result of a duplication, so **the *dark purple* *K. waltii* genes are duplicated in *S. cerevisiae*.**

   b. Both *S. cerevisiae* and *K. waltii* are descendants of a common ancestor. **At some time after the evolutionary lines for these two species separated, a portion of the *S. cerevisiae* genome was duplicated in a progenitor of *S. cerevisiae*. Over time one copy was lost of many of the duplicated genes. Occasionally both copies of a gene were retained**, probably because they had changed by mutation into genes with slightly different functions that were both valuable to the organism. This hypothesis explains both the presence of the duplicated genes in *S. cerevisiae* and the interleaving pattern of genes from *K. waltii* that are found in the same order on the two different *S. cerevisiae* chromosomes.

   The following figure diagrams the processes described. Individual genes are indicated with arrows. The duplication event that occurred in the *S. cerevisiae* lineage is represented by the appearance of new copies of these genes (hatched arrows). Some of the duplicated genes became lost as their nucleotide sequences diverged; probably they first became nonfunctional pseudogenes (shown with Xs) and then changed further so that any relationship to the ancestral gene was obscured. But in the cases of the genes marked in the problem as *dark purple*, the *S. cerevisiae* genome retained both copies of the duplicated gene.

chapter 12

52. You would focus your attention on the genes that are still duplicated in *S. cerevisiae* (the two genes in *purple*) and compare the sequences of both copies with that of *K. waltii*. From this information you could infer the sequence of a gene ancestral to all three copies. The question you would attempt to answer is whether the two *S. cerevisiae* genes have diverged from the ancestral gene at similar rates, or whether one looks similar to the ancestral gene while the other has diverged must faster.

    The sequence of the *K. waltii* gene would serve as a control. Because only one copy of the *K. waltii* gene exists, it likely fulfills a function similar to that of the ancestral gene and thus should have evolved from the ancestor at a (relatively slow) rate indicative of the degree of selection against changes that would disrupt this function.

53. **A Robertsonian translocation that fused two acrocentric chromosomes in the great apes into a large metacentric chromosome appears to have given rise to human chromosome 2.** This translocation must have occurred in the lineage leading to humans at some time since we last shared a common ancestor with any of the great apes. The small reciprocal product of the translocation must have been lost from the human genome.

chapter 13

Bacterial Genetics

Synopsis

The lives of bacteria are intricately involved with our own lives. Human bodies are roughly 90% bacteria (based on cell number), and bacteria play essential roles in human physiology. Some bacteria are pathogens that cause human disease. Bacteria also shape our environment, for example, by metabolizing nitrogen to allow growth of the plants we depend on. The study of bacterial genetics thus has important practical applications and is necessary for a basic understanding of the biology of life on the earth. Investigating the genetics of pathogenesis is essential if we are to develop new, effective antibiotics; research on the genetic control of bacterial metabolism is helping us to understand how bacteria can disperse oil spills or produce ethanol from inedible biomass. Modern-day geneticists use bacteria as vehicles for cloning recombinant DNA molecules, and also as tools to study gene functions universal to all cells.

Two topics introduced in this chapter are of particular interest. First, *horizontal gene transfer* of DNA between different strains and species of bacteria has allowed these organisms to evolve rapidly. Bacteria can thus adjust their metabolic functions to live in a variety of extreme environments. A second fascinating topic covered in this chapter is *gene targeting*. Scientists have harnessed the process of horizontal gene transfer, and through artificial transformation of linear double-stranded DNAs, they can change any gene in the genome of many bacterial species at will – delete it, add base pairs to it, or alter one or more of its base pairs. You will see later in the book (Chapter 18) how similar techniques can be applied to manipulate the genomes of eukaryotes.

Key terms

prokaryote – single-celled organism whose genome is not enclosed in a nucleus

pathogen – a microorganism that causes disease in its host

core genome and **pangenome** – Genes shared by all strains of a given bacterial species are the *core genome* of that species; the core genome plus all the genes found in some strains and not others of the species constitutes the *pangenome* of that species.

IS elements and **Tn elements** – Small bacterial transposons that include genes needed for their own transposition but do not carry any genes for other functions in the host cell are called *insertion sequences*, or *IS elements*; larger bacterial transposons carrying transposase and drug resistance genes and flanked by IS elements are called *Tn elements*.

plasmids and **episomes** – Small circles of double-stranded DNA that can replicate in bacteria independently of the bacterial chromosome are *plasmids*; plasmids that can integrate into the bacterial chromosome (e.g. the F factor) are episomes.

metagenomics – collective analysis of genomic DNA from natural communities of microorganisms

bacteriophages (or **phages**) – viruses that infect and propagate in bacteria. Some phages can insert into the bacterial chromosome; the integrated phage genome is called a **prophage,** and the bacterium housing the prophage is called a **lysogen.**

lysis – the process during phage propagation in a host bacterium where the bacterial cell breaks open, releasing progeny phage. Fluid containing lysed bacterial cells and thus progeny phage are called **lysates.**

minimal medium – an aqueous solution containing a sugar and inorganic salts that can support growth of prototrophic (wild-type) bacteria

prototrophs and **auxotrophs** – Wild-type bacteria that can grow on minimal medium because all of their biosynthetic pathway genes are functional are *prototrophs*; bacteria that require a supplement to minimal medium because they have a mutation in a biosynthetic pathway gene are *auxotrophs*.

screens and **selections** – In genetic *screens*, individual organisms (or clones) are tested for the phenotype in question; in genetic *selections*, conditions are devised so that only those organisms with the phenotype in question can survive.

horizontal gene transfer – the introduction and incorporation of DNA into a recipient from a different individual or from a different species

transformation – one of the ways in which bacteria share genes; occurs when DNA – either linear DNA fragments or circular plasmids - from the donor is taken up by the recipient; the recipient is called a **transformant.** Spontaneous transformation is called **natural transformation.** To increase transformation efficiency and to make it happen in bacterial species that do not undergo natural transformation, researchers can use various techniques to make bacterial cell walls and membranes permeable to DNA – a process called **artificial transformation.**

conjugation – another way in which bacteria share genes. In conjugation, the temporary connection of two bacterial cytoplasms allows DNA transfer, through replication, from the donor to the recipient. After conjugation, the recipient is called the **exconjugant.**

transduction – yet another way in which bacteria share genes. Transduction is a mechanism of horizontal gene transfer where bacterial donor DNA is packaged in phage during lysis, and this DNA is then transferred to a recipient bacterium upon subsequent infection; recipients are known as **transductants.** In **generalized transduction,** any portion of a bacterial genome can be transferred; in **specialized transduction,** only bacterial genes located next to insertion sites of prophages are transferred.

F factor (**F plasmid, F episome**) – a large plasmid that carries genes required for conjugation and gene transfer. Bacteria that have an F factor are called **F+ bacteria** and serve as the donors during conjugation; the **F-** recipients lack an F factor. Strains in which the F factor has integrated into the bacterial chromosome are known as **Hfr bacteria** because they transfer genes from the donor bacterial chromosome at <u>h</u>igh <u>f</u>requency.

chapter 13

> **merodiploid** – bacteria that are partially diploid because they house an extra copy of some bacterial genes on a plasmid such as the F factor
>
> **gene targeting** – using cloned, genetically engineered DNA to alter the base pair sequence of a genome in a particular manner at a specific location through homologous recombination

Four Key Generalizations

1. **Bacterial genomes are monoploid.** This fact means that the sole allele of a gene carried in the genome dictates the phenotype directly, without complications of dominance/recessiveness. Also, because each gene is present in only one copy, and because bacteria multiply by binary fission, all of the cells in a colony are genetically identical (excepting rare mutants).

2. **Bacterial genomes are circular.** Linear fragments of DNA cannot be maintained in bacterial cells. For a DNA sequence from a linear fragment to remain in all the cells of a colony, homologous recombination must occur. Two crossovers are required, one on either side of the sequence from the linear fragment that will now replace a piece of the original circular chromosome.

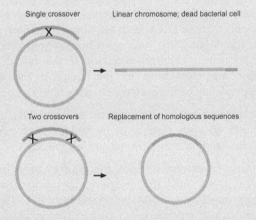

3. **Plasmids are small circles of DNA that can replicate independently of the bacterial genome.** One consequence of this fact is that researchers can make *merodiploids* (partial diploids) in which one copy of a gene is located on the large bacterial chromosome, and a second copy is found on a plasmid. The construction of merodiploids makes complementation analysis in bacteria possible. A second consequence of the fact that plasmids are small circles is that a single crossover by homologous recombination between the bacterial chromosome and a plasmid will incorporate the plasmid into the chromosome, in effect making a larger circle such as an Hfr chromosome.

chapter 13

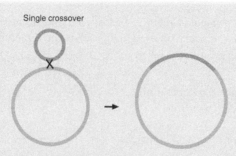

4. **Horizontal gene transfer between bacteria is asymmetric.** One strain is the donor and the other is the recipient. In most cases, the cell resulting from the transfer has only a few genes from the donor cell; most of the genes come from the recipient's genome.

Problem Solving

Most of the problems in this chapter are about horizontal gene transfer – the remarkable ability that bacteria have to transfer genes between individuals of the same species or between different species. The horizontal gene transfer mechanisms are *transformation, conjugation,* and *transduction.* A few key points to keep in mind:

- Alteration of the bacterial chromosome through horizontal gene transfer requires not only that the DNA from the donor bacterium enter the recipient cell, but also that the donor DNA undergoes homologous recombination with the recipient's chromosome. (The only exception to this rule occurs when the genes from the donor are transferred to the recipient on a plasmid that can replicate independently of the bacterial chromosome.)

- Incorporation of a linear fragment of donor DNA into the bacterial chromosome requires an even number of crossovers – an odd number would linearize the circular bacterial chromosome and kill the cell.

- Because crossovers are rare, the number of crossovers required to incorporate donor DNA into the recipient bacterial chromosome is usually 2; DNA incorporated by 4 crossovers is relatively rare. This fact will help you determine gene order when three genes are involved in a cross; see Solved Problem I on pp. 484-485 for an example.

Transformation and transduction: To solve problems involving these forms of horizontal gene transfer, you need to consider three factors:

(i) A physical limit exists on the amount of donor DNA that is transferred into a recipient cell. This limit depends on the average size of the fragments used in transformation, and on the size of the DNA that can be packaged into a bacteriophage head in transduction. If two genes are further apart than these limits, they cannot be cotransformed or cotransduced into a single recipient cell at any frequency.

(ii) Within these physical limits, the frequency of cotransformation or cotransduction of two genes from the donor is inversely proportional to the distance between the genes. (The closer the genes are located with respect to each other, the higher the frequency of cotransformation or cotransduction.)

chapter 13

(iii) Information about the relative order of three genes from a transformation or transduction experiment can be obtained by considering whether two or four crossovers would be needed to form a particular class of transformant or transductant. (Again, see Solved Problem I on pp. 484-485.) This is conceptually similar to performing a three-point cross in an organism like *Drosophila*.

Conjugation: Depending on how the experiment is set up, conjugation can provide several forms of information:

(i) *Distance from the origin of transfer.* The further a gene is located from the origin of transfer in a given Hfr donor strain, the later in time will be the first evidence of its appearance in exconjugants. Also, because mating pairs can break their connection by chance at any time, the further a gene is located from the origin of transfer, the lower will be the fraction of exconjugants containing the donor gene ultimately obtained. (See Fig. 13.20 on p. 474 for an explanation of these points.)

(ii) If genes are close together, they cannot be mapped accurately with respect to each other by trying to measure their distances from the origin of transfer. In such cases, consider again whether two or four crossovers would be needed to obtain particular classes of exconjugants.

Vocabulary

1.

a.	transformation	8.	transfer of naked DNA
b.	conjugation	5.	transfer of DNA requiring direct physical contact
c.	transduction	9.	transfer of DNA between bacteria via virus particles
d.	lytic cycle	7.	infection by phages in which lysis of cells releases new virus particles
e.	lysogeny	6.	integration of phage DNA into the chromosome
f.	episome	3.	small circular DNA molecule that can integrate into the chromosome
g.	auxotroph	1.	requires supplements in medium for growth
h.	pangenome	4.	the core genes that define a bacterial species plus all of the genes unique to individual strains
i.	gene targeting	2.	a method for mutagenizing genes in bacterial genomes

chapter 13

Section 13.1

2. It is not easy to discriminate unicellular organisms as eukaryotes, bacteria, or archaea. As the problem states, neither cell shape nor cell size is diagnostic. Eukaryotes have nuclei and mitochondria enclosed in their own membranes, while bacteria and archaea (being prokaryotes) do not. However, because many unicellular organisms have cell walls it is not always possible to look through the cell walls to visualize the presence or absence of nuclei, and mitochondria are often hard to see. Views of these structures in the electron microscope could discriminate eukaryotes from the other two groups, but not bacteria from archaea.

    The nature of the genome is not completely predictive. Most bacteria have a single circular chromosome while eukaryotes have multiple linear chromosomes, but some bacteria have linear chromosomes or multiple circular chromosomes. All known archaea have circular chromosomes, but there is no particular reason a currently uncharacterized species of archaea might have a linear chromosome.

    There are many other interesting comparisons that could be made between these three groups: for example, eukaryotes and archaea have translational systems with methionine as the initiating amino acid, not formylmethionine as in bacteria; eukaryotes and bacteria have lipids of a particular type while those in archaea are of another type; all three groups have genes interrupted by introns although these are far more common in eukaryotes than the other two groups.

    The most reliable measure for characterizing unicellular organisms is actually DNA sequence comparisons with known members of each group. Historically bacteria and archaea were in fact grouped together until DNA sequence comparisons made it clear that they comprised two groupings that were separated a very long time ago in evolution.

Section 13.2

3. a. In order to determine the number of nucleotides necessary to identify a gene, you must calculate how many bases represent a unique sequence in a DNA molecule of the size of the *E. coli* genome (4.6 Mb or 4,600,000 bases), assuming that the sequence is random. There are 4 bases possible at each position in a sequence, so 4^n represents the number of combinations found in a sequence n bases long. For example, $4^2 = 16$ is the number of unique sequence combinations that could be made with 2 positions. Therefore, if you were looking for a unique 2-nucleotide sequence, you might expect to find it, on average, every 1/16 nucleotides. A sequence of 11 nucleotides would appear $1/4^{11}$ or once in every 4×10^6 bases (4 Mb); a sequence of 12 nucleotides would appear uniquely $1/4^{12}$ or one in 16.8 Mb. Therefore **you need a sequence of about 12 nucleotides in order to define a unique position in the *E. coli* genome**.

    This problem can also be solved using the equation $4^n = 5 \times 10^6$ (= 5 Mb). To solve for n, rearrange the equation: $n\log 4 = \log(5 \times 10^6)$; $n = 11.1$. Thus you need at least 12 nucleotides to find a unique nucleotide sequence in a random sequence of 5 Mb.

chapter 13

b. This part of the problem assumes that you have determined a sequence of contiguous amino acids within a protein. The 12 nucleotides shown in part (a) define a unique position in the genome that would encode 4 amino acids. However, because of the genetic code's degeneracy, you actually know the identity of only about 8 of these nucleotides. (For amino acids specified by 6 codons, you would potentially know one less nucleotide per codon; for tryptophan and methionine, which are specified by only a single codon, you would know all three nucleotides in the codon.) **If you had a sequence of 6 amino acids, you would probably know at least 12 unique nucleotides**. Because many genes evolved through a pattern of duplication followed by divergence, some protein domains of 6 amino acids might appear in more than one protein. As a result, knowing a few more than 6 amino acids would make the case that you have identified the correct gene even stronger.

c. The gene you found might be strain-specific; that is, **the gene is a part of the *E. coli* pangenome, but it is not part of in the core genome.**

4. a. Because replication from a single origin is bidirectional (there are two replication forks), the rate of synthesis is 2000 nt/sec. As the *E.coli* genome size is about 4.6 Mb, the minimum time to replicate the *E. coli* chromosome is 4,600,000 nt / 2000 nt/sec = 2300 sec; 2300 sec / 60 sec/min = **38.33 minutes.**

b. Under optimal conditions, the origin of replication can "fire" before the previous round of replication and cell division are complete. This fact means that *E. coli* cells growing under these conditions can have multiple copies of regions around the origin of replication, as shown in the following figure. You can think of this phenomenon as the existence of replication bubbles within replication bubbles. Immediately after cell division, the daughter cells that are "born" have genomes that are already partially replicated, so they can finish replicating such genomes in less than ~38 minutes required to copy the whole chromosome.

The figure that follows compares the replication of the *E. coli* chromosome under slow growth (left) and rapid growth (right) conditions. *oriC* denotes the origin of replication of the chromosome, from which the replication bubble extends in both directions by extension of the two replication forks. *terC* signifies the position of replication termination (that is, where the replication forks going in opposite directions from *oriC* would eventually meet with each other). *Gene A* is located close to *oriC*, while *gene B* is located near *terC*. You can see that the origin of the parental chromosome fired once and part of this molecule has been replicated; then, before the forks met each other at *terC*, the origins of the two partial daughter molecules then fired again. One interesting result of the rapid firing of origins in high growth conditions is that the dosage, and thus the expression level, of genes near *oriC* (like *gene A*) is higher than those of genes (like *gene B*) near *terC*.

chapter 13

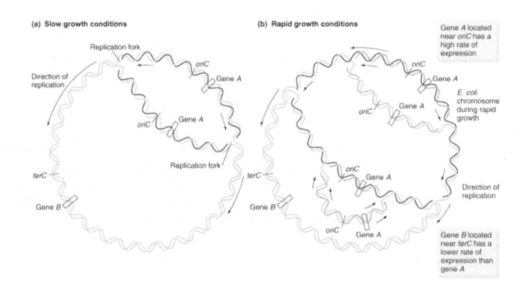

5. Genome features unique to eukaryotes include: **multiple chromosomes, linear chromosomes, centromeres and telomeres, introns and exons, large intergenic regions, enhancers, methylation of Cs; chromatin (packaging into nucleosomes).**

6. Genes can move between plasmids and the bacterial chromosome and *vice versa* by **transposition;** a gene flanked by IS elements can become a composite transposon.

7. Bacteria that grow in high saline environments are likely make proteins that don't aggregate under high salt conditions. A **metagenomics analysis of bacteria in salt lakes** could identify genes that potentially encode such proteins.

   To test the function of the candidate proteins you find, you could proceed in any of three ways: (i) You would knockout individual genes encoding the candidate proteins using the *gene targeting* methods described in Fig. 13.31 on p. 481. If the gene product was involved in helping proteins retain their solubility in high salt, then the mutant bacteria would no longer be able to grow in a high salt environment. (ii) You could construct recombinant DNA plasmids that place genes encoding candidate proteins into a plasmid vector, and use these to transform bacteria that could not normally grow in high salt. You would then select for transformants that could grow in high salt. This approach would not work if more that one such protein were needed to allow this growth. (iii) You could construct recombinant DNA plasmids that encode fusion proteins in which a protein tag is added to the sequences coding for a candidate protein. You could then purify the fusion proteins from *E. coli*, and add them to solutions containing protein aggregates formed in high salt. It is possible that the fusion protein might reverse the aggregation.

chapter 13

Section 13.3

8. a. To find the original linezolid-resistant strains, you would **take a wild-type (linezolid-sensitive) isolate of pneumococcus and place it on petri plates containing linezolid. This is a direct selection** for the desired linezolid-resistant bacteria. You will find very rare resistant colonies amidst a very large number of originally sensitive bacteria without using replica plating or treatment with mutagens.

    Finding linezolid-sensitive derivatives of the linezolid-resistant mutants is much harder. **You must somehow screen to find the rare desired bacteria, because they will not grow in the presence of the antibiotic as the resistant bacteria do. You would likely use replica plating** to identify the desired bacterial colonies that grow the absence of the drug and not in its presence. **You would probably use mutagenesis with chemical mutagens** to increase the frequency of desired mutations.

    b. In general, mechanisms that affect the bacterial toxicity of an antibiotic include the import of the antibiotic into the cell or its removal from the cell; chemical modifications of the antibiotic and thus its activation or detoxification; and the interactions of the antibiotic with its cellular targets.

    In terms of linezolid-resistant mutants, you could imagine either loss-of-function or gain-of-function mutations that would affect antibiotic import. For example, **a loss-of-function mutation in a gene encoding a receptor or pump that imports the antibiotic into the cell would prevent the antibiotic from entering the cell. Less likely, a gain-of-function mutation that would increase the efficiency of a system that exports the antibiotic out of the cell would lower the intracellular concentration of linezolid, making mutant cells more resistant to the drug. Similarly, gain-of-function mutations in genes encoding proteins that participate in a system that normally detoxifies the antibiotic would increase the cell's resistance. Loss-of-function mutations in genes whose protein products could modify linezolid in ways that increase its toxicity are also possible ways to increase linezolid resistance. Finally, it is possible that a mutation might disrupt an RNA or protein component of the 50S ribosomal subunit so that it could no longer be bound by linezolid.** Such mutations could not be loss-of-function mutations in terms of the function of the component in translation, because the cell would die if it could not synthesize proteins. These are only some of the possibilities.

    The linezolid-sensitive derivatives could be due to a reversion of the original mutation. Alternatively, they could be caused by the opposite of any of the types of mutations described above. For example, a gain-of-function mutation in a gene encoding a protein that helps import the antibiotic into the cell would raise the intracellular concentration of linezolid and make the cells more sensitive to the drug's action at lower extracellular concentrations of linezolid.

9. The initial tube of bacteria has 2×10^8 cells/ml. The first step of the dilution series is a 10^{-2} dilution (0.1 ml of the initial tube into 9.9 ml of diluent = 0.1 ml/10.0 ml = 1/100). The second step of the dilution is again 10^{-2}, so at this point the total dilution is 10^{-4}. The third step is a 10^{-1} dilution (1 ml/10.0 ml = 1/10) for a total dilution of 10^{-5}. You

then put 0.1 ml (10^{-1} ml) of this dilution on the petri plate. **Therefore, you are putting $10^{-5} \times 10^{-1}$ ml $\times$ (2×10^8 cells/ml) = 2×10^2 cells on the first petri plate, which should grow into 200 colonies.** The fourth step of the series is another 10^{-1} dilution, for a total dilution of 10^{-6}. You again plate 0.1 ml of this dilution, so you expect $10^{-6} \times 10^{-1}$ ml $\times$ (2×10^8 cells/ml) = 2×10^1 cells = **20 colonies on this second plate.**

10. In this problem, "minimal media" contain no carbon sources, so growth of bacteria on minimal medium requires supplementation with the indicated sugars. In rich medium (e.g. rich medium + X-gal), it is easiest to assume that the sugar is glucose. Note that X-gal is a substrate for the ß-galactosidase enzyme, but it cannot serve as a carbon source.

   a. **(iv)** Lac+ cells are able to use lactose as the sole carbon source for growth, while Lac- cells would not be able to grow if the only sugar in the media were lactose. Medium iv is thus selective, because Lac+ cells can grow on it but Lac- cells cannot. Note that this selective medium requires supplementation with the amino acid methionine because the original bacterial strain is Lac- Met-.

   b. **(iii)** A screen is different from a selection. For a genetic screen, you need to be able to examine the phenotype of each individual cell or colony. To screen for Lac+ cells, you therefore need a medium on which both Lac+ and Lac- cells can grow but on which they have different visible phenotypes. In a selection, only certain genotype(s) of cells can grow (min + lac selects for Lac+ cells and against Lac- cells). The rich medium (iii) has glucose, so it allows both Lac+ and Lac- cells to grow; medium iii is therefore <u>not</u> selective. However, the X-gal in these plates distinguishes between the two phenotypes (Lac+ cells are blue; Lac- cells are white).

   c. **(ii)** To select for Met+ cells, the medium should lack methionine, demanding that the bacteria must be able to synthesize methionine in order to grow. Medium ii is the only choice that lacks methionine. This medium also contains glucose so Met+ cells would grow whether they are Lac+ or Lac-.

Section 13.4

11. a. As expected, no strain A or strain B revertants were recovered in 10^8 cells. If the reversion rate is 1 in 10^7 cells, **the chance that two independent reversions would occur in the same bacterial chromosome in strain A is $(1 \times 10^{-7})^2 = 1 \times 10^{-14}$. In strain B, the chance of three independent reversions is $(1 \times 10^{-7})^3 = 1 \times 10^{-21}$. In the experiment, the cultures were grown only long enough to generate 10^8 cells** – not long enough statistically for reversion. (That is, to have a reasonable chance of obtaining just one revertant of either strain, you would have to grow a minimum of 10^{14} cells. Because wild-type colonies were recovered when strains A and B were grown together, and not when they were grown separately, the wild types must have come about through some mechanism other than reversion that involved gene transfer.

   b. In 1953, about a year after Lederberg and Tatum demonstrated the existence of gene transfer between bacterial strains A and B, William Hayes demonstrated that this transfer was *asymmetrical*, meaning that one strain was the donor and the other was the recipient. He established which strain was which by **including antibiotic**

chapter 13

sensitivity/resistance markers to strain A and strain B. For example, suppose strain A is <u>resistant</u> to the drug streptomycin, while strain B is <u>sensitive</u> to the same drug. Hays then checked the Met+ Bio+ Thr+ Leu+ Thi+ exconjugants that he obtained to see if they were resistant to streptomycin. None were. But if he now made strain A sensitive to streptomycin and strain B resistant, all of the exconjugants were streptomycin resistant. Taken together, these two results implied that strain A was the donor and strain B was the recipient.

12. a. In order to be sure that gene transfer is taking place, you would require other markers in addition to ampicillin resistance; only in this way could you discriminate among donor cells, recipient cells, and the cells that result from gene transfer. For example, suppose the donor cells were sensitive to the drug streptomycin as well as being resistant to ampicillin, and that the recipient cells were resistant to streptomycin as well as being sensitive to ampicillin. Cells that grew on plates containing both antibiotics must be the result of gene transfer, as long as the frequency of gene transfer is higher than the frequency of reversion.

    b. Transduction is the transfer of bacterial genes mediated by phage. The DNA is protected from exposure to enzymes when it resides inside the protein head of the phage. Transformation is gene transfer using naked DNA, which is not enclosed in any protective structure. The DNA being transferred by transformation will therefore be susceptible to degradation by DNase. **If the transfer of the amp^r allele occurs after DNase treatment, transduction must be occurring**.

    c. This famous so-called "U-tube experiment," is diagrammed in part (b) of the figure that follows. [Part (a) essentially reiterates Fig. 13.14 on p. 471.] The membrane separates solutions containing strain A and strain B; the membrane allows small molecules like sugars to flow between the two halves of the "U", but cells or molecules larger than the size of the pores could not. The membrane would prevent conjugation, because the cells could not touch each other but the membrane would allow transformation and transduction to proceed. Thus, **if no recombinant colonies were formed, conjugation would be ruled out, but transformation and transduction would be possible**. To distinguish between these two latter possibilities, you could use a membrane in which the pores are smaller than the size of a bacteriophage but larger than a fragment of DNA.

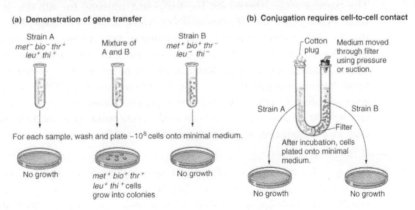

13. Do a mating between the mutant cell with 3-4 copies of F and a wild-type F- recipient. If the mutation responsible for the increased copy number is in the F plasmid, you expect the exconjugant (the recipient strain into which the plasmid is transferred so that it now becomes F+) to have the higher copy number. If the mutation is in a chromosomal gene, the higher copy number phenotype would not be transferred into the recipient.

    There is however a potential complication with this experiment. When you do an F+ × F- mating, some of the F+ donor cells will have converted to Hfr cells that can transfer bacterial DNA (possibly including a gene that influences the copy number of the F plasmid). An alternative experiment could avoid this complication. **You could isolate the F plasmid DNA from the mutant cell, and then transform this plasmid into new recipient cells. By examining the number of copies of the F factor in the transformed cells, you could tell whether the trait was carried by the plasmid.**

14. **The donor DNA is fragmented into small pieces of about 20 kb during transformation.** Furthermore, in the transformant, the donor DNA replaces only a small percentage of the recipient's chromosome. Thus only genes that are close together on the chromosome can be cotransformed. The entire *E. coli* chromosome is about 4.6 Mb long. If *purC* and *pyrB* are located half way around the chromosome from each other, they are roughly 2.3 Mb apart, making it impossible for them to be cotransformed.

15. The *purE* and *pepN* genes will be cotransformed at a higher frequency if the *H. influenzae* Rd non-pathogenic strain was used as a host donor strain rather than the pathogenic *b* strain. There is a lower likelihood that the two genes will be on the same piece of DNA in the pathogenic strain because they are separated by 8 more genes-worth of DNA (~ 8 kb of DNA) that is not present on the chromosome in the *H. influenzae* Rd non-pathogenic strain.

16. **Transfer the plasmid (by transformation) into a non-toxin producing recipient strain. If the gene is encoded on the plasmid, the transformed cells will produce the toxin.** This approach requires that the plasmid carry all of the genes needed to produce the toxin that are not found in the recipient strain.

17. a. **The partner strain should be F-, Str^r, and mutant for all the markers to be transferred from the Hfr strain (Pyr-, Met-, Xyl-, Tyr-, Arg-, His-, Mal-).** Note that to perform this conjugation experiment, you require one marker (Str^r in this case) that allows you to kill off the donor cells and only to examine exconjugants in which other genes are transferred into the recipient. It is always a good idea to know beforehand that the gene conferring streptomycin (Str) resistance/sensitivity is located further from the origin of transfer than any of the genes whose transfer you will be measuring. The reason is that you want to make sure you are looking only at exconjugants in which genes from the donor are transferred into the Str^r recipient.

    b. Selecting for Pyr+ exconjugants selects for an early marker transferred from the donor into the recipient. The frequency with which genes beyond *pyrE* are transferred decreases with distance from this early marker; this phenomenon is called the *gradient of transfer*. The gradient of transfer is due to the fragility of conjugation bridge between the cells, and the fragility of the DNA being transferred as well. The further a gene is from *pyrE*, the more likely it is that the connection between the donor and recipient or the DNA will be broken before that gene can be transferred. Recall that this phenomenon explains the different plateaus of transfer in Fig. 13.20b on p. 474. **The order of genes is:** *pyrE xyl mal arg met tyr his*.

18. As shown in the diagram that follows, transfer of Pyr+ and Arg+ requires at least 2 crossovers – one to the left of *pyrE* (closer to the F factor origin of transfer) and one to the right of *arg* (further from the F factor origin). [Note that in all of the following diagrams, the donor DNA (*orange*) is linear, and the recipient chromosome (*blue*) is a circle. The exconjugant chromosome with *blue* and *orange* regions is also a circle.]

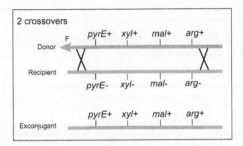

Additional crossovers within those two points, as long as they occur in even numbers, will also result in Pyr+ Arg+ exconjugants. For example, as shown in the following diagram, Pyr+ Arg+ Xyl- Mal- exconjugants can be generated by 4 crossovers.

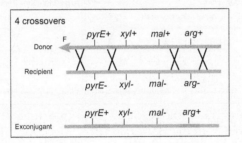

Two other kinds of exconjugants – Pyr+ Arg+ Xyl- Mal+ and Pyr+ Arg+ Xyl+ Mal- – can also be generated by 4 crossovers:

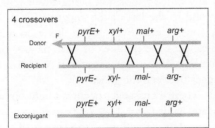 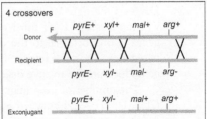

As 4 crossovers are much less likely to occur than 2 crossovers, by far **most of the Pyr+ Arg+ exconjugants will also be Xyl+ Mal+.**

19. a. The exconjugants were all selected to be Arg+, so 1 crossover occurred to the right of *arg* (that is, further from the F factor origin of transfer). **In the four most frequent exconjugant types (with 80, 40, 20 and 20 colonies), a second crossover occurred** in different locations closer to the F factor, while **the two infrequent exconjugant types (one colony each) required** 3 additional crossovers closer to the F factor – for **a total of 4 crossovers**. These various crossover events are depicted in the diagrams that follow. [Note that in all of the following diagrams, the donor DNA (*orange*) is linear, and the recipient chromosome (*blue*) is a circle, as is the exconjugant chromosome.]

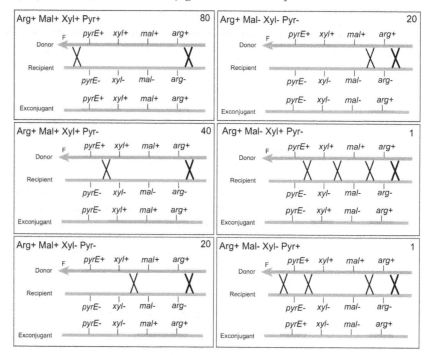

b. Because all the exconjugants are Arg+, the entire length of the orange donor DNA shown in the figures in part (a) – at least through the *arg* gene – had to be transferred into the recipient. The other crossover necessary for transfer of the *arg+* allele is between the *arg* gene and the F factor origin (F in the diagrams above). This second crossover can occur at any random position along the length of the *orange* DNA to the right of the origin and to the left of the *arg* gene. The frequency with which the second (or third or fourth) crossover occurs between any of the two genes is proportional to the distance – the number of base pairs – between those two genes. Thus, the relative numbers of the crossovers that occur between any two genes indicate the relative distances between the genes.

The number of exconjugants where a crossover occurred between *arg* and *mal* is (20 + 1 + 1) = 22; between *mal* and *xyl* is (20 + 1) = 21; between *xyl* and *pyrE* is (40 + 1 + 1) = 42. Therefore, expressed as a fraction of the largest of three distances,

13-14

the relative distances between the genes are: 22/42 = 0.52 (*arg* ↔ *mal*); 21/42 = 0.50 (*mal* ↔ *xyl*); 42/42 = 1.0 (*xyl* ↔ *pyrE*). In other words, **the *arg* ↔ *mal* and *mal* ↔ *xyl* distances are nearly the same, and each is about half the *xyl* ↔ *pyrE* distance.**

c. The relative distance between *pyrE* and the F factor origin can also be determined in the same manner. You can see from the diagram in part (a) that the number of exconjugants where a crossover occurred between *pyrE* and the F factor origin is (80 + 1) = 81. Normalized to the *xyl* ↔ *pyrE* distance, 81/42 = 1.9. Thus, **the *pyrE* ↔ F distance is 1.9× the *xyl* ↔ *pyrE* distance.**

Relative distances

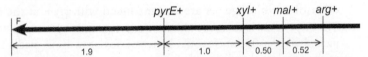

20. a. Arbitrarily place and orient the first Hfr insertion site (HfrA) on the bacterial chromosome and then order the genes that are transferred by that Hfr. When you place HfrB on the same map, notice that the first gene transferred by this Hfr is *lys*. HfrB could be placed on either side of this gene. However, the second gene transferred determines on which side of the first gene the F factor is inserted and the directionality of transfer. The time of transfer for four of the markers [*gly*, *phe*, *tyr*, *ura*] is indistinguishable, so we cannot put them in an order on the map, but this cluster of four genes can be placed relative to *lys* and *nic*. To introduce a convention used in several other problems in this chapter, the tip of the arrowhead in the following indicates the first DNA sequence transferred from the Hfr donor to the F- recipient. So for Hfr B, *lys* is the first marker transferred, followed by [*gly*, *phe*, *tyr*, *ura*], and then *nic*. For the conjugation with HfrA, 60 minutes is insufficient time to transfer the *lys* gene from the donor to the recipient.

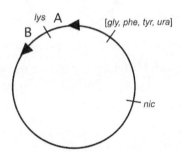

b. Note that *phe* was cotransduced with *ura* more frequently than with *tyr*, and that all of the Tyr+ transductants are also Ura+. These two facts both indicate that the order is *phe-ura-tyr*. None of the cotransduction classes is very rare, so none of these result from quadruple crossovers. The relationship of these three genes to other markers and the order of *gly* gene is still unknown, and this uncertainty is represented in the figure that follows by placing *gly* in parenthesis and the flanking marker *lys* and *nic* in brackets.

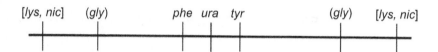

c. **To map the *gly* gene with respect to other markers, select for Gly+ transductants on min + lys + phe + tyr + ura+ nic. Then score the other markers to determine which genes are cotransduced with *gly+* at the highest frequency.**
The Gly+ transductants would be selected on min + lys + phe + tyr + ura, and then individual transductants would be tested for growth on media lacking one or more of the unselected amino acids.

21. a. **The order of the genes is either *pab ilv met arg nic (trp pyr cys) his lys* or *pab ilv met arg nic (cys pyr trp) his lys*.** (See the following diagram.) This order is established only by looking at the time of entry after the start of mating; this parameter is directly related to the distance of a gene from the origin of transfer in the given Hfr strain. Note the *arrowheads* that indicate the location and orientation of the origin of transfer in the Hfr strains A-E (see the answer to Problem 20 for an explanation of the directionality of the arrowheads).

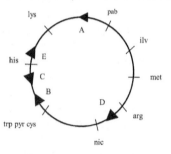

b. After mating with HfrB long enough for transfer of *nic,* and then selecting for Trp+, most of the exconjugants (790/1,000) also recombined in the *pyr+ cys+* alleles from the donor. This result reinforces the conclusion from the interrupted mating experiment in part (a) that the three genes (*trp, pyr,* and *cys*) are closely linked. The reason is that all Trp+ exconjugants must have had a crossover on either side of the *trp* gene, and the frequent transfer of *pyr+* and *cys+* along with *trp+* indicates that most of these crossovers must have occurred outside the *pyr-trp-cys* region. The low crossover frequency within the *pyr-trp-cys* region means that most of the base pairs of DNA transferred into the recipient from the donor are to the left and right of this three-gene region. This situation is diagrammed in the figure that follows. Note that although we know sufficient time elapsed to allow transfer of the *nic+* gene from the donor to the recipient, we did not score the phenotype associated with this gene. Thus, the crossover to the right in the figure could have been to the left or the right of the *nic* gene; it is shown arbitrarily to the left of *nic*.

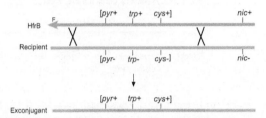

In most conjugation experiments, the smallest class of recombinants reveals the gene in the middle. Usually, the smallest class of recombinants represents those produced by four crossovers instead of two. In these kinds of exconjugants, after the four crossovers have occurred, the two outside genes will have the genotype of the Hfr parent and the gene in the middle will have the genotype of the F- parent. However, this problem has an interesting twist because the smallest exconjugant class cannot be produced by four crossovers. In the data in the table, neither of the two exconjugant classes that have two genes from the Hfr recombined into the F- chromosome [(Trp+ Pyr+ Cys-) or (Trp+ Pyr- Cys+)] is the smallest class of exconjugants. Instead, the smallest class is represented by Trp+ Pyr- Cys-, in which only the *trp+* gene from the Hfr was transferred into the F- chromosome by recombination.

This result indicates that the *trp* gene is the one in the middle. The exconjugant class produced by four crossovers was not recovered because it was Trp- and exconjugants were selected on the basis of their being Trp+. The following figure illustrates this point. The quadruple crossover class from this cross is Pyr+ Trp- Cys+, but such exconjugants could never be seen because the original selection was for Trp+. (In the following figure and all subsequent diagrams, the placement of *pyr* and *cys* with respect to *nic* was chosen randomly.)

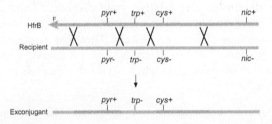

Instead, the smallest exconjugant class (Pyr- Trp+ Cys-) is the one where the two crossovers required to produce it are the least likely (lowest frequency) crossovers.

Because *pyr*, *trp*, and *cys* are all closely linked, the lowest frequency crossovers will be those that occur between the two outside genes and the middle gene of the three [that is, between (*pyr* and *trp*) and (*trp* and *cys*)], resulting in transfer of the Hfr allele of the middle gene (*trp+*) into the F- strain:

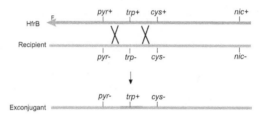

Crossovers between these closely linked gene pairs are less frequent than crossovers to the left of all three genes (between the F factor origin of transfer and the leftmost of the three genes) or crossovers to the right of all three genes (between rightmost of the three genes and a position somewhere to the right of *cys*. This must be the case because, as explained above, the data tell us that less DNA (a smaller number of base pairs) exists between the three tightly linked genes than exists in the regions to the left or the right. Exconjugants in which one rare crossover occurred within the *pyr-trp-cys* region, and one more frequent crossover outside that region occurred, represent the two more frequent classes:

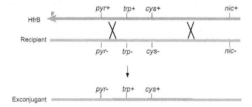 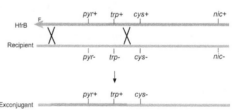

Therefore, the data indicate that **the order is *pyr trp cys*** (that is, *trp* is the gene in the middle). We do not know whether *pyr* or *cys* is closer to *nic*. Surprisingly, **we cannot determine any relative gene distances from these data.** The reason is that the three double crossover classes whose origins were just diagrammed all vary in the positions of both crossovers. Thus, the frequencies of each of the classes reflect crossovers in two intervals, and we cannot compare them. We can only determine relative distances when one of the intervals in two different classes is the same; see Problem 19.)

22. To get a stable exconjugant cell from an Hfr mating requires *recA*-mediated recombination of some of the donor genes into the chromosome in the F- recipient cell. Therefore, **this assay would detect *recA-* mutants in the F- cell based on the inability to form stable exconjugants** on the selective medium. Note that selective

medium means here that only exconjugants that have obtained markers from the donor strain would grow, but neither donor cells nor the original recipient cells would be able to multiply under these conditions.

23. Transfer into bacteria of virulence genes next to phage genes suggests that the transfer occurred through **specialized transduction** (see Fig. 13.26b on p. 478). This mechanism of horizontal gene transfer would begin with the infection of a pathogenic bacterial host cell with a *temperate bacteriophage* (see Fig. 13.24 on p. 477). The bacteriophage DNA becomes incorporated as a *prophage* into the bacterial chromosome, making the bacterial cell into a *lysogen*. Suppose that the site of prophage integration is next to a *pathogenic island* of virulence genes in the chromosome of the pathogenic host.

    When the prophage DNA is excised from the lysogen's chromosome, a mistake happens, creating a DNA molecule that has some genes from the bacteriophage as well as some genes from the host bacterium's pathogenic island. This DNA molecule can be incorporated into a new bacteriophage particle. When this bacteriophage infects a new non-pathogenic host, recombination between this DNA molecule and the genome of the new host would transfer both phage genes and pathogenic genes into the host. The host cell would then acquire pathogenic properties, and the phage genes and pathogenic genes would be adjacent to each other.

24. Generalized transduction occurs during the lytic cycle, when a phage packages into its head a segment of the fragmented bacterial chromosome from lysed cells instead of its own (phage) genome, and then transfers that DNA fragment into another bacterium that it infects. Specialized transduction occurs when, upon induction, a lysogen excises from the bacterial chromosome aberrantly, such that a segment of adjacent bacterial DNA is attached to a partial phage genome, and is packaged into the phage particle. **In generalized transduction, any fragment of the bacterial chromosome can be packaged in the phage head as long as it is the correct size. In specialized transduction, only DNA adjacent to the lysogen insertion site can be packaged into the phage particle.**

Section 13.5

25. In both parts of this question, you will be transforming bacterial cells with a plasmid library. Each clone that results will be a *merodiploid*, that is, a partial diploid cell that contains a full bacterial chromosome and a plasmid that has only one or a few genes.

    a. A nonsense mutation is likely to cause a loss of gene function. **Transform the mutant strain with a plasmid library made from a wild-type strain; colonies in which the mutant phenotype is rescued to wild-type should contain a plasmid with a wild-type copy of the gene in which the nonsense mutation had been found.** Note that this strategy is based on the idea that in almost all cases, loss-of-function mutations on the chromosome will be recessive to the wild-type allele on the plasmid.

    b. **Generate a plasmid library of genomic DNA from the mutant strain, and use it to transform wild-type bacteria. Transformants with the mutant phenotype should contain a plasmid with the mutant gene.** Here, the underlying assumption is that a gain-of-function allele on the plasmid is dominant to the wild-type allele on the bacterial chromosome.

26. (1) The Trp- phenotype in the original mutant strain could be mapped to a position on the bacterial chromosome relative to other genes using any one of the various methods described, particularly conjugation or cotransformation/cotransduction with other markers at known locations. The mutation should map to either *gene X* or *gene Y*.

(2) The two plasmids could each be used to generate a loss-of-function mutations in *gene X* or *gene Y* by gene targeting (see Fig. 13.31 on p. 481). Whichever new loss-of-function mutant has a Trp- mutant phenotype is the gene corresponding to the original mutant.

(3) Complementation tests in merodiploids with single copy plasmids (for example, the F factor) containing either *gene X* or *gene Y* could determine which gene corresponds to the Trp- mutant phenotype. (The idea here is that the reason the "wrong" plasmid corrected the Trp- phenotype is that the plasmid was present in many copies in each bacterium; you assume that this effect would not be seen if the "wrong" gene was present on a plasmid found in only a single copy per bacterial cell.)

(4) Both *gene X* and *gene Y* could be amplified by PCR from the Trp- mutant, and the sequences of the genes determined. By comparison to the gene sequences in a wild-type *E. coli* strain, mutations in one or the other gene that would be likely to inactivate gene function could be detected.

27. a. The easiest way to get the plasmid into *S. parasanguis* cells is by transformation. *Streptococci* take up DNA naturally, as seen in the early experiments of Griffiths with *Streptococcus pneumoniae*, but **artificial transformation could be used to increase the efficiency of cells that uptake the plasmid. It is critical that the transformation be done at low temperature, so that the plasmid with the temperature-sensitive origin ($repA^{ts}$) can replicate in transformed cells. Select for transformed cells containing the plasmid by plating on media containing kanamycin and erythromycin.**

Next, raise the temperature to the restrictive conditions and allow the cells to grow. The plasmid is unable to replicate at the high temperature, but the transposon will insert into random positions in the bacterial genome. These insertions will contain the *erm^r* gene. **To identify the bacteria with insertions plate them at high temperature on medium containing erythromycin but not kanamycin. Use replica plating to make sure that individual colonies are susceptible to kanamycin. Each individual Ery^r Kan^s colony should have an *IS256* insertion in a different genomic location.**

b. Using this plasmid as the basis for transposon-based mutagenesis has several advantages.

(1) As explained in part (a), **you can screen easily for strains that have a potentially mutagenic insertion of the transposon into the bacterial chromosome.**

(2) The frequency of strains with transposon insertions will be high among the cells that you screen because (i) the plasmid cannot be maintained at high temperature due to the temperature-sensitive origin, and (ii) cells that are Ery^r

chapter 13

Kan^s are most likely to have formed by transposition of *IS256* into the bacterial chromosome.

(3) **The presence of the transposon allows the researchers to identify rapidly new mutations that might disrupt the ability of the bacteria to cause dental plaque** because the transposon "tags" the mutant gene (see part [c] below).

c. **Inverse PCR (Fig. 13.30, p. 481) could be used to purify DNA adjacent to the *IS256* in the key mutant strains, and the amplification product could be sequenced to determine the particular gene disrupted by the insertion.**

28. a. The *Mbo*I site would be cleaved by the restriction enzyme as indicated by the *red lines* in the diagram that follows. You would then use DNA ligase to a dilute solution of the resultant DNA fragments. Among the products of the ligation is a small circle of DNA in which the *Mbo*I sticky ends shown in the diagram a joined together. (The low concentration of DNA fragments in the ligation mix reduces the probability that separate restriction fragments will be connected.) You now use PCR to amplify the region shown in the figure. The PCR primers within the Mariner element that would amplify a product including adjacent genomic DNA are: **5' CTCCCTTCCGCCTTTTT 3' and 5' GATATTTGCCAAACTAA 3'** (see the following diagram).

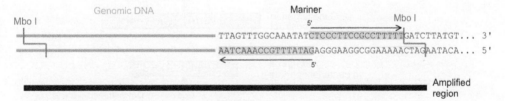

b. The DNA fragment that the primers would amplify is shown as a thick black line on the diagram above. The sequence of *blue* genomic DNA region would identify the insertion site of the transposon.

29. a. The DNA construct that would be used to exchange the *yodA* coding sequences for those of *GFP* is shown in the following figure. The coding region for GFP in *green* in the figure would start with the initiation codon for this protein and end with the stop codon. The **GFP sequences would be cDNA** because *E. coli* genes don't have introns and so *E. coli* does not have splicing machinery. A few thousand base pairs 5' and 3' to the coding would be required for efficient gene targeting.

Note that the regions needed for gene expression, particularly the *yodA* gene promoter and the Shine-Dalgarno sequences involved in the initiation of translation, will need to remain in place in the bacterial chromosome after the gene exchange. Ultimately, the cells containing the altered *yodA* gene will glow green when cadmium is present because transcription from the *yodA* promoter takes place only when the environment contains this heavy metal.

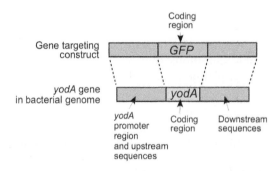

b. The *yodA* gene is nonfunctional after gene targeting because the ORF of the gene is completely replaced by that of GFP; no YodA protein is made.

c. If the gene targeting construct was altered so that *GFP* coding sequences were fused in-frame with *yodA* coding sequences – either at the 5' end or the 3' end - the fusion protein could potentially function like YodA and also like GFP (fluoresce). An example of *yodA-GFP* fusion gene construct for gene targeting (*GFP* fusion at the 3' end of *yodA*) is shown in the following diagram:

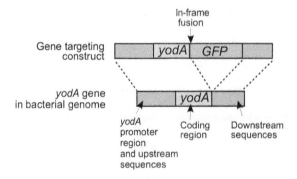

Section 13.6

30. a. Penicillin kills bacteria by preventing them from resynthesizing their cell walls after they divide. In other words, **penicillin only kills cells that are growing, but not cells that are not growing.** Rare cysteine auxotrophs will not divide in medium lacking cysteine, and so they will survive for some time in the presence of penicillin. These cells will not grow, but they also will not die. Prototrophic cells will grow in the absence of cysteine, and the penicillin will kill them.

b. To screen for cysteine auxotrophs, the cells that survive penicillin selection would be plated on a master plate consisting of minimal medium supplemented with cysteine, and then replica plated onto minimal medium. Colonies that grow on the master plate but do not grow on the minimal medium are cysteine auxotrophs.

c. The enrichment procedure would not work if the starting strain contained a plasmid with the *pen^r* gene because all of the cells would be penicillin-resistant, whether they were cysteine auxotrophs or not. The penicillinase enzyme encoded by this gene cleaves penicillin and inactivates it.

chapter 14

Organellar Genetics

Synopsis

Mitochondria and chloroplasts, organelles key to providing energy for cellular metabolism, have their own genomes. Because scientists think that these organelles originated from bacterial cells that established a symbiotic relationship with the ancient precursors of eukaryotic cells, it makes sense that mtDNA (mitochondrial DNA) and cpDNA (chloroplast DNA) are circular molecules. Many of the genes in organellar genomes encode components of protein complexes that participate in the energy metabolism occurring in the organelles, while other organellar genes encode RNAs and proteins that function in transcription and translation that take place in the organelles.

Just like mutations in nuclear genes, mutations in organellar genes can affect an organism's phenotype. Organelles are usually, but not always, inherited maternally – in the egg cytoplasm. Human mitochondrial DNAs are maternally inherited, so mutations in mitochondrial genomes can result in a variety of human diseases that display non-Mendelian maternal inheritance. Critical to understanding the inheritance of organellar traits are the facts that : (i) each cell contains multiple mitochondria and/or chloroplasts; (ii) each organelle houses several copies of its genome; and (iii) replicated organelles are distributed randomly into daughter cells after mitosis. Because of these facts, different cells in an individual may have different proportions of wild-type and mutant organellar genomes, adding complexity to the relationship of organellar genotype and phenotype.

Key terms

mitochondria – organelles that convert energy derived from nutrient molecules into ATP via the process of *oxidative phosphorylation*

chloroplasts – plant organelles that capture solar energy and store it in the chemical bonds of carbohydrates through the process of *photosynthesis*

biolistic transformation – the use of a gene gun to generate plants with transgenic chloroplasts, or **transplastomic plants**

endosymbiont theory – proposes that chloroplasts and mitochondria originated when free-living bacteria were engulfed by primitive nucleated cells; host and guest adapted to the group arrangement and each derived benefit (*symbiosis*).

non-Mendelian inheritance – pattern of inheritance that does not follow Mendel's laws and does not produce Mendelian ratios among the progeny of various crosses

maternal inheritance – the most frequent mechanism for inheritance of organellar genomes, in which they are transferred to progeny in the egg

chapter 14

> **paternal inheritance** – transfer of organellar genomes to progeny via the male gamete
>
> **uniparental** and **biparental inheritance** – modes of transmission of organellar DNAs in which they come only from one parent (*uniparental inheritance*, which can be either maternal or paternal) or from both parents (*biparental inheritance*).
>
> **heteroplasmic** – describes a cell or an organism that has organellar DNAs of different genotypes
>
> **homoplasmic** – describes a cell or an organism who organellar DNAs all have the same genotype
>
> **cytoplasmic segregation** – chance distribution of all of one type of organellar DNA in a heteroplasmic cell into a single daughter cell during mitosis
>
> **threshold effect** – a phenomenon where a particular fraction of wild-type organellar DNAs is sufficient for the normal phenotype
>
> **mitochondrial gene therapy** – nuclear transfer from an oocyte with mutant mitochondria to an oocyte with normal mitochondria

Problem Solving

When attempting to solve the problems at the end of Chapter 14, keep in mind that many factors govern the relationship between organellar genotype and phenotype:

1. Each organelle has multiple copies of its genome.
2. Organelles (and at least some of their genome copies) replicate prior to cell division, and the organelles are distributed randomly to daughter cells during mitosis and meiosis.
3. For a specific organellar gene mutation, a threshold exists for the proportion of wild-type gene copies required for normal function at the level of the organelle, the cell, and the tissue or organ. Even for a specific gene mutation, the thresholds for normal cellular function can vary in different tissues and organs.

These three points are of particular importance when considering a *heteroplasmic* cell or organism that has some wild type and some mutant organellar genomes. Cell division of a heteroplasmic cell can produce daughter cells or gametes that have different ratios of the wild type and mutant genomes. Depending on the ratio and the threshold needed for normal function, one daughter cell or progeny may have a mutant phenotype and the other cell or other progeny may have a wild type phenotype. In fact, after several rounds of cell division one type or the other type of organellar genome can be lost through the process of *cytoplasmic segregation*, producing a *homoplasmic* cell that has only type of organellar genome.

Some problems in this chapter focus on issues relating to the process of gene expression within organelles. The genes in organellar genomes need to be transcribed, and in some cases, translated. Some of the components needed for gene expression are encoded in the organellar genome itself, while other components are encoded by nuclear genes; these latter components must be imported into the organelle. The gene expression systems within organelles have several interesting wrinkles; the most important of these is the fact that the genetic codes used by mitochondria can differ from the normal genetic code (see Table 14.1

chapter 14

on p. 495). As a result, it is often the case that mRNA transcripts of nuclear genes cannot be translated properly within the mitochondrion, while the mRNAs made from mtDNA cannot be translated correctly on ribosomes in the cell cytoplasm.

Vocabulary

1.

a.	cytoplasmic segregation	4.	a cell with a mixture of different mtDNAs generates a daughter cell with only one kind
b.	heteroplasmic	6.	cell with mtDNAs or cpDNAs with different genotypes
c.	homoplasmic	2.	cell that has mtDNAs or cpDNAs that are all of one genotype
d.	maternal inheritance	1.	transmission of gene through maternal gamete only
e.	uniparental inheritance	7.	transmission of genes through either a maternal or a paternal gamete – not both
f.	isogamous	3.	having gametes of a similar size
g.	threshold effect	5.	a specific fraction of wild-type organellar DNAs is required for a wild-type phenotype

Section 14.1

2. Human mitochondrial DNA is 16,500 bp; each cell has on average 1000 mitochondria; each mitochondrion has on average 6 copies of its genome. Therefore, a human cell has on average 16,500 bp/mtDNA × 1000 mitochondria × 6 mtDNAs/mitochondrion = 990,000,000 bp of mtDNA in total. The diploid nuclear genome has 6,000,000,000 bp (6 billion bp). Thus, 990,000,000 bp/ 6,000,000,000 bp = 0.165 = **16.5% mtDNA.**

3. a. Based on the universal genetic code (see Fig. 8.2 on p. 256), the human DNA sequence that could encode Trp His Ile Met is (RNA-like strand):

   5' TGG CA(T/C) AT(T/C/A) ATG 3'.

   b. Based on the human mitochondrial genetic code (see Table 4.1 on p. 495), the human mtDNA sequence that could encode Trp His Ile Met is (RNA-like strand):

   5' TG(G/A) CA(T/C) AT(T/C) AT(G/A) 3'.

4. Fewer tRNAs are needed to translate mitochondrial mRNAs than to translate mRNAs transcribed from nuclear genes because the wobble rules are different in each case. **The wobble bases of mitochondrial tRNAs can generally recognize more different bases than the wobble bases of the cytoplasmic tRNAs that help translate the transcripts of nuclear genes.** For example, unmodified U present in the wobble positions of eight different mitochondrial tRNA anticodons [(1) 5' UAG; (2) 5' UAC;

(3) 5' UGA; (4) 5' UGG; (5) 5' UGU; (6) 5' UGC; (7) 5' UCG; and (8) 5' UCC] enables each of these tRNAs to recognize four different codons [(N = A, G, C, or U): (1) 5' CUN (Leu); (2) 5'GUN (Val); (3) 5' UCN (Ser); (4) 5' CCN (Pro); (5) 5' ACN (Thr); (6) 5' GCN (Ala); (7) 5' CGN (Arg); and (8) 5' GGN (Gly)]. At least two different cytoplasmic tRNA species are usually required to cover the four codons corresponding to each of these amino acids using the genetic code wobble rules for translation of nuclear gene mRNAs. (Note in Fig. 8.21b on p. 275 that in the wobble rules used for translation on cytoplasmic ribosomes, xo^{5}U-modified tRNAs usually cannot cover codons with C in the wobble position.)

5. The tRNA synthetases for making charged mitochondrial tRNAs are encoded by genes in the human nuclear genome. These genes are transcribed in the nucleus, and the resultant mRNAs are translated on ribosomes in the cell cytoplasm. The tRNA synthetase proteins are then imported into the mitochondria, where they can function to add the correct amino acid to the mitochondrial tRNA.

    It turns out that in humans, separate nuclear genes exist for each type of cytoplasmic and mitochondrial tRNA synthetase, with two exceptions: A single glycine–tRNA synthetase and a single lysine-tRNA synthetase aminoacylate both cytoplasmic and mitochondrial tRNA^{Gly} and tRNA^{Lys}, respectively. Also, mitochondria need only 19 tRNA synthetase enzymes because tRNA^{Gln} is aminoacylated by glutamic acid (Glu) tRNA synthetase; the glutamic acid on the charged tRNA^{Gln} is subsequently transamidated (enzymatically modified) to glutamine (Gln).

Section 14.2

6. a. **Both** cpDNA and mtDNA encode tRNAs.

    b. **Both** cpDNA and mtDNA encode electron transport proteins.

    c. In **neither** cpDNA nor mtDNA are all genes necessary for organellar function present; both organelles import important proteins encoded in the nuclear genome.

    d. **Only mtDNA varies greatly in size depending on the species**; cpDNA is fairly uniform in size across species.

7. a. Each cell has 2 copies of the nuclear genome. Each cell also has 1000 mitochondria × 10 mtDNAs/mitochondrion = 10,000 copies of mtDNA; each cell has 50 chloroplasts × 20 cpDNAs/chloroplast = 1000 copies of cpDNA. For every copy of the nuclear genome, there are 5000 copies of mtDNA and 500 copies of cpDNA. Therefore, if you obtained 100 reads of a single-copy nuclear DNA sequence, you would get 5000 × 100 = **500,000 reads of a mtDNA sequence** and 500 × 100 = **50,000 reads of a cpDNA sequence**.

    b. The answer to part (a) demonstrates that one simple criterion for determining whether a sequence is part of the nuclear genome or of an organellar genome is the number of reads; each part of an organellar genome will be represented in many more reads that each part of the nuclear genome. However, other independent criteria can help distinguish these sequences:

chapter 14

(i) The organellar DNAs of a previously uncharacterized plant species should display similarity with the organellar DNAs of other plant species. This criterion, though simple and usually effective, is not absolute because during evolution, some organellar genes have been relocated to the nucleus. Thus, a gene found in mtDNA or cpDNA in one species is not always guaranteed to be found in the organellar genomes of a different species.

(ii) The organization of organellar and nuclear genes differ. Mitochondrial and chloroplast genes are much more densely packed than nuclear genome genes; mitochondrial genes in most species have no introns, and mitochondrial genes in many species differ from nuclear genes in terms of the start and stop codons employed in their genetic codes.

(iii) If enough reads are made, computer analysis will assemble nuclear DNA sequences into long, linear contigs. The situation is clearly different in the case of cpDNA, because the reads of cpDNA will assemble into circles that will probably be no longer than ~200 kb in length. Some plant mtDNAs are also circular (as is also the case in all animals), but other plant mtDNAs are actually linear. The lengths of mtDNAs may also be quite variable, from less than 10 kb to more than 2400 kb. If the reads of mtDNA do not assemble into a circle and instead are seen only as linear contigs, it may be difficult to differentiate them from nuclear DNA.

8. a. spectinomycin resistance gene — 2. gene used to select choloroplast transformants
    b. chloroplast DNA — 1. homologous DNA that mediates integration
    c. unique restriction site — 4. site at which DNA can be inserted
    d. *ori* — 3. sequence for replication in *E. coli*

Section 14.3

9. **(a) and (d)** are characteristics of chloroplasts and mitochondria that are similar to the characteristics of bacteria but dissimilar to those of eukaryotic cells. While alternate codons [choice (b)] are used in the mitochondria of many species, such variations from the 'universal' code are also found in the nuclear genes of unicellular eukaryotes like ciliates (see p. 264) and certain yeasts. Most bacteria use the universal genetic code, with the exception of some species that use UAG stop codons to specify pyrrolysine (see Problem 8.20 on p. 294). Introns [choice (c)] are found in genes in chloroplast genomes and are also found in the mitochondrial genomes of some species like yeast but not in the mtDNA of animals; in contrast, introns occur only very rarely in bacterial genomes.

10. a. Several changes must be made to a nuclear gene like *ARG8* in order for it to be transcribed and its mRNA translated in mitochondria.
    First, any **introns would have to be removed**. (Some genes in yeast mtDNA do have introns, but the mechanisms of splicing in mitochondria and in the nucleus are not the same and depend on different kinds of sequences in the primary transcripts.) Removal of introns can be accomplished by starting with cDNA, rather than genomic sequences.

Next, **some of the codons in the nuclear gene would have to be changed because the genetic code is somewhat different in the nucleus than in mitochondria**. Table 14.1 (p. 495) shows the changes in the genetic code specific to human mitochondria, not yeast mitochondria. The actual changes in the yeast mitochondrial genetic code are: AUA encodes Met, not Ile; UGA encodes Trp, rather than signaling "stop"; and any codon whose first two letters are CU encodes Thr, not Leu. Thus, an AUA codon in the nuclear gene would have to be changed to a different Ile codon; UGA could not be used as a stop codon, but rather would cause Trp to be inserted into the protein. Using UGA for Trp codons in the *ARG8* gene has an additional experimental advantage: this would prevent expression of the novel gene in the event it somehow escaped from the mitochondria back into the nucleus. Any codon starting with CU would have to be changed to an alternative Leu codon.

Third, **the open reading frame of the altered nuclear gene would have to be placed under the control of a promoter, a translational start site, and a transcriptional termination site that work in mitochondria**. The nature of these regulatory sites differ for mitochondrial and nuclear genes.

The researchers who actually conducted these experiments used a DNA synthesizer to make an *ARG8* open reading frame without introns and with the proper genetic code alterations. This achievement was possible because the *ARG8* gene is quite small, but it would also be possible to do this by altering an *ARG8* cDNA using *in vitro* mutagenesis. The investigators then replaced the open reading frame of an actual cloned mitochondrial gene with the open reading of the altered *ARG8* that they made. This step put the *ARG8* sequences in the proper position with respect to mitochondrial regulatory sequences. Next, **the researchers introduced the cloned gene into yeast mitochondria using microprojectile bombardment (the biolistic gun)**.

b. A yeast strain with *ARG8*-expressing mitochondria has at least two advantages. First, transfer of *ARG8* into mtDNA provides a phenotype that depends mitochondrial gene expression. **Such a yeast strain allows one to select for function of the mitochondrial genetic system in mutants that are unable to respire** (grow on glycerol). If the nuclear *ARG8* gene was deleted or otherwise mutated, then the yeast cells could survive on medium lacking arginine only if they had mitochondria making this arginine biosynthetic enzyme. Second, because the expression of this *ARG8* gene was now controlled by authentic mitochondrial regulatory regions like promoters, scientists could use this strain to study mitochondrial regulatory sequences. For example, they could try to **find arginine auxotrophs that could no longer make arginine because there was a DNA change that obliterated the function of the promoter**. By sequencing such mutations in mtDNA, **they could figure out a lot about the function of regulatory elements in the mitochondrial genome**.

11. a. Human cells contain only two copies of nuclear genomic DNA, but a typical cell contains hundreds of copies of mtDNA. Therefore, **in situations where the DNA sample is substantially degraded or where there are very few cells available** (both of which occur often in forensics), mtDNA is used to identify individuals because it is possible to recover enough of it.

chapter 14

b. The disadvantage of typing individuals with mtDNA versus nuclear DNA is that **it is impossible to distinguish individuals with the same maternal lineage,** such as siblings. A grandmother, mother, daughter, and son would all have the same mtDNA sequence (excepting rare mutations or unusual cases of heteroplasmy).

Section 14.3

12. For a newly induced mitochondrial mutation to express itself phenotypically at the level of a single mitochondrion, it would have to be present in a sufficient number of mtDNAs; to cause a mutant phenotype at the level of a cell, a sufficient number of mitochondria expressing the phenotype would need to be present in the cell; to express itself phenotypically within a tissue or an organ, a sufficient number of cells expressing the mutant phenotype would need to be present to affect the function of that tissue or organ. The definition of a "sufficient number" would depend on the specific mutation in the particular gene and on the kind of cell being observed.

    Newly induced mitochondrial mutations have the best chance of affecting phenotype at the level of a tissue or organ when they either occur in or segregate into female germ-line cell precursors (oogonia). Each oocyte has about 100,000 mitochondria that all descend from only about 10-20 mitochondria present in each oogonium. During germ cell proliferation, a mutant mtDNA could easily proliferate and become a large fraction of the mtDNA population in an oocyte, and therefore in the resulting zygote after fertilization.

13. Many mechanisms exclude the mitochondrial DNA from one parent in the zygotes. **(i) Most often, the small size of the sperm excludes organelles. (ii) Cells can degrade organelles or organellar DNA from the male parent;** for example, in some species the zygote destroys the paternal organelle after fertilization. **(iii) Early zygotic mitoses can distribute the male organelles to cells that will not become part of the embryo. (iv) The fertilization process itself may prevent the paternal cell from contributing any organelles (only the sperm nucleus is allowed into the egg).**

14. Heteroplasmic cells have a mixture of the two genotypes in their organelles. Homoplasmic cells have only one type of genome. Thus, homoplasmic cells can either be totally normal or totally mutant. **If the mutation is very debilitating to the cell, either because of the loss of energy metabolism in the case of mitochondria or of photosynthetic capability in the case of chloroplasts, a cell that is homoplasmic for the mutant genome will die.** Therefore, you will find the mutant organellar genome only in heteroplasmic cells.

15. a. **Paternal inheritance of mtDNA is excluded** because three of the four seedlings inherited mtDNA from the oocyte (plant 1). Maternal inheritance is possible because seedlings 1-3 have both maternal (plant 1) alleles, and although seedling 4 has only one of the maternal (plant 1) alleles, that outcome could be due to segregation into an oocyte of only mtDNA containing this allele. Biparental inheritance is possible because seedlings 1-4 all have the allele common to both parents, and so some of their mtDNAs could have come from the pollen.

chapter 14

b. To distinguish between between maternal and biparental inheritance of mtDNA, **make plant 2 the maternal parent and plant 1 the paternal parent.** If inheritance is strictly maternal, all of the progeny plants should have only the variant of mtDNA seen in plant 2; if biparental, the progeny plants should have both alleles.

c. **To determine if plant 2 is in fact homoplasmic, would need to isolate mtDNA from all the tissues of plant 2, and make sure that only the single variant is found in PCR analysis.** This experiment would be more sensitive if you sequenced both variants of mtDNA and were able to design a primer pair that could amplify only the variant not seen in the leaf of plant 2. The reason this experiment is necessary is that **the single leaf assayed originally could have come from a region of the plant that was homoplasmic for a single variant due to cytoplasmic segregation.**

d. **Differing proportions of the two mtDNA variants in the four seedlings is due to cytoplasmic segregation either in the germ line of plant 1, or during growth of the seedlings.** Whether mtDNA inheritance is maternal or biparental, the differing proportions of the two variants in the seedlings could reflect the proportions present in individual oocytes of plant 1. [As explained in part (a), this proportion could be 100% one type in the oocyte that gave rise to seedling 4.] Alternatively or in addition, cytoplasmic segregation during growth of the seedlings could explain the results. In the zygote, different cells with different proportions of the mtDNAs due to early cytoplasmic segregation events could proliferate at different rates. Because mitochondria divide when cells divide, the result could be a change in the proportion of mtDNA variants that were present originally in the fertilized egg.

16. In all of the diagrams in this problem, *unfilled black outlined squares* = mtDNA; *black filled circles* = cpDNA; *red lines* = male sterile nuclear DNA; *blue lines* = male fertile nuclear DNA.

    a. Remember that mitochondria and chloroplasts are inherited maternally in plants. Therefore, in the first cross diagrammed below, **the female parent, which is the male sterile strain, provides progeny zygotes with both the organelle genomes and a haploid set of chromosomes, while the male parent, which is the male fertile strain, gives only a haploid set of chromosomes.** In the diagram, only one homologous pair of nuclear chromosomes is shown, and it is meant to symbolize all nuclear chromosomes.

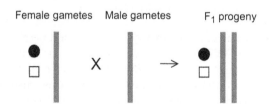

chapter 14

b. During meiosis in the germ line of the F_1 in part (a), recombination occurs between the nuclear homologs from its two parents. In the female gametes, the organellar genomes come from the original male sterile strain. Two generations of F_1 backcrosses are diagrammed below. First the F_1, and then female backcross progeny (BC_1) are backcrossed to the male fertile strain. Recombination also occurs between the nuclear homologs during meiosis in the BC_1 germ line cells.

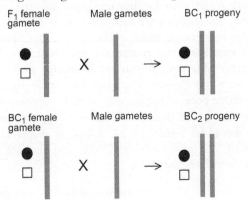

Note that in each generation, only one of the many different possible recombinant chromosomes is pictured.

c. **Each generation of backcrossing increases the fraction of the nuclear genome from the male fertile strain** (*blue*). **On the other hand,** as the hybrid progeny were always the female parent in the backcrosses, **the organellar genomes from the original male sterile strain have been inherited by the backcross progeny in an unbroken line. If male sterility continued to occur in all the progeny in each successive backcross generation,** despite the dwindling proportion of the original male sterile strain genome present, **the implication is that an organellar gene, as opposed to a nuclear gene, is responsible for the male sterile phenotype.**

17. a. Remember that this type of CMS is caused by mutant mitochondrial genomes that prevent pollen formation. You must also realize that the line used as the female parent produces both eggs and pollen (it can self-fertilize), but the line used as the male parent provides only pollen – the pollen is collected and applied to the female line. **If the female parental inbred line is male sterile, then this line cannot self-fertilize and the seed companies would not have to do anything more to prevent self-fertilization.** If the female parent instead had normal mtDNA (was male fertile), then much of the corn produced by the cross would not be hybrid corn, but would be instead simply more of the inbred line used as the female in the cross. Prior to the use of the CMS technique, seed companies would prevent self-fertilization by hiring high school and college students to go through the fields and manually remove the tassels from the plants used as the females in the cross before they produced pollen. This procedure is very labor-intensive and expensive.

b. Mitochondrial inheritance in corn is uniparental from the female parent. Thus, the F_1 corn plants in part (a) inherit mutant CMS mitochondrial DNA. These F_1 corn plants must nonetheless be fertile so that kernels form in the F_1 ears of corn. (Each corn kernel is like one of Mendel's peas – each comes from a different fertilized egg that has developed into an embryonic plant.) If the *Rf* allele of the nuclear *Restorer* gene suppresses the mitochondrial sterility mutation, then **the sterile inbred line in part (a) must also be homozygous for the recessive for the *rf* allele of *Restorer*.** These plants are then male sterile and would thus be the female parent in the hybrid-generating cross, as discussed above. **The other inbred line, the male parent supplying the pollen for the cross, would have to have at least one (and preferable two) dominant *Rf* alleles of *Restorer*.** (Note that as long as the male parent was *Rf/–*, it could, like the female parent, have CMS mitochondria and yet be fertile because the nuclear *Rf* allele suppresses the sterile phenotype.) The F_1 hybrid corn would thus have CMS mitochondria but also carry *Rf*, so the F_1 corn plants would be fertile. In the following diagram, *filled **black squares*** represent mutant (CMS) mitochondrial DNA, and unfilled *black outlined squares* represent normal mitochondrial DNA.

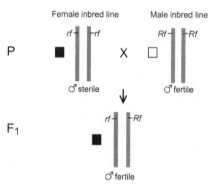

Note that if the inbred male line was *Rf/rf*, then half the F_1 hybrid plants would be male sterile (*rf/rf*) and the other half fertile (*Rf/rf*). This would not be a practical difficulty because if many F_1 seeds were planted in the same field, the male fertile plants would produce enough pollen to fertilize the ovules from the male sterile plants.

c. One method to produce CMS plants that are also *rf/rf* is to **make a fertile "Maintainer" line that has mitochondria with a normal (non-CMS) genome but whose nuclear genomes are the same as the CMS plants and are also *rf/rf*. The *rf/rf* CMS plants are used as the female parent** (they cannot produce pollen). **When pollen from Maintainer plants fertilizes the CMS plants, the progeny will have CMS mitochondria and will also be *rf/rf*; in other words, these progeny will be identical to their maternal parents.** In the following diagram, *filled **black squares*** represent mutant (CMS) mitochondrial DNA, and unfilled *black outlined squares* represent normal mitochondrial DNA.

chapter 14

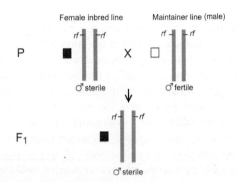

A second, more complicated method (not shown) uses CMS Rf/rf plants of the same inbred background. One fourth of the progeny would be CMS rf/rf plants that could be used to make hybrid seed, but you would need some way to identify these plants and isolate them from their male fertile CMS Rf/rf siblings.

d. There are two potential issues with using hybrid corn, both of which revolve around the idea of *monoculture*. As hybrid seed is made by seed companies, **farmers now depend on the seed companies to provide their seed**. In the past farmers saved part of their harvest and used this for corn seed the next season. However, the fact that most farmers now pay to buy hybrid seed shows the economic advantages of the increased yield due to hybrid corn. Second, farmers now grow only the few varieties of hybrid corn sold by the seed companies. **The genetic variation of the corn crop overall is reduced**. If the hybrid corn is susceptible to a newly emerging disease, this reduced genetic variation could have disastrous consequences. In 1970, a new mutant strain of a fungus causing leaf blight emerged that was particularly lethal to corn with a CMS cytoplasm called *Texas* (*T*). This fungus decimated the hybrid corn crop in the southern U.S. each year until new fungus-resistant forms of hybrid corn could be developed.

18. a. Fermentation generates ATP independent of mitochondria. **The *cox2-1* mutant yeast cannot grow on glycerol because glycerol is nonfermentable and the mutants lack mitochondrial function (electron transport), so they must rely on fermentation for energy.**

b. To answer this question, remember that the inheritance of mtDNA in yeast is biparental, but that cytoplasmic segregation will occur rapidly after a few rounds of mitotic division of heteroplasmic yeast because the mitotic buds only inherit a few mtDNAs (Fig. 14.14 on p. 505). As a result, yeast cells are usually homoplasmic.

The data suggest that *cox2-1* is a mitochondrial gene. The diploids formed by a mating of *cox2-1* and wild-type strains have roughly half *cox2-1* and half normal mitochondria (they are thus originally heteroplasmic). After growing mitotically for some time, about half of the diploid cells have normal mitochondria, and half have mutant mitochondria; each type is now homoplasmic. The diploids that have *cox2-1* mutant mitochondria cannot grow on nonfermentable glycerol and cannot sporulate. The diploids with the normal mitochondria can grow on glycerol, and when they sporulate, all 4 spores have normal mitochondria, so they can all grow on glycerol. Because mating type is controlled by nuclear genes, the fact that half the spores are mating type *a* and half are mating type α underscores that the nuclear genes are segregating normally in the spores.

The observations are inconsistent with *cox2-1* as a nuclear gene. If this were the case, you would expect that all of the heterozygous *cox2-1*/+ diploids would remain heterozygous, and either all of the diploid cells would be able to grow on glycerol if *cox2-1* is recessive to +, or none of them would grow on glycerol if *cox2-1* is dominant to +. Additionally, under this scenario all of the *cox2-1*/+ diploids should upon meiosis segregate 2 *cox2-1* (cannot grow on glycerol) : 2 + (can grow on glycerol) spores. Instead, all the haploid spores could grow on glycerol, which is consistent with the alternative hypothesis that *cox2-1* is a mitochondrial gene.

c. **The *pet111-1* mutants behave exactly the way we would expect for a recessive mutation in a nuclear gene** [see part (b)].

19. a. **The *petites* can grow on glucose plates because glucose is fermentable—it can be used to generate ATP without mitochondrial function (oxidative phosphorylation); they cannot grow on either glycerol or ethanol because using these molecules to generate ATP requires mitochondrial function, which they lack. The *grande* yeast (wild-types) can grow on glucose, glycerol, or ethanol because they have functional mitochondria.** The reason that *petite* colonies are small is that fermentation is less efficient than oxidative phosphorylation (cells generate fewer ATP molecules from a given number of glucose molecules by fermentation as compared with oxidative phosphorylation in mitochondria).

    b. **The odd uniparental inheritance pattern of the *grande* phenotype in *grande* × *petite* crosses is due to the fact that the *petite* strains used by Ephrussi have no mtDNA at all.** Thus, in the diploid cells resulting from the mating, all of the mtDNA comes from the *grande* parent that has wild type mtDNA. This situation differs from the cytoplasmic segregation that occurs in heteroplasmic diploids (as in Problem 18b) because in this case, one parent supplies all the mtDNA.

Section 14.5

20. One characteristic of a mitochondrial mutation in many organisms is **maternal inheritance**. Most of the offspring (both male and female) of an affected female are affected. None of the offspring of affected males are affected. Another indication of mitochondrial inheritance is **differing levels of expression of the mutant phenotype in different progeny** due to differing amounts of heteroplasmy in either the egg or the cells of the embryo.

21. a. **The mother (I-1) may have been heteroplasmic with a very small proportion of mutant mtDNA. Alternatively, a spontaneous mutation could have occurred** either in the mitochondrial genome of the mother's oogonia, her egg that gave rise to individual II-2, or in the early zygote of individual II-2. Due to cytoplasmic segregation, a mutation that occurred in one mtDNA genome could come to represent a large proportion of mtDNAs in the some mitotic descendants of the cell in which the mutation happened.

b. To distinguish between heteroplasmy and new mutation, **you could analyze the mtDNA of somatic cells from various tissues in the mother (I-1). If the mother was heteroplasmic, most of her tissues should show at least a small fraction of mutant mtDNA.** If the mutation occurred in her germ line and was inherited by II-2 (or if it occurred in the zygote) the mother's somatic cells would not show any defective mtDNA.

22. **The zygote that formed these twins was heteroplasmic, with both wild-type and mutant mitochondrial genomes. Early in embryonic development, two cell masses separated to become the two identical twins. The ratio of wild-type to mutant mtDNAs in the two cell masses may have differed. Furthermore, during the individual development of the twins the ratios of wild-type to mutant mtDNAs may vary in different tissues.** The more affected twin probably had a higher proportion of mutant mitochondria in tissues such as muscles and brain that particularly depend upon energy supplied by mitochondria.

23. **If the patient is male then you could reassure him that none of his children would be affected by the disease** (assuming his mate is unaffected by MERRF). **If the patient is female then she is most likely heteroplasmic. There is a strong chance that her child might be affected by MERRF, but there is no way to determine either the numerical probability that the child would be affected or the severity of the disease**, because of the random events in mtDNA distribution through many rounds of cell division that will influence the ratio of wild-type and mutant mtDNAs in the egg and in various tissues in the fetus. It is relatively easy by **amniocentesis and PCR to check for the mutant mitochondrial DNA in the fetus.** However, **these results are not diagnostic** because the distribution of wild-type and mutant mtDNA can vary considerably in various tissues. For example, the fetal cells shed into the amniotic fluid may be devoid of mutant mtDNA, but cells in other fetal tissues might still have some mutant mitochondrial genomes.

24. a. **The three diseases can be caused by deletion of different mitochondrial genes that affect mitochondrial function to different degrees. It is even possible that all three syndromes could have the same genetic cause, but the differences could be due to heteroplasmy and cytoplasmic segregation.**

    b. **The origin of replication must be outside of the DNA deleted in these syndromes.** (That is, outside of the 7.6 kb removed in the largest of these deletions.) Deletion of the mtDNA replication origin would result in a mutant mtDNA that is unable to replicate; it would be lost and no cells would end up heteroplasmic.

    c. **One possibility to explain the lack of maternal inheritance of PEO is that germ-line cells carrying the mtDNA deletions are at a disadvantage either for multiplication or for oogenesis relative to homoplasmic segregants that do not have the deleted DNA. As a result, either the germ line would be populated mostly by cells without the deletions, or only the homoplasmic segregants would produce viable eggs. A woman who has the deleted mtDNAs in her somatic cells (and thus has PEO symptoms) might not have mutant mtDNA in her germ-line cells or in the eggs she makes.**

25. a. The product of the nuclear gene is imported into the mitochondria and it works in the same pathway as the product of the mitochondrial gene. For example, some protein complexes that operate in oxidative phosphorylation include one subunit encoded by a nuclear gene and another subunit encoded by a mitochondrial gene. Loss of either gene function results in the failure of electron transport and thus in the symptoms of Leigh syndrome.

   b. **No**, the pedigree shown does not enable you to distinguish between nuclear versus mitochondrial gene inheritance. One possibility consistent with the data is that Leigh syndrome is caused by a recessive mutation in an autosomal nuclear gene (*SURF1*); both parents would be heterozygous. The other possibility is a mtDNA mutation (*MT-ATP6*) and the mother is heteroplasmic; her overall or tissue-specific proportion of mutant mtDNA is below the threshold for disease symptoms. The tissue distribution and/or overall proportion of mutant mtDNA is high enough in 2 of her 3 children for them to display disease symptoms.

   c. Knowing that the disease is mitochondrial would inform the affected male that he cannot pass it on to his children, and would inform both the affected female and her unaffected sister that potentially they both could do so. If the disease were caused by a rare recessive allele, the unaffected sibling could test her genomic DNA to determine whether or not she is a carrier. If so, all three siblings could test their partners for a mutation in the gene before they decide to have children.

26. Scientists could use a PCR assay that distinguishes SNPs in the mtDNAs from the nuclear donor and the cytoplasm donor oocyte to test cells from different tissues of Mito and Tracker. Likely these SNPs would lie in the hypervariable regions of mtDNA found in humans and other primates (see Problem 11). These PCR assays would not however be absolutely conclusive. Without testing every single cell in Mito and Tracker, the researchers cannot be completely certain that a few mtDNA molecules from the nuclear donor oocyte are present.

chapter 15

Gene Regulation in Prokaryotes

Synopsis

The genetic experiments of Jacob and Monod provided the first big ideas about how genes are turned and off in response to changes in the cellular environment. A basic principle derived from these *lac* operon experiments is that proteins (*trans*-acting factors) bind to specific DNA sequences in genes to regulate transcription initiation. This binding can be influenced by small molecules that are present within cells or that come into cells from the external environment. Later genetic studies of the *trp* operon that led to discovery of the *trp* attenuator showed that bacteria regulate gene expression not only at the point of transcriptional initiation, but also during transcriptional elongation. Again, the function of the attenuator reflects the concentrations of small molecules, allowing the cells to respond to environmental conditions.

The *trp* attenuator is a sequence of RNA that can assume alternative configurations. More recently, scientists have discovered that RNA can mediate gene expression in many ways. *Cis*-acting RNA leader sequences can regulate not only transcription elongation as does the attenuator, but also the initiation of translation (*riboswitches*). Small *trans*-acting RNAs encoded in bacterial genomes also regulate translation initiation. Finally, *antisense RNAs* can influence either the transcription of a gene or the stability or the translation of the gene's RNA.

The work of Jacob and Monod also revealed another key idea: Genes in bacterial cells whose products cooperate in the same biochemical pathway are often clustered together in units called *operons*. The genes in the operon are expressed into a single, *multicistronic RNA*. This fact provides a means for the *coordinate regulation* of the genes in the operon: By controlling the transcription of the operon, the cell can regulate all of the genes comprising the operon simultaneously.

After you read Chapter 16, which describes the mechanisms that regulate genes in eukaryotic organisms, it will be interesting to compare and contrast what you learn in that chapter with the material in this chapter. You will see that some regulatory mechanisms are unique to prokaryotes, while some are found only in eukaryotes; nonetheless, many of the basic principles of gene regulation are shared by all organisms.

Key terms

gene expression – the process by which a gene's information is converted into RNA (transcription) and then, for protein-coding genes, into a polypeptide (translation).

RNA polymerase – the enzyme that transcribes a DNA sequence into an RNA transcript.

promoter – DNA sequences near to (and usually upstream of) a gene's transcription start site to which RNA polymerase binds directly to initiate transcription

ribosome binding site – region on prokaryotic mRNAs containing both an initiation codon and a *Shine-Dalgarno box*; ribosomes bind to these elements to start translation of protein-coding sequences on the mRNAs.

catabolic pathway – a metabolic pathway that breaks down complex molecules to produce smaller molecules (and often, energy). Lactose utilization, which breaks down lactose into glucose and galactose, is an example of a catabolic pathway. (See Fig. 15.2 on p. 517.)

anabolic pathway – a metabolic pathway that synthesizes a complex molecule from simpler ones. The genes of the *trp* operon encode enzymes that cooperate in the anabolic pathway that leads to the synthesis of the amino acid tryptophan.

transcription factor – a protein that binds DNA sequence-specifically and regulates transcription. In prokaryotes, two types of transcription factors exist: **positive regulators** activate transcription, and **negative regulators** (also called **repressors**) inhibit transcription.

inducible regulation – a mechanism of gene control where transcription occurs only in the presence of a molecule called an *inducer*. An **inducer** is a small molecule that binds to a positive regulator, altering the positive regulator's conformation so that it can bind DNA. Inducible regulation is particularly important for the regulation of catabolic pathways.

repressible regulation – a mechanism of gene control where transcription occurs only in the absence of a *corepressor*. In prokaryotes, a **corepressor** is a small molecule that binds to a **repressor**, thereby altering repressor conformation so that it can bind DNA. Repressible regulation is particularly important for the regulation of anabolic pathways.

induction – the process by which an inducer causes transcription of a gene or a set of genes

operon – a unit of DNA composed of specific *structural genes* (encoding functional products), plus a *promoter* and an *operator*. These elements work in unison to regulate the coordinated response of expression of the structural genes to environmental changes.

allosteric protein – a protein that changes its conformation reversibly when bound to a specific *effector* molecule

constitutive mutants – cause gene transcription to occur all the time in an unregulated manner; that is, regardless of environmental conditions

effector – a small molecule that binds to an allosteric protein or to an RNA molecule and causes a conformational change. As examples, allolactose (an inducer), cAMP, and tryptophan (a corepressor) are both effectors.

in *trans* – describes the action of a protein or RNA that can bind to target sites on any DNA or RNA in the cell; a protein or RNA that acts in *trans* is said to be ***trans*-acting.** Molecules that act in *trans* diffuse throughout the cell to find their target sites.

in *cis* – describes the action of a DNA site or a region in an RNA molecule that acts only on the DNA or RNA to which it is connected physically; a DNA site or an RNA that acts in *cis* is said to be ***cis*-acting.**

domain – a discrete region of a protein that has a particular function. For example, the *lac* repressor polypeptide has a domain that binds to the *operator*, a domain that binds to the *inducer* allolactose, and a domain that allows polypeptides to associate into multimers.

catabolite repression – inhibition of the expression of genes or *operons* like the *lac* operon when glucose or another preferred catabolite is present

RNA leader sequence – the 5' UTR of an mRNA; in prokaryotes, RNA leader sequences can act in *cis* to regulate transcription elongation or the transcript's translation.

hairpin loops (stem loops) – structures formed when a single strand of RNA folds back on itself because of complementary base pairing between different regions in the same molecule

attenuation – a type of gene regulation that responds to translation of an ORF in the RNA leader; the RNA leader is called the **attenuator.** When this ORF is translated efficiently, transcription of a gene or operon terminates in the RNA leader sequence before a complete mRNA transcript is made.

terminator – in prokaryotes, stem-loop structures that cause RNA polymerase to stop downstream transcription. Terminator formation in an RNA leader sequence (such as an attenuator or a riboswitch) can be prevented by formation of a different, competing stem-loop structure called an **antiterminator.**

riboswitch – an allosteric RNA leader that binds a small molecule *effector* to control gene expression

RNA thermometer – an allosteric RNA leader that regulates translation in response to temperature through a stem-loop structure whose stability is temperature-dependent.

sRNAs – small RNA molecules that regulate translation in *trans* by base pairing with sites on mRNAs. This base-pairing of sRNA and mRNA can either hide or expose the ribosome binding site on the mRNA.

antisense RNAs – regulatory RNAs that are complementary in sequence to the mRNAs they regulate because they are transcribed using the opposite strand of DNA as a template. Antisense RNAs can block transcription or translation of their target mRNAs; the target mRNAs are sometimes called *sense RNAs.*

fusion gene – a gene constructed using recombinant DNA technology that is made up of parts of two or more different genes. For example, a fusion gene may contain the 5' regulatory region of one gene and the ORF of a different gene.

reporter gene – a fusion gene whose ORF encodes a protein that is easy to detect; expression of that protein "reports" activity of the fusion gene's promoter and also verifies that the fusion gene transcript was translated. (The term *reporter gene* also sometimes refers to the ORF of the fusion gene only.)

RNA-Seq (cDNA deep sequencing) – method for analysis of all of the transcripts made in an organism (the **transcriptome**) under a specific set of conditions. The data are obtained as approximately one billion sequence reads each corresponding to an individual cDNA molecule. With current technology, each read is about 150 nt long.

chapter 15

> **quorum sensing** – a communication system whereby bacteria sense their population density to regulate the transcription of particular genes

Structures of prokaryotic genes and mRNAs

Solving problems about the regulation of prokaryotic genes requires that you have a thorough knowledge of prokaryotic gene and mRNA structure. Be sure that you understand the functions of all of the parts of the typical operon and the mRNA in the diagram that follows. Critical to this diagram is the point that an operon's mRNA is *polycistronic,* meaning that it contains the ORFs for multiple *structural genes*. The expression of all the structural genes in the operon can be coordinated by events that occur near the promoter and thus that affect the amount of the polycistronic mRNA. Even though they are on a single mRNA, these ORFs can be translated independently, because they each have a separate translation initiation signal [made up of a Shine-Dalgarno sequence and a ribosome binding site (RBS)].

Another important feature of prokaryotic gene expression is that transcription of the gene or operon and translation of the mRNA occur is the same space in the cell. Furthermore, translation can begin before the mRNA has been completely transcribed. These facts are central to phenomena like attenuation, where the structure of the 5'UTR (the upstream leader) of the operon mRNA can influence whether the rest of the operon will be transcribed.

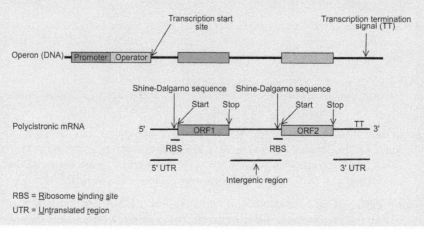

Fusion genes and reporter genes

Our understanding of prokaryotic gene structure has enabled scientists to cut and paste different parts of genes together – to make **fusion genes**. Such fusion genes are not only important experimental tools, but they also allow pharmaceutical companies to employ bacteria as factories to make clinically useful proteins. The fusion genes you saw in this chapter (Fig. 15.25 on p. 533 and Fig. 15.27a on p. 534) contain the regulatory region of one gene, and the open reading frame (ORF) of a different gene. Such recombinant genes are sometimes called *transcriptional fusions* because transcription of an ORF is controlled by foreign regulatory sequences. If the regulatory region of the first gene is well-characterized and easy to manipulate, it can be used to make in bacterial cells large amounts of the protein encoded by the ORF.

Some fusion genes are **reporter genes;** the ORF of a reporter gene (e.g. *lacZ*) encodes a protein (e.g. β-galactosidase) for which there is a convenient assay. By monitoring reporter protein levels, researchers can assess the activity of the promoter and previously uncharacterized regulatory sequences that drive transcription of the reporter gene. The reporter can be used to test different base pairs in the putative regulatory elements for function through mutagenesis; to identify the *trans*-acting factors that bind the regulatory elements; or to test the transcriptional response of the regulatory region to various environmental factors.

Figure 15.26 on p. 533 shows the special case of a **"promoterless" reporter gene**. By incorporating this promoterless reporter gene into a transposon, researchers can isolate many strains in which the reporter is inserted into different, random positions in a bacterial genome. Reporter expression under a specific condition means that the reporter inserted adjacent to the regulatory region of a gene that responds to that stimulus.

In later chapters, you will see that reporter genes and other fusion genes are invaluable in studies of eukaryotic as well as prokaryotic organisms.

Protein/DNA interactions

Gene regulation involves *cis*-acting DNA elements and *trans*-acting factors that bind them. The following list clarifies key points how about transcription factors interact with DNA.

- In general, transcription factors bind to double-stranded DNA rather than single-stranded DNA; the DNA strands do not denature in order to bind transcription factors. Atoms at the edges of the base pairs (that are not involved in forming the hydrogen bonds between complementary base pairs) are available to form noncovalent bonds with atoms of the amino acid R groups within the DNA-binding domain of the transcription factor. Most of these interactions between amino acids and bases take place in the major groove of the DNA (review Fig. 6.15 on p. 185).

- The specificity of particular transcription factors for specific DNA sequences stems from the particular sequence of amino acids in the DNA-binding domain and the particular base pair sequence of the binding site. The sum of all of these attractive forces adds up the *affinity* of the transcription factor for the binding site.

- All transcription factors, no matter how high their affinity is for a site on DNA, are in *equilibrium* between bound and unbound states. How frequently a particular transcription factor is bound to a particular site on DNA depends on the affinity of the DNA-binding domain of the protein for that base pair sequence, and also on the concentrations of the protein and the DNA in the nucleus.

chapter 15

Problem Solving

One of the most consistent themes in this chapter is that regulation of prokaryotic gene expression in response to the environment often occurs during the initiation of transcription and requires three different components: (1) a functional structural gene (e.g. *lacZ*$^+$); (2) intact *cis*-acting DNA sites [e.g. a promoter (*P*) and an operator (*o*)]; (3) functional *trans*-acting factor gene(s) (e.g. *lacI*$^+$).

Related concepts to keep in mind include:

- *cis* and *trans*: When analyzing a genotype to determine how particular genes will be expressed under specific environmental conditions, keep in mind the difference between *cis*-acting DNA sites and *trans*-acting factors. The promoter and operator are *cis*-acting sites at the 5' end of every operon; they function only with that operon – not with other genes or operons on unconnected DNA molecules. Transcription factors, on the other hand, are *trans*-acting proteins that can work on any operon in the cell; it does not matter whether the gene that encodes a *trans*-acting factor is on the same DNA molecule as the operon the factor regulates. The reason is that once it is made, the transcription factor protein can diffuse in the cell cytoplasm and find its binding site on any DNA molecule present there.

- **Prokaryotic promoters:** A prokaryotic promoter is simply a specific base pair sequence that attracts RNA polymerase to initiate transcription of a gene or operon. Amino acids that form the DNA-binding domain of RNA polymerase have an affinity for the particular base pair sequence in the promoter that is higher than the general affinity of RNA polymerase for binding DNA at random sequences.

- **Prokaryotic transcription factors can be positive regulators or negative regulators:** Positive regulators increase the affinity of RNA polymerase for the promoter. Positive factors do so because they bind DNA sites near the promoter, where the factors make contact with RNA polymerase. These contacts serve as attractive forces: A promoter with a positive regulator bound near it has a higher affinity for RNA polymerase than when the positive regulator is not bound there. Negative regulators can bind to DNA sites (operators) that overlap with the promoter and thus prevent RNA polymerase from binding there or from initiating transcription.

Vocabulary

1.

a.	induction	10.	stimulation of protein synthesis by a specific molecule
b.	repressor	8.	negative regulator
c.	operator	5.	site to which repressor binds
d.	allostery	2.	protein or RNA undergoes a reversible conformational change
e.	operon	7.	group of genes transcribed into one mRNA
f.	catabolite repression	1.	glucose prevents expression of catabolic operons

g. reporter gene   9. often fused to regulatory regions of genes whose expression is being monitored

h. attenuation   6. termination of transcription elongation in response to translation

i. sRNA   3. regulates translation of mRNAs in *trans*

j. riboswitch   4. RNA leader that regulates gene expression in response to a small molecule or ion

Section 15.1

2. The first step of gene expression is the binding of RNA polymerase to the promoter. **If the main controls of gene expression were later in transcription or in translation, the cells would sometimes waste energy making RNA that would not be used. Bacteria are clearly evolving under the constraint of conserving the energy they obtain from nutrients in their environments.**

3. a. **i, ii, iii.** The first two, i and ii, reflect the rate at which the transcription of different operons is initiated. This rate will depend on the intrinsic affinity of the promoter of each operon for RNA polymerases - some promoters are stronger than others. (Although not in the list, the binding of sequence-specific DNA binding-proteins, like repressors or positive regulators, to DNA sequences near promoters also regulates levels of mRNAs). In addition, promoters are recognized by different RNA polymerases - some sigma factors may be needed for RNA polymerase to recognize some promoters and not others. Point iii indicates that the level to which an mRNA accumulates in the cell depends not only on its rate of transcription but also on its rate of degradation. Different mRNAs can be degraded at different rates due to differences in their structures that are recognized by various ribonucleases in the cell.

   b. **iv, v, vi.** Point iv is quite interesting. In operons the "distal" genes that are further from the promoter may be less efficiently transcribed or transcribed later than the "proximal" genes that are closer to the promoter. If the mRNA of the operon is long it will take RNA polymerase some minutes to reach the distal genes. The RNA polymerase might sometimes fall off the DNA or encounter transcription termination signals before transcribing the distal genes. As transcription and translation are coupled in bacteria, the transcription of distal genes might actually be regulated by the rate at which the proximal genes in the same mRNA are translated. Point v indicates that the efficiency of translational initiation for genes in the same operon can vary because of differences in Shine-Dalgarno sequences and because of differences in the way the various ribosome binding sites fold in three-dimensional space. Point vi is important because the amount of protein that accumulates in a cell is not only due to the rate it is translated, but also to the rate at which that protein is degraded.

4. The lack of *rho* function must be lethal for the cell, so **the *rho* gene function is essential**. Conditional mutations are the only sort of mutation that can be isolated for essential genes. Given that the Rho protein is needed to terminate correctly the

transcription of many genes and operons in the bacterial genome, it is not surprising that *rho* gene function is essential for survival.

Section 15.2

5. Mutations in the promoter region can act only in *cis* to the structural genes immediately adjacent to this regulatory sequence. A promoter mutation in one operon will not affect the expression of a second, normal operon, in *trans*.

6.

Structure of the *hup* operon mRNA

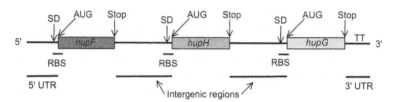

Transcription (Feature Fig. 8.10, pp.265-266) begins with the first base at the 5' end of the mRNA shown above. RNA polymerase transcribes the template DNA strand until it reaches the transcription termination signal (TT). The transcription termination signal could be a hairpin loop (an intrinsic termination signal) near but not directly at the 3' end of the mRNA, or alternatively it could be a sequence near the 3' end of the mRNA that is bound by a protein factor like Rho (an extrinsic termination sequence).

Translation (Feature Fig. 8.25, pp. 278-279) begins when a ribosome binds to the ribosome binding site (RBS) in the mRNA, which is made up of the Shine-Dalgarno sequence (SD) and, a short distance downstream, the initiation codon (AUG). The rectangles in different shades of *pink* represent the translated regions (open reading frames or ORFs) for each gene. Note that translation starts at three separate locations, one for each gene in the operon. In prokaryotes the initiating amino acid is fMet. Translation terminates at a nonsense codon (Stop), leading to the release of the polypeptide and the dissociation of the ribosomal subunits.

Note that several regions in the mRNA are *not* translated. These include the sequences upstream of the first SD sequence, known as the *5' UTR* (untranslated region), also called the *RNA leader sequence*. Other nontranslated regions of the mRNA include the sequences between each ORF (the *intergenic regions*) and the sequences downstream of the last ORF (the *3' UTR*) that include the transcription termination signal TT.

7. **Statement (b) will be true**. When a positive regulator is inactivated, the expression of the operon will be much reduced.

8. a. When the gene encoding the DNA binding protein has a loss-of-function mutation, the result is constitutive expression of the *emu* operon. Therefore, **the wild-type regulatory protein blocks transcription; in other words, it is a negative regulator of *emu* operon expression**. You can imagine for example that the DNA binding protein could be a repressor like the Lac repressor. You could similarly

imagine that the regulatory site to which the DNA binding protein binds is similar to the operator site in the Lac operon.

b. **Strain (i) will have inducible *emu1* and constitutive *emu2* expression, while strain (ii) will have inducible *emu1* and *emu2* expression.** In both strains, the Emu2 protein can be made only from the structural gene on the F′ plasmid, while the Emu1 protein can only be made from the structural gene on the bacterial chromosome. Both strains should have enough Reg2 binding protein to repress transcription of *emu* operons on either the plasmid or the chromosome, because wild-type alleles of the *reg2* gene should be dominant to mutant loss-of-function alleles, and because binding proteins such as the Lac repressor work in *trans*.

   In strain (i), the only functional *emu1*$^+$ gene (on the bacterial chromosome) is normally regulated (that is, it is inducible) because the *cis*-acting *reg1* DNA binding site (operator) is intact and can interact with the Reg2 repressor. However, the F′ plasmid, with the only functional *emu2*$^+$ gene, lacks a functional *reg1* operator. The Reg2 repressor cannot regulate the operon on the F′ plasmid, so *emu1* expression is constitutive (the operon is transcribed all the time).

   In strain (ii), the *emu* operons on both the F′ plasmid and the bacterial chromosome have functioning *reg1*$^+$ operators. Binding of the Reg2 repressor protein to these operators makes the expression of both the *emu1* and *emu2* structural genes inducible.

c. **Strain (i) will have inducible *emu1* and *emu2* expression while strain (ii) will have inducible *emu1* and constitutive *emu2* expression.** In this case, both strains will have sufficient *trans*-acting Reg1 repressor. Strain (i) has functional operators on both the F′ plasmid and the bacterial chromosome, while the only functional operator in strain (ii) is on the bacterial chromosome. Thus, the expression of *emu1*$^+$ from the operon on the bacterial chromosome will be normally inducible, while expression of *emu2*$^+$ from the operon on the F′ plasmid will be constitutive because the Reg1 repressor cannot bind to the mutant operator.

9. When the λ phage infects a cell with an integrated λ phage (that is, the cell is a lysogen), the *c*I repressor protein is already present in the cytoplasm so can bind and repress the transcription of the genes on the incoming phage chromosome. Thus the incoming phage is repressed and no progeny can be made. The nonlysogenic recipient cell has no the *c*I repressor protein in the cytoplasm, so the incoming infecting phage goes into the lytic cycle, producing progeny phage.

10. These revertants of the constitutive expression of the *lac* operon could be changes in the base sequence of the operator site that compensate for the missense mutation in *lacI*. For example, if an amino acid necessary for recognition of operator DNA is changed in the *lacI* mutant, a compensating mutation changing a recognition base in the operator could now allow the mutant LacI protein to recognize and bind to the mutant operator site. Note that this explanation depends on the fact that the original *lacI* mutation was a missense mutation changing the identity of a single amino acid. If the *lacI* mutation instead lead to more severe changes in the Lac repressor protein (for example, if the mutation was a nonsense or frameshift mutation early in the

coding region), this scenario for suppressing the constitutive phenotype would not be possible.

11. The *lacZ* gene codes for the β-galactosidase enzyme. The *lacY* gene codes for a permease.

   a. ***lacZ* is constitutive; *lacY* is constitutive.** The o^C mutation acts in *cis* on *lacZ*$^+$ and *lacY*$^+$; because the repressor cannot bind to the o^C site, the two genes are expressed at high levels whether or not the inducer allolactose is present.

   b. ***lacZ* is constitutive; *lacY* is inducible.** The o^C mutation acts in *cis* on *lacZ*$^+$. Even in the presence of I^+, which works both in *cis* and in *trans*, *lacZ*$^+$ is constitutive because repressor made from the I^+ allele cannot bind the o^C operator. Expression of *lacY*$^+$ is inducible because it is downstream of a wild-type operator.

   c. ***lacZ* is constitutive; *lacY* is inducible.** As in part (b), *lacZ*$^+$ is constitutive because the o^C mutation acts in *cis* and cannot respond to I^+, while *lacY*$^+$ is inducible because it is downstream of a wild-type operator.

   d. **no expression of *lacZ*; *lacY* is constitutive.** *lacZ*+ cannot be transcribed without a functional promoter, and *lacY*+ is consititutive because it is connected physically to o^C.

   e. **no expression of *lacZ*; no expression of *lacY*;** I^S makes a *trans*-acting superrepressor that binds o^+ on both copies of the operon. The superrepressor cannot bind to the inducer allolactose, so expression of the structural genes cannot be induced.

12. a. In the presence of glucose alone, **the repressor protein (LacI) is bound** to the operator because no inducer (allolactose) is present. Because glucose is present, the levels of cAMP are low, so the CRP protein cannot bind to its binding site adjacent to the operon's promoter.

    b. In the presence of glucose + lactose there will be **no proteins bound to the regulatory region** of the *lac* operon. Under this condition, the LacI repressor protein binds to the inducer allolactose and cannot bind to the operator. As in part (a), the presence of glucose prevents the CRP protein from binding to the regulatory region because the levels of cAMP are low.

    c. In the presence of lactose only, **the CRP-cAMP complex will bind to the promoter region** of the *lac* operon. The LacI repressor binds to the inducer rather than to the operator. In the absence of glucose, the activity of adenyl cyclase is high and thus the level of cAMP is high, allowing the CRP-cAMP complex to bind to its target site adjacent to the operon's promoter.

13. a.

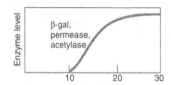

Because one I^+ gene is present and both operons are o^+ P^+, both copies of the *lac* operon are inducible. Neither copy of the *lac* operon is transcribed in the first 10

minutes in the absence of lactose, because the LacI repressor is bound to the operators. Starting at 10 minutes when lactose is added as the sole carbon source, all three proteins will be induced because at least one wild-type structural gene is present to encode each protein and the LacI repressor can no longer bind to the operators.

b.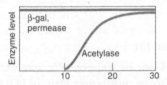

The $lacZ^+$ and $lacY^+$ genes are constitutive (they are expressed even if lactose is absent) because they are physically connected to the *cis*-acting o^C mutation. On the other hand, $lacA^+$ is inducible because the only functional (+) $lacA$ gene copy is physically connected to o^+, and an I^+ gene is present.

c.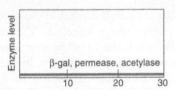

All three genes are off even in the presence of lactose because a functional copy of each is connected to a functional operator (o^+), and I^S makes a *trans*-acting superrepressor that prevents induction of either o^+ operon. (The superrepressor protein cannot bind to the inducer so it always binds to o^+ sites.)

d.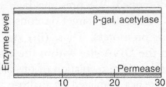

Both copies of the I gene are non-functional (I^-). In the absence of repressor protein, any wild-type structural gene downstream of a functional promoter (P^+) will be expressed constitutively. Therefore, $lacZ^+$ and $lacA^+$ are constitutive, while $lacY^+$ is not expressed at all because the promoter of the operon in which it is located is nonfunctional.

e.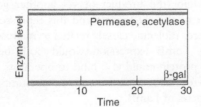

In the absence of repressor protein, any wild-type structural gene with a functional promoter (P^+) will be expressed constitutively. Therefore, $lacY^+$ and $lacA^+$ are constitutive, while $lacZ^+$ is not expressed at all because the promoter of the operon in which it is located is nonfunctional.

14. a. The *malT* gene codes for a positive regulator that increases the expression of all three operons. A loss-of-function mutant in *malT* would be unable to transcribe the genes in the three operons, so **the bacterial strain would be unable to utilize the sugar maltose as a carbon source (Mal-).**

    b. **No, the maltose operons would not be expected to have CRP binding sites. The expression of the MalT positive regulatory protein is catabolite sensitive**, meaning that in the presence of the preferred carbon source glucose, the MalT protein would not be synthesized. This fact allows catabolite sensitive expression of all the other maltose structural genes; cells would not use maltose if glucose is present. (Although it is possible that the three *mal* operons could have binding sites for the CRP-cAMP complex, these sites would not be required.)

    c. Bacteriophage lambda (λ) binds to the LamB protein in the outer membrane of the host bacterial cell in order to initiate phage infection. Therefore, *E. coli* will be sensitive to infection with lambda only if the LamB protein is expressed. **The cells will be sensitive to infection when they are grown in media with maltose** (activating *malT* which will in turn increase expression of *lamB*) **and without glucose** (so there is no catabolite repression to reduce *lamB* expression).

    d. Any mutations that affect presence of a functional LamB protein would be resistant to bacteriophage λ. These will include: **(i) *lamB-* point mutations or deletions; and (ii) promoter mutations affecting transcription of the *malK lamB* operon.**

    The problem defined the MalT protein as a positive regulator although the mechanism of action was not described. We presume here that MalT works by binding sites near the promoters of the three target operons in a fashion that helps RNA polymerase recognize these promoters. Thus, **(iii) *malT-* loss-of-function mutations; or (iv) mutations in the DNA site within the *malK lamB* operon to which *malT* binds** would prevent positive regulation of the *lamB* gene.

    It is possible that among the genes regulated by maltose is a gene encoding a protein that helps import maltose into the cell. **(v) A loss-of-function mutation in a gene whose product is needed for maltose import** would block the cell's ability to synthesize LamB protein and would thus make the cell resistant to bacteriophage λ.

    The problem further states that maltose induces the synthesis of LamB protein via expression of MalT, but the mechanism was not described. The most likely scenario is that a repressor protein (the product of yet another gene) binds to operator sites near the promoter of the *malT* gene, and this repressor protein can associate with maltose or an inducer molecule closely related to maltose. In this way, MalT expression and indirectly LamB expression would both be induced by maltose. (The putative repressor protein could also bind to operators located at the three target operons.) (vi) Superrepressor mutations in the presumptive repressor gene could also prevent the synthesis of LamB.

chapter 15

15. This problem can be approached in either of two ways: (i) by starting with the expression data and assessing what types of mutations could produce that particular phenotype; or (ii) by starting with each single mutant and matching it with an expression pattern. Using the latter approach, the superrepressor mutant (a) will show no expression of β-galactosidase under any conditions and therefore is either mutant 3 or 4. The operator deletion (b) will give constitutively high levels of β-galactosidase expression in the presence of either glycerol or lactose, but it will show low expression with lactose + glucose because it will be subject to catabolite repression. Therefore the operator deletion is mutant 5 or 6. The amber suppressor tRNA (c) would have no effect on its own and is mutant 7. The defective CRP-cAMP binding site (d) causes the same low levels of β-galactosidase expression with lactose or lactose + glucose because the CRP-cAMP complex cannot bind to the promoter region to increase expression. Thus mutant 1 or 2 contains a mutation in the CRP-cAMP binding site. The nonsense mutation in β-galactosidase (e) will give no expression, so it is mutant 3 or 4. The nonsense mutation in the repressor gene (f) leads to constitutive expression of β-galactosidase, as in mutants 5 and 6. A defective *crp* gene (g) means the CRP-cAMP complex cannot form, so it cannot bind to the promoter region to augment gene expression; this phenotype is seen in mutants 1 and 2. To summarize, mutations 1 and 2 are the mutant CAP-cAMP binding site and the defective *crp* gene; mutations 3 and 4 are a superrepressor and a nonsense mutation in the β-galactosidase gene; mutations 5 and 6 are an operator deletion and the nonsense mutation in the repressor gene; and mutation 7 is an amber suppressor tRNA.

The next stage of the analysis involves understanding the results of the double mutant and merodiploid genotypes. An amber nonsense mutation of the *lacZ* gene would be suppressed by mutation 7 (amber suppressor tRNA) in the same cell. Therefore, mutant 3 contains the nonsense mutation in β-galactosidase (*lacZ*) gene and mutant 4 must be a superrepressor mutation. Likewise, a nonsense mutation in the repressor gene would be suppressed by mutation 7 in the same cell, so mutant 5 is the mutant repressor gene and mutant 6 is the operator deletion. The defective *crp* gene can be distinguished from the defective CRP-cAMP binding site by the merodiploid genotypes presented (the genotypes with an F' plasmid in addition to the bacterial chromosome). In both merodiploids, the bacterial chromosome has a mutant gene for β-galactosidase, but all other parts of the *lac* operon expression system are wild type. The F' element has either a mutant CRP-cAMP binding site or a defective *crp* gene. In the latter case, the presence of the *trans*-acting wild type crp^+ gene on the bacterial chromosome will give normal regulation of the *lac* operon. However if the mutation on the F' element is in the *cis*-acting CRP-cAMP binding site, the phenotype of the merodiploid will still be mutant. Thus, mutant 1 is the *crp* mutation and mutant 2 is the altered binding site. In summary: **a=4; b=6; c=7; d=2; e=3; f=5; and g=1.**

16. a. **If you screen only for Lac+ revertants, many of the revertants you find will be mutations in just the *lac* operon.** These could be compensating mutations in the *lac operon's* CRP-cAMP binding site (that is, mutations altering the binding site's sequence so that it can bind to mutant CRP-cAMP complexes) or mutations in the operon's promoter that would allow strong recognition by RNA polymerase even in the absence of the CRP-cAMP complex.

Demanding that the suppressor mutations yield both Mal+ and Lac+ phenotypes will limit the revertants you find to more general cases that affect **all** CRP-cAMP binding or activity. These could be second mutations in the CRP gene (such that a *crp* gene with the original mutation and the second mutation would now be functional), or mutations in genes that affect either CRP or cAMP production or the function of the CRP-cAMP complex.

You should note that the kinds of suppressors you isolate in either screen depend in large part on the kind of mutation in the *crp* gene you started with. The answers above assume that the *crp* gene had a missense mutation, so that mutant gene product would be made but has an altered amino acid sequence that would compromise some aspect of its function. If you started with a nonsense or frameshift mutant that could not make full-length CRP protein, the suppressors you find (excluding nonsense or frameshift suppressor mutations in tRNA genes) would have to bypass or circumvent the entire catabolite repression mechanisms. Very few ways can be envisioned to accomplish this kind of bypass.

b. **Most likely, the α subunit of RNA polymerase interacts directly with the CRP protein**. If the original *crp* mutation was missense, then the gene could encode a full-length CRP protein that could no longer interact with the α subunit of RNA polymerase. If the gene encoding the α subunit had a compensating missense mutation, then the mutant CRP protein might be able to form contacts with the mutant RNA polymerase.

Much less likely is the hypothesis that the mutant RNA polymerase might recognize the promoters of the operons involved in metabolizing sugars more strongly than normal, so that positive control through the CRP-cAMP complex would no longer be needed for high levels of the expression of these operons. First, you know that promoter recognition is mostly mediated by the σ subunit of RNA polymerase rather than the α subunit. Second, a mutation that altered RNA polymerase's recognition of many promoters would likely be lethal to the cell.

17. a. Mutations **i, iii, v and vi** would all prevent the strain from utilizing lactose. (i) Deletion of *lacY* would prevent importation of lactose into the cell because Lac permease would be missing. (iii) A loss-of-function missense mutation in *lacZ* would abolish the ability of β-galactosidase to cleave lactose into galactose + glucose. (v) If the cell had a superrepressor mutation, transcription of the Lac operon would not take place even if lactose was present in the medium because the superrepressor protein would always occupy the operator and could not be removed by it. (vi) By inverting the structural genes but not the regulatory region including the promoter, the structural genes could not be transcribed properly (the resultant message would be an RNA with the same sequence as the DNA's template strand, not the same sequence as the RNA-like strand, and the polarity would be incorrect).

b. The *lacY* deletion will be complemented by any other mutation that expresses the LacY (permease) protein. These will include **mutations ii, iii and iv**. (In the strains with a Lac+ phenotype, the *lacY+* gene will be expressed from the operon on the plasmid, not from the operon on the bacterial chromosome.) Cells carrying plasmids with mutations v and vi will be Lac- in phenotype because the *lacY* genes on the plasmids could not be expressed as explained in part (a), while the *lacY* gene on the bacterial chromosome has a *lacY* deletion.

chapter 15

c. First look for striking patterns that might indicate a specific mutation. For example, mutant 6 shows expression of *lacZ* when combined with any of the other mutations. Of the possible choices, mutation 6 could only be an o^C mutation. Strains containing mutation 5 and any of the mutations other than mutation 6 also have a very consistent pattern: *lacZ* is never expressed, except when combined with mutation 6 (the o^C mutation). Mutation 5 therefore shuts down the other copy of the operon in addition to its own copy, which can be explained by a superrepressor (I^s) mutation.

Looking at the remaining mutations, the inversion of the entire *lac* operon (iv) should not have an effect on expression and should not influence expression from other copies of the operon. Mutation 4 leads to inducible *lac* expression except when it is combined with mutation 5, the superrepressor. Thus, mutation 4 is equated with (iv).

The inversion that does not include *lacI, P,* and *o* should not show expression because the regulatory region is now in the opposite orientation from the genes (see the answer to part [a]). This inversion also should not influence expression from the other mutant copy. Looking at the patterns for mutations 2 and 3, mutation 3 does not show expression except when combined with 4, so mutation 3 fits with this inversion.

By process of elimination, mutation 2 is a loss-of-function *lacZ* missense mutation. This conclusion makes sense because a cell with one operon containing this mutation can synthesize β-galactosidase only if the other operon has a *lacZ+* gene that can be expressed.

To summarize: **mutation 1 is i; mutation 6 is ii; mutation 2 is iii; mutation 4 is iv; mutation 5 is v; and mutation 3 is vi.**

18. a. **Mutations in either o_2 or o_3 alone have only small effects on Lac repressor binding.** Note on Fig. 15.17 on p. 526 that one dimer in the Lac repressor tetramer can bind to *either o_2 or o_3*. If a *lac* operon had a mutation that disrupts one of these sites, the repressor still could bind to the either site. **A mutation in either element would have only a small effect on transcription of the operon** and would therefore be difficult to detect in screens for mutations that affect *lac* operon regulation.

b. **A mutagen that causes small insertions would produce an o^C mutant phenotype if the insertion disrupted the operator such that the Lac repressor could no longer bind to it efficiently.** There are two ways in which a small insertion could disrupt the operator sequence. First, the insertion could disrupt the specific base pair sequence at o_1 so that the repressor could not recognize it. A second way that an o^C mutant phenotype could be produced is if the insertion alters the spacing between binding sites. An insertion could change the spacing between the binding sites for the two Lac repressor monomers at o_1, or it could alter the spacing between the two operators (o_1 and o_2/o_3).

The reason spacing is important is that the Lac repressor is a tetrameric protein. A functional Lac operator must have the four binding sites for the four subunits of the repressor protein on the same "face" of the DNA. (It takes 10.5 bp to achieve one full turn of the double-helix; binding sites that are multiples of ~11 bp apart will be on the same "face" of the DNA.) Two subunits of the tetramer bind to the two recognition sequences at the o_1 site and the other two subunits of the tetramer

15-15

bind to either the o_2 or o_3 site: the DNA in between the two operators must bend to allow simultaneous interactions with two sites. Small DNA insertions may change the way that the binding sites face one another.

c. The o^C mutations described in part (b) that are caused by small insertions of DNA would most likely be insensitive to the I^S superrepressor protein, but *lac* operons with some such mutations might show some sensitivity to I^S depending on the nature and location of the particular DNA insertion. ("Insensitive" here means that a double mutant strain with both an o^C mutation and an I^S superrepressor mutation would show high levels of constitutive expression of the *lac* operon. "Sensitive" means that the level of *lac* operon expression in the double mutant would be lower than in a strain with the o^C mutation alone; this low level would be seen in the presence or absence of lactose.)

An o^C mutation caused by disruption of the base pair sequence of the o_1 binding site, or alteration of the spacing between the monomer binding site at o_1, will be insensitive to an I^S superrepressor protein because the superrepressor would not be able to bind o_1, the site most critical for repression. The effect on superrepressor binding of an o^C mutation caused by altering the spacing between the two operators (o_1 and o_2/o_3) is more difficult to predict. LacI dimers might still be able to bind to the o_1 site, although the efficiency would be lower than normal because the full tetramer could not contact o_1 and o_2/o_3 simultaneously. In such cases, the level of *lac* operon expression would be reduced by the superrepressor; this expression level would be unaffected by the presence or absence of lactose because the superrepressor cannot bind to the inducer.

19. a. **At its left end, the operator begins at a position after the endpoint of deletion 1 and before the endpoint of deletion 5.** (The operator is intact in deletion 1 because the operon is still inducible, but part of the operator has been removed in deletion 5 because the operon is now constitutively expressed.) **The right endpoint of the operator cannot be determined by these data.**

b. **The deletions may have removed bases within the promoter that are necessary for the initiation of transcription.**

20. **DNA-binding sites show rotational symmetry when they are bound by homodimers.** Each monomer in the dimer has the same DNA-binding domain that recognizes the same base pair sequence. The two DNA-binding domains in the homodimer are positioned as mirror-images with respect to one another (see the chapter's opening figure on p. 514), and so must be the base pair sequences that they bind.

The promoter is not rotationally symmetric because RNA polymerase is not a homodimer. It is instructive to ask why the promoter site cannot be rotationally symmetrical. The answer is that if this was the case, RNA polymerase would not "know" in which direction it should travel in order to transcribe the adjacent gene or operon.

chapter 15

Section 15.3

21. Attenuation, as in the *trp* operon, is unique to prokaryotes because it depends on translation of a nascent mRNA while the transcript is being elongated. **Transcription and translation occur simultaneously in prokaryotes, but this is not so in eukaryotes.** In eukaryotes, transcription occurs in the nucleus and transcripts are exported to the cytoplasm where they are translated.

22. a. **At least three ribosomes** are required for the translation of *trpE* and *trpC* from one mRNA molecule. The fact that there is a full length mRNA means that early in transcription one ribosome must have initiated translation of the leader and stalled at the tryptophan codons in the attenuator, resulting in antiterminator formation. A second ribosome must have bound to the ribosome binding site at the beginning of the *trpE* open reading frame, and a third ribosome must bind to the ribosome binding site at the beginning of the *trpC* open reading frame.

    b. If the two tryptophan codons in the leader were deleted, the terminator stem loop would form causing transcription termination. (With respect to Fig. 15.21 on p. 530, the ribosome would translate the short open reading frame in the RNA leader rapidly, allowing the formation of the 3-4 stem-loop structure that terminates transcription.) **The transcription of the *trpE* and *trpC* genes would be rare** regardless of tryptophan concentration, because the level of full length mRNA from the *trp* operon would be very low.

23. One open reading frame exists in this mRNA leader sequence, beginning with the first nucleotide. The predicted amino acid sequence is:

    N Met Thr Arg Val Gln Phe Lys His His His His His His His Pro Asp C

    Notice the 7 histidines in a row out of a total of 16 amino acids. Thus, if the cell is starving for histidine, the ribosomes will pause at the His codons (CAC or CAU) in the leader sequence because the tRNA^{His} molecules will not be completely charged with histidine. This ribosome pausing presumably allows the leader to fold up into an antiterminator configuration similar to that shown in Fig. 15.21b on p. 530 for the *trp* operon. If the leader adopts this antiterminator configuration, RNA polymerase can complete transcription of the operon, giving maximal production of the polycistronic mRNA and maximal expression of the structural gene proteins that synthesize histidine.

24. a. Because the promoter is deleted, transcription cannot occur and **there is no expression of either *trpE* or *trpC* under any conditions**.

    b. The repressor cannot bind the operator, so transcription initiates constitutively. However, the attenuator is functional, so transcription often terminates prematurely (before RNA polymerase reaches the structural genes) when tryptophan is present. There will be **lower (but measurable) levels of expression of both *trpE* and *trpC* when tryptophan is present, and high levels of expression of both genes when there is no tryptophan**.

    c. The repressor cannot bind tryptophan and thus cannot bind to the operator, so transcription of the structural genes is initiated constitutively. However, as in part (b), the attenuator is normal so transcription often terminates prematurely when tryptophan is present. Thus, **there will be lower (but measurable) levels of expression of both *trpE* and *trpC* when tryptophan is present, and high levels of expression of both genes when there is no tryptophan.

d. The repressor cannot bind to the operator, so expression of *trpC* and *trpE* is constitutive. The attenuator is nonfunctional, so the operon will always be transcribed at the maximal level. Thus, ***trpC* and *trpE* will show completely constitutive expression, with high levels of the gene products made whether or not tryptophan is present**.

e. The **expression of *trpC* is repressible by tryptophan** because the wild-type allele of this gene is controlled by a normal operator and because the repressor is *trans*-acting. When tryptophan is added to the medium, it will act as a corepressor that interacts with the repressor; the repressor/corepressor complex will bind to the operator and shut down transcription. The **expression of *trpE* is partially constitutive** (meaning that this gene is always transcribed, but the levels of expression are lower when tryptophan is present) because the *cis*-acting operator is defective but the attenuator is still functional.

f. The **expression of *trpC* is repressible by tryptophan** for the same reasons just described in part (e). There will be **no expression of *trpE*** because one *trpE* gene has a null mutation while the wild-type $trpE^+$ gene is downstream of a defective *cis*-acting promoter.

g. Expression of ***trpE* is fully constitutive** ("on" at high levels all the time) because the *cis*-acting operator is defective and the *cis*-acting attenuator site is defective. ***trpC* expression is partially constitutive** (this gene is always transcribed, but the levels of expression are lower when tryptophan is present) because the $trpC^+$ gene is in *cis* with a defective operator and a functional attenuator.

25. To distinguish between the transcription interference and translation inhibition mechanisms for antisense RNA action, **you can test whether or not the antisense RNA can act in *trans* to inhibit expression of the gene it regulates; action in *trans* would rule out transcription interference**. Assuming that you have a method for detecting the protein product of the gene, a simple way to do this experiment would be to **use recombinant DNA technology to make a plasmid construct in which a strong bacterial promoter drives transcription of the antisense RNA**. You would then grow bacteria transformed with the plasmid under conditions where the gene product in question is normally expressed from the bacterial chromosome. **If the antisense RNA functions by interfering with sense transcription (in *cis*), then the antisense RNA from the plasmid should have no effect on the levels of the gene product. If the antisense RNA functions in *trans* by interfering with translation of the sense RNA, then bacteria containing the plasmid should synthesize little or no protein product.**

26. a. Lac repressor: **protein; negative regulator; transcription initiation; in *trans***

    b. *lac* operator: **DNA; negative regulator; transcription initiation; in *cis***

    c. CRP: **protein; positive regulator; transcription initiation; in *trans***

    d. CRP-binding site: **DNA; positive regulator; transcription initiation; in *cis***

    e. Trp repressor: **protein; negative regulator; transcription initiation; in *trans***

f. charged tRNA^Trp: **RNA + small molecule; negative regulator; transcription elongation; in** *trans*

g. *trp* operon antiterminator: **RNA; positive regulator; transcription elongation; in** *cis*

h. riboswitch terminator: **RNA; negative regulator; transcription elongation; in** *cis*

i. sRNA that blocks translation: **RNA; negative regulator; translation initiation; in** *trans*

27. a. *CsrA*: **protein; negative regulator; translation initiation; in** *trans*. *CsrB*: **RNA; positive regulator; translation initiation; in** *trans*.

    b. The CsrA/CsrB system is likely to respond to glucose levels. Target genes for glycogen synthesis (anabolic) could be induced by high glucose if a repressor of *CsrB* transcription cannot bind its operator when the repressor is bound to glucose. Target genes for glycogen breakdown (catabolic) could be repressed by high glucose if a positive regulator of *CsrB* transcription binds DNA only when it is not bound to glucose.

    You should note that the underlying logic for regulation of target genes is opposite to that governing the *lac* and *trp* operons. The *lac* operon is catabolic and induced by the substrate (allolactose), while the *trp* operon is anabolic and repressed by the synthesis product (tryptophan). The opposite is the case for the genes targeted by the CsrA/CsrB system: An anabolic operon is induced by the substrate (glucose), while a catabolic operon is repressed by the synthesis product (glucose). The reason for the reverse is the control by glucose rather than by glycogen. This makes biological sense because: (i) glycogen is a large molecule that cannot easily be imported into cells; (ii) the anabolic pathway generates glycogen as a means of storing excess glucose; and (iii) the catabolic pathway breaks down glycogen when glucose levels are low.

Section 15.4

28. The loss of LexA function leads to the new expression of many genes. **Therefore, the wild-type LexA protein acts as a negative regulator that shuts off the expression of these target genes.** One possible model is that the LexA protein is a repressor that binds to the operators of target genes or operons. However, given the information in the problem, LexA could also suppress target gene expression in other ways, for example by preventing translation of target gene/operon mRNAs.

    It turns out that LexA is indeed a repressor that turns off target gene transcription by binding to an operator sequence. The mechanism by which DNA damage induces expression of the target genes is very interesting. In the presence of DNA damage, another protein called RecA interacts with LexA in a fashion that causes the LexA protein to cleave itself, disrupting LexA's DNA binding domain. LexA can no longer bind to operators, so target gene/operon expression is induced.

    This system becomes even more interesting when you consider that among the genes targeted by LexA are the genes encoding proteins that participate in the SOS system of DNA repair (described in Chapter 7 on p. 233). When DNA damage occurs, LexA is inactivated, leading to induction of the genes encoding these SOS system enzymes. These enzymes are now synthesized and the DNA damage can be corrected.

Because the SOS system is error-prone, cells need to repress the synthesis of these enzymes if the DNA is not damaged; this is the function of the LexA repressor.

29. **Fuse the *lacZ* coding sequences to the promoter and regulatory elements of one of the motility genes whose expression was increased during cell growth in poor carbon sources. Then introduce this fusion gene into a *lacZ⁻* *E. coli* strain.** This bacterial strain will make high levels of ß-galactosidase in media containing poor carbon sources but lower levels media with rich carbon sources. **Next, mutagenize this strain and look for mutants in which *lacZ* expression does not increase under poor growth conditions.** Such mutants can be assayed with a ß-galactosidase substrate like X-gal that changes color when cleaved. Mutant cells would not produce ß-galactosidase and so could not cleave X-gal and would give rise to white colonies on petri plates that were made with poor carbon source medium. The cells that do produce ß-galactosidase will cleave the X-gal and give rise to blue colonies on the same plates.

    The above experiment would allow you to find mutants in genes involved in the foraging response. **You would next have to identify the gene that was altered in each mutant strain.** One approach would be to map the genes; this procedure would be much easier if you mutagenized the cells with a transposon and then identified the transposon insertion site with inverse PCR (review Figs. 13.29 and 13.30 on pp. 480-481). Another approach would be to "rescue" the mutant phenotype by transforming mutant cells with a plasmid library containing random fragments of the bacterial genome and looking for transformants that now give rise to blue colonies on poor carbon source petri plates (review Fig. 13.28 on p. 480).

30. a. **If the X-gal plates contained maltose, the colonies could be blue because transcription of the *lacZ* reporter would be induced. Plates without maltose would have white colonies.** (Note: these colonies are wild-type in the sense that all of the genes encoding regulators of the maltose-inducible operon are wild-type.)

    b. In a blue colony, the reporter is transcribed in the absence of maltose. (Presumably, these colonies would also be blue in the presence of maltose.) Blue colonies could result either from a gain-of function mutation in a gene encoding a positive regulator of *mal* operon expression, or from a loss-of-function mutation in a gene encoding a negative regulator of this operon. The wild-type regulator, whether positive or negative, would have to respond to the presence or absence of maltose.

      If the mutant gene in this colony encodes a positive regulator of the maltose-inducible operon: (i) The wild-type protein encoded by this gene would be activated by maltose. In the likely case that the protein is a transcriptional regulator, the wild-type protein might bind DNA more efficiently in the presence of maltose. **(ii) The mutation would result in a hypermorphic or neomorphic gain-of-function. (iii) The mutation would either increase the amount of the protein or alter the protein's amino acid sequence to make the protein work more efficiently or in a new way.** For example, if the wild-type protein binds to DNA weakly in the absence of maltose, the mutation might result in more of the protein being produced; alternatively, mutant cells might produce normal amounts of a mutant protein that could bind DNA efficiently in the absence of maltose. The reporter gene would thus be expressed (the colony would be blue) even if maltose is not present.

If the mutant gene in a blue colony encodes a repressor of the maltose-inducible operon: (i) **The wild-type protein encoded by this would function in the absence of maltose but not in its presence.** In the most likely scenario, this protein would be a transcriptional repressor like the Lac repressor that could bind DNA in the absence, but not in the presence, of an inducer related to maltose. (ii) **The mutation would be loss-of-function [either an amorph (null) or a hypomorph].** (iii) **The mutagen is not specified, so the mutation could be a nonsense mutation, a missense mutation in a key amino acid for protein function, a frameshift mutation near the start of the ORF, or a deletion of the gene.** If the negative regulator function is missing, the reporter gene would be expressed (the colony would be blue) regardless of the presence of maltose.

c. If the mutation is gain-of-function, transformation of mutant bacteria with a wild-type genomic library will not rescue the mutant phenotype (constitutive expression, seen as blue colonies on plates lacking maltose) to wild-type (inducible expression, seen as white colonies on plates lacking maltose). However, transformation of a wild-type strain with a genomic library made from mutant genomic DNA would cause the mutant phenotype: A colony transformed with a plasmid containing the gain-of-function mutant gene would be blue in the absence of maltose.

If the mutation is loss-of-function, the opposite results would be observed. Transformation of mutant bacteria with a wild-type genomic plasmid library would result in rescue of the mutant phenotype to wild type (in colonies transformed with a wild-type copy of the gene corresponding to the mutation), while transformation of mutant bacteria with a mutant genomic library would not – the genomic library would not contain a wild-type copy of the mutant gene.

31. a. **The plasmid diagrammed below** also contains an origin of replication and a drug resistance gene that are not shown. The *lac* operon regulatory region must contain the operon's promoter, operator, CRP-cAMP binding site, and the ribosome binding site. Note that the *lac* regulatory region must be followed by sequences encoding MBP, DDDDK, and erythropoietin (EPO) in that order because the N terminus of the fusion protein is the N terminus of MBP, while the C terminus of the fusion protein is the C terminus of EPO.

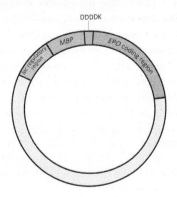

b. The ribosome binding site is in the *lac* regulatory sequences **(ii)**.

c. In order for the plasmid to encode the desired fusion protein, MBP, DDDDK, and the EPO coding region must all be in frame with one another: **(i, iii, iv)**.

d. The EPO coding sequences should come from a **cDNA** because bacterial genes do not have introns and therefore bacteria cannot splice eukaryotic RNAs that contain introns.

e. Based on the information given in the text, the most logical choices here would be to use either **lactose or allolactose** as the inducers to "turn on" expression of the fusion gene. In actuality, these compounds are metabolized (cleaved by β-galactosidase) so quickly that the induction would only be temporary. Genetic engineers can instead use a compound called isopropyl β-D-1-thiogalactopyranoside (IPTG) as the inducer for fusion genes under the control of the *lac* operon regulatory region. This compound is a molecular mimic of allolactose, so it can bind in the same way to the Lac repressor, but it is not cleaved at all by β-galactosidase. **The inducer should be added to the culture after the population has reached high density** because the fusion protein may be harmful to the bacterial cells. If the inducer was added too early, the culture might stop growing so that too few cells would be available, and too little fusion protein could be produced.

f. As its name implies, the maltose binding protein (MBP) binds tightly to the sugar maltose. **The fusion protein could be purified by passing an extract from MBP-DDDDK-EPO-expressing bacteria over column containing maltose molecules coupled covalently to an insoluble resin.** You would then wash off proteins that are not bound tightly to the column, leaving the fusion proteins on the resin.

g. **You would add enterokinase enzyme to the column in part (f)** that has bound MBP-DDDDK-EPO. This enzyme cleaves polypeptides just after (C terminal to) DDDDK. Thus after cleavage, N MBP-DDDDK C will remain bound to the column through the association of MBP with maltose, while EPO freed of MBP and DDDDK will elute from the column and can be collected.

32. Genes transcribed in response to osmolarity changes are likely to include genes transcribed in response to other stresses like heat shock as well as genes that are turned on specifically in response to the osmolarity change. Compare RNA-Seq data from bacteria grown under high and low osmolarity with bacteria grown in normal temperatures and in heat shock conditions. **RNAs present specifically in the heat shocked culture and also specifically in either high or low osmolarity conditions are likely to be general stress response genes.** (*Note:* In this usage, "present specifically" does not necessarily mean that no transcripts of the gene are seen in a particular condition, but instead the proportion of "reads" from the total mRNA would be significantly higher under one condition or the other. It will also be interesting to look at the RNA-Seq data to see if the expression of any genes is turned down in one condition versus the other.)

33. a. There are **19 genes** depicted in Fig. 15.29; one gene in light blue just upstream of *modA* is clearly shown but is unnamed. The average gene density is **1 gene per 1000 bp**, which is **characteristic of bacterial genomes**. The figure depicts 5

multicistronic operons; however, if you counted each "singleton" genes as a separate operon, you could say that 10 operons exist in this region [or even 11 as discussed in part (e) below].

b. No transcripts were found for the gene *t2110*, but the RNA-Seq analysis was performed under a single environmental condition; **under different conditions, it is possible that *t2110* transcripts could be detected**. In databases, *t2110* is curated as a "hypothetical gene" because it has a long ORF, but it has never been demonstrated that the polypeptide encoded by this ORF is ever made in cells.

c. Premature termination of transcription due to regulation by an attenuator or a riboswitch could be detected as **fewer sequence reads at the end of an operon that at the beginning**. The *hutUH* operon may be regulated this way. Note on Fig. 15.29 that RNA-Seq detects significantly more reads for *hutU* than for *hutH*.

d. Antisense transcription would be detected as **sequence reads in both directions in a region of the genome**. (You would see some reads represented in *green* and some reads in *purple* in a particular position of the DNA sequence.) No such examples are seen in the figure.

e. The **figure shows more reads of the *galM* sequence than reads of the *galETK* genes**. This is unexpected because *galM* is furthest from the promoter; if anything, you would expect RNA polymerase to terminate transcription before it reached *galM*. This finding suggests that *galM* might be a single gene controlled by its own promoter rather than part of the same operon with *galETK*.

f. **No**. What is recorded by RNA-Seq is the relative number of transcripts corresponding to different genes. Translational regulation would not be apparent in the data.

34. (i) The sRNA sequences should not contain any ORFs. (ii) In many cases, sRNAs could be excluded as being fragments of a larger mRNA because the same RNA-Seq experiment may not reveal any nearby transcripts in the same orientation. (iii) The reverse complement of the sRNA sequence should exist in the sense strand of one or more mRNAs, specifically in the 5' UTR.

Section 15.5

35. Using RNA-Seq analysis, the transcriptome of *V. fischeri* growing at high density and making light could be compared with that of non-bioluminescent *V. fischeri* growing at low density. The *luxICDABE* operon would be identified as a transcript present at much higher levels in bioluminescent bacteria than in those not making light. However, *luxR* would not be identified as it is transcribed at the same levels in *V. fischeri* under both low and high density conditions.

36. The key to solving this problem is to remember that loss-of-function mutations in the same gene do not complement (−), while mutations in different genes do complement (+). In addition, mutations that inactivate an entire operon will fail to complement mutations in single genes within that operon. The complementation matrix is shown in the following diagram. Along the diagonal, recall that mutations cannot complement themselves.

	1	2	3	4	5	6	7	8	9
1									
2	−								
3	+	+							
4	+	+	+						
5	+	+	+	+					
6	+	+	+	+	−				
7	+	+	+	+	−	−			
8	−	−	−	−	+	+	+		
9	−	−	−	−	+	+	+	−	

You can use this complementation matrix to sort the 9 bioluminescence mutants into four complementation groups: group I = 1, 2, 8, 9; group II = 3, 8, 9; group III = 4, 8, 9; group IV = 5, 6, 7. These complementation results are unusual compared with those you saw in Chapter 7 in that mutants 8 and 9 are members of three distinct complementation groups. The reason is that the mutations in strains 8 and 9 prevent expression of the entire *luxICDABE* operon. In contrast, the mutants that define the three individual complementation groups each affect only one of the genes in the operon: Mutants 1 and 2 are *luxA-*, mutant 3 is *luxB-*, and mutant 4 is *luxI-*. Note that mutants 8 and 9 are not included in complementation group IV (mutants 5, 6, 7) because the group IV mutants are *luxR-*, and the *luxR* gene is not part of the *luxICDABE* operon.

37. a. The *luxR* gene is transcribed at low levels – not high enough to make the amount of *lacZ* transcript required for blue colonies.

    b. The *luxICDABE* operon is transcribed at low levels without autoinducer.

    c. Transcription of *luxR* is not controlled by autoinducer; Fig. 15.32 on p. 538 does not indicate any way in which this could occur.

    d. LuxR bound to autoinducer activates transcription of the *luxICDABE* operon. The *luxI* gene encodes Synthase, which makes more autoinducer. The operon needed to be transcribed and translated in order to make enough autoinducer to produce enough *lacZ* transcript for the colonies to turn blue.

    e. *luxR* transcription is constitutive but at a low level.

38. Molecules that mimic the ligand for the quorum-sensing receptor would repress *V. cholera* toxin expression and could potentially be used as drugs to prevent cholera symptoms.

chapter 16

Gene Regulation in Eukaryotes

Synopsis

As with prokaryotic genes, important steps in the regulation of eukaryotic genes occur at the level of transcription initiation. Because eukaryotic chromosomes are not naked DNA, but instead are protein/DNA complexes called chromatin, eukaryotic transcriptional activators and repressors function differently than those of prokaryotes. In multicellular eukaryotes, the multiplicity of diverse differentiated cell types requires that the control regions of the genes interact with many more transcription factors than is the case with the control regions of genes in prokaryotes. A fascinating feature of the control of eukaryotic transcription initiation discussed in this chapter is the existence of epigenetic phenomena such as *genomic imprinting* in which patterns of cytosine methylation in DNA that modulate gene expression can be inherited both through cell division and also through the gametes. (Interestingly, although it was not presented in Chapter 15, imprinting exists also in bacteria, but through adenine methylation.) The existence of epigenetic phenomena has many implications for both inheritance and evolution, and the study of events such as genomic imprinting is therefore an extremely active area of research.

Gene expression in eukaryotes is not regulated only at the level of transcriptional initiation, but also at subsequent steps. The housing of DNA in a nucleus precludes certain prokaryotic post-initiation control mechanisms like transcription attenuation that depend on transcription and translation occurring in a single cellular compartment. Nonetheless, some post-transcriptional gene control mechanisms that are of great importance in eukaryotes are either not available to prokaryotes or are employed in different ways. This chapter focuses on two such mechanisms. One is the ability of RNA-binding proteins to determine patterns of alternative splicing, thus allowing primary transcripts to be processed into different mature mRNAs. Surprisingly, the control of alternative splicing plays a key role in determining the sex of *Drosophila*. A second post-transcriptional mechanism described in the chapter concerns the existence of several classes of small RNA molecules that affect the stability and translation of eukaryotic mRNAs.

Key terms

promoter – a DNA sequence near the transcription start site that attracts RNA polymerase to the gene. Eukaryotic pol II promoters bind RNA polymerase indirectly, and contain a **TATA box,** a sequence of roughly seven nucleotides which are all As and Ts.

basal factors – transcription factors that bind the promoter, such as TATA-box binding protein (TBP), and their associated proteins, such as TBP-associated factors (TAFs); together, these proteins form a *basal factor complex* that attracts RNA polymerase to the promoter

enhancer – a *cis*-acting DNA element that regulates transcription of one or more genes and that may be located thousands of base pairs distant from the promoter. Enhancers contain binding sites for transcription factors and are responsible for the spatiotemporal specificity of transcription.

activators and **repressors** – eukaryotic transcription factors that bind sites within enhancers. *Activators* are positively acting factors that either stabilize the basal factor complex at the promoter, or recruit **coactivators**, which are proteins that displace nucleosomes from the promoter. *Repressors* are negatively acting factors that recruit **corepressors**, which are proteins that either disrupt the basal factor complex at the promoter, or close chromatin at the promoter.

indirect repressors – proteins that prevent transcription of a gene by inhibiting the function of an activator rather than by binding to DNA

GFP (green fluorescent protein) – a jellyfish protein used as a reporter protein by researchers; when activated by UV light, GFP glows in live cells

insulators – DNA sites that bind proteins in order to organize chromosomes into loops; an enhancer can interact only with promoters in the same loop

DNA methylation – in vertebrates, the addition of a methyl group to the cytosine base in a CpG sequence (5' CG 3') to form 5-methylcytosine.

CpG island – a *cis*-acting transcriptional regulatory element found upstream of some eukaryotic genes and that is rich in CpG sequences; the activity of the CpG island depends on the extent of cytosine methylation within it. Nonmethylated CpG islands bind activators, while methylated CpG islands exclude activator binding and may bind repressors.

epigenetic phenomenon – a heritable (directly through cell division or through gametes) change in gene expression not caused by a mutation in the gene's base pair sequence

genomic imprinting – epigenetic phenomena due to methylation of transcriptional regulatory elements, where the maternal or paternal origin of an allele affects its expression in the progeny. **Imprinted alleles** are transcriptionally silenced.

miRNAs (micro-RNAs) and **siRNAs (small interfering RNAs)** – small *trans*-acting RNAs that regulate the stability or translation of specific mRNA targets through complementary base pairing. miRNAs are produced by processing of long primary transcripts (*pri-miRNAs*), while siRNAs are produced by the processing of double-stranded RNAs. A given mature miRNA or siRNA binds a protein complex called **RISC (RNA induced silencing complex)**, and guides it to the target mRNA.

piRNAs (Piwi-interacting RNAs) – small *trans*-acting RNAs that guide complexes containing *Piwi proteins* to transposable elements (TEs) or to TE transcripts. piRNAs are important for limiting the mobilization and transposition of TEs.

RNA interference – a research technique for reducing or turning off gene expression that exploits the siRNA pathway. Double-stranded RNA (dsRNA) corresponding to the base pair sequence of a gene is introduced into a cell or organism. Enzymes in the cell or organism process the dsRNA into functional siRNA, which then guides the RISC to the target gene's mRNA, leading to degradation of that mRNA.

chapter 16

Structures of eukaryotic genes and mRNAs

To think about gene regulation, you must be clear about the component parts of eukaryotic genes and mRNAs. In the figure below, note the following: (1) An enhancer can be thousands of base pairs away from the promoter. Most often, as shown in the figure, enhancers are located upstream of the promoter, but enhancers are sometimes found downstream of the entire transcriptional unit or even in the introns of the gene. (2) A gene is typically flanked by insulators if that gene's enhancer(s) do not activate expression of other genes also. (3) The start and stop codons can be in any exons; in other words, several exons can be devoted to the 5' UTR and/or the 3' UTR.

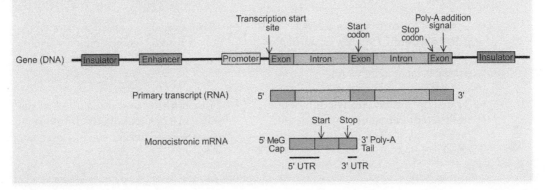

Eukaryotic vs. prokaryotic gene regulation

Eukaryotes	Prokaryotes
• **Promoters** bind basal factor complex, which attracts RNA polymerase	• **Promoters** bind RNA polymerase directly
• **Enhancers** determine temporal and spatial specificity of transcription	• **Operators** and **σ factors** control transcriptional response to environment
• **Insulators** organize genomic DNA to control enhancer/promoter interactions	• Neither **enhancers** nor **insulators** exist
• **Activators** bind enhancers and either stabilize the interaction of the basal complex with the promoter, or recruit **coactivators** that clear the promoter of nucleosomes	• **Activators** bind DNA near promoters and interact directly with RNA polymerase to stabilize the interaction between RNA polymerase and the promoter

- **Repressors** bind enhancers and recruit **corepressors** that either destabilize the interaction of RNA polymerase with the basal complex, or recruit nucleosomes to the promoter

- **Alternate RNA splicing** can generate different proteins from a single gene and can be regulated by cell type-specific splicing factors

- Small RNAs (**miRNAs, siRNA, piRNAs**) control stability or translation of mRNA targets through complementary base pairing

- Transcription and translation occur in different cellular compartments; mechanisms such as attenuation cannot occur

- **Repressors** bind the operator and either inhibit RNA polymerase binding to the promoter, or prevent RNA polymerase from initiating transcription; **corepressors** are effectors that bind repressors and alter their conformation so that they can bind DNA

- **No RNA splicing** occurs

- **sRNAs** regulate translation of mRNA targets through complementary base pairing

- Transcription and translation of the same mRNA can occur simultaneously, allowing mechanisms such as attenuation to exist.

Fusion genes in eukaryotes and prokaryotes

In Chapter 15, you were introduced to the idea that scientists employ recombinant DNA technology to make gene fusions. These gene fusions are useful for studying the regulation of gene expression and are also valuable for producing protein drugs in bacteria. Because the regulatory elements of prokaryotic and eukaryotic genes are somewhat different, gene fusions have different structures in bacteria and eukaryotes. The diagrams that follow summarize these differences and also shows two kinds of eukaryotic fusion genes that you will encounter in later chapters. In these diagrams, P = promoter; o = operator; colored boxes are DNA sequences from different genes; arrows indicate transcription start sites.

Two kinds of reporter genes, *lacZ* (encoding β-galactosidase) and *GFP* (encoding jellyfish green fluorescent protein) can be used both in prokaryotes and eukaryotes. Historically, *lacZ* was used first. The *lacZ* gene product can be detected in living bacterial colonies, but eukaryotic cells have to be killed in order to detect β-galactosidase. A major advantage of GFP as a reporter protein is the ability to detect GFP in living cells of both prokaryotes and eukaryotes.

chapter 16

Prokaryotes

- Reporter fusion to analyze regulatory elements

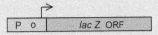

- Fusion to produce large amounts of Protein X

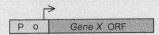

- Promoterless reporter inserted randomly in genome to identify regulatory elements and thereby genes with specific regulation

Eukaryotes

- Reporter fusion to study enhancer properties

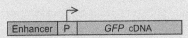

- Fusion to produce large amounts of Protein X (Chapter 17)

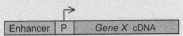

- Enhancerless reporter inserted in genome randomly to identify enhancers and thereby genes with specific expression patterns
  Problem 11 - "enhancer traps"

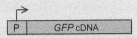

- In frame fusion of ORFs generates tagged hybrid protein (Chapter 18)

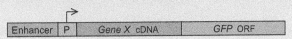

Problem Solving

The problems in this chapter are mostly straightforward, but they require you to synthesize material from several earlier chapters in the book – particularly information about gene expression (Chapter 8), the molecular analysis of gene structure (Chapters 9 and 10), and chromatin (Chapter 11). Keep in mind the structure of a eukaryotic gene and how it compares with a prokaryotic operon, and the other differences between prokaryotic and eukaryotic gene expression outlined on previous pages of this study guide.

Vocabulary

1.
   a. basal factors — 7. bind to promoters
   b. repressors — 6. bind to enhancers
   c. CpG — 4. site of DNA methylation
   d. imprinting — 2. pattern of expression depends on which parent transmitted the allele
   e. miRNA — 9. prevents or reduces gene expression posttranscriptionally
   f. coactivators — 8. bind to activators
   g. epigenetic effect — 10. change in gene expression caused by DNA methylation
   h. insulator — 1. organizes enhancer/promoter interactions
   i. enhancer — 3. activates gene transcription temporal- and tissue-specifically
   j. ChIP-Seq — 5. identifies DNA binding sites of transcription factors

Section 16.1

2.
   a. differential splicing: RNA splicing of any kind occurs in **eukaryotes** only
   b. positive regulation: occurs in **both** prokaryotes and eukaryotes; *positive regulators* in prokaryotes, and *activators* in eukaryotes
   c. chromatin compaction: only **eukaryotes** have chromatin
   d. attenuation of transcription through translation of the RNA leader: **prokaryotes** only. Transcripts are translated during transcription elongation only in prokaryotes; in eukaryotes, transcription in the nucleus is physically separated from translation in the cytoplasm.
   e. negative regulation: occurs in **both** prokaryotes and eukaryotes through *repressors*
   f. translational regulation by small RNAs: occurs in **both** prokaryotes and eukaryotes; in prokaryotes through sRNAs, and through miRNAs and siRNAs and in eukaryotes

3. Events that can affect the type or amount of protein produced in a eukaryotic cell **include: transcript processing (including alternate splicing of the RNA), export of mRNA from the nucleus, stability of the mRNA, changes in the efficiency of translation (including regulation by miRNAs), stability of the protein product, chemical modification of the gene products, and localization of the protein product in specific organelles.**

chapter 16

Section 16.2

4. a. tRNAs; **pol III**

    b. mRNAs; **pol II**

    c. rRNAs; **pol I**

    d. miRNAs; **pol II**

5. a. **(i)** To identify a gene's enhancer(s), fuse the reporter to **(z)**, fragments of genomic DNA around the gene. A genomic fragment that includes an enhancer will cause GFP expression in at least some of the tissues in which the endogenous gene is normally expressed in mice. (The particular pattern of reporter expression observed will depend on how many enhancers the endogenous gene contains and how many of these enhancers were fused to the reporter gene.) **(ii)** To express GFP tissue-specifically, fuse the reporter to **(y)**, a known kidney-specific enhancer. GFP will be expressed in the kidney. An alternative answer is **(z)**, fragments of genomic DNA surrounding a gene. This approach will work if the gene is known to be expressed tissue-specifically; one or more fragments of the DNA in the vicinity of the gene are likely to contain an enhancer that determines this tissue-specific expression. **(iii)** To identify genes expressed in neurons, fuse the reporter to **(x)**, random mouse genome sequences. Those that contain enhancers active in neurons will cause GFP expression in neurons; these enhancers can be used to identify the genes that they control normally.

    b. As one example of ectopic expression, assume that the gene of interest is not expressed in the kidneys normally. If you fused the mouse gene to **(y)**, a known kidney-specific enhancer, the gene should then be ectopically expressed in the kidneys. This approach should work with any characterized enhancer because you could then predict where the fusion gene will be expressed.

       You might want to express a gene ectopically to determine whether a mutant phenotype will result; the nature of that phenotype can sometimes reveal information about the gene's normal function.

6. First, let's number the different parts of the regulatory region so that we can refer to them:

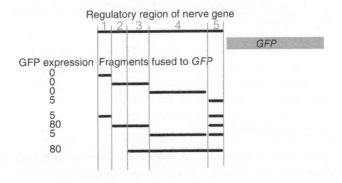

Maximal levels of transcription occur when both an enhancer and a promoter are present. Without a promoter, no transcription occurs; with only a promoter (and no enhancer), only low (basal) levels of transcription will occur. **The promoter must be located in fragment 5** because reporters with at least fragment 5 produce basal levels of GFP, and reporters without fragment 5 produce no GFP. This localization is expected because promoters generally are found very close to the start of transcription. **The enhancer must be in fragment 3** because reporters with at least fragment 3 + fragment 5 have maximal levels of GFP. Note that in this example, as in most cases, the enhancer and the promoter are not located adjacent to each other.

7. Constitutive expression of the galactose genes in yeast will be observed if GAL4 binding to UAS_G cannot be prevented. Therefore, **a *GAL80* mutation in which the protein is either not made or is made but cannot bind GAL4 protein** will prevent indirect repression of the galactose genes and lead to constitutive synthesis. **A *GAL4* mutation that prevents GAL4 protein from binding GAL80 protein** would also lead to constitutive expression of the galactose genes.

8. a. As yeast is a eukaryote, the expectation is that **each gene would be transcribed individually.** Eukaryotic mRNAs are monocistronic.

   b. If the three genes (*GAL7, GAL10,* and *GAL1*) were somehow cotranscribed in a single mRNA, analysis of their ORFs in the genomic DNA should indicate that all three genes are transcribed from the same template strand – in the same direction. In fact, *GAL1* is transcribed from a different template strand than are *GAL7* and *GAL10*, so it would be impossible for a single transcript to contain all three coding sequences.

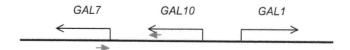

To test if *GAL10* and *GAL7* are cotranscribed, generate cDNAs made from yeast cells that express both genes (for example, wild-type yeast cells grown in the presence of galactose or a strain containing a null mutation in *GAL80* that expresses *GAL7* and *GAL10* constitutively). Only if the two genes are cotranscribed will they be present on a single cDNA. You could establish this point by performing "deep sequencing" (RNA-Seq) on the cDNAs from this cell. Another way to answer this question is to perform PCR on the cDNA library (not on genomic DNA) from these cells using the primer pair shown above (*red arrows*). If the two genes were cotranscribed you would see a PCR product. However, because *GAL7* and *GAL10* are actually expressed from different transcripts, you would not find any cDNAs containing the coding sequences for both of these genes by either method.

9. a. **If Id were a repressor, it would likely bind DNA at the same site on an enhancer as MyoD, or at an overlapping site. Id would inhibit MyoD function indirectly by competing for enhancer binding. Id could also function directly as a repressor by recruiting a corepressor to the enhancer.**

The answer to part (b) of this problem below discusses other scenarios by which Id could function indirectly as a repressor even if it does not bind to an enhancer.

b. **As an indirect repressor, Id could bind to MyoD's activation domain (quenching), or it could form inactive Id/MyoD heterodimers (heterodimerization, also called titration). Another possibility is that Id could interact with MyoD in the cytoplasm and prevent MyoD from entering the nucleus (sequestration).**

c. **If hypothesis (a) were the case, Id ought to have a DNA-binding domain that would be recognizable in the amino acid sequence. If the heterodimerization/titration sub-hypothesis in (b) were the case, Id would have the same dimerization domain as MyoD.** The quenching and cytoplasmic sequestration hypotheses would be more difficult to discern from the Id amino acid sequence.

10. Let gene *A* encode transcription factor 1 and gene *B* encode transcription factor 2.

    a. If both transcription factors are required for expression (complementary gene action), only the *A– B–* genotype will produce blue flowers. Diagram the cross:

    Aa Bb × *Aa Bb* →
    9/16 *A– B–* (blue) : 3/16 *A– bb* (white) : 3/16 *aa B–* (white) : 1/16 *aa bb* (white)
    = **9/16 blue : 7/16 white flowered plants.**

    b. If either transcriptional factor is sufficient for blue color (redundant gene action), then three of the four genotypic classes from a dihybrid cross will produce blue flowers:

    9/16 *A– B–* (blue) : 3/16 *A– bb* (blue) : 3/16 *aa B–* (blue) : 1/16 *aa bb* (white)
    = **15/16 blue : 1/16 white flowers.**

11. a. **In addition to *GFP* coding sequences, a gene construct that would express GFP in *Drosophila* needs to have a promoter, an enhancer, a transcription start site, a 5' UTR, and a transcription termination sequence within a 3' UTR.** The promoter, transcription start site, and 5' UTR need to be directly upstream of the coding region and in that order; the enhancer can be upstream of the promoter or downstream of the 3' UTR; the transcription termination sequences and 3' UTR are directly downstream of the coding sequences. Wing-specific expression would be achieved with a wing-specific enhancer.

    b. **The "enhancerless" gene integrated in a different genomic location in each strain. GFP is expressed only if the construct integrated near an enhancer; the tissue and temporal specificity of that enhancer dictates the GFP expression pattern.** Reporter gene constructs like this are called *enhancer traps;* they can be used to identify genes expressed in particular cell types and/or in specific times during development when the constructs integrate near the enhancers for such genes.

12. You would start with a *Drosophila* strain from part (b) of Problem 11 in which the "enhancerless" GFP reporter is expressed in a tissue-specific pattern suggesting that it has integrated near an enhancer. **You would fuse wild-type *Drosophila* genomic DNA fragments from the region near the integration site to the same**

enhancerless promoter – *GFP* construct. You would then transform the various constructs into the genome of wild-type flies, and look for animals expressing GFP in the same tissue-specific pattern. In this way, you could find a genomic fragment that contains the putative enhancer.

Remember that an enhancer should function upstream or downstream of a promoter, in either orientation. To verify that the genomic fragment indeed contains an enhancer, you should thus make recombinant DNA constructs in which the DNA fragment is again added to the promoter-GFP construct, but the location and orientation of the fragment relative to the coding sequence should be varied. After transformation into flies, GFP expression should again occur in the same tissue-specific pattern. Similarly, you could make constructs that include subfragments of the genomic piece in order to determine the precise boundaries of the enhancer within the original fragment.

13. a. **In order to block enhancer function, an insulator needs to be located** *between* **an enhancer and a promoter.** Insulators prevent the expression of a given gene to be affected by enhancers associated with other genes that are located on the other side of the insulator.

   b. **To identify insulators, you could make an enhancer-containing construct like the one in Fig. 16.15 on p.555, and insert random genomic sequences between the enhancer and the promoter driving** *RFP* **expression. Flies transformed with the construct will express GFP, but not RFP, if the DNA fragment contains an insulator.**

14. a. The target genes of Myc-Max and Mad-Max are the same, but Myc-Max heterodimers are activators while Mad-Max heterodimers are repressors. The implication is that in order to drive the cell cycle forward, Myc competes with Mad for Max binding. Myc expression at the G_0-to-S transition means that **target genes with enhancers that bind Myc-Max and Mad-Max would drive the cell cycle forward.** These target genes would be repressed in the resting state (during G_0) when the *mad* gene is expressed, but they would be transcribed when the *mad* gene is turned on just before S phase. The target genes could encode proteins needed for the cell to progress into S phase and then into other phases of cell division.

   b.

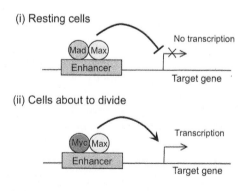

chapter 16

c. Expression of Mad during G_0 keeps genes repressed that would otherwise advance the cell cycle. When Myc is expressed, it competes with Mad for Max binding, and the target genes that advance the cell cycle to S are activated instead of repressed. As the cell gets closer to the transition between G_0 and S, the *mad* gene would presumably no longer be expressed and the Mad protein in the cell would be degraded and disappear. This latter property would act as an autocatalytic switch so that once a cell begins the transition, the decision to divide would become irreversible.

d. As will be explained in Chapter 19, Myc is a proto-oncogene; that is, a normal gene whose product promotes cell proliferation. **Cancer-causing mutations in Myc would be gain-of-function alleles** that in some way caused Myc to be overactive (hypermorphic). The more Myc function present in the cell, the more likely that the cell will proliferate in the out-of-control fashion typical of tumor cells.

15. a. Prokaryotic repressors, like the Lac repressor, directly block RNA polymerase from binding the promoter and/or block the movement of RNA polymerase bound at the promoter to the transcription start site (see Fig. 15.4(b) on p. 519). Eukaryotic repressors recruit corepressor proteins that either close chromatin or prevent pol II from binding the promoter (see Fig. 16.9 on p. 552).

b. Prokaryotic activators, like CAP, attract RNA polymerase to the promoter by touching the polymerase directly. Eukaryotic activators bind coactivators that displace nucleosomes open chromatin, or recruit RNA pol II to the promoter through interactions with other proteins such as those in the Mediator complex (see Fig. 16.6 on p. 551).

16. The UAS_G-*lacZ* construct is a reporter gene that "reports" the interaction between the protein of interest (the "bait") and a protein encoded by a cDNA (the "prey"). The Gal4 DNA-binding domain brings the bait protein to UAS_G. If a prey cDNA encodes a protein that interacts with the bait protein, the Gal4 activation domain will also be recruited to UAS_G. The two fusion proteins bound together constitute a complete activator – a DNA-binding domain and an activation domain. Expression of β-galactosidase will signal the presence of a prey protein that interacts with the bait protein:

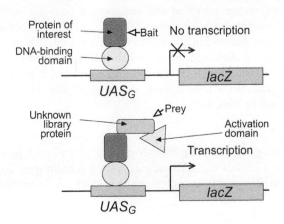

To identify proteins that interact with a protein of interest (the bait), you would take yeast cells previously transformed with the reporter plasmid (UAS_G-$lacZ$) and the bait plasmid (encoding the fusion between the protein of interest and the GAL4 DNA-binding domain) and now transform these cells with a prey cDNA library (in which each clone can express a fusion between the protein encoded by the cDNA and the activation domain of Gal4). Each yeast colony will contain a different prey plasmid from the library.

If the bait and the prey proteins interact with each other, the activation domain will be brought to UAS_G and $lacZ$ will be expressed. Colonies that express β-galactosidase will be blue on X-gal indicator plates. The prey plasmid can be purified and its cDNA identified, allowing you to find a protein that interacts with the protein of interest.

Section 16.3

17. Remember that Prader-Willi syndrome is caused by a mutation in an autosomal maternally imprinted gene. In order to answer this question, you must be able to figure out the genotype of the original affected parent. Remember that all individuals have only one transcriptionally active allele of this gene, and that is the allele they inherited from their father.

    a. **True.** An affected male has a mutant allele of the gene that he inherited from his father and an imprinted (transcriptionally silenced) and presumably normal allele of the gene that he inherited from his mother. In the affected male's gametes, somatic imprints are removed and sex-specific ones are applied. As this gene is maternally imprinted, neither allele is imprinted in his germ line. Therefore, half of his sperm will have the non-imprinted normal allele and the other half will have the non-imprinted mutant allele. All of his children will receive an inactivated (presumably wild-type) allele of this maternally imprinted gene from their mother and so half of his sons (and also half of his daughters) will be affected.

    b. **True.** [See part (a).]

    c. **False.** An affected female has a mutant allele of the gene that she inherited from her father and an imprinted (transcriptionally silenced) and presumably normal allele of the gene that she inherited from her mother. In the affected female's gametes, somatic imprints are removed and sex-specific ones are applied. As this gene is maternally imprinted, both of her alleles are imprinted, meaning that this gene is transcriptionally silenced in all of her eggs. Therefore, half of her eggs will have the imprinted normal allele and the other half will have the imprinted mutant allele. All of her children will receive an active (presumably wild-type) allele of this maternally imprinted gene from their father, so none of her children (whether sons or daughters) will be affected.

    d. **False.** [See part (c).]

18. Draw out the pedigree, including the information given in the problem. In the diagram that follows, the alleles of the maternally imprinted *IGF-2R* gene are *60K* and *50K*, while { } represents an allele that is transcriptionally inactivated (imprinted).

chapter 16

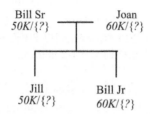

a. Both of Joan's alleles will be imprinted in her gametes, so Jill and Bill Jr each express the allele they received from Bill. Therefore, **Bill Sr's genotype is 50K/{60K}**. With this information, nothing further can be said about **Joan's genotype (60K/{?})**, and there is no way of telling which of Joan's alleles Jill and Bill Jr received.

b. Bill Sr's complete genotype and the further information about Pat and Tim are presented in the pedigree that follows:

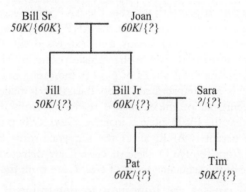

As in part (a) for Joan, both of Sara's alleles will be imprinted in her gametes. Therefore, Pat and Tim express the transcriptionally active alleles they received from Bill Jr. Because Tim's phenotype is 50K, Bill Jr must be heterozygous 60K/{50K}, with his imprinted {50K} allele having come from Joan. Therefore, **Joan's genotype is 60K/{50K} and Bill Sr's genotype is 50K/{60K}**.

19. It may be helpful to refer Fig. 16.20a on p. 559, which shows a typical pedigree diagram for a phenotype controlled by a paternally imprinted autosomal gene.

   a. **No.** The alleles of the gene are not expressed in the germ cells of male I-2. The gene is paternally imprinted, and sex-specific imprints are applied in the germ line. Therefore, by the end of meiosis, both alleles of this gene are imprinted (transcriptionally inactivated).

   b. **No.** The allele of the gene from male I-2 will not be expressed in the somatic cells of II-2. The gene inherited from her father was turned off in his germ cells, so the somatic cells arising after fertilization will contain an inactive copy from I-2.

   c. **Yes.** The allele of the gene from male I-2 will be expressed in the germ cells of II-2. The somatic cell imprinting inherited from the I-2 (I-2 allele inactivated) is erased during gametogenesis, and sex-specific imprints applied. A female will not inactivate either of her alleles of this gene in any of her eggs, so both alleles of this gene will be active in the germ cells of II-2.

d. **No.** The allele of the gene from male I-2 will not be expressed in the somatic cells of II-3. The allele inherited from I-2 is imprinted, so the son's somatic cells will not express that allele.

e. **No.** The allele of the gene from male I-2 will not be expressed in the germ cells of II-3. The imprinting from I-2 is erased during gametogenesis, but the male will re-imprint both alleles of the gene in his germ cells.

f. **Yes.** The allele of the gene from male I-2 (should it be the one inherited) would be expressed in the somatic cells of III-1. The grandson (III-1) would have inherited the non-imprinted allele of the gene from his mother (II-2), so the allele from I-2 is expressed in the somatic cells of III-1.

g. **No.** The allele of the gene from male I-2 (should it be the one inherited) would not be expressed in the germ cells of III-1. Male III-1 will have imprinted both alleles of this gene in his germ cells.

20. a. *Gene A* **is not imprinted;** whether the AKR strain was the female or male parent, half the sequence reads were from the AKR allele, meaning that the *Gene A* alleles from the AKR and PWD strains were expressed equally in the fetus. *Gene B* **is maternally imprinted:** When the female parent is AKR, no sequence reads are obtained from the AKR alleles of *Gene B*, indicating that the *Gene B* allele inherited from AKR is not transcribed; when the AKR is male parent (and PWD is the female parent), all of the sequence reads are from the AKR allele of *Gene B*, indicating that the PWD allele is inactive. *Gene C* **is paternally imprinted:** When the female parent is AKR, all of the sequence reads from *Gene C* are AKR (no transcripts from the PWD allele of *Gene C* are detected); when the male parent is AKR, none of the sequence reads of *Gene C* are from the AKR allele.

b. **Reciprocal crosses are important to determine if a gene is imprinted as opposed simply to having a mutation that prevents its expression irrespective of which parent the allele came from.**

c. A diagram of the cross looks like this:

♀ $A^{AKR} B^{AKR} C^{AKR} / A^{PWD} B^{PWD} C^{PWD}$ × ♂ $A^{PWD} B^{PWD} C^{PWD} / A^{PWD} B^{PWD} C^{PWD}$

At each gene (*A*, *B*, and *C*), the progeny of this cross could be heterozygous *AKR/PWD*, or homozygous *PWD/PWD*. We determined in part (a) that *Gene A* is not imprinted, that *Gene B* is maternally imprinted, and *Gene C* is paternally imprinted.

If a progeny mouse is heterozygous (*AKR/PWD*) for any one of the genes, the expected expression patterns are as shown in the diagram below at left. The key insight is that in the progeny, the *AKR* alleles come from the female parent and the *PWD* alleles from the male parent. If a progeny mouse is homozygous (*PWD/PWD*) for any one of the genes, that mouse will have no *AKR* alleles so the percentage of mRNA reads that are *AKR*-specific will be zero.

chapter 16

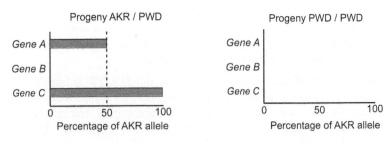

21. As shown in the diagram below, **an insulator that controls imprinting could be located between separate placenta and fetal heart enhancers, such that the insulator is between the placenta enhancer and the promoter, but not between the heart enhancer and promoter.** In this configuration, imprinting could be tissue-specific: The gene could be maternally imprinted in the placenta but not the fetal heart.

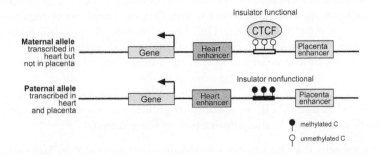

22. An antibody that binds only to DNA containing 5-methylcytosine could be used to identify genomic DNA fragments that are sex-specifically methylated; these fragments could include regulatory sequences of imprinted genes. The procedure you would use is similar to that outlined in Fig. 16.17 on p. 556, but you would start with genomic DNA rather than chromatin. **You would make separate genomic DNA preparations from male and female cells. You would then fragment the DNA and incubate it with the antibody; the antibody would immunoprecipitate (IP) DNA fragments only if they are methylated. You would then sequence the IPed fragments and determine if any sequences are recovered only from male or only from female cells. You would then query the genome sequence to determine if these sex-specifically methylated sequences are near or within genes. Finally, you would use RNA-Seq to verify whether those genes are imprinted.** Allele differences in the copies of the gene inherited from each parent would allow you to determine if the transcripts from the gene in the progeny correspond only to one parental allele.

Section 16.4

23. a. A single eukaryotic gene can give rise to several different types of mRNA molecules through **alternate splicing of the primary transcript.** In addition, some genes have **multiple promoters,** and so different transcription start sites can also generate different transcripts.

b. A single mRNA species can produce proteins with different activities through **posttranslational protein processing (e.g. cleavage of a zymogen) or posttranslational protein modifications (e.g. phosphorylation or ubiquitination).** (Review Fig. 8.26 on p. 280.)

24. Experiments to determine whether or not the 5' UTR or the 3' UTR of *hunchback* regulate its translation would involve making transgene constructs and introducing them into the fly. You will learn the details of transforming transgenes into the *Drosophila* genome in Chapter 18, but assume for now that engineered DNAs can in fact be integrated into fly chromosomes.

    You could perform three kinds of experiments. **(1) You could replace the 5' UTR or 3' UTR of a cloned copy of the *hunchback* gene with the corresponding part of another gene. (2) You could alter the base pair sequences of 5' UTR or 3' UTR of the cloned copy of the *hunchback* gene.** In either case, you would ask whether the manipulated *hunchback* gene you put into the fly genome could rescue the phenotype associated with *hunchback* loss-of-function mutations.

    A third approach relies on the fact that the translational regulation of *hunchback* affects the distribution of the Hunchback protein. As you will learn in Chapter 18, although the *hunchback* mRNA is initially found everywhere in the embryo, the Hunchback protein is present only in the anterior half of the embryo. (A protein called Nanos prevents Hunchback translation in the posterior half of the embryo.) **(3) You could thus attach the 5' UTR or 3' UTR of *hunchback* to a reporter gene like GFP. If these UTR sequences regulate translation, you might see GFP fluorescence only in the anterior part of the embryo.**

25. **The protein may be posttranslationally modified only in fat cells (or in all cells except for fat cells) so that it is only active in fat cells. Alternatively, the protein may need a cofactor to be activated, and this cofactor is present only in fat cells.**

26. a. Transcription, processing of the primary transcript, and export of the mature mRNA to the cytoplasm take place before translation, but these events should only take a few minutes rather than 6 hours (1/4 of a day). Let's assume that all of these processes take about the same time for the interesting gene mRNA and the *GFP* mRNA. Therefore **the difference in first detection of the mRNA of the interesting gene and the GFP protein could be the result of some kind of translational regulation.** For example, translation of the mRNA could depend on a factor that is present in mouse embryos only starting at day 8.75. **A plausible alternative explanation is that the delay is caused by different sensitivities in detecting mRNA versus protein.** In other words, a lot of GFP protein has to accumulate before fluorescence is detectable.

    The difference between the disappearance times of the mRNA and the GFP protein can be explained in at least two ways. **It is possible that GFP protein is more stable than the mRNA, so the protein remains in the cells for several days longer than the mRNA.** A second scenario involves the fact that the mRNA expressing GFP is not identical to the mRNA for the endogenous gene of interest. **It is thus also possible that *GFP* mRNAs are more stable than those of the interesting gene.**

    b. **GFP protein expression should more accurately indicate the onset of gene activation than the cessation of gene activity.** Presumably the *cis*-acting DNA

chapter 16

elements controlling the transcription and translation of the *GFP* reporter gene allow reporter expression to be turned "on" and "off" with the same schedule as the transcription of the interesting gene. However, as explained in part (a) above, if mRNA or protein stability for the two genes differs then GFP detection may cease sooner or continue longer than the activity of the protein product of the gene of interest. The data presented in the problem indeed indicate that the timing of GFP fluorescence aligns closer to the start than to the cessation of gene transcription. However, the investigator also needs to remember that the observation of GFP fluorescence lags the appearance of gene transcripts by a few hours.

27. a. **Raise an antibody to the Argonaute-like protein, and use it to purify complexes between this protein and RNAs that exist within human cells.** To do this, you would first isolate RNA-protein complexes from cells and try to stabilize these complexes by crosslinking them covalently with a reagent like formaldehyde. You would then immunoprecipitate (IP) the complexes containing the Argonaute-like protein using the antibody you prepared. After removing the proteins in these complexes with proteases, you would convert the IPed RNAs into cDNAs using reverse transcriptase. You would then sequence the cDNAs recovered and compare the sequences to the those of known human miRNAs.

b. In a mouse that lacks the Argonaute-like protein, any mRNAs that it down-regulates would be more abundant than in a wild-type mouse. You would thus **perform an RNA-Seq analysis comparing the mRNAs of the mutant mouse and its most closely related wild-type counterpart.** You would then **look for mRNAs that are overrepresented in the mutant relative to the wild-type transcriptome.**

28. a. **The gene construct should express both a sense and an antisense RNA corresponding to the *gene X* ORF** (indicated as *"Gene X"* in the diagram below). The idea is that the bacteria should make double-stranded (ds) RNA. When nematodes (worms) eat the bacteria, enzymes in the worms will process the dsRNA into an siRNA, and this will lead to the degradation of the target gene's mRNA.

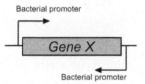

b. To determine if *gene X* mRNAs have been degraded by siRNA, **purify mRNAs from worms that have and have not eaten bacteria transformed with the *gene X* construct and use the two samples to make cDNAs. Using a primer pair within *gene X* and the cDNAs as the template, try to PCR-amplify a product. If the *gene X* mRNAs have been degraded by siRNA, you will be able to detect a product only from the worms that have *not* eaten the transformed bacteria.** Alternatively, you could analyze the cDNAs by RNA-Seq and determine if *gene X* mRNAs are missing from the sample obtained from animals that have eaten the transformed bacteria.

If you know that null mutants in *gene X* cause a particular phenotype, another way to test if *gene X* expression is obliterated (or at least is greatly reduced) is simply to examine the worms that have eaten the bacteria transformed with the construct and ask whether these animals show the mutant phenotype. In general, scientists would not perform this kind of experiment because the purpose of doing RNA interference is usually to ask whether knockdowns of *gene X* cause some unknown phenotype. (That is, these experiments would be done only in cases where null or hypomorphic mutations in the gene were not available.)

c. **Sometimes siRNAs affect expression of genes other than those intended. So-called "off-target effects" occur when some of the siRNA sequences are at least partially complementary to other genes.** This problem can be overcome at least in part by choosing carefully the sequences in the *gene X* construct shown in the answer to part (a). You would select those sequences within *gene X* that have the least similarity to sequences anywhere else in the worm genome.

Section 16.5

29. a. **Mutation of the Tra- or Tra2-binding site that prevents Tra or Tra2 binding but does not alter key Fru-M amino acids** would result in Fru-M expression in females. Even though these proteins are present in females, they could not associate with the *fru* primary transcript so the male-specific mRNA encoding Fru-M would be obtained.

    b. **A construct in which a *fru-M* cDNA is downstream of the *fru* promoter** would express Fru-M in females. In effect, the primary transcript of this construct is already spliced into the male-specific pattern.

30. a. **Null mutations in the *Sxl* gene are of no consequence to XY males, but are lethal to homozygous XX females.** Males don't normally express *Sxl*. The reason that the absence of Sxl kills XX animals (females) rather than transforming them into males is the dosage compensation pathway (which equalizes the transcriptional activity of X-linked genes in males and females) is downstream of *Sxl*. Sxl protein in normal XX females leads to reduction in the transcription of X-linked genes. In contrast, the X-linked genes of XX flies that lack Sxl are over-expressed and this is lethal.

    Constitutively active *Sxl* alleles are transcribed in both XY and XX flies. As XX flies express Sxl anyway, **constitutively active *Sxl* would have no effect in females, but males (XY) would die.** In XY flies with a constitutively active allele, Sxl protein would reduce the amount of X-linked gene expression to about half or what it would be in normal XY flies; the result is lethality.

    b. **As explained in part (a), the lethal effects of *Sxl* mutations are all about dosage compensation.**

    c. **Null alleles of *tra* transform XX flies (females) into morphological and behavioral males.** Without Tra protein, Dsx-M is produced instead of Dsx-F, and Fru-M is made instead of Fru-F. These male flies are sterile because they lack the Y chromosome and thus the Y chromosome genes that contribute to male fertility.

chapter 16

Constitutively active *tra* alleles transform XY flies (males) into morphological and behavioral females because Dsx-F and Fru-F would be expressed instead of Dsx-M and Fru-M.

d. The *tra* mutants survive whereas the *Sxl* mutants don't because **Sxl controls dosage compensation, while *tra* does not**; the *tra* gene controls morphological sex and behavior. In parts (a) and (b), you learned that the sex-specific lethality associated with *Sxl* mutants reflects aberrations in dosage compensation.

e. As Tra-2 is required for Tra function, **the effects of *tra-2* loss-of-function would be equivalent to *tra* loss-of-function**: XX flies are transformed into morphological and behavioral males (that are sterile because of the absence of the Y chromosome), whereas XY flies would be normal males.

f. XX animals without Tra function would be transformed into morphological and behavioral males [see part (c)]. **As the *tra⁻* XX males express Fru-M, they would display male sexual behavior and would try to mate with females.**

31. a. Because Sxl protein in made only in females and because MSL-2 is made only in males, the Sxl protein must repress translation of the *msl-2* mRNA from the proper (fourth) AUG. One way in which this could be accomplished is if Sxl instead promotes translation initiation from one of the wrong (out-of-frame) AUGs. **A reasonable model would be that Sxl binds the 5' UTR of *msl-2* mRNA near one of the three upstream AUGs, promoting its usage by the ribosome.** In fact, Sxl binds near the third AUG and promotes translational initiation at this site. The ribosome is then inhibited from scanning the mRNA for the correct downstream (fourth) AUG, and MSL-2 protein will not be made in females.

b. **Sxl protein is only made in females (XX), so the Sxl protein will bind to the *msl-2* mRNA in females but not in males.** It makes sense for *msl-2* translation to be inhibited only in females, as MSL-2 protein promotes dosage compensation by doubling the level of X-linked gene expression in males.

c. *Sxl* **loss-of-function is lethal to XX animals due to the failure to repress MSL-2 expression in the absence of Sxl**: Both X chromosomes will be improperly hypertranscribed in such XX animals. *Sxl* **gain-of-function alleles that are constitutively active are lethal to XY animals because Sxl will inhibit MSL-2 expression.** Because MSL-2 is not made, dosage compensation does not take place: The X chromosome in these animals will not be hypertranscribed (relative to the X chromosomes in XX females) as it should be.

d. **Loss of *msl-2* gene function would have no effect on females (XX) at all** because Sxl normally inhibits *msl-2* expression so that dosage compensation does not occur. **Without *msl-2* expression, males (XY) would die because they don't dosage compensate.** (That is, the X chromosome will not be hypertranscribed and will thus not produce the correct amount of X-linked gene products.)

chapter 17

Manipulating the Genomes of Eukaryotes

Synopsis

Scientists desire to manipulate genomes for several important reasons, ranging from the study of biological phenomena, to the improvement of agricultural crops and domesticated animals, to the production of protein drugs, to the treatment of genetic syndromes by *gene therapy*. Chapter 17 presents two general strategies for altering genomes. The first is the insertion of *transgenes* into random positions in plant and animal genomes. The second strategy is alternatively called *gene targeting* or *targeted mutagenesis*. Armed only with a gene's nucleotide sequence and cloned copies of a gene, researchers can now change at will any base pair of any gene in the genome of virtually any organism. It is thus possible, at least in theory, to knock out the function of any gene of interest, or to replace an allele of a gene with a different allele that could be created *in vitro* by recombinant DNA technology.

One goal of the chapter is to introduce you to the methodologies scientists use to create transgenic organisms and to perform targeted mutagenesis. As you will see, several of these strategies rely on biological events that are specific to the species whose genome is to be manipulated. Although it is useful for you to learn the details of some of these methods, it is more important for you to think hard about what kind of genome manipulation will allow researchers to achieve the ultimate goals of the work. What kinds of changes would you need to make in genomes in order to improve a crop or to make a mouse model of a human genetic disease? The techniques you would employ depend upon the end result you desire.

The chapter concludes with a discussion of experimental gene therapy in humans, in which potentially therapeutic genes are introduced into a patient's somatic cells in order to fight a the effects of a disease. To date, the successes of gene therapy in humans have been limited, but in certain cases, such as with an inherited form of blindness, partial cures have been obtained. Gene therapy in humans has not yet lived up to its potential, but that potential for alleviating many diseases is vast.

Key terms

transgenic organism – a plant or animal whose genome contains a **transgene** transferred into the organism by a scientist. The source of the transgene may be an individual of the same species or of a different species; researchers may also create novel transgenes by using recombinant DNA technology.

pronuclear injection – a method for making transgenic mammals in which DNA is injected into a zygote's sperm or egg nucleus (called a **pronucleus**) just after fertilization

***P* element transformation** – a method for generating transgenic *Drosophila* in which a recombinant *P* element containing a transgene is injected into embryos; the recombinant transposon integrates into a chromosome of a germ-line cell.

***Agrobacterium*-mediated T-DNA transfer** – method for generating transgenic plants in which bacteria containing a recombinant plasmid infect plants, and the part of the plasmid containing the transgene integrates into the plant genome. The transformation vector in the recombinant plasmid is **T-DNA,** which is the part of a so-called *Ti (tumor-inducing) plasmid* in *Agrobacterium* that is transferred to the host plant cell genome during infection.

pharming – the use of transgenic animals and plants to produce protein drugs

reproductive cloning – creation of a cloned embryo by insertion of the nucleus of a somatic cell into an enucleated egg cell. The hybrid egg is transplanted into the uterus of a foster mother and allowed to develop to term.

GM (genetically modified) crops – agricultural plants containing transgenes that enhance their properties, such as size, nutritional value, or shelf-life, or impart resistance to environmental stresses or insect pests

targeted mutagenesis –technologies that enable scientists to alter specific sites in the genome of virtually any organism in any particular way that they desire

ES (embryonic stem) cells – cultured embryonic cells that continue to divide without differentiating and that are **totipotent** – capable of becoming any cell type

chimeras – organisms made of cells from two or more different individuals

knockout mice – mice with an induced mutation that destroys (knocks out) the function of the targeted gene

conditional knockout mice –animals whose genome contains an engineered gene that can be made nonfunctional through deletion of an exon only in specific tissues or only under particular conditions. Such a gene is said to **floxed;** the removable exon is flanked by DNA elements called **loxP sites** that bind **Cre** recombinase protein. Cre mediates site-specific recombination between two *loxP* sites inserted on either side of an exon and thereby delete the exon.

knockin mice – mice in which a gene has been altered (but not knocked out) by targeted mutagenesis; the alteration can be a point mutation or a large insertion of DNA. A synonym for the process of creating a knockin allele is *gene replacement*.

TALENs – short for TALE (transcription activator-like effector) nuclease; synthetic restriction enzymes that can be designed to recognize and cut any DNA sequence in a genome for the purposes of targeted mutagenesis. TALENs can be used to remove or add base pairs at any site in the genome.

gene therapy – manipulation of the genome in order to cure a disease. In humans, ethical and legal considerations limit the practice of gene therapy to a patient's somatic cells.

therapeutic gene – a cloned gene that is introduced into a patient's somatic cells and whose product is meant to cure a disease. For *in vivo* **gene therapy,** the cloned gene is introduced into cells in the patient by injection or inhalation. In *ex vivo*

chapter 17

> **gene therapy,** somatic cells are removed from the patient, the therapeutic gene is introduced in cell culture, and the cells are reintroduced into the patient.
>
> **retroviral vectors** – partial retrovirus genomes genetically engineered for use as vehicles to introduce therapeutic genes into patient cells; the therapeutic gene becomes stably integrated into the genome of the patient's somatic cells
>
> **AAV vectors (<u>a</u>deno-<u>a</u>ssociated <u>v</u>iral vectors)** – partial AAV genomes genetically engineered for use as vehicles to introduce therapeutic genes into patient cells; the therapeutic gene does not integrate into patient chromosomes

Problem Solving

The problems in this chapter ask you to think creatively about how to use cloned DNA to modify the genomes of organisms. A few key points to keep in mind:

Transgenes

- **Transgenes can be inserted into random locations** in an organism's genome using the following methods:
  - Pronuclear injection (random recombination)
  - Transposon vectors
  - Retroviral vectors
  - T-DNA vectors (plants)

- **To be useful, the protein encoded by transgenes must have a dominant effect** because the genome of the transgenic organism is otherwise usually intact. However, it should be remembered that as a side product of the integration process, transgene insertions into random locations can disrupt the structure and/or expression of nearby genes.

Gene targeting

- **Linear double-stranded DNAs introduced into cells can be targeted to specific locations in the genome through homologous recombination.** The free ends of the DNA introduced determine the location where homologous recombination takes place. Gene targeting through homologous recombination is extremely inefficient. It is thus necessary to include a selectable marker like neo^r to identify cells in which a potential targeting event occurred. Furthermore, because the introduced DNA can also insert at random locations, it is necessary to screen the cells with the selectable marker for those in which the targeting construct indeed underwent homologous recombination with the intended target locus.

- **Knockouts are genes that have been targeted by homologous recombination so as to disrupt their function.** Typically, the selectable marker is inserted into a protein-coding exon. Knockouts for most genes are recessive alleles, so phenotypic effects are observed only in homozygotes for the knockout allele. However, if the targeted gene is haploinsufficient, the knockout will cause a dominant mutant phenotype.

- **Knockins are genes that have been targeted by homologous recombination so as to alter their structure in specific ways.** Often, knockin mice are designed to mimic point mutations in human disease genes in the hope that mouse models for the disease can be used to understand the biochemical and physiological basis for the disease and to screen for drugs that might ultimately treat the condition in humans.

- **TALENS are synthetic restriction enzymes used to cut genomes at specific locations to improve the efficiency of gene targeting.** Microhomology-mediated end-joining to repair the DNA ends produced by TALEN cleavage is imprecise and can thus lead to a knockout. Homologous recombination near these DNA ends with a linear fragment of DNA can lead to gene replacement (a knockin).

Conditional knockouts and knockins

- **Conditional knockouts/knockins allow researchers to study the effects of gene alteration in a specific tissue or under certain conditions; they are particularrly useful when whole organisms with the knockout/knockin cannot survive.**

- **Making mouse conditional knockouts and knockins involves the Cre/loxP recombination system.** Cre is a site-specific recombinase that performs crossing-over at two small (34 bp) loxP sites. The base pair sequence of a loxP site is asymmetrical – it has direction. When two loxP sites on the same DNA molecule are in the same orientation, Cre-mediated recombination between the two sites deletes the DNA between them. If the loxP are in opposite orientation, Cre-mediated recombination between two loxP sites inverts the orientation of the DNA between them. The diagram at the top of the next page illustrates these two possible outcomes of Cre-mediated recombination:

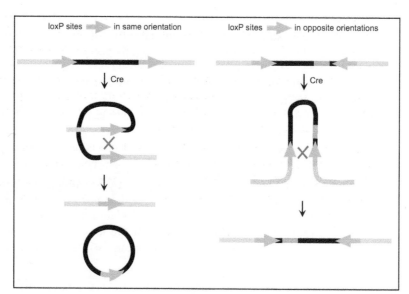

chapter 17

Vocabulary

1.
a.	transgene	7.	gene transferred by a scientist into an organism's genome
b.	pronuclear injection	9.	method of DNA transfer used for many vertebrates
c.	floxed gene	3.	useful for making conditional knockouts
d.	T-DNA	8.	vector of bacterial origin used for constructing transgenic plants
e.	AAV vector	1.	genetically engineered viral genome that transfers a therapy gene
f.	packaging cells	12.	generate viral particles for gene therapy
g.	TALEN	10.	synthetic restriction enzyme
h.	knockout mouse	11.	loss-of-function mutant through gene targeting
i.	knockin mouse	2.	contains additional or altered DNA through gene targeting
j.	Cre recombinase	6.	causes crossovers at loxP sites
k.	ES cells	4.	can develop into any cell type
l.	GM organism	5.	plant or animal that carries a transgene

Sections 17.1 and 17.2

2. a. Most likely the missense mutation is a loss-of-function allele. **Make transgene constructs that express wild-type cDNAs corresponding to each of the three different splice forms.** These constructs would need to include sequences allowing them to be expressed properly in hair follicles [see part (b) below]. **Using pronuclear injection, transform white mutant mice with each construct and test if any one of them rescues the white mutant phenotype to wild-type (gray).** It could also be possible that a combination of two or all three splice forms is required for the wild-type phenotype (gray).

    b. Because the constructs are based on cDNAs, they will require sequences not found in the cDNAs that allow the mRNAs to be expressed and processed correctly. Thus, as seen in the figure that follows, the constructs will need to include a hair follicle enhancer and a promoter that can interact with the enhancer; in addition, a poly-A addition site is needed to make sure that the mRNAs produced from these transgenes will have a poly-A tail to ensure their stability and translation.

chapter 17

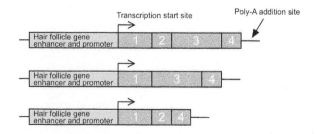

3. a. Cross the transgenic strain to wild type, and cross F$_1$ animals heterozygous for the transgene insertion to each other. If limb deformity is not caused by the transgene insertion, some F$_2$ animals homozygous for the transgene insertion will not have the mutant phenotype. If the limb deformity is due to the transgene insertion, homozygosity for the transgene should correlate with limb deformity.

   b. The transgene has inserted into the gene and caused a loss-of-function mutation. **Inverse PCR** (see Fig. 13.30 on p. 481) should enable you to identify the gene into which the transgene inserted.

   c. **Conduct a complementation test.** Cross the homozygous transgenic mutant mice with mice homozygous for the *ld* mutation. The progeny are *trans*-heterozygotes for the transgene insertion and *ld*. If the transgene insertion is a loss-of-function mutation in the same gene that is altered in *ld* (that is, if the mutations are allelic), the mutations will fail to complement each other and the progeny will show the limb deformity. The mutations will complement each other (no mutant phenotype will be seen in the progeny) if they are mutations in different genes.

4. a. Make a fusion gene construct in which the *HoxA1* promoter and enhancer drive expression of a *HoxD4* cDNA.

   b. If the absence of HoxD4 protein is what distinguishes the occipital bones (E) from the cervical vertebrae (C1 and C2), **in animals transformed with the *HoxA1-HoxD4* fusion gene, the occipital bones should be transformed into cervical vertebrae.**

5. a. Transform *faf-* flies with a construct in which the *faf* gene regulatory sequences from *Drosophila* drive expression of a *Usp9* cDNA; if the rough eye phenotype is rescued to wild-type, the human USP9 protein can substitute functionally for fly Faf protein.

   b. If mice with loss-of-function *Fam-* mutations have mutant eyes, then *Fam* must be required for mouse eye development. In Section 17.3, you will learn how to use cloned *Fam* DNA to generate mice with loss-of-function *Fam-* mutants; these would be knockouts or conditional knockouts.

chapter 17

6. a. **Fly lines in which the enhancer trap construct integrated next to an enhancer will express GFP so you will see green fluorescent tissue(s).** Depending on the nature of the enhancer in a particular GFP-expressing line, expression of GFP could be ubiquitous in every cell, or limited to specific cell types at particular stages of development.

    b. **Screen for lines that express GFP only in the wing. Use inverse PCR** (see Fig. 13.30 on p. 481) **to amplify the genomic DNA adjacent to the enhancer trap transgene insertion site.** Sequence the amplified genomic DNA and compare it to the *Drosophila* genome reference sequence to identify the gene at the insertion site. This gene should be expressed in the fly wing because it is located near a wing-specific enhancer.

    c. **Enhancer trap insertions can sometimes insert within the gene normally regulated by the "trapped" enhancer, disrupt gene function, and cause a loss-of-function mutation.** Homozygotes for the transgene insertion could, therefore, have a mutant phenotype. (Review Problem #3 above.)

7. a. **First, identify the gene that encodes the fish antifreeze protein.** One way you could accomplish this goal is to purify the antifreeze protein, determine the amino acid sequence of this protein by chemical means, and then examine the fish genome for a gene that could encode it. **Next, you would obtain a cDNA for the gene.** Several techniques would allow you to do this; one method is to use primers specific for the gene and reverse transcriptase to amplify the specific cDNA from a pool of total fish mRNA. **Next, insert the cDNA into a T-DNA vector that contains a high-efficiency plant promoter and enhancer, and transform the recombinant T-DNA into *Agrobacterium*** (see Fig. 17.5 on p. 579). **Finally, spray the transformed *Agrobacterium* onto plant cells and then grow embryos from single cells under selection for T-DNA insertion.** (If the T-DNA vector included a gene for herbicide resistance, you would add the herbicide to the cultures of the plant cells.) **Embryos that survive selection should express the fish antifreeze gene cDNA under control of a plant promoter.**

    b. Researchers did find fish antifreeze protein in plants, but it did not seem to function. **The most likely possibility is that key posttranslational modifications were not made to the fish protein in plant cells.** A second possibility to consider is that the antifreeze protein in fish does not work by itself, but instead it functions only in conjunction with a cofactor (another protein or even a small molecule) that is found in fish but not in plants.

8. a. The most straightforward way to make a transgene that would inhibit the function of a particular gene is through RNA interference. **An RNA interference transgene should have a promoter at both the 5' end and the 3' end of a complete or partial cDNA for the gene;** the 5' end promoter makes the sense transcript, and the 3' end promoter makes an antisense transcript. The sense and antisense transcripts hybridize to form double-stranded RNAs that will be processed into siRNAs, which will bind RISC and target the endogenous gene transcripts for degradation.

chapter 17

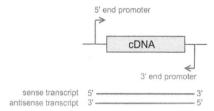

Another possibility is to **construct a transgene that expresses a dominant negative allele of the gene.** This strategy would be more difficult; you would have to understand a lot about the structure and function of the gene product to design a dominant negative allele, and probably high levels of overexpression of the transgene would be required for it to have a chance to knock down gene function. Moreover, even when they work, dominant negative alleles rarely inhibit the function of the wild-type allele completely.

b. **An RNAi construct for the *CFTR* gene could be made that consists of one or a few of the large number of exons of the gene, and two copies of a strong lung-specific promoter oriented as shown in the answer to part (a). The transgene could be incorporated into the mouse genome by pronuclear injection.** Because the insertion sites and copy numbers vary in different lines, the levels of interfering RNA expressed are line-specific. Thus, in different transgenic lines, *CFTR* gene expression would be inhibited to different extents. To determine whether or not *CFTR* gene mRNA is effectively degraded, lung cell mRNA from normal mice and different lines of RNAi transgenic mice would be purified and used to make cDNA. The cDNA would be used as a template for PCR with primer pairs that amplify exons of *CFTR*. If *CFTR* mRNA is degraded by RNAi in a specific line, an amplification product should be detected when the cDNA from the normal mice is used as a template, but no amplification product (or much less) should be obtained using the cDNA from that RNAi line. (As an alternative to PCR, you could also perform RNA-Seq to determine the residual level of expression.) Mouse lines in which the *CFTR* transcripts are degraded successfully would be analyzed to determine if they have a phenotype similar to humans with cystic fibrosis.

For the *CFTR* gene, it may be possible to generate a dominant negative by using a very strong lung promoter to overexpress an allele that makes a mutant protein that cannot function as a chloride ion regulator (due to a deletion in the ORF or a missense mutation), but can insert into the cell membrane. As above, the transgene would be transferred to the mouse genome by pronuclear injection. The mutant protein, if expressed at much higher levels than normal CFTR, potentially could out-compete the normal protein for space at the cell membrane. You could then determine if the mice with the transgene have a mutant phenotype similar to humans with cystic fibrosis.

9. a. In the diagrams that follow, IR = inverted repeat; *ori* = origin of replication that allows the plasmid to be maintained in *E. coli*. A key point is that the plasmid on the left that contains the transgene has *Tol2* inverted repeats, and the plasmid on the right that supplies *Tol2* transposase protein does not.

Study Guide Solutions Manual for Genetics, Fifth Edition   413

chapter 17

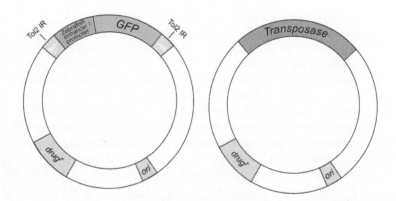

b. Using a medaka fish transposon (*Tol2*) in zebrafish means that there is no endogenous source of Tol2 transposase. Thus, once inserted in the zebrafish genome, **the transgene will be stable.**

Section 17.3

10. a. Some genes in the mouse genome might be haploinsufficient for an essential function in undifferentiated ES cells. For these genes, cells heterozygous for knockouts will die and cannot be maintained as a cell line.

    b. Some knockouts that are viable as heterozygotes in totipotent ES cells could be haploinsufficient for the function of key tissues later in mouse development. Mice with these knockouts would be inviable.

    c. *Fam* is haploinsufficient in ES cells, and so it is not possible to use ES cells to make an entire mouse that is *Fam-/ Fam+*, let alone a homozygous knockout mouse (*Fam-/ Fam-*). To ask if *Fam* is required for mouse eye development, **a researcher could make a conditional knockout mouse using the Cre/loxP recombination system.** A construct that could be introduced into the mouse *Fam* gene by homologous recombination in ES cells is shown in the following diagram (see Fig. 17.10a on p. 587 for additional details):

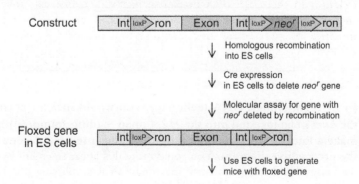

chapter 17

The mouse would also have a *cre* gene (expresses Cre recombinase protein) controlled by an eye-specific enhancer, as shown in Fig. 17.10b on p. 587. Eye-specific Cre expression would result in deletion of the floxed exon of *Fam* (the *purple* exon in the preceding diagram) only in eye cells. Thus in a mouse heterozygous for a wild-type *Fam* gene and a floxed *Fam* gene (*Fam+/Fam floxed*), the eyes would be *Fam+/Fam-* (exon deleted), while all of the other cells in the animal are *Fam+/Fam floxed* (exon not deleted). The eyes of these mice might display a mutant phenotype if *Fam+* is haploinsufficient in eyes. Mice that lack all *Fam+* function in their eyes could be generated by crossing two *Fam+/Fam floxed* heterozygotes to obtain a *Fam floxed /Fam floxed* homozygote. Eye-specific Cre expression in such a homozygote would result in deletion of the floxed exon in both *Fam floxed* gene copies, and homozygous *Fam- / Fam-* eye cells.

d. **As *Fam* is haploinsufficient for viability in ES cells, it is likely also haploinsufficient in most mouse cells. Thus, it seems likely that two copies of *Fam+* are required for viability of eye cells and thus that *Fam+/Fam-* eyes would be absent because the heterozygous eye cells would die.** Another possible outcome is that *Fam+* levels are less crucial in the eye than in ES cells, and so the cells live but the eyes don't develop properly and are malformed. Yet a third result could be that only parts of the eye are missing or formed abnormally, meaning that *Fam* is required for proper development of specific parts of the eye. Finally, it is possible that the eye would be transformed into another organ – for example, an antenna. You will learn in Chapter 18 that mutations resulting in transformations of one body part into another are called *homeotic mutations*. This would mean that *Fam* is *homeotic gene* – a gene pivotal in making cell fate decisions.

11. a. **The diagram below shows a construct for knocking out the yeast *H2A* gene.** The blue rectangles can be sequences within the *H2A* gene and at each end as shown, or 5' or 3' to the *H2A* gene (not shown). Homologous recombination between the *H2A* gene in the yeast genome and the linear construct lead to replacement of the normal allele with the allele containing the *kan^r* gene in the middle of the open reading frame. This allele should be nonfunctional (a null allele).

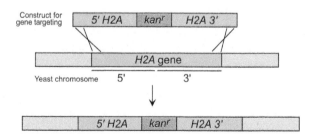

b. **(i) By recombinant DNA technology you would make a plasmid containing the *kan^r* gene inserted into the *H2A* open reading frame. (ii) You would then make a linear fragment containing the interrupted *H2A* gene as indicated in the preceeding diagram.** You could make this linear fragment by cutting it out of the plasmid with restriction enzymes or by PCR amplifying the desired sequences from the plasmid. **(iii) You would transform yeast with the linear DNA** fragment and select for kanamycin-resistant colonies. (iv) You would analyze the *H2A* genomic region of single colony using PCR and sequencing of the PCR product to check for insertion via homologous recombination.

c. You would expect that the *H2A* gene is essential because its protein product is a critical part of nucleosomes (review Chapter 11). You would thus perform the gene targeting in a **diploid strain** and would plan to knock out only a single allele. (Gene targeting is sufficiently inefficient that it would be unlikely to knock out both copies of the gene simultaneously.) In a diploid, a wild-type gene copy will remain to prevent the successfully targeted yeast cell from dying.

d. If haploid *H2A* knockouts are viable, such a result would mean that, **counter to expectation, *H2A* is not an essential gene**. The most likely explanation would that **the yeast genome contains another gene that makes a protein very similar to H2A and that can substitute functionally for the H2A protein in nucleosomes**.

12. The idea is to synthesize a PCR primer pair that will amplify the neo^r gene template and at the same time add loxP sites to the two ends of the gene. This can be done by synthesizing long primers that contain the 34 nt loxP site, and also ~20 nt of the neo^r gene. In addition, the very 5' end of each primer will include a restriction site (RE) that is not found elsewhere in the PCR amplification product but is found in the intron into which this DNA fragment will be ligated. As shown in the following diagram, the PCR product will have the same restriction site at both ends (RE), and loxP sites flanking the neo^r gene.

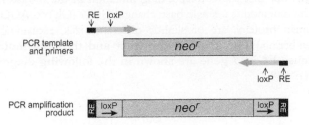

You can now cut the PCR product with the restriction enzyme and ligate it into a plasmid containing the intron that has been cut with the same restriction enzyme that leaves the same sticky ends:

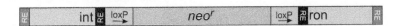

13. a. Because LCA is due to loss of *RPE65* gene function, a mouse **knockout** could be used to generate a model for LCA. As is true of most loss-of-function mutations, the LCA phenotype is recessive and appears only in homozygotes or in trans-heterozygotes for two different alleles. Thus, the mouse model would need to be homozygous for the knockout allele. If these knockout mice are blind, then *RPE65* likely functions similarly in mice and humans.

b. The procedure you would use to create a mouse model for fragile X syndrome would depend on which aspect of the disease you wanted to study. Because the disease is caused by loss of *FMR-1* gene function, **a knockout could model the disease phenotype.** It turns out that the *FMR-1* gene in mice is located on the murine X chromosome as well. Male mice hemizygous for the knockout and female mice homozygous for the knockout in fact show phenotypes that are highly reminiscent of the symptoms of fragile-X syndrome seen in human children. On the other hand, **if you want to study how new mutant alleles arise from pre-mutation alleles, or how the expansion of CGG repeat number inhibits *FMR-1* expression, you could generate a knockin with additional CGG repeats.**

c. Huntington disease is caused by a gain-of-function allele of *Htt* that makes a protein with an abnormally long polyQ region. To study the disease phenotype, how the mutant protein functions, or the process by which a pre-mutation allele becomes a disease allele, a **knockin that adds CAG repeats** to the mouse gene's coding region at the same position these repeats are found in the human *Htt* would be needed. Such knockin mouse models with 150 CAG repeats have indeed been created, and they show late onset neurodegeneration. Pharmaceutical companies can use these mice to establish whether candidate drugs will alleviate the symptoms, although the models are difficult to use because the onset is quite late compared to the life span of the mouse.

14. a. **The object is to knockin into the mouse *FGFR3* gene the missense mutation (G380R) that causes achondroplasia when present in the human gene.** It turns out that the corresponding mutation in the mouse *FGFR3* gene is G374R. [The mutation is a single base change: **G**GG (Gly) to **A**GG (Arg). The mouse and human proteins are of slightly different sizes, explaining the difference in the numbering.] **The knockin construct and the final mutation it makes in the mouse *FGFR3* gene are shown in the following diagram**, based on Fig. 17.11 on p. 588:

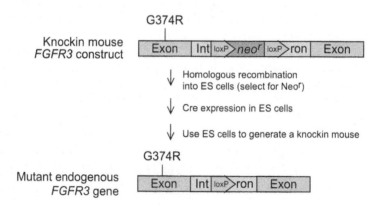

b. TALENs could also be used to generate the same mutation in the mouse *FGFR3* gene. **First, design synthetic genes for two TALEN proteins that will bind to base pair sequences flanking the GGG codon for G374 (see Fig. 17.13, p. 589).**

chapter 17

Next, use *in vitro* transcription to make mRNAs from each TALEN gene. Inject the two different kinds of TALEN mRNAs into the cytoplasm of one-cell stage mouse embryos (zygotes), along with linear DNAs containing the **G374R point mutation.** The TALEN proteins will be synthesized in embryonic cells by translation of the mRNAs, and the TALEN proteins will cut mouse chromosomes at the position of the nucleotides encoding G374 of the *FGFR3* gene. The ends produced by this cleavage promote homologous recombination between the mouse chromosome and the linear DNA.

You will next need to implant the injected embryos into surrogate mothers. When these animals are born, you then test the mouse pups for cells with the mutation using a PCR assay. Because the injected mice that test positive for the mutation are likely mosaic for mutant and normal cells, you need to backcross these mice to normal mice. If cells in the germ line have the mutation, the gametes they produce will generate non-mosaic heterozygotes in the next generation.

15. a. **To express GFP, the gene trapping construct must integrate within an intron of a gene.** The gene trapping construct is in essence an exon, and if it inserts between the exons of a mouse gene, it can potentially be incorporated into a spliced mRNA. Because of its poly-A addition site, the construct will become the final exon of the altered gene.

   b. **Only a fraction of mice with the gene trapping transgene integrated into an intron will express GFP.** This statement is true even if we assume that the transgene will always be spliced into the mRNA made from the primary transcript of the altered gene.

   First, the construct can insert in two different orientations with respect to the promoter. Only the orientation where the *GFP* open reading frame ends up in the transcript could possibly result in GFP protein expression.

   Second, the position of the transgene with respect to other exons matters. If the transgene inserts between exons that both contain 5'UTR sequences, then the ATG of the transgene should signal the beginning of GFP translation, and GFP should be made. If the transgene inserts between coding exons, then translation will have begun upstream and the ATG of the GFP cDNA must be in frame with the ORF of the preceding exon in order for GFP protein to be translated; the chances of this occurring are only 1 in 3 (if the orientation is correct).

   Finally, if ORFs are in frame, GFP amino acids will be attached at the C terminus of the protein encoded by the preceding exon(s). As part of a hybrid protein, GFP may or may not fold correctly, and thus fluorescence may or may not be detectable.

   c. If the transgene inserts such that GFP protein is made without other amino acids attached to it [see part (b)], then the information from enhancer trapping and gene trapping is substantially similar; GFP would be reporting the pattern of expression of the regulatory elements of the gene that trapped it. However, one subtle difference between the patterns of expression revealed by the two methods exists. If the gene trapping construct inserts between exons that both contain 5'UTR sequences, then the mRNA expressing GFP will have 5'UTR sequences from the mouse gene. The 5'UTR sequences could influence the translation or stability of the mRNA for GFP, perhaps in a tissue-specific way. An enhancer trap construct has its own 5'UTR, so such effects could not be observed.

If GFP inserts such that it is made as the C terminus of a hybrid protein [see part (b)], then information about protein subcellular localization could potentially be revealed, especially if the N-terminal portion of the hybrid protein contains all or most of the normal protein encoding by the gene at the insertion site. If the N-terminal part of the protein is able to go to its normal location in the cell (for example, a particular organelle), the attached GFP will fluoresce there. In an enhancer trap line, GFP would not be expressed as a hybrid protein, so this information would not be available.

d. Both enhancer trap insertions and gene trap insertions can generate mutations in genes at their insertion sites. However, **a successfully trapped <u>gene</u> is more likely than a trapped <u>enhancer</u> to be associated with a mutation in the gene.** An enhancer trap construct can express GFP without the transgene inserting into the gene. A gene trap construct, however, must insert into an intron of the gene in order for GFP to be made. The stop codon that ends the GFP ORF will stop further translation of the mRNA and thus will prevent translation of any amino acids encoded by exons downstream of the insertion site.

16. a. **A transgene in which the B cell regulatory region is fused to a *myc* cDNA introduced into the mouse genome by pronuclear injection should increase the frequency of tumor formation in the mouse immune system.**

b. The transgenic mice described in part (a) would express the *myc* cDNA in B cells throughout the life of the organism. In this part of the problem, we are exploring how to turn on *myc* overexpression starting at a defined point in the organism's development.

One way to overexpress Myc in B cells starting at 1 week of age is to generate a transgene like the one shown in the following diagram, which can be introduced into the mouse genome by pronuclear injection. (In the diagram, the B cell regulatory region contains both a promoter and a B cell-specific enhancer; T= transcription termination sequence.) If the mice also contain a *hs-cre* construct and are heat-shocked when they are one week old, Cre expression will cause the B cell enhancer to flip, so that now it expresses *myc* instead of *GFP*. (See the diagram on p. 17-4 of this chapter of the Solutions Manual that shows the effects of Cre-mediated recombination when loxP sites are in inverted orientation.) The GFP gene in this construct serves two purposes. First, it is a positive control to ensure that mice have in fact been transformed by the construct. Second, you can make sure that the heat shock worked as desired if the B cells stop expressing GFP and thus lose their fluorescence.

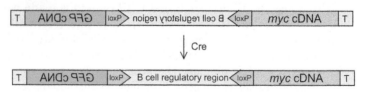

chapter 17

Another construct that would work has loxP sites in the same orientation flanking *GFP*; the *GFP* cDNA, initially transcribed in B cells, will be deleted when Cre is expressed:

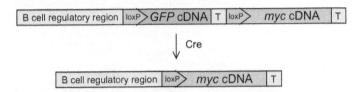

17. a. To generate a mouse that expresses Pax6 everywhere it is expressed normally except in its eyes, introduce **a transgene that expresses the reverse complement of *Pax6* cDNA from an eye-specific enhancer. The latter construct will inhibit endogenous *Pax6* expression in the eye by RNAi.** That is, eye cells will have double-stranded *Pax6* RNA that can be processed into *Pax6* siRNAs; these in turn will lead to degradation of *Pax6* mRNAs.

b. **A mouse strain with a floxed *Pax6* gene and *hs-cre* transgene would enable the scientist to turn off *Pax6* expression by heat-shocking the mice at any age.** The mouse would need to be homozygous for the floxed *Pax6* because if a normal copy of *Pax6* remained in the genome, its expression would be unaffected by the heat shock. As Cre expression is not eye-specific, the mice would likely die at some point after heat shock due to the loss of Pax6 function in essential organs. One way around this problem would be to express the reverse complement of *Cre* cDNA from a transgene with a pancreas-specific promoter [see part (a)].

c. **The knockin construct shown at the top of the figure that follows would generate the floxed *Pax6* gene presented at the bottom of the figure.** Note that the *GFP* gene within the floxed *Pax6* gene is divided into two pieces (a 5' piece containing a mouse basal promoter and the 5' UTR, and a 3' piece with the codons for the GFP amino acids), and a transcript made with the intervening sequences. Prior to heat shock, a mouse with the floxed *Pax6* gene would make *Pax6* as normal, because all the alterations of the gene are within introns, so the *GFP* and loxP sequences would be spliced out of the primary transcript. However, such a mouse would be unable to make GFP prior to heat shock because the *GFP* sequences are interrupted.

The mice homozygous for the floxed allele also have a transgene that expresses Cre only in the eye. Recombination catalyzed by Cre will delete exon 2 from the floxed gene only in the eye. *Pax6* will become inactivated in eye cells. Simultaneously, the promoter and ORF of *GFP* are brought together (the endogenous *Pax6* enhancer will work with the mouse promoter that drives *GFP* expression) so the cells with no *Pax6* function will fluoresce in green. (See Fig. 17.10a on p. 587 for more details.)

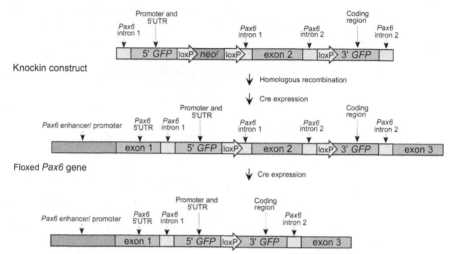

18. a. Some human diseases do not show symptoms until an advanced age. As the lifespan of the mouse is shorter, mice age faster. Thus, depending on the biochemistry and physiology underlying the disease, **it is possible that the mice homozygous for the knockout will die of natural causes before the effects of the knockout will become manifest.** As discussed in Problem 13(c) in this chapter, this phenomenon has in fact lessened the utility of mouse models for Huntington disease.

b. Homozygosity for some loss-of-function mutations in humans may not be lethal, yet it may cause disease phenotypes in individuals who are born. Homozygosity for knockouts of the corresponding gene might be lethal *in utero* in mice. Thus, **homozygous mice could not suffer from the same condition because these mice could never be born.** One way to circumvent this issue is to generate a mouse model for the disease using a conditional knockout where the gene can be inactivated in a specific tissue and time after birth.

c. **If mice have several copies of a gene, a knockout for only one of these copies is likely to be without effect, because the other copies could supply the gene product.** Two strategies could allow investigators to deal with this issue. First, all the copies of the gene could be knocked out sequentially to model the loss-of-function disease in humans. Alternatively, an RNAi transgene could potentially knock down or knock out expression of all of the copies if a single siRNA could be designed that was homologous to all of these genes.

d. **Mice from different inbred lines could have distinct alleles of modifier genes (suppressors and enhancers) resulting in differing expressivity of a knockout mutant phenotype; in fact, the mutant phenotype might not be observed at all in some genetic backgrounds.** If a particular knockout mouse does not show a mutant phenotype, it might be useful to outcross it to a different strain and then regenerate the homozygotes. Mixing two different genetic backgrounds might eliminate some suppressors and reveal a human disease-like phenotype.

chapter 17

e. **Knockout mice in which the expression of nearby genes is disrupted might display phenotypes completely unrelated to the human disease being studied.** It would thus be difficult to know which phenotypes are associated with the knocked-out gene, and which with the nearby gene. Possible ways to deal with this situation include making different kinds of knockouts that should not affect adjacent genes (for example, using gene targeting to introduce a nonsense mutation into the gene of interest) or knocking down the function of this gene by RNA interference.

19. a. **The following figure diagrams a knockin construct that would allow you to tag the full-length protein product of the gene in question with GFP.** You would obtain the GFP coding sequences from a cDNA, and you would use recombinant DNA technology to insert the GFP sequences in frame with the gene's ORF just upstream of the stop codon. In this way, the altered gene will encode a fusion protein in which all the amino acids of the gene product are followed at the C terminus by GFP. The *neo*^{r} gene is required as a selectable marker to identify ES cells that have integrated the knockin construct by homologous recombination.

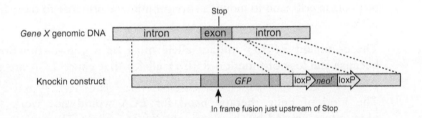

(Note that in the diagram above, if the Stop codon is located in the final exon, the *light blue* rectangle to the right of the "exon" labeled as "intron" would simply be genomic DNA 3' to the gene. In this case, it might not be necessary to remove the *neo*^{r} gene.)

b. **The knockin strategy has the advantage that the *gene* X regulatory sequences do not need to be characterized because they are already present in *gene* X on the mouse chromosome.** Suppose that you employed instead a transgene that would express a Protein X-GFP fusion protein. To ensure that the fusion protein is made in those cells in which *gene* X is normally expressed, the transgene would have to contain a DNA fragment with the *gene* X regulatory sequences. You would have to conduct several experiments to identify such a DNA fragment.

Section 17.4

20. Alteration of the genome of human germ-line cells raises so many ethical issues that it is illegal in most countries. **One major concern is that genome alterations passed through the germ line would affect every cell in the body of resulting progeny, and could easily produce unanticipated phenotypic consequences in these people and in their descendants for generations to come. These changes may not easily be reversed.** In addition, many ethicists are concerned about the ramifications of altering genomes to enhance specific characteristics of their children

and their descendants: Will the alterations truly aid the progeny, or would they instead simply reflect what the parents currently think about the desirability of particular traits? How can one draw the line?

21. a. **Retroviral vectors deliver to the target cells double-stranded DNA copies of the therapeutic gene to the nucleus, and these copies are then incorporated by reverse transcriptase/endonuclease into a chromosome. AAV vectors deliver to the target cells double-stranded DNA gene copies that remain extrachromosomal.**

   b. **The advantage of retroviral vectors is that the therapeutic gene is stable in a chromosome. By contrast, when using AAV vectors, because the therapy gene is extrachromosomal, the DNA degrades eventually and the gene therapy must be repeated. However, the main drawback of retroviral vectors is that they can cause mutation when the therapeutic gene integrates into the chromosome.** If the integration event interrupts a gene, the result would be a loss of function mutation. The integration event could alternatively be a gain-of-function mutation if the strong enhancer in the retroviral vector long terminal repeat (LTR) causes a nearby gene to be overexpressed. Another drawback of retroviral vectors is that human cells tend to mount a stronger immune response to retroviruses than to AAVs.

22. a. **The dominant *RPE65* mutant allele must be a gain-of-function mutation because the loss-of-function *RPE65* alleles that cause LCA are recessive to wild-type; *RPE65* is not haploinsufficient.**

   b. **The kind of gene therapy used for LCA would not work for retinitis pigmentosa caused by the dominant *RPE65* allele.** The gene therapy strategy for LCA was replacement of the loss-of-function alleles with a wild-type gene copy. A therapy gene for retinitis pigmentosa would somehow have to inhibit the function of the dominant mutant *RPE65* allele or gene product.

23. a. **The issue for designing an siRNA that will prevent the expression of the mutant allele specifically is that siRNAs can bind to and inhibit expression of mRNAs without perfect sequence complementarity.** The dominant mutations in disease alleles often are single base changes that alter a codon. Thus, it is possible that an siRNA with perfect complementarity for the disease allele would still be able to inhibit expression of the normal allele because only a single base pair is different.

   One way that scientists try to get around this problem is to look for silent base pair differences in the normal and mutant allele that an interfering RNA could discriminate between. If such polymorphisms were found, then it might be possible to design interfering RNAs that differ from the wild-type allele in the patient at more than one position.

   The problem is even more acute in the case of rare dominant alleles that cause overexpression of a normal protein product; here, the mRNAs produced by the mutant and wild-type allele could be identical. The problem could be circumvented only if the mRNAs produced by the two alleles were not in fact identical because of silent polymorphisms.

   b. **Triplet repeat disorders like Huntington disease always present a challenge to design of an interfering RNA that would target only the mutant allele.** The reason is that

chapter 17

the difference between the normal and disease allele is not a base substitution, but a difference in copy number of a triplet repeat. Because the triplet repeat region in normal alleles is usually longer than an interfering RNA (~21-30 nt), it is impossible to design an interfering RNA that would base pair only with a mutant allele that has an expanded triplet repeat region. As described in part (a), one potential way around this problem is to target silent sequence differences in the wild-type and mutant alleles.

chapter 18

The Genetic Analysis of Development

Synopsis

Chapter 18 explains the methods used to achieve one of the biggest successes of modern-day science – the use of genetics as a tool for understanding development. Largely through genetic analysis, scientists now understand in great mechanistic detail how a fertilized egg becomes a multicellular organism. Much of what has been learned about developmental mechanisms from *Drosophila* genetic screens and the subsequent cloning and analysis of developmental genes is transferable to humans; evolution tends to repurpose and embellish developmental pathways, rather than invent new ones from scratch.

Key terms

developmental genetics – the use of genetic analysis as a tool to study the molecular mechanisms of development – that is, how a zygote becomes a multicellular organism

mutant screen – a process where researchers examine a large number of mutagenized organisms and identify rare individuals with a mutant phenotype of interest

signal transduction cascade (or **pathway**) – form of molecular communication in which the binding of proteins to receptors on cell surfaces constitutes a signal that is converted through a series of intermediate steps to a final intracellular regulatory response, usually the activation or repression of transcription of target genes

developmental pathway – a description of the interactions between genes and gene products to produce a particular outcome in development

sensitized phenotype – a mutant phenotype that is sensitive to alteration by changes in the activities of genes that work in the same (or a parallel) pathway with the mutant gene causing the phenotype

dominant enhancers and **dominant suppressors** – mutations that, when heterozygous with a wild-type allele, display no mutant phenotype in an otherwise wild-type background (they are recessive to wild-type alleles). However, in double mutants, these mutations make the sensitized phenotype caused by mutation in a different gene appear more mutant (*enhancers*) or less mutant (*suppressors*).

RNA *in situ* hybridization – an experimental approach to determining the expression pattern of a particular mRNA in the context of an entire organism or tissue. Labeled cDNA sequences corresponding to the gene's mRNA are used as a probe on preparations of thinly sectioned tissues, or in some cases whole organisms or tissues. Signals where the probe is retained through complementary base pairing (*hybridization*) indicate cells containing the gene's mRNA.

RNA-Seq – method for analysis of the transcriptome of an organism in which millions of cDNAs are sequenced

antibody staining – method for monitoring expression of a particular protein in a cell, tissue, or whole organism. Protein synthesized *in vitro* is an **immunogen**; it is injected into an animal, the animal generates antibodies to the protein, and the antibodies are purified and tagged with a fluor. When applied to the tissue, the antibody binds the immunogen protein and fluorescently labels it.

fusion gene – gene made up of parts of two or more genes

fusion protein – protein encoded by open reading frames from more than one gene

genetic mosaics – an organism made of cells with different genotypes

mosaic analysis – observation of mosaic (*gene+/gene−*) tissues in order to determine the cells in which a particular gene must be active to allow the animal to develop or function normally - the **focus of action** of a gene

FLP/FRT recombination system – the use in transgenic organisms of FLP recombinase enzyme from yeast and the DNA site that it binds (FRT) to perform site-specific recombination; one use in flies is to cause mitotic recombination for mosaic analysis. This site-specific recombination method can be used interchangeably with the Cre/loxP recombination system.

temperature-sensitive (ts) mutations – mutant alleles that function like wild type at low temperatures **(permissive temperature)** and are loss-of-function alleles at high temperatures **(restrictive temperature)**

epistasis – a gene interaction in which the effects of alleles at one gene hide the effects of alleles at another gene. Epistatic interactions are sometimes observed in **double mutants** – organisms with mutant alleles of two different genes – when the two genes work in the same pathway.

switch/regulation pathway – a linear developmental pathway that conforms to a particular model in which a signal controls a switch that can be either "on" or "off" in any particular cell. An "on" switch initiates a cascade of gene regulation events (genes are either activated or inactivated) that lead to a change in the cell's fate. The outcome of the pathway is thus one of two developmental states, depending on whether the switch is "on" or "off".

maternal effect genes – genes encoding maternal components (those supplied in the egg by the mother) that enable the development of her progeny

zygotic segmentation genes – in *Drosophila*, three groups of genes transcribed from the genome of the zygote that function in a hierarchy to divide the organism into an array of identical segments. At the top of the hierarchy are the **gap genes**; they express transcription factors in broad regions along the AP axis. Next are **pair-rule genes,** which also encode transcription factors, and are expressed in a periodic pattern of seven rings, or stripes, in the early embryo. Finally, the **segment polarity genes** are expressed in periodic pattern of fourteen rings, one in each segment. The segment polarity genes encode cell communication proteins that enable cells within a segment to detect their position in that segment.

homeotic gene – a gene that gives originally identical groups of cells individual identities during development

morphogen – a substance that defines different cell fates in a concentration-dependent manner

homeobox – in a family of transcription factors, many of which are encoded by homeotic genes, a region of DNA (~180 bp) that encodes the DNA-binding domain, called the **homeodomain**

bithorax complex (BX-C) – in *Drosophila*, a cluster of homeotic genes encoding homeodomain transcription factors that control the identity of segments in the abdomen and posterior thorax

Antennapedia complex (ANT-C) – in *Drosophila*, a region containing several homeotic genes encoding homeodomain transcription factors that specify the identity of segments in the head and anterior thorax

Hox **genes** – constitute a gene superfamily in *Drosophila* and humans encoding homeodomain transcription factors. The *Hox* genes pattern flies and people along the AP body axis during development.

ectopic expression – gene expression that occurs outside the cell or tissue where the gene is normally expressed

Mutant screens

Several different kinds of mutant screens in *Drosophila* were presented in this chapter. The details of how researchers perform mutant screens depend on the biology, especially the reproductive biology, of the particular organism. However, the general principles of mutant screens are illustrated well in the *Drosophila* screens presented. A few important points follow:

- Mutant screens are called F_1, F_2, F_3, or F_4 screens, depending on the generation, post-mutagenesis, which is screened for the mutant phenotype.

- In all of the screens described in Chapter 18, sperm are mutagenized by feeding mutagen to males of the Parental (P) generation. The F_1 flies will each contain a single mutagenized homolog (from the male parent) and a normal homolog (from the female parent) of each chromosome.

- It is important to realize that in the F_1, any mutation of interest exists initially only on a single chromosome in one organism. Further crosses are performed to generate additional copies of that chromosome with a potentially interesting mutation, in additional organisms. This is essential either to keep a stock of that mutation for further study, or to make animals homozygous for the mutation for screening.

- The modifier screens described in Fig. 18.5 (p. 603) and Fig. 18.7 (p. 605) were F₁ screens. Researchers prefer to perform F₁ screens when possible: Only one cross (the Parental cross) needs to be set up, and so thousands or potentially even millions of F₁ may be screened for a phenotype. However, F₁ screens are possible only when the mutant phenotype can be observed in organisms with only one copy of the mutation ($m-/m+$).

- The mutant screen shown in Fig. 18.18 (p. 613, lefthand side) is an F₃ screen for zygotic segmentation genes. To screen for zygotic lethals, it was necessary to screen the F₃ generation because the mutant phenotype was observed in homozygotes, and animals with the mutant phenotype could not survive. The F₁ and F₂ generations were needed to homzygose individual mutagenized chromosomes. The F₂ generation was also required to make balanced stocks of the different mutagenized chromosomes, so that mutations of interest could be recovered in heterozygotes.

- The mutant screen shown in Fig. 18.18 (p. 613, righthand side) is an F₄ screen for maternal effect segmentation genes. In this case the F₄ embryos had to be screened because their mothers had to be homozygous for a particular mutagenized chromosome.

Epistasis analysis

In Chapter 3, you were first introduced to gene interactions, and the phenomena of recessive and dominant epistasis. Back then, you were meant to understand why epistasis occurs between alleles of genes that function in the same pathway, and how epistatic interactions can alter Mendelian ratios. Here, you are asked to think about epistasis in a deeper, more analytical way. Sometimes, epistatic interactions can be used to order genes in a developmental pathway. Although this sort of double mutant analysis can be extremely useful, several problems in this chapter ask you think about how very complex this kind of analysis is in reality and, therefore, how many assumptions a scientist must make to draw conclusions from epistatic interactions. A few important points to keep in mind:

- Single mutants must cause different phenotypes: Researchers make double mutants to ask many different kinds of questions. However, when double mutants are made to test for epistatic interactions, it's necessary that each single mutant cause a different phenotype. In epistasis analysis, a scientist asks: Do the double mutants ($a-$ $b-$) have the $a-$ phenotype, or the $b-$ phenotype? Obviously, $a-$ and $b-$ have to cause different mutant phenotypes in order for that question to be answerable.

- Not all double mutants will show epistasis: Even if two mutants cause different phenotypes, they may not display epistatic interactions. Epistasis occurs only when the two genes function in the same pathway. Other kinds of results include additive interactions, enhancement, and suppression.

- Mutant alleles must be nulls or constitutively active: The reason why only "full off" (null) or "full on" (constitutively active) alleles must be used for

chapter 18

epistasis analysis is that if hypomorphic or hypermorphic (but not fully on) alleles are used, the interaction you are observing may not be epistasis, but simply additive effects. This is best illustrated with an example. Suppose a pathway exists where in response to UV light, an organism makes a pigment; the pigment-making pathway requires $a+$ and $b+$ in that order.

Signal	Switch			Outcome
UV	ON	gene $a+$ → gene $b+$ →		Pigment
	OFF	gene $a+$ ✗→ gene $b+$ ✗→		No pigment

Null alleles ($a-$ or $b-$) would generate the same mutant phenotype – no pigment whether the signal is on or off – and so they cannot be used for epistasis analysis. However, if $a-$ is hypomorphic, then some pigment will be produced in response to UV light in $a-$ mutants, but in $a-$ $b-$ double mutants; $b-$ will appear to be epistatic to $a-$. However, if $b-$ is the hypomorph, and $a-$ is the null, $a-$ $b-$ double mutants would not make pigment; $a-$ would appear to be epistatic to $b-$. You can see that, in this case, the double mutant phenotype will be the phenotype of whichever mutation is null, regardless of gene order in the pathway. If a researcher thought that both $a-$ and $b-$ were nulls, but one was actually hypomorphic, a different result would be obtained, depending on whether $a-$ or $b-$ was the true null allele, and neither result would mean anything about gene order in the pathway.

				Result when switch ON
Wild-type	gene $a+$	→	gene $b+$	→ Pigment
a^{null}	gene $a-$	✗→	gene $b+$	→ No pigment
a^{hypo}	gene a^{hypo}	→	gene $b+$	→ Less pigment
b^{null}	gene $a+$	→	gene $b-$	✗→ No pigment
b^{hypo}	gene $a+$	→	gene b^{hypo}	→ Less pigment
a^{null} b^{hypo}	gene $a-$	✗→	gene b^{hypo}	→ No pigment
a^{hypo} b^{null}	gene a^{hypo}	→	gene $b-$	✗→ No pigment

Problem Solving

The problems in Chapter 18 draw upon all that you learned in previous chapters. In addition, the problems ask you to test hypotheses by designing and interpreting experiments. Two points not covered above to keep in mind:

chapter 18

- FLP/FRT recombination system: You might be wondering why *Drosophila* geneticists used FLP/FRT to make mosaics instead of Cre/LoxP. The answer is simply serendipity; the fly geneticist who had the idea to use site-specific recombination to cause efficient mitotic recombination set up the FLP/FRT system, and the mouse geneticist who had the idea for generating conditional knockouts with site-specific recombination chose Cre/loxP. The two systems are interchangeable.

- Maternal effect genes and mutations: The idea that the genotype of an organism's mother, rather than the genotype of the organism itself, controls phenotype is not entirely new to you. You saw in Chapter 14 that in humans, for example, mitochondria come from the maternal gamete exclusively. However, keep in mind that a big difference exists between organelle-base maternal inheritance and maternal effect inheritance: In mitochondrial inheritance, it is the mtDNA itself that is being passed from the egg to the progeny, while in maternal effect inheritance, rather than maternal genomic DNA, gene products (RNAs and proteins) made from the maternal nuclear genome are put into the egg.

Vocabulary

1.

a.	epistatic interaction	11.	double mutant has phenotype of one of the two mutants
b.	regulative determination	4.	the fate of early embryonic cells can be altered by the environment
c.	modifier screen	8.	method for identifying pleiotropic genes
d.	RNAi	7.	suppression of gene expression by double-stranded RNA
e.	ectopic expression	12.	a gene is turned on in an inappropriate tissue or at the wrong time
f.	homeodomain	9.	a DNA-binding motif found in certain transcription factors
g.	green fluorescent protein	13.	a tag used to follow proteins in living cells
h.	genetic mosaics	3.	individuals with cells of more than one genotype
i.	segmentation genes	1.	divide the body into identical units (segments)
j.	homeotic genes	5.	assign identity to body segments
k.	morphogen	6.	substance whose concentration determines cell fate
l.	maternal effect genes	10.	encode proteins that accumulate in unfertilized eggs and are needed for embryo development
m.	signal transduction pathway	2.	initiated by binding of ligand to receptor

chapter 18

Section 18.1

2. a. To analyze the role of a particular gene in heart development, you would want to: make loss-of-function and maybe also gain-of-function mutations; analyze the mutant phenotypes; monitor the mRNA and protein expression patterns. **These experiments could not be done with humans for ethical reasons. The best model organism to use would probably be a mouse.** The reason is that mice have hearts that are similar to those of humans and the manipulations outlined above can all be done with mice. A primate would have a heart closest to that of a human, but primarily for ethical reasons, most people would avoid primate models as much as possible.

    b. To find genes required for heart development, you would want to perform some kind of genetic or molecular screen. In choosing a model organism, the considerations include: How similar is the organism's heart to the human heart? (The more similar the two are, the more likely the genes you find will have similar functions in humans.) How easy is it to perform the desired screen? (For example, will complex dissections be required?) How many mutants will I be able to screen? (The number of mutants you will be able to screen depends on the generation time of the organism and how many progeny a mating pair produce.) **Mice, zebrafish, and *Drosophila* would all be good choices for different reasons.**

    Mice would probably be the best choice for a molecular screen, such as RNA-Seq analysis of heart tissue; mouse and human hearts are the most similar, and the dissections required to isolate heart tissue are probably easiest in mice because they are the largest of the animals.

    If using a genetic screen approach, each of the three models would have advantages and disadvantages. The main advantage of mice is their similarity to humans. Genetic screens are possible with mice, but they are expensive and time-consuming. In addition, loss-of-function mutations affecting heart development are likely to be homozygous lethal, and balancer chromosomes that help to maintain lines of lethal mutations exist only for a few mouse chromosomes. Zebrafish genetic screens are also labor-intensive and time-consuming. One advantage of zebrafish is that the embryos are transparent so the heart could be visible in live fish. **Even though *Drosophila* hearts are quite different from human hearts, *Drosophila* has major advantages over mice or zebrafish for a genetic screen approach.** The generation time of *Drosophila* is less than two weeks, each female produces hundreds of progeny, balancer chromosomes enable stocks of homozygous lethal mutations to be maintained easily, and flies are relatively cheap and easy to culture. This means that many more potential mutants could be screened in *Drosophila* than in the other two models, and more genes identified. Some of the genes, at least, should be relevant to human heart development.

3. a. **In *C. elegans*, laser ablation at this early stage of development would almost certainly be lethal**, because you would destroy a large proportion of the cell types that would eventually develop from the descendants of the ablated cell, and there are no other cells that could replace these. **In mice, the loss of one out of four early embryonic cells would have no effect**, because the descendants of the cells

that were not destroyed could take the place of the descendants of the cell that was destroyed.

b. Separating the four cells would be **lethal to *C. elegans*** because none of these cells would be able to develop into a complete organism. **In mice it is possible that the separated cells could develop into a mouse.**

c. Fusing two four-cell embryos **in *C. elegans* would likely be lethal**, because proper development depends upon signals between the various cells and this would almost certainly be disrupted by the fusion event. **In mice, such a fusion would be tolerated giving rise to a chimeric animal.** Recall from Chapter 17 that similar fusions are in fact an important step in the procedure for making knockout mice.

Section 18.2

4. a. **Screen homozygous mutants for failure to move toward UV light.** A screen like this was, in fact, how the first *sevenless* mutants were identified.

   b. **Mutations in every gene required for R7 development would <u>not</u> be recovered in the screen described.** A screen like this one requires that the homozygous loss-of-function mutants are viable, which precludes the recovery of mutations in **pleiotropic genes** required for R7 development and also for essential developmental pathways. (Hypomorphic alleles of pleiotropic genes that are homozygous viable potentially could be identified.) Also, mutations in **genes with redundant functions** in the R7 developmental pathway would not be identified.

   c. **Complementation tests** could be performed to determine if newly identified mutants ($m-$) are alleles of *sev* or *boss*. Simply cross the mutants with *sev* or *boss* homozygotes and determine whether $m-/sev$ or $m-/boss$ flies lack R7s (no complementation – mutations in the same gene) or are wild-type (complementation – mutations in different genes).

   d. **Additional mutant alleles of a gene (m) required specifically for eye development could be identified in a screen for noncomplementing mutations.** Mutagenize male flies, cross them to $m-/m-$ females, and screen the F$_1$ (mutagenized chromosome/$m-$) for the $m-/m-$ mutant phenotype; F$_1$ with the characteristic $m-/m-$ mutant phenotype have new mutant alleles of gene m on the mutagenized chromosome. A researcher might want to generate many different $m-$ alleles to help him identify the gene at the molecular level; gene identification often involves comparing the base pair sequences of wild-type and mutant alleles and this is much easier to do when many different mutant alleles are available. The base pair sequences of many different mutant alleles can also help scientists analyze the relationship of a gene's structure to its function; the mutant base pairs could reveal specific amino acids or regulatory elements key to function of the gene.

5. a. **The rg^{73} allele** causes the more severe phenotype (very rough eyes) and is therefore the stronger allele.

chapter 18

b. As the rg^{41} phenotype is more wild type (less severe) this allele directs either the production of more Rugose protein, or the production of the same amount of a more active Rugose protein than the rg^{73} allele.

c. **Muller's test compares the phenotype associated with homozygosity for a recessive mutation with that associated with heterozygosity for the mutation and a deletion of the gene. If the phenotypes are identical in these two genotypes, then the mutation is a null mutation. If the mutation/deletion has a more severe phenotype, then the mutation is not null but hypomorphic.** The answers to parts (a) and (b) above establish that rg^{41} **is a weak allele that makes at least some Rugose protein**, so it cannot be a null (no activity) allele. This leaves rg^{73} as the only candidate for a null allele. Muller's test suggests that rg^{73} **is likely a null allele** because the phenotype of a rg^{73} homozygote is the same as that of rg^{73} / $Df(1)JC70$. We know that the deletion must be null for the rg gene because the gene is missing. If rg^{73} encoded some Rugose protein then you would expect that rg^{73} / $Df(1)JC70$ would make half the amount of gene product of rg^{73} homozygotes, resulting in a more severe phenotype.

d. A researcher who wants to understand the function of a gene would want to know the consequence to the organism of a complete loss of that gene's function – the null phenotype.

e. Muller's test could be used to determine if the dominant allele is hypermorphic. The mutant phenotype of $rg^{hyper}/rg+$ flies is due to levels of $rg+$ activity in excess of wild-type levels. **Heterozygous $rg^{hyper}/rg+$ flies should have more $rg+$ activity and thus a <u>more severe</u> mutant phenotype than $rg^{hyper}/Df(1)JC70$ flies.**

f. The mutant phenotype of $rg^{anti}/rg+$ flies is due to levels of $rg+$ activity that are lower than wild-type levels. **Heterozygous $rg^{anti}/Dp(rg+)$ flies should have more $rg+$ activity and thus a <u>less severe</u> mutant phenotype than $rg^{anti}/rg+$ flies.**

6. a. You could use the myo-2::GFP trangenics as the genetic background in which to perform a mutagenesis screen for worms with a morphologically abnormal pharynx. Expression of GFP in the pharynx will make the structure easily visible in living worms.

b. Generate a fusion gene construct where $pha4$ coding sequences are fused to the regulatory region of a gene expressed in an organ other than the pharynx, for example, the vulva. If $pha4$ is a master regulator of pharynx formation, then worms transformed with the $pha4$ fusion gene should transform vulval cells into pharynx. If the worms also contain the myo-2::GFP transgene, the transformation would be obvious in living worms because GFP would be expressed not only in the pharynx, but also in cells located where the vulva would normally be.

7. You want to determine if $gene\ X$ is important in specification of the pharynx so you want to disrupt expression there; however you want to maintain normal expression of $gene\ X$ in all other tissues. Therefore **express a double-stranded RNA (RNAi) corresponding to gene X in the pharynx**. One way to do this is to put gene X between two regulatory regions for myo-2 so that gene X is transcribed in opposite

directions (the figure on the left that follows). Alternatively you can place two copies of gene *X* oriented in opposite directions downstream of a single regulatory region for *myo-2*; when these genes are transcribed, you will get a double-stranded hairpin RNA (the figure on the right that follows). **Transform these constructs into worms containing *myo-2::GFP* and examine the pattern of green fluorescence for indications of the effects of RNAi for gene *X* on pharyngeal development.**

8. a. The screen for modifiers of the hypomorphic *sev⁻* phenotype relied on the viability and fertility of the starting strain flies. And to identify and recover the modifier mutants, the F₁ flies with a modified (enhanced or suppressed) *sev⁻* phenotype also had to be viable and fertile. **It might be possible for homozygotes for a hypomorphic *Egfr* allele to be viable and fertile and have a morphological mutant phenotype useful for a modifier screen. But even if so, enhanced F₁ flies would likely be dead, making it impossible to recover enhancers in the same sort of modifier screen.**

b. Using transgenic flies in which the *Egfr* mutation is confined to an organ that is non-essential in the laboratory, like the eye, is one way to get around the lethality problem noted in part (a). For example, **a fusion gene could be constructed where eye-specific transcriptional regulatory sequences drive expression of an *Egfr* cDNA encoding a hypermorphic receptor protein.** (This is similar to the transgene described in Fig. 18.7 on p. 605.) Because EGFR signaling is used in a variety of developmental decisions in the eye, the Ras/MAPK pathway proteins are present in essentially all eye cells. Therefore, hypermorphic EGFR would alter many developmental pathways and cause eye malformations that could be used as the background for a modifier screen. With this transgene, enhancers would identify genes that antagonize EGFR signaling, and suppressors would be mutations in genes that help relay the EGFR signal. **Alternatively, an RNAi transgene could be constructed in which eye-specific regulatory sequences drive expression of an *Egfr* exon in both directions.** A transgenic line could be identified in which EGFR protein expression is knocked down, but not completely gone; these flies would essentially have an eye-specific hypomorphic *Egfr* allele. If such an RNAi transgene were used as a screen background, enhancer mutations would identify genes that help relay EGFR signals, and suppressors would be genes that antagonize EGFR signaling.

9. a. A dominant negative Ras protein expressed in R7 precursors would prevent Sevenless receptor signaling; **some ommatidia in the fly eyes would lack R7s.**

b. Using a *sev-Ras^{S17N}* (Ras dominant negative) transgenic line as a screen background, the mutant phenotype is due to loss of Ras function. This means that loss-of-function mutations in genes that relay the Sevenless signal would be recovered as enhancers, while loss-of-function mutations in genes that antagonize Sevenless signaling would be suppressors. Therefore, **the genes encoding Ras, Drk, Sos, Raf, Mek, MAPK, and Pnt could be identified as enhancers, and the genes for Gap and Yan as suppressors.**

c. If the screen background were $sev\text{-}Ras^{G12V}$ (hypermorphic Ras), the mutant phenotype is due to too much Ras function. This means that loss-of-function mutations in genes that relay the Sevenless signal would be recovered as suppressors, while loss-of-function mutations in genes that antagonize Sevenless signaling would be enhancers. Therefore, **the genes encoding Ras, Drk, Sos, Raf, Mek, MAPK, and Pnt could be identified as suppressors, and the genes for Gap and Yan as enhancers.**

10. a. Diagram the testcross: $w/w\ ;\ ro-\ /\ ro+\ P[w^+]\ ♀\ \times\ w/w\ ;\ ro-\ /\ ro-\ ♂\ \rightarrow$
The parental chromosomes are $ro-$ and $ro+\ P[w^+]$; the recombinant chromosomes are $ro-\ P[w^+]$ and $ro+$. The $ro-$ allele from the male parent reveals the rough eye phenotype if the chromosome from the female parent is $ro-$. The $P[w^+]$ transgene is a dominant allele; in the w/w background, flies with one copy of $P[w^+]$ have red eyes, and flies with no $P[w^+]$ have white eyes. Therefore, the parental phenotypes are 145 red, smooth eyes ($ro+\ P[w^+]$) and 152 white, rough eyes ($ro-$), and the recombinant phenotypes are 2 white, smooth eyes ($ro+$) and 1 red, rough eyes ($ro-\ P[w^+]$). RF = $(1 + 2)/(145 + 152 + 1 + 2) = 3/300 = 1\%$. **The ro gene and the $P[w^+]$ insertion are linked, and the $ro \leftrightarrow P[w^+]$ distance is 1 m.u.**

b. The data in part (a) do not indicate whether ro is centromeric or telomeric to the $P[w^+]$ element. **To determine the relative positions of the centromere, ro, and the $P[w^+]$, use a $P[w^+]$ chromosome for the testcross that has another mutation (m-) that is very close to the centromere which has a scorable mutant phenotype.** (The mutation can be either dominant or recessive; in this example, m- is recessive to wild type.) The testcross should be:
$w/w\ ;\ ro-,\ m+/\ ro+,\ m-\ P[w^+]\ ♀\ \times\ w/w\ ;\ ro-,\ m-\ /\ ro-,\ m-\ ♂\ \rightarrow$
If ro is centromeric to $P[w^+]$ (left side in the following diagram), then single crossovers between ro the $P[w^+]$ yield $m+\ ro-\ P[w^+]$ chromosomes. [Note that if ro were closer to the centromere than m (not shown) most of the crossovers would be between $m+$ and the $P[w^+]$.] If ro is telomeric to $P[w^+]$ (right side in the following diagram), then single crossovers between between ro the $P[w^+]$ yield $m-\ ro-\ P[w^+]$ chromosomes.

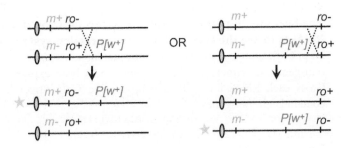

Therefore, in the testcross, analyze the progeny with rough, red eyes; if most of them have the m+ phenotype, then *ro* is centromeric to *P[w+]*; if most of them have the m– phenotype, then *ro* is telomeric to *P[w+]*.

c. The insertion sites of the *P[w+]* elements have been determined using inverse PCR and this information is available on the *Drosophila* genome project website, as is the genomic reference sequence of *Drosophila*. **In a region of 5000 bp, likely only one gene is present.** Use the genomic DNA sequence of this gene to design PCR primer pairs that allow you to amplify exons of the candidate gene from one of more *ro–* mutant strains. Sequence the PCR products to find mutations likely to alter gene function. Verify the identity of the candidate gene by introducing a transgene containing the suspect gene's genomic DNA into *ro–* flies. If the *ro–* mutant phenotype is rescued to wild type by the transgene, then the gene corresponding to the *ro–* mutation is in the transgene.

d. The *Drosophila* genome project database will contain a virtual translation of the *rough* gene ORF, and an analysis of the putative amino acid sequence of the protein. A computer program will have searched the amino acid sequence for common sequence motifs that signify particular functions, such as **DNA-binding, insertion in the cell membrane**, etc. In fact, *rough* encodes a transcription factor with a DNA-binding domain called a *homeodomain*.

11. a. The diagrams that follow show three different ways to generate double-stranded RNA corresponding to a gene that should knock out or knock down the gene's expression by RNAi. In the diagrams, the purple rectangle corresponds to a regulatory region (enhancer and promoter) and the blue rectangles are all or part of a gene's cDNA.

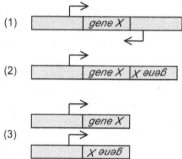

b. **The best way to perform a mutant screen for wing development genes using RNAi is to use the Gal/UAS binary expression system** (see Solved Problem III on p. 623). Obtain transgenic fly lines that already exist, each of which has an RNAi transgene that targets a different gene, and whose expression is driven by UAS_G. Cross each RNAi line with a wing-specific Gal4 driver line to obtain flies that contain both transgenes:

chapter **18**

Screen the flies for abnormal wings. **Pleiotropic** *gene X*s **with roles in wing development that would cause lethality if their expression were knocked out everywhere can be identified because the interfering RNA will be expressed only in the wing.**

Section 18.3

12. a. The pattern of *rugose* mRNA expression in whole fly embryos or dissected tissues from different developmental stages can be monitored in RNA *in situ* hybridization experiments. Protein expression can be detected by staining whole embryos or tissues with antibodies. Gene expression in a particular celltype <u>does not</u> mean that the gene product is necessarily required there; genes can have redundant roles.

    b. Hypomorphic alleles can make lower amounts of the normal gene product, or normal amounts of a mutant gene product that functions less efficiently than wild type. Therefore, homozygotes for a hypomorphic *rugose* allele may or may not have less mRNA than wild type (detected by RNA *in situ* hybridization), and may or may not have less protein (detected by antibody staining). By the strict definition of a null allele (no gene product at all), a null allele should make no mRNA and no protein. By the looser definition where a null allele makes no *functional* gene product, null alleles could make normal levels of mRNA and protein with alterations in base pair and amino acid sequence. Therefore, like hypomorphic alleles, the gene products of null alleles may or may not show differences from wild type in assays of RNA or protein quantity. **In summary, if lower levels of mRNA or protein are detected, then the allele could be either hypomorphic or null. If no mRNA or protein are detected, it could mean that no gene product is made, or that low levels of gene product are made that are beyond the limits of detection. Therefore, neither RNA** *in situ* **hybridization nor antibody staining is particularly useful for distinguishing between hypomorphic and null alleles.**

    For a dominant *rugose* allele, RNA *in situ* hybridization and antibody staining could sometimes be useful to discriminate between the different types of dominant alleles. Antimorphic alleles should have elevated levels of mRNA and protein relative to wild type; the reason is that in order to antagonize the function of the wild-type gene product, the mutant gene product needs to be highly overexpressed. Hypermorphic alleles can make excessive amounts of the normal gene product, or altered gene products that function more efficiently than wild type. Thus, a hypermorphic allele may or may not produce elevated levels of mRNA or protein. Neomorphic alleles may make mutant gene products that function differently from wild type, or they make wild-type gene products in abnormal expression patterns (ectopic expression). In summary, **if the RNA and protein expression patterns appear wild type, the allele could be hypermorphic or neomorphic; if ectopic expression is observed, the allele is likely neomorphic; if more mRNA or** protein than wild-type amounts are observed, then the allele is likely hypermorphic or antimorphic.

13. a. Boss is the ligand, expressed in R8, for the Sevenless receptor, expressed in five R7 precursor cells. **Mosaic ommatidia with a wild-type phenotype (R7 present) must have R8s that are w^+ $boss^+$; mosaic ommatidia with a mutant phenotype (R7 absent) must have R8s that are w^- $boss^-$.**

   b.

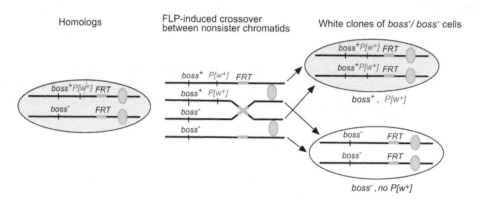

14. a. The flies also contain a UAS_G-GFP transgene and a FLP transgene elsewhere in the genome:

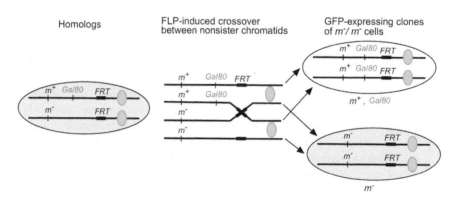

   b. Clones could be restricted to the adult nervous system by having **FLP expression driven by an adult nervous system-specific enhancer.**

15.

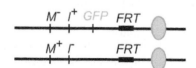

chapter 18

16. Mutant alleles whose finished protein products misfold when the temperature is raised (activity ts alleles) are more useful for temperature shift experiments than mutant alleles whose protein products misfold only if the temperature is raised during translation (translation ts alleles). The reason is that the protein of the activity ts allele will become inactive immediately when the temperature raised, while for translation ts alleles, only protein synthesized after the temperature shift will be affected; previously synthesized protein could be stable for a very long time after the temperature shift and its presence could complicate the determination of when a protein is required.

17. a. The data indicate that *zyg-9* is required starting **at about 52 minutes up until 70 minutes after fertilization.** (We are assuming that the *zyg-9* ts allele used is ts for activity.)

    b. A wild-type (*zyg-9+*) allele provided to the early embryo by the sperm has no effect on the polarity of the of the embryo; it doesn't rescue the *zyg-9* mutant phenotype to wild type. **One possibility is that some sort of imprinting is occurring; the allele from the male is not expressed. In this case, the ts Zyg-9 protein in the egg comes from the embryo's maternal allele. Another possibility is that *zyg-9* is a maternal effect gene;** in this case, neither the maternal nor the paternal allele in the embryo genome is expressed at this early stage of embryogenesis. **The ts Zyg-9 protein would have been transcribed from the mother's genome and deposited in the egg.**

18. A comparison of the two leftmost columns in the diagram shows that after 4 hours at restrictive temperature, the morphogenetic furrow has advanced anteriorly by 2 rows. During these 4 hours, each ommatidium in the very left column of the diagram has matured into the form to the right of it along the horizontal (row). Five examples of such maturation are shown in the diagram, including two examples where unpatterned cells anterior to the morphogenetic furrow have begun to assemble into ommatidia. To determine the different requirements for Notch protein, compare the development of wild-type ommatidia (middle column in the diagram) with N^{ts} mutants (rightmost column in the diagram) after 4 hours at restrictive temperature. Starting with the most anterior ommatidium, and thus the earliest developmental event (row 1), and moving posteriorly to row 5, we can determine that **Notch is required for: (row 1) R8 determination; (row 2) spacing of separate ommatidia within rows; (row 3) preventing extra photoreceptors from joining the ommatidium between R3 and R4; (row 5) R7 determination.**

Section 18.4

19. a. The *sev−* mutant phenotype is no R7; the *yan−* mutant phenotype is up to five R7s. As the *sev− yan−* double mutants have the *yan−* phenotype, ***yan−* is epistatic to *sev−*.**

    b. Yan represses transcription of target genes that promote R7 development, and modulation of Yan function happens downstream of Sevenless. **Activation of Sevenless results in Yan inactivation, and therefore derepression of genes that promote R7 cell determination. If Yan is absent, Sevenless activation is** no longer required to derepress the target genes. Therefore, it's expected that *yan−* would be epistatic to *sev−*.

c. Two null mutations, like *sev−* and *yan−*, can elicit different mutant phenotypes if they function antagonistically in the same pathway.

d. **Ras and Yan both function downstream of Sevenless, but Ras works in the same direction in the pathway as Sevenless, while Yan and Sevenless functions are antagonistic.** It therefore makes sense that a constitutively active *Ras* mutant allele and a null allele of *yan* could both be epistatic to a *sev* null allele.

e. A constitutively active Yan protein made by *yan^D* would always repress the genes that promote R7 development, whether Sevenless was activated or not. Therefore, **no R7s would be present in *yan^D/yan+* eyes and *yan^D* would be epistatic to both *sev−* and *Ras^{G12V}*.** Because *yan^D* functions downstream of both *sev* and *Ras*, neither the absence of Sevenless nor the presence of constitutively active Ras^{G12V} prevents Yan^D protein from repressing its target genes and inhibiting R7 development.

20. a. The *Drosophila* sex determination cascade is a switch/regulation pathway. **The switch is the activation of *Sxl* transcription; the signal that controls the switch is the number of X chromosomes. In cells with two X chromosomes (XX), the switch is flipped ON; in cells with one X chromosome (XY), the switch is flipped OFF.**

b. **In terms of morphology, XY animals that are either *tra−* or *ix−* are normal males.** XY flies normally make neither Tra protein nor Dsx-F, and Ix is required only for Dsx-F function. **XX animals that are *tra−* are morphological males;** without Tra, the *dsx* primary transcript is spliced to produce Dsx-M, which causes male morphology. **XX animals that are *ix−* will be morphologically intersex;** Dsx-F is made and it can activate female morphology genes, but without Ix, Dsx-F cannot repress male morphology genes. Thus, both male and female morphology genes are active.

c. In XY animals, *tra− ix−* double mutants should have no mutant phenotype [see part(b)]. **In XX animals, *tra− ix−* double mutants would be expected to have the *tra−* mutant phenotype – transformation into morphological males.** The reason is that in the absence of Tra, Dsx-M is produced instead of Dsx-F. As Ix is required only for Dsx-F function, its absence has no effect in a *tra−* background; therefore, *tra−* is epistatic to *ix−*.

d. In the previous examples of *sev− Ras^{G12V}*, and *sev− yan−*, the epistatic gene was the downstream gene. But as you can see from part (c) above, this is not always the case. Why? The answer has to do with the state of the signal in the cells affected by the mutation – whether the cells that are expressing the mutant phenotype are cells in which the signal has turned the switch ON, or cells in which the signal has turned the switched OFF.

The *sev− Ras^{G12V}* and *sev− yan−* examples have a feature in common; in both cases, one of the mutations (*sev−*) causes cells in which the signal has flipped the switch ON to have a mutant phenotype, and the other mutation (*Ras^{G12V}* or *yan−*) causes cells in which the signal has flipped the switch OFF to have a mutant

chapter 18

phenotype. In *sev−* eyes, the R7 precursor cell that touches R8 is switched ON (it touches R8 with Boss on its surface) but without a functional Sevenless receptor, it cannot respond. On the other hand, in Ras^{G12V} or *yan−* mutants, the mutant phenotype happens in cells that in which the switch is OFF – cells that don't touch R8 and whose Sevenless receptors are not activated by Boss ligand become R7s inappropriately. In cases like this, **when in double mutants the two mutations cause a mutant phenotype in cells with opposite signal states, the epistatic gene is the one downstream in the pathway.**

In *tra− ix−* double mutants, the cells in which each mutant phenotype occurs are XX cells – in both cases the switch is flipped ON. In parts (a) and (b) above, we determined that the signal is the number of X chromosomes, and that in XX cells, the switch is flipped ON, meaning that transcription of the *Sxl* gene is activated. We also determined that both *tra−* and *ix−* cause mutant phenotypes only in XX animals – cells in which the switch is flipped ON. In cases like this, **when both mutations in double mutants cause phenotypes in cells in which the switch is flipped ON, the epistatic mutation is upstream.** (The same is true if both mutations have phenotypes in cells where the switch is flipped OFF.)

The purpose of this question is to show you just how complex epistasis analysis is, and why we said in the text that scientists have to verify any conclusions they draw from epistatic interactions about the order of genes in pathways with biochemical experiments. It's important to realize that even when scientists think that they know enough about a pathway and the mutant alleles of the genes to draw conclusions from epistasis analysis, it is always possible that there is something about the pathway that they don't know – or that the mutant alleles are not really null or not completely constitutively active. (We apologize for not having a problem that shows you why hypomorphic alleles or alleles that are hypermorphic by not completely unregulated can confuse epistasis analysis – see the example given above on p. 18-5.)

Section 18.5

21. a. **Maternal inheritance of organelle DNAs means that all of the organelles containing their own genomes (mitochondria or chloroplasts) are provided in the egg cytoplasm. In maternal effect inheritance, the products of the genes (RNAs and/or protein) expressed by the maternal genome – not the DNA itself – are inherited in the egg cytoplasm** and affect the phenotype of progeny.

    b. **The pattern of mitochondrial inheritance is that mothers with the trait transmit the trait to <u>all</u> (or *some*, in the case of heteroplasmy) of their progeny (males and females) in every generation, but <u>none</u> of the male progeny transmit the trait. For maternal effect genes (recessive traits), a female will transmit the trait to <u>all</u> of her progeny (males and females), but <u>none</u> of her progeny will transmit the trait unless (in rare instances) female progeny inherited the maternal effect mutant allele from their paternal gamete also.**

22. a. In a pure-breeding line of sinistral snails, all of the snails are *s−/s−*; in a pure-breeding line of dextral snails, the genotype is *s+/s+*. In a cross of *s−/s−* hermaphrodites to *s+/s+* males, all of the progeny will be sinistral; the

hermaphrodites provide the eggs and because *s* is a maternal effect gene, **all of the progeny have the *s−* (sinistral) phenotype.**

b. In a cross of *s+/s+* hermaphrodites and *s−/s−* males, **all of the progeny have the *s+* (dextral) phenotype.** The *s+/s+* hermaphrodites provide the eggs, and as *s* is a maternal effect gene, all of the progeny will have the *s+* (dextral) phenotype.

c. **The F$_1$ from part (a) and part (b) are both *s+/s−*; if they self-fertilize, all of the F$_2$ progeny will be dextral** because the gene product from the maternal *s+* allele will be present in all of the eggs, regardless of their genotype.

d. The F$_2$ from both part (a) and part (b) are 1/4 *s+/s+* : 1/2 *s+/s−*: 1/4 *s−/s−*. **The selfed *s+/s+* and *s+/s−* F$_2$ will produce dextral progeny** (the maternal *s+* allele will provide *s* gene product to the eggs), **and the selfed *s−/s−* F$_2$ will produce sinistral progeny** (no *s+* allele is present in the maternal genome so no *s* gene product is put into the eggs).

23. a. **The females were homozygous for random loss-of-function mutations.** If the gene products were essential outside of the germ line, the homozygous females would be dead rather than sterile.

b. **Analyze F$_3$ males and determine if they have a mutant phenotype;** if they do, the maternal effect mutation also plays a role in males.

c. **The balancer chromosome enabled the production and identification of homozygous mutants.** The balancer prevented recovery of recombined second chromosomes in the F$_3$, thereby keeping the *Cy* marker and the *m−* mutation separate. This means that *m−/m−* homozygotes could be identified as non-curly-winged F$_3$, and the *m−* chromosome could be propagated in *m−/Balancer* flies.

24. a. If *gene X* is located on the X chromosome, females heterozygous for a *gene X−* mutation and an *ovo^D* mutation, both on chromosomes with FRTs near the centromere, could be used to generate FLP-induced *gene X−* homozygous clones in the germ line. As *ovo^D/ovo+* germ-line cells do not produce eggs, any eggs produced must result from mitotic recombination and therefore must have come from *gene X−/gene X−* germ-line cells:

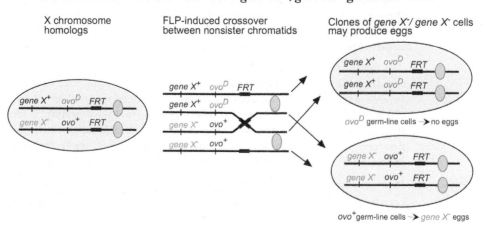

(i) To determine if *gene X* RNA and/or protein are normally supplied to eggs, **eggs from homozygous *gene X–* germ-line cells may be useful as negative controls in experiments to monitor RNA (RNA *in situ* hybridization) or protein (antibody staining).** The *gene X–* allele would be most useful for this purpose if it were a true null that produced no RNA and no protein. (ii) If maternally supplied *gene X* products are not required for proper embryogenesis, then the fertilized eggs from *gene X–* germ-line cells, if fertilized with wild-type (*gene X+*) sperm, should develop normally. **If maternal supplies of *gene X* products are required for embryogenesis (*gene X* is a maternal effect gene), the *gene X–* fertilized eggs will not develop properly, even if fertilized by *gene X+* sperm.**

b. As described in part (a), if *gene X* is a maternal effect gene, *gene X–* eggs will be produced by homozygous *gene X–/gene X–* germ-line cells; but even if they are fertilized by *gene X+* sperm, they will not undergo embryogenesis normally. If *gene X* is needed for oogenesis, even the *gene X–/gene X–* germ-line cell clones that have lost the *ovo*^{D} mutation [see diagram in part (a)] will **not produce eggs.** If a gene is required for oogenesis and is also a maternal effect gene, the maternal effect function would be obscured.

25. **Human females homozygous for mutations in maternal effect genes would be sterile** – their eggs would lack an RNA or protein required for early development.

26. **Injection of *bicoid* mRNA synthesized *in vitro* into the posterior pole of normal eggs results in embryos with two, mirror-image anterior poles;** heads develop at both end of the embryo.

27. a. The *hunchback* gene is transcribed in both the female germ line, and also in the early embryo. In addition to a promoter, **either a single enhancer that is active in both the female germ line and the early embryo must be present, or separate female germ line and early embryo enhancers.**

b. Recall that *hunchback* is both a maternal effect gene and a gap gene. The *hunchback* open reading frame encodes a transcription factor that is recognizable by its **Zinc finger DNA-binding domain.**

c. **The 3' UTR of** *hunchback* mRNA has binding sites for Nanos protein; Nanos prevents translation of maternally supplied *hunchback* mRNA in the posterior half of the embryo.

28. a. The observation is that if no *hunchback* mRNA is supplied maternally, Nanos protein (which is also maternally supplied) is not needed. The implication is that **the only function of Nanos is to inhibit *hunchback* function.**

b. It may seem puzzling to you that the fly has a gene, *nanos*, whose sole function is to inhibit the maternal function of another gene, *hunchback*. Why doesn't the fly just not make maternal *hunchback* in the first place, and then it wouldn't need the *nanos* gene at all? **Developmental biologists have discovered many double negative mechanisms like this one because the process of evolution is messy and it results in inefficiencies.** How could this particular situation with *hunchback* and *nanos* have come about? In ancestors of *Drosophila,* maternal stores of *hunchback* may have been required; or perhaps Nanos was present in the egg because it had a different target gene, and its presence allowed the regulatory region of *hunchback* to express the gene maternally (with no harm done) as well as zygotically. Hypotheses like these can be tested by analyzing the expression of *hunchback* and *nanos* in fly species that are the ancestors of *Drosophila*.

29. a. The absence of *knirps* function has no effect on Hunchback protein distribution but does affect Kruppel protein localization. More specifically, these results show that **Knirps is needed to restrict the posterior limit of the zone of Kruppel expression.** Because these gap gene proteins are all transcription factors, the implication is that Knirps represses Kruppel transcription, at least in this part of the embryo.

    b. In the absence of Nanos, **Hunchback protein would be seen throughout the embryo** because there is no Nanos protein to inhibit the translation of maternally supplied *hunchback* mRNA.

30. a. The *eve* stripe 2 enhancer is active only in the stripe 2 cells because only in that region of the embryo are the enhancer binding sites filled often enough with activator proteins (Bcd and Hb) as opposed to repressor proteins (Kni and Gt) for transcription to be activated. **The reason is that the levels of Bcd and Hb are high in the stripe 2 cells, while the levels of Kni and Gt are low.**

    b. Two different scenarios could account for the ability of a single enhancer to function in two nonadjacent stripes. **One possibility is the enhancer responds to the same transcription factors in those two stripes**; this would require that the pattern of transcription factor expression is similar in these the two nonadjacent regions. **Alternatively, different transcription factors could regulate *eve* transcription in the two nonadjacent stripes; different sites on the enhancer could be occupied to activate transcription in each of the two stripes.**

    c. **Transcription of secondary pair-rule genes, like *ftz*, is regulated (in part) by primary pair-rule gene proteins. As the primary pair-rule proteins are present in a periodic pattern of seven stripes along the AP axis of the embryo, a single enhancer with binding sites for primary pair-rule proteins could activate *ftz* transcription in a seven-stripe pattern.** The enhancer's binding sites would be occupied (or empty) in a pattern that repeated seven times along the AP axis.

    This concept is illustrated in the diagram that follows. In the diagram, each circle represents a cell along the AP axis of the embryo, and the numbers indicate 4 of 14 divisions, each 4-cells wide, along the AP axis. Each color represents a different primary pair-rule protein (black represents no protein).

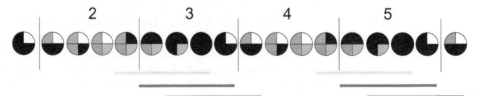

chapter 18

In the model above, each primary pair-rule protein's expression pattern is in a stripe 4 cells wide, and the starting point of each protein's expression stripe is one cell out-of-register with the start of the next protein's expression stripe. The colored lines at the bottom of the preceding diagram represent 4 different periodic expression patterns (other than those of the primary pair-rule genes themselves) generated by primary pair-rule gene expression. If the enhancer of a secondary pair-rule gene can activate transcription in response to any one of the four combinations of primary pair-rule protein patterns in the circles above the line, but not in the patterns indicated by the other four circles, then that secondary pair-rule gene will be transcribed in seven stripes along the AP axis.

To determine if a single *ftz* enhancer exists, **different fragments of genomic DNA surrounding the *ftz* gene could be fused to an "enhancerless" GFP reporter gene.** GFP expression would be monitored in *Drosophila* lines transformed with the different fusion genes. **If the genomic DNA fragment in the reporter fusion contains a single *ftz* enhancer with the properties described above, embryos with that transgene should express GFP in seven stripes corresponding to the pattern of *ftz* expression.**

d. **Segment polarity gene transcription is controlled by pair-rule gene proteins. As explained for *ftz* transcriptional regulation above, because the pair-rule proteins are expressed in a striped pattern along the AP axis, one enhancer that binds pair-rule proteins could activate *engrailed* transcription in a striped pattern.** The pair-rule proteins are each expressed in seven stripes, but it's easy to imagine how a seven-stripe pattern of transcription factors could activate transcription from a single *engrailed* enhancer in fourteen stripes. The diagram that follows illustrates the idea. In the diagram, each circle represents a cell along the AP axis of the embryo, and the numbers indicate 4 of 14 divisions, each 4-cells wide, along the AP axis. Each color represents a different pair-rule protein.

In the preceding diagram, each pair-rule protein's expression pattern is in a stripe 4 cells wide, and the starting point of each protein's expression stripe is one cell out-of-register with the start of the next protein's expression stripe. Each cell in a 2-division wide (8 cell wide) stripe expresses a unique combination of pair-rule proteins. If the *engrailed* enhancer is activated by either of the two combinations of proteins indicated by the arrows in the preceding diagram, then *engrailed* would be transcribed in fourteen stripes.

31. The easiest way to think about how the BX-C controls the identities of segments T3-A8 is that the cells in each segment have a "codeword" – a particular combination of BX-C genes are expressed, and in a segment-specific pattern and level. (By pattern is meant the particular array of cells in the segment that express the gene.) The diagram that follows illustrates this concept.

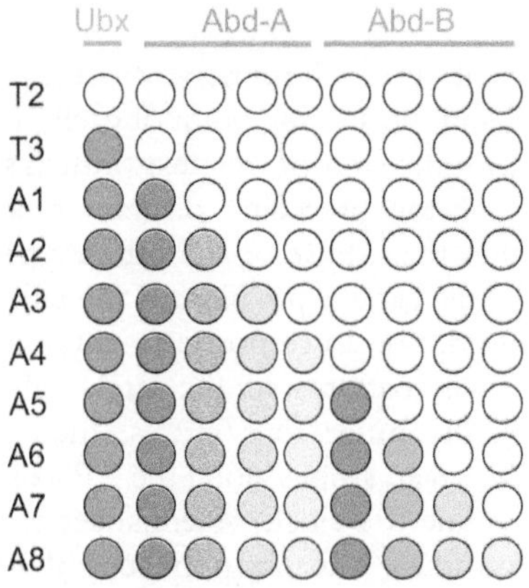

In the preceding diagram, the codeword for T3 is *Ubx* expression, the codeword for A1 is the T3 codeword plus a particular level and pattern of *Abd-A* expression, the codeword for A2 is the A1 codeword plus additional *Abd-2* expression, etc. The most important element of each codeword is the final element – that which makes the segment different from the one anterior to it.

Using this model, it's easy to see (as shown in the following diagram) that in the absence of *Ubx,* T3 is transformed into T2; the cells that would normally form T3 now have the T2 codeword – no BX-C expression.

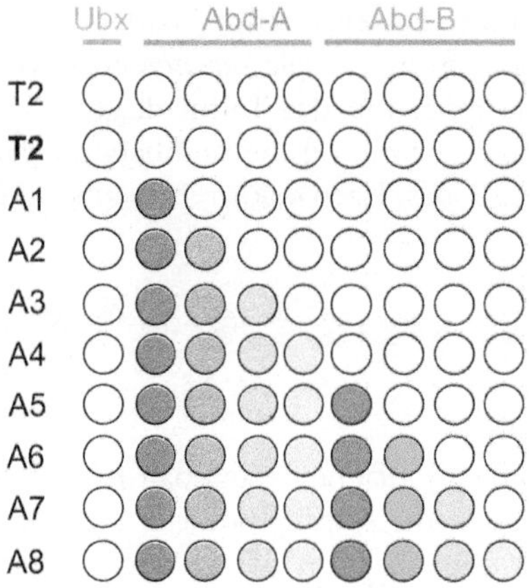

a. In embryos lacking all *Abd-B* gene function, there would be an anterior-directed transformation of the segments in which *Abd-B* is expressed. The result would be that **the segments A5-A8 would all be transformed to look like A4,** which is the next most anterior segment in which *Abd-B* is <u>not</u> expressed. According to the model, in the absence of Abd-B, A5-A8 would all have the A4 codeword:

chapter 18

	Ubx	Abd-A	Abd-B
T2	○○○	○○○	○○○
T3	●○○	○○○	○○○
A1	●●○	○○○	○○○
A2	●●●	●○○	○○○
A3	●●●	●●○	○○○
A4	●●●	●●●	○○○
A4	●●●	●●○	○○○
A4	●●●	●●●	○○○
A4	●●●	●●●○	○○○
A4	●●●	●●●	○○○

b. Segments A5-A8 are morphologically distinguishable from each other, even through they each express all three BX-C genes. Each has a different codeword because the level and pattern of *Abd-B* expression is different in each segment A5-A8.

c. Without any BX-C expression, every segment from T2-A8 would have the codeword for T2; all of the segments that normally express BX-C genes would be transformed to T2:

	Ubx	Abd-A	Abd-B
T2	○○○	○○○	○○○
T2	○○○	○○○	○○○
T2	○○○	○○○	○○○
T2	○○○	○○○	○○○
T2	○○○	○○○	○○○
T2	○○○	○○○	○○○
T2	○○○	○○○	○○○
T2	○○○	○○○	○○○
T2	○○○	○○○	○○○
T2	○○○	○○○	○○○

d. In wild-type flies, the wing is in segment T2 while the haltere is in segment T3 (Fig. 18.26, p. 619), so *Contrabithorax* mutations result in a transformation from an anterior structure to a posterior structure. As posterior-to-anterior transformations (T3-to-T2) are associated with loss-of-function mutations of the *Ubx* gene, **it is likely that *Contrabithorax* mutations are gain-of-function mutations in which the *Ubx* gene is expressed ectopically in T2**. Remember that *Ubx* is ordinarily expressed in T3 but not in T2. Most gain-of-function mutations are dominant. The majority of *Contrabithorax* mutations are chromosomal rearrangements that place the *Ubx* gene near an enhancer element that turns on gene expression in T2. This situation is similar to that of gain-of-function mutations in *Antennapedia* that place a leg on the head of the fly (Fig. 8.32, p. 287).

chapter 18

32. a. The *PcG* genes encode **repressors and corepressors;** the repressors bind specific DNA sequences within the silencers and recruit complexes of corepressors such as histone methyltransferases and histone deacetylases that close chromatin.

b. When any one of PcG proteins is absent, the silencers cannot function and the BX-C genes are derepressed: segments T2-A8 all have the A8 codeword:

chapter 19

The Genetics of Cancer

Synopsis

This chapter presents a straightforward thesis: Cancers are genetic diseases caused by the accumulation in somatic cells of mutations in certain key genes; these mutations lead to uncontrolled proliferation and other phenotypes such as metastasis. The clinical properties of different cancers depend on which key genes have been changed by mutation and the nature of the cells that have accumulated these mutations. All cancers involve an insidious positive feedback pathway in which cell proliferation increases the chances that new mutations will occur, and these new mutations lead to increased rates of cell proliferation.

Researchers have identified two major classes of genes whose mutant alleles can contribute to cancer. **Proto-oncogenes** are genes whose products promote cell-cycle progression; dominant gain-of-function alleles of proto-oncogenes are called **oncogenes**. In contrast, **tumor-suppressor genes** encode products that either inhibit cell proliferation or that protect the genome from changes. Loss-of-function mutations in tumor-suppressor genes contribute to cancer by removing these important safeguards. It is critical to note that normal genomes contain wild-type alleles of both proto-oncogenes and tumor-suppressor genes, because the products of these genes are required both for cells to proliferate at the proper rate and for genomes to remain intact; it is mutations in these genes that promote cancer.

Recent advances in DNA sequencing technologies have led to the advent of one of the most hopeful trends in the fight against cancer: *personalized cancer treatment*. Clinicians can now determine the genome sequences of a patient's tumor cells and normal cells; comparisons of these genomic DNA sequences can help scientists pinpoint the particular critical mutations in proto-oncogenes and tumor suppressor genes that contributed to the patient's particular cancer. Armed with this knowledge, physicians can sometimes prescribe treatments effective against that specific cancer. This strategy is imperfect, in part because the genomes of tumor cells are constantly accumulating new mutations. Personalized cancer genetics is nonetheless becoming increasingly powerful as researchers gain new information about the response of tumors with particular mutations to an increasing armament of anti-cancer drugs.

Key terms

contact inhibition – the process by which normal cells stop dividing when they come into physical contact; normal cells cultured on petri plates thus proliferate until they form a *monolayer* one cell thick at the bottom of the plate. Cancerous cells lose contact inhibition; in culture, tumor cells form piles called *transformed foci* that are many cells thick.

apoptosis/programmed cell death –When the genome of a normal cell becomes damaged past the point where it can be repaired, a genetically-directed program of cell death called *apoptosis* occurs, leading to the destruction of that cell. Cancerous cells often have lost their ability to undergo apoptosis, and this loss is dangerous because it allows cells with damaged genomes to proliferate.

autocrine stimulation – Most normal cells will not proliferate unless they receive the proper molecular signals from their environment. Cells that exhibit *autocrine stimulation* can instead divide in the absence of these external signals (that is, the tumor cells provide their own division-stimulating signals).

senescence versus immortality – Most normal somatic cells undergo *senescence* – an end to proliferation – after they have undergone a limited number of cell divisions; in contrast, many cancer cells continue to proliferate with no such limitation – they are *immortalized*. An important facet of immortalized cells is that they express the enzyme *telomerase*, whereas most normal somatic cells lack telomerase so their chromosomes lose DNA from the telomeres during each round of cell division.

genomic instability – describes the tendency of cancer cells to accumulate genomic changes at a rapid rate. These changes can include increased frequencies of point mutations, chromosomal rearrangements, aneuploidy, and even polyploidy.

angiogenesis – a process whereby tumors stimulate the growth of blood vessels into the tumor so that the cancerous cells can be supplied with oxygen and nutrients

metastasis/malignancy – a property of cancerous cells that allows them to escape into the bloodstream and thus to start secondary tumors in other parts of the body

immune surveillance – describes the ability of a person's immune system to identify newly formed cancer cells and to destroy them. The cells in successful tumors accumulate mutations that somehow allow them to evade this normal process of immune surveillance.

tumor progression – the process by which pre-cancerous and cancerous cells progressively accumulate (through mutation) additional properties that make the cancers increasingly invasive and malignant

growth factors – molecules that promote or inhibit cell proliferation; those that promote cell division are **mitogens.** Those mitogens that are produced by one cell type but act on other cell types are called *hormones*.

growth factor receptors – molecules embedded in the membranes of receptive cells that are targeted by growth factors. The binding of a growth factor to a growth factor receptor at the surface of a cell initiates a **signal transduction pathway/cascade** that eventually influences the activity of transcription factors within the cell's nucleus. These transcription factors regulate the transcription of genes whose products either promote or inhibit cell division.

signal transducers – molecules that act as intermediaries within signal transduction pathways to transmit the signal (the binding of a growth factor to a growth factor receptor at the cell surface) to the effectors of the pathway (the transcription factors that regulate the expression of genes whose products are involved in cell division).

cyclin-dependent kinases (CDKs) – the key enzymes that drive cell cycle progression. CDKs generally consist of two subunits: One subunit is a *kinase*, an enzyme that phosphorylates (adds phosphate groups to) other proteins. The other subunit is a *cyclin*, a polypeptide whose abundance varies with stages of the cell cycle and which functions as a specificity factor to direct the kinase to phosphorylate particular protein substrates.

checkpoint – a molecular pathway that prevents cell cycle progression until necessary preconditions have been fulfilled. For example, the G_1-to-S *checkpoint* prevents cells from replicating their chromosomes until any DNA damage has been repaired.

gene amplification – a type of *genomic instability* in which the number of copies of particular chromosomal regions is increased. Gene amplification can result either in *homogenously staining regions* (*HSRs*) in which the additional copies are positioned in tandem with respect to each other at one location in the genome, or in multiple *double minutes,* small chromosome-like bodies lacking centromeres and telomeres. Gene amplification is particularly important to tumor progression if the region now found in additional copies includes a *proto-oncogene.*

oncogene/proto-oncogene – An *oncogene* is a cancer-inducing, gain-of-function mutant in a *proto-oncogene*, which is a normal gene whose product promotes cell cycle progression.

tumor-suppressor gene – A tumor-suppressor gene encodes a product that impedes cell cycle progression or that protects the genome from alterations. Loss-of-function mutations in tumor-suppressor genes promote cancer.

loss-of-heterozygosity (LOH) – If a cell is heterozygous for a loss-of-function mutation in a tumor-suppressor gene, then one allele of the gene provides wild-type function while the other allele is nonfunctional. Any event that removes or inactivates the remaining wild-type allele results in a *loss-of-heterozygosity*; the cell is left with no functional copies of the tumor-suppressor gene, and this can contribute to cancer.

chemotherapy – treating cancer patients with drugs that kill cancer cells

personalized cancer treatment – tailoring cancer treatments to the individual based upon the suite of mutations in oncogenes and tumor-suppressor genes found by comparing the genomic sequences of tumor tissues and normal tissues in that patient

driver mutations – the mutations found in the genomes of the cancerous cells in the patient that are most likely to contribute to tumor formation. Other mutations in the genomes of cancer cells that are unlikely to contribute to tumor formation are often called *passenger mutations*. Because of *genome instability,* cancer cell genomes are likely to accumulate several driver mutations and many passenger mutations.

druggable targets – the protein products of genes that have been mutated in cancer cells; most druggable targets are the proteins encoded by oncogenes

cancer landscapes – tabulations of the mutations in the cancer cells from many patients that help scientists determine both what phenotypes are likely to occur when particular suites of oncogenes and tumor-suppressor genes are mutant and what kinds of drugs are most likely to be successful in treating specific subclasses of tumors

chapter 19

Problem Solving

The most critical figures in this chapter are:

- Fig. 19.6 on p. 635, which illustrates the hypothesis that cancers result from the accumulation, within a line of somatic cells, of mutations within certain key genes (proto-oncogenes and tumor-suppressor genes)
- Fig. 19.19 on p. 644, which shows the key differences between oncogenes and tumor-suppressor genes
- Fig. 19.24 on p. 648, which depicts the various types of events than can cause a *loss-of-heterozygosity* by removing or inactivating the remaining functional copy of a tumor-suppressor gene from a cell in which the other copy of the gene was non-functional

Oncogenes versus tumor-suppressor genes:

This distinction is central to solving many problems in this chapter.

Oncogenes are gain-of-function mutations in normal genes called **proto-oncogenes**. The proto-oncogenes often encode proteins that are involved in *signal transduction pathways* through which cells respond to mitogenic signals. The gain-of-function mutations may increase the amount of the normal gene product (for example, duplications, HSRs, and double minutes can increase the dosage of proto-oncogenes and therefore the amount of gene product), or they can be mutations that increase the efficiency of the gene product or that remove a domain of the protein that could inhibit its function. As is the case for most gain-of-function mutations, oncogenic alleles are dominant to wild-type alleles of the proto-oncogenes, so one copy of an oncogene (that is, one mutant allele) is sufficient to contribute to tumor progression.

Tumor-suppressor genes are normal genes whose products either retard the cell cycle (for example, they may participate in cell cycle checkpoints) or protect the genome from mutation and other instabilities. The mutations of tumor-suppressor genes that matter are loss-of-function mutations.

A point critical for understanding cancer is that *mutations of tumor-suppressor genes have effects that are recessive at the level of the cell but dominant at the level of the organism*. This statement means first that two "hits" (mutations) are necessary to knock out the function of the tumor-suppressor gene so that its product can no longer protect a cell from acquiring cancerous properties. However, if an individual inherits one wild-type allele of a tumor-suppressor gene from one parent and one loss-of-function allele from the other parent, this inheritance constitutes a head start for cancer formation as one of the "hits" is always present. The reason predilection for cancer is dominant at the level of the organism (as would be seen in pedigree analysis) is that it is highly likely that in a single cell, the one remaining wild-type copy of the tumor-suppressor gene will be inactivated or lost. Many opportunities exist for the second "hit" to occur, and as Fig. 19.24 on p. 648 shows, in any of the appropriate cells, many types of potential events could happen that would provide the second "hit."

chapter 19

Vocabulary

1.

   a. mitogenic growth factor — 7. signals a cell to leave G_0 and enter G_1

   b. tumor-suppressor genes — 6. mutations in these genes are recessive at the cellular level for cancer formation

   c. cyclin-dependent protein kinases — 8. cell-cycle enzymes that phosphorylate proteins

   d. apoptosis — 2. programmed cell death

   e. oncogenes — 1. mutations in these genes are dominant for cancer formation

   f. growth factor receptor — 9. protein that binds a hormone

   g. signal transduction — 3. series of steps by which a message is transmitted

   h. checkpoints — 5. control progress in the cell cycle in response to DNA damage

   i. cyclins — 4. proteins that are active cyclically during the cell cycle

Section 19.1

2. a. Contact inhibition: Normal cells stop growing when they come into contact with each other. This property is observed mostly in culture, where normal cells form a *monolayer* one cell thick; they stop growing when they take up the entire surface of the petri plate. Cancer cells do not have this property, so they continue to grow after contacting each other, leading to the formation of piles of cells called *transformed foci*. The loss of contact inhibition contributes to cancer because cells no longer have one of the controls on their growth. In addition, recall that the more cells divide, the more mutations they can accumulate that can contribute further to the aggressiveness of the tumor. [This latter point is also part of the answers to parts (b)-(e) of this question, but it will not be repeated in the answers below.]

    b. Autocrine stimulation: Normal cells will divide and grow only when they are stimulated to do so by the binding of a growth factor such as a hormone to a receptor protein on the cell surface. Most cancer cells exhibit *autocrine stimulation*, meaning that they can divide rapidly in the absence of a growth factor. Autocrine stimulation contributes to cancer because the organism cannot limit the growth of cells that have this property.

    c. Apoptosis: When the DNA of normal cells is damaged beyond the capacity of repair systems to deal with the damage, the cells undergo *apoptosis* (programmed cell death). As a result, cells with damaged DNA cannot grow and replicate. Many cancer cells do not undergo apoptosis in response to cell damage. As a result, cells

d. Telomerase expression: Normal somatic cells do not express the enzyme telomerase. Thus, the chromosomes of these cells become shorter each time the cells replicate, providing somatic cells with an upper limit to the number of times they can divide before essential DNA is lost. Many lines of cancer cells express telomerase, which counteracts this shortening. As a result, such cancer cells become effectively immortalized; they can continue to divide indefinitely. Again, these cancer cells have lost one of the normal controls on their division and growth.

e. Senescence due to telomere shortening: This is the flip side to part (d). *Senescence* is the property of normal somatic cells that results in their failure to divide further once telomeres have shortened so far that essential DNA is lost. Many cancer cells are instead immortalized because they express telomerase; they do not undergo senescence.

f. Genomic stability: Normal mitotic cell division ensures that (with the exception of rare mistakes) progeny cells have the same genome as the parental cell. The number of chromosomes remains the same, and chromosomal rearrangements do not occur. Cancer cells have unstable genomes that accumulate many alterations, including point mutations, chromosomal rearrangements, and changes in chromosome number. These alterations can mutate or alter the expression of oncogenes and tumor-suppressor genes, contributing to cancer.

g. Angiogenesis: Solid tumors require nourishment from blood; large tumors thus generate molecular signals that stimulate the growth of blood vessels into the tumor.

h. Metastasis: Normal cells grow within particular regions of the organisms that are usually defined by membranous structures. If cancer cells *metastasize*, they break these boundaries and can start to colonize other parts of the body, forming new tumors at new sites.

i. Susceptibility to immune surveillance: Our bodies normally can identify cancerous cells and destroy them by a variety of immune system measures. For a tumor to become established in the body, the tumor cells must accumulate mutations that allow them to evade this immune surveillance.

Section 19.2

3. You need to do epidemiological studies of the colon cancer profile looking for possible correlations between either diet or genetic differences with the incidence of cancer. **To assess the role of diet, you need to control (as best possible) the genetic make-up of the population examined and vary the diet. You could set up studies contrasting the cancer rates between a population of recent immigrants from India now residing in the United States and a population of people of the same ethnic background who have remained in India.** If possible, you might also be able to distinguish within each group subpopulations who favor either a Westernized diet or the historical diet of the ethnic group in India.

chapter 19

To assess the role of genetic differences in influencing cancer rates, you need to keep other factors, most importantly the diet, as constant as possible. You would thus compare the incidence of cancer between genetically dissimilar groups who have similar diets. Researchers often accomplish this task by studying cancer rates among different ethnic groups living in the same area. For example, people of Indian heritage who have lived in Brooklyn for a long period could be compared with Brooklynites from other ethnic groups, as long as all of the individuals surveyed had similar diets.

4. These 'conflicting views' are easily reconciled. We know that **some of the environmental agents that are implicated in increased cancer risk cause increased level of mutations**, so this fact is consistent with the idea that mutations in genes are necessary to cause cancer. **The inherited mutations that lead to predisposition to cancer inactivate one allele of a gene (often a tumor-suppressor gene) that inhibits cell growth.** This inherited mutation is just one of several mutations that must occur within a cell to lead to cancer. The environmental and inherited factors are therefore both affecting genes that are important for regulating the cell cycle.

5. **The development of tumors depends upon random mutational events that must "hit" several oncogenes and tumor-suppressor genes within a cell lineage.** A single mutation, even if it hits one such gene, is rarely enough by itself to cause a cell to become cancerous; instead, among the descendants of that cell, additional mutations in other key genes must occur, and it takes time for these mutations to accumulate through random, rare events. In some cases, tumors will not develop either because the carcinogen (presumably, a mutagen) has not caused a mutation in any key gene, or because not enough such mutations occur among the descendants of a cell that has sustained a mutation in a single cancer gene. Tumors do sometimes develop at the site of carcinogen application because the chemical induces a mutation in a key gene, and this gives the cell a "head start" on becoming cancerous.

6. a. The facts that many affected individuals are observed over three generations and that all the patients develop the disease at an early age are both clues that an inherited mutation is an important factor in the development of colon cancer in this family. Examination of the pedigree suggests that **predisposition to colon cancer in this family could be an autosomal dominant trait. If this is true, then individual II-2 must have the mutation but not express it**. Two possibilities exist with regard to individuals I-1 and I-2: Either one of them has the mutation but does not express it, or the mutation arose anew in one of their germ lines.

   Although the evidence is strong that genetics plays a major role in the development of colon cancer in this pedigree, the data do not prove that the prevalence of colon cancer in this family is entirely due to inheritance. As discussed in part (b), it is also conceivable that some family members have been exposed to an environmental factor that contributes to development of the disease.

   b. **Individuals I-1, I-2, and II-2 are not among the high coffee consumers. Perhaps the predisposition to colon cancer is a combination of a particular genotype and the environmental factor of consumption of the special coffee.** Thus, the interaction of a certain genotype and an environmental factor results in the phenotype of colon cancer. In this scenario, the genotype alone is not enough to predispose you to cancer: Individuals II-2 and either I-1 or I-2 might have the cancer-promoting mutation, but they do not develop the disease because they are not coffee drinkers. On the other hand, note that a substantial number of individuals who drink coffee are not affected (for example III-8, III-9, III-10, III-11, III-12) so coffee alone does not appear to have a major effect on the early onset of colon cancer.

7. a. **The reason that the blood of patients with B cell lymphomas has mostly one type of antibody molecule reflects the fact that a single B cell (or B cell precursor), whose antibody genes have already undergone rearrangements to produce a particular antibody protein, now becomes cancerous and starts to divide out of control.** A large percentage of the B cells in the patient will be making this one single type of antibody protein. In contrast, in normal individuals, B cells can only divide a discrete number of times, and the blood will be contain millions of different kinds of antibody proteins made from millions of different B cells; no single B cell will have enough descendants for its particular antibody to predominate.

    Of interest, the existence of B cell lymphomas allowed researchers (who subsequently won a Nobel Prize) to study the structure of antibody molecules. By obtaining the antibody proteins from the blood of such patients, the researchers started with substantially pure preparations of a single type of antibody. Their studies would have been impossible if they started with samples from normal blood composed of millions of different kinds of molecules.

    b. **B cell lymphomas provide support for the clonal theory of cancer because it is clearly the case that the cancerous cells in the blood all derive from a single B cell** (whose antibody genes had previously undergone rearrangement, so that the cell produces only one kind of antibody). This cell lineage divides uncontrollably, producing a huge clone of B cells all making the same antibody molecule.

Section 19.3

8. a. Molecules outside the cell that regulate cell cycle include **hormones and growth factors** (see Figs. 19.10 and 19.11 on p. 638). The external parts of growth factor receptors would also qualify, although other parts of these receptors are found embedded in the cell membrane or are located in the cytoplasm.

    b. Most of Section 19.2 is a discussion of the molecules inside the cell that regulate the cell cycle. These include **cyclins, cyclin dependent protein kinases, molecules in signal transduction pathways (the internal domains of growth factor receptors, intermediaries such as Ras and MAP kinases, and the transcription factors targeted by the pathway), and the proteins that participate in various cell-cycle checkpoints.**

9. c, e, a, b, d. (See Fig. 19.11 on p. 638 and Fig. 19.12 on p. 639.)

chapter 19

10. a. Ras is inactive when it is bound to GDP (Ras-GDP; Fig. 17.3d). When RAS is bound to GTP it is activated and in turn activates three protein kinases (the MAP kinase cascade). This cascade activates a transcription factor which causes cells to divide. Therefore, **a Ras mutant protein that stays in the GTP-bound state is permanently activated and will cause the cell to continue dividing**.

b. This second Ras mutant protein is blocked in the Ras-GDP form at the restrictive temperature. Therefore, **under the restrictive conditions the cells will not divide**.

11. a. Figure 19.14 on p. 640 is reproduced below, with two additional *red arrows* indicating about when the SCF and the APC become activated. **The SCF complex couples ubiquitin to cyclins E (and perhaps cyclin D), because those are the cyclins that are present during S phase. The APC couples ubiquitin to cyclin B (and perhaps cyclin A), because those are the cyclins that are present during M phase.** By coupling ubiquitin to cyclins that will then be degraded, the SCF and APC ensure the end of S phase and M phase, respectively.

b. The simplest hypothesis for the activation of the SCF and the APC at the proper times is that this activation depends on the cyclin-dependent kinases (CDKs) present during a particular cell cycle phase. Thus, CDKs containing cyclins D and E would activate SCF (possibly by adding phosphate groups to protein components of SCF), and SCF in turn would cause the degradation of those cyclins, the loss of the activity of these CDKs, and the exit from S phase. CDKs containing cyclins A and B would activate the APC (again through phosphorylation of APC subunits), and the APC would in turn cause degradation of those cyclins, loss of the activity of these CDKs, and the exit from M phase (that is, the beginning of anaphase).

This mechanism for the progression of the cell cycle is fundamentally elegant. The transcription and translation of particular cyclins initiate the process. These cyclins, in the context of CDKs, promote the events that must occur in each cell cycle stage. But in addition, the cyclins ensure that the stage will eventually end because they activate the enzyme complexes (SCF and APC) that lead to cyclin destruction.

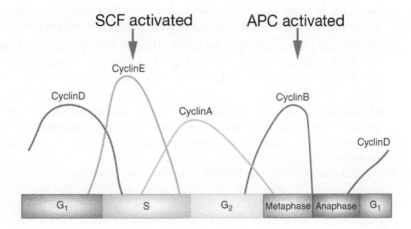

12. As explained on in Chapter 4 (on p. 98) and in Chapter 11 (p. 400-401), kinetochores that are not properly attached to spindle fibers generate molecular checkpoint signals that prevent anaphase onset. The M phase checkpoint ensures that cohesin remains intact until all the chromosomes are connected to the spindle. **In the M phase checkpoint, molecules made by unattached kinetochores prevent activation of the anaphase promoting complex (APC).** If the APC is not active then M phase cyclins remain at high levels and cohesin remains intact. In other words, the cells stay in prometaphase or metaphase while unattached kinetochores are present and the checkpoint is signaling.

As seen in Problem 11 above, **the APC must become activated at the beginning of anaphase to destroy M phase cyclin, allowing cells to leave M phase. The activated APC adds ubiquitin to protein substrates. When this happens the ubiquitylated proteins are rapidly destroyed by the proteosome. One simple hypothesis is that cohesin is also targeted by the APC, as it must be destroyed at the beginning of anaphase**, thus ensuring that sister chromatids separate at the beginning of anaphase.

Activated APC does in fact lead to cohesin degradation and thus sister chromatid separation, but the actual situation is slightly more complicated than you could have predicted from the given information. The APC-directed degradation of cohesin turns out to be indirect. Active APC adds ubiquitin to another protein called Securin which is found in a complex with a proteolytic enzyme named Separase during M phase. (Separase is depicted in Fig. 11.26 on p. 401.) When Securin and Separase are together in the complex, the Separase is nonfunctional. When Securin is destroyed, then Separase cleaves cohesin and separates the sister chromatids.

13. a. You would treat a culture of haploid yeast cells with a mutagen, and then plate out the yeast so that individual cells will form colonies. **You would grow the master petri plate at the lower (permissive temperature) so that colonies would grow whether or not they had a *ts* mutation in a gene needed for survival. You would then make a replica of this plate and grow the cells at high (restrictive) temperature.** If the cells in the colony had a *ts* mutation in an essential gene, then a colony would not appear on this replica plate. You would then retrieve the colony with the *ts* mutation from the original master plate.

b. **You would grow cells from the colony first at low (permissive) temperature, and then shift the culture to the high (permissive) temperature. You would then look at the cells to see if they arrested their growth at a particular point in the cell cycle.** Do the cells all stop growing when they have a single large bud, as was seen in Fig. B on p. 641? Do they have no bud at all? Do they have buds of a particular size? Cell cycle mutants will arrest at a particular time in the cell cycle, while mutants in other essential genes will die or stop growing at random phases of the cell cycle.

You could then go further and look at structures like the spindle and chromosomes to see if all the cells arrest with M phase-like spindles or with contracted chromosomes. In fact, using new technology based on the Green Fluorescent Protein, you could even make movies of these structures to see when arrest occurs.

chapter 19

c. **Once you have obtained a collection of *ts* mutants in cell cycle genes, you would determine which mutants are allelic by complementation analysis.** Assuming that the mutations are recessive and represent loss-of-function, you would make diploid yeast cells that combine two different *ts* mutations. If the diploid cells still exhibit temperature-sensitive growth, you would conclude the mutations do not complement each other and therefore are in the same gene. If the diploid cells can grow normally at the restrictive (high) temperature, you would conclude the mutations complement each other because they are in different genes.

   Note that to make diploid yeast cells, you need to mate haploid cells of opposite mating types. This can be done if you found mutations in haploid cells that were mating type a and others in haploid cells that were mating type α. Other methods exist for converting cells of one mating type into another, but these were not explained in the text.

d. Note that cells with large buds just before the temperature shift arrest as two cells with large buds after the temperature shift. This finding indicates that the protein encoded by this *ts* gene is not required for any process that occurs after some stage of large bud formation nor for the initiation of the cell cycle (because new buds can still form at the restrictive temperature. Thus, **the protein product of the gene is essential for some process that occurs after the cell cycle begins (after new buds form) but prior to cytokinesis (because cells with large buds that were already past the arrest point at the time of the temperature shift can produce two daughter cells).** It turns out that cells with disrupted DNA synthesis during S phase can arrest with large buds, but the same is true for cells that have defects in M phase.

   To distinguish among these possibilities, you would need to look more carefully at the arrested cells. For example, if the cells had only one but not two copies of all their chromosomes, the gene product would be needed for S phase. If the cells had two copies of all their chromosomes but had not formed spindles, the gene product would be needed for an early step in M phase.

e. **You would take yeast cells that have a *ts* mutation in the gene encoding the yeast CDK1, and then add to these cells a gene that has the corresponding human gene.** One way to do this is to take advantage of the fact that yeast plasmids are known to exist. You would make a recombinant DNA molecule in a yeast plasmid vector that contains a cDNA for the wild-type human CDK1 kinase. Ideally, you would place this cDNA sequence downstream of a the promoter sequence for the yeast *CDK1* gene (so that the cells would make approximately the same amount of the human protein as would be found in a normal yeast cell, and according to the same cell cycle schedule. You would then transform this plasmid into yeast cells. If the growth of the cells was normal even at restrictive (high) temperature, then the human gene could replace the function of the homologous gene in yeast. Historically, this experiment demonstrated that all eukaryotic cells share important cell cycle regulators.

chapter 19

Section 19.4

14. **Genome and karyotype instabilities are both causes and consequences of cancer because of the intimate relationship between mutation and cell division.** Consider for example the case of mutations in a tumor-suppressor gene encoding an enzyme that helps repair DNA damage. The lack of the DNA repair enzyme causes genome instability (because it will cause new mutations to accumulate in other genes), and some of these new mutations will promote cancerous cell growth. On the other hand, unrestricted cancerous cell growth creates new opportunities for mutation; that is, more genome instability.

15. Proto-oncogenes are genes that code for proteins that stimulate cell division. Therefore, oncogenic mutations in such genes usually increase the level of expression of the gene product. Increasing the level of expression of the proto-oncogene results in cells that divide too frequently or at the wrong time, increasing the probability of cancer. A deletion of a proto-oncogene will decrease the level of expression, not increase it. Therefore, **choice c. will not be associated with cancer**. All of the other changes will increase the probability that the affected cell, or its mitotic descendants, will become cancerous.

16. Note: In Problems 16 and 17, we make the reasonable assumption that the normal (non-translocated) homologs of the chromosomes involved in the translocation do not have any mutations in any cancer-related genes near the translocation breakpoints.

    a. **False.** Notice that the leukemic cells of the patient are heterozygous for the translocation. This fact means that the cancer-promoting properties of the translocation must be dominant at the level of the cell. This fact is thus not consistent with the characteristics of a tumor-suppressor gene, but makes more sense if some kind of dominant gain-of-function mutation has occurred that involves a proto-oncogene near the translocation breakpoint that is converted into an oncogene because of the translocation.

    The actual situation for chronic myelogenous leukemia (CML) was presented in the text on Fig. 12.16 on p. 423. The dominant gain-of-function mutation associated with the translocation converts the *c-abl* proto-oncogene into a fused *bcr/c-abl* oncogene. The protein encoded by the *c-abl* proto-oncogene is a kinase that participates in a signal transduction pathway; the kinase is regulated so that it is active only when it is needed. The protein encoded by the fused *bcr/c-abl* oncogene still functions as a kinase, but its activity can no longer be turned off. As a result, the signal transduction pathway leading to cell proliferation is constitutive: It is "on" all of the time, even when it should be "off".

    b. **False.** As just stated in part (a) the mutation must involve a gain-of-function, not a loss-of-function, of a proto-oncogene so that it is converted into an oncogene.

    c. **False.** The germ-line cells of the patient are still normal and do not have the translocation. The event producing the translocation must have occurred in some somatic cell in the lineage of white blood cells.

    d. **True.** Most of the cells in this individual have a normal karyotype and are not cancerous, while the leukemic cells are heterozygous for the translocation.

chapter 19

e. **True.** Fragments of DNA from the leukemic cells will include the region around the breakpoint that has the mutant oncogene. When normal mouse tissue culture cells are transformed with this DNA, they will form transformed foci because the effect of the mutant oncogene is dominant. (That is, the mouse cells will have two copies of the normal proto-oncogene, but if one copy of the human mutant oncogene is added, this can cause the cells to acquire cancerous properties.) If cells from the transformed foci are injected into mice, the mice can in turn develop tumors. This type of experiment was shown on Fig. 19.21 on p. 646 of the text.

f. **False.** As stated in part (a), the gene near the translocation breakpoint must have been a proto-oncogene. Proto-oncogenes encode gene products that drive the cell cycle forward rather than gene products that restrain the cell cycle (as do tumor-suppressor genes).

g. **False.** One (gain-of-function) mutational event that converts a proto-oncogene into an oncogene is sufficient to promote cancer. Two loss-of-function mutational "hits" in a tumor-suppressor gene are needed to promote cancer, but the gene at the breakpoint is clearly a proto-oncogene rather than a tumor-suppressor gene.

h. **False.** A cancer treatment would need to turn off the activity of the oncogene at the translocation breakpoint. A drug that turns on the oncogene's activity even more would have exactly the wrong effect; it would promote cancer rather than alleviating the condition.

17. The genesis of the chronic myelogenous leukemia (CML) in Problem 16 is a reciprocal translocation between nonhomologous chromosomes that is found in the cancerous cells of this patient, while normal cells do not contain this rearrangement. **You would diagnose whether the patient's blood is free of leukemic cells by performing a PCR that would amplify only DNA from the translocation. If one PCR primer binds to one of the chromosomes at one side of the translocation while the other primer binds to the other chromosome on the other side of the breakpoint, then the PCR primers will span the translocation.** Fig. 12.6b on p. 423 shows the arrangement of the requisite PCR primers with respect to the translocated chromosomes). One advantage of this method is its sensitivity, because PCR could detect only one or a few leukemic cells that remain after treatment among a large population of normal cells.

18. a. **All of these genes (i-v) are potential oncogenes** because their normal function is to promote cell division and growth. None of them could be tumor-suppressor genes.

    b. **If adding a phosphate inhibited the phosphatase, then all the actions of Kinase A would ensure that the transcription factor was activated.** That is, Kinase A would make the transcription factor active both by phosphorylating the transcription factor and by preventing the phosphatase from removing the activating phosphorylation. By contrast, if adding a phosphate activated the phosphatase, then Kinase A would simultaneously add and remove phosphates from the transcription factor; in such a circumstance, Kinase A functions would cancel each other out.

c. **The phosphatase gene is likely to be a tumor-suppressor** because its protein product inhibits the production of growth factors. The phosphatase would keep the transcription factor inactive. In the absence of active transcription factor, the mitosis-promoting factors would not be synthesized.

d. In the table that follows, homozygous means homozygous for the particular mutation for the row, heterozygous means one mutant allele and one wild type allele, E = excessive cell growth, D = decreased cell growth, N = normal cell growth.

Mutation	Homozygous	Heterozygous
i	E	N
ii	D	N
iii	D	N
iv	D	N
v	E	E
vi	E	E
vii	E	E
viii	D	D
ix	E	N

19. a. **A tumor-suppressor gene** - most inherited tumors are due to the inheritance of an inactive allele of a tumor-suppressor gene. The inheritance is dominant because it is highly likely that the wild-type allele of the tumor-suppressor gene obtained from the nonaffected parent will be inactivated or lost in some of the progeny's cells.

b. As implied in part (a), the allele inherited from the affected parent is a **loss-of-function** mutation that inactivates the *NF1* gene.

c. **Ras-GDP**. If neurofibromin is a tumor-suppressor protein, its presence would slow down or restrict the cell cycle. This could be accomplished if neurofibromin favors the formation of the inactive form of Ras, Ras-GDP.

d. **ii, iv, v** and **vi** could all be "second hits" that would inactivate or cause the loss of the functional allele of *NF1* inherited from the normal parent. The allele from the affected parent is already inactivated or deleted, so additional mutations in that allele would have no effect. Mitotic chromosome nondisjunction or chromosome loss (v) might produce an aneuploid cell without a functional copy of the *NF1* gene. Mitotic recombination in a heterozygote could produce a cell homozygous for the inherited *NF1* mutation, but the recombination event would have to occur between the *NF1* gene and the centromere of the chromosome carrying it (Fig. 5.24).

e. The neurofibromas in these patients are **sporadic**. In other words, these patients inherit wild-type alleles of *NF1* from both parents. Then **there must be a clone of cells in which one of the copies of *NF1* is inactivated by some rare event then the other allele of *NF1* is inactivated or lost**. These tumors are extremely rare because two rare events must happen in the same line of cells. In contrast, the inherited disease requires just one rare event as the patient starts out with a

chapter 19

defective allele. Sporadic tumors are restricted to one part of the body because these clones of cells with no *NF1* function are so rare; in addition, the second inactivating event is likely to occur late in the patient's life and only in one tissue. Patients with the inherited form of neurofibromatosis instead have many pre-cancerous growths (see Fig. 10.19 on p. 358).

20. *Note:* Throughout this problem, the identities of the three point mutations vi, vii, and viii cannot be distinguished.

   a. **iii (trisomy), iv (duplication of a chromosomal region), and [vi, vii, or viii] (a rare hypermorphic or neomorphic point mutation).** Genetic alterations that could convert proto-oncogenes into tumor-promoting oncogenes would be gain-of-function mutations (vi, vii, or viii) or increases in gene dosage (iii and iv).

   b. **i (mitotic recombination), ii (deletion of a chromosomal region), v (uniparental disomy), and [vi, vii, or viii] (a null or hypomorphic point mutation).** [If one point mutation was assigned previously in part (a) as a gain-of-function mutation, it cannot also be a gain-of-function mutation in part (b).] Genetic alterations of tumor-suppressor genes that promote cancer would be loss-of-function mutations (vi, vii, or viii) or decreases in gene dosage (ii). In heterozygotes for a loss-of-function mutation of a tumor-suppressor gene, mitotic recombination (i) or uniparental disomy (v) could result in loss of the remaining wild-type copy of the gene.

   c. **i (mitotic recombination) and v (uniparental disomy).** The idea here is that prior to the event listed, a cell has two copies of a tumor-suppressor gene but one copy is inactive (that is, the cell is heterozygous for a loss-of-function mutation – but not a deletion – in a tumor-suppressor gene). Mitotic recombination (i) or uniparental disomy (v) would make the cell homozygous for the mutant allele, so there would still be two copies of the gene but both would be mutant.

   A new null point mutation (vi, vii, or viii) in the previously wild-type copy of the tumor-suppressor gene would likewise result in a cell with two copies of the gene, both of which are inactive. This situation is technically not a loss-of-heterozygosity (because the cell would be a heterozygote for two different loss-of-function mutations), but you are still essentially on the right track if you also answered vii.

   For parts (d) – (h), you first need to understand the difference between the representation of a locus and a specific allele. Each column is a specific locus (nucleotide pair) along the chromosome; each row represents the possible alleles of that locus (that is, A, G, C, and T). It is next crucial to pay attention to the number of copies of each SNP analyzed in the microarray. *White* means no copies of the indicated allele, *orange* means one copy, and *red* means two copies. You then add up the number of copies of all the alleles of a locus in the sample; for example, the tumor has 3 total copies of locus *a* on Chromosome 15 (1 A and 2 Gs).

d.

i (mitotic recombination)	ii (deletion)	iii (trisomy)	iv (duplication)	v (uniparental disomy)
16 *a-g*	16 *s-w*	15 *a-z*	14 *n-t*	17 *a-z*

You were not asked to locate any point mutations (vi-viii) because this is impossible [with one special exception noted in part (f) below]. The possibility that any of the SNP polymorphisms examined on the microarray contributes to cancer is extremely remote given that these are constitute only a very small sample of all possible polymorphisms on these chromosomes. The SNPs examined are instead anonymous in the sense that they would have no effect on phenotypes including cancer.

e. **The mitotic recombination event must take place in the interval between SNPs *g* and *h* on Chromosome 16.** Note that in the tumor tissue, the sequences distal to the mitotic recombination site (that is, further from the centromere) become homozygous, while those proximal to the mitotic recombination site remain heterozygous as in the normal tissue.

f. Although none of the SNP polymorphisms examined on the microarray is likely to promote cancer, the data do suggest that there are three places in the genome in which the patient might have inherited a point mutation in a tumor-suppressor gene or developed a somatic point mutation in a liver cell. **One location is region *a-g* of Chromosome 16.** If a point mutation existed in a tumor-suppressor gene in this region, then mitotic recombination in region *g-h* of this chromosome could generate a cell homozygous for this point mutation. **A second possibility is anywhere on Chromosome 17** (regions *a-z*), where uniparental disomy could generate a cell homozygous for a point mutation in a tumor-suppressor gene. **The third region is Chromosome 16 region *s-w*,** where the patient inherited a deletion, and a point mutation in a tumor-suppressor gene may have developed subsequently in a liver cell.

g. **Alteration ii.** The most clear-cut evidence that the patient inherited a mutation that could promote cancer is that the interval containing SNPs *s-w* on Chromosome 16 is deleted (the normal tissue has only one copy of these loci). On a normal chromosome, this region presumably contains a tumor-suppressor gene, so the patient starts out (as a zygote) with only one copy of this gene.

As just discussed in part (f), the possibility exists that the patient may have inherited a point mutation in region *g-h* of Chromosome 16 and/or somewhere on Chromosome 17 (*a-z*), but the evidence for the existence of such point mutations is only indirect.

h. Consistent with the answers to parts (f) and (g), the three scenarios are as follows. The three point mutations (vi-viii) are interchangeable.

Scenario 1: The first hit (ii) is that the patient inherited from a parent a deletion of a tumor-suppressor gene in region *s-w* of Chromosome 16. The second hit (vi) would be a point mutation in the remaining wild-type copy of the tumor-suppressor gene.

Scenario 2: The first hit is that the patient inherited from a parent a loss-of-function point mutation in a tumor-suppressor gene lying in region *a-g* of Chromosome 16 (vii). (This mutation could also have occurred somatically in a liver cell; the data do not address this point.) **The second hit would be the mitotic recombination event (i) that occurred in region *g-h* in a liver cell originally heterozygous for the point mutation.** The mitotic recombination would make a daughter cell homozygous for the point mutation.

Scenario 3: The first hit is that the patient inherited from a parent a loss-of-function point mutation in a tumor-suppressor gene lying somewhere on Chromosome 17 (viii). (This mutation could also have occurred somatically in a liver cell; the data do not address this point.) **The second hit would be the uniparental disomy event (v) in a liver cell originally heterozygous for the point mutation.** The result would be a liver cell that is now homozygous for the point mutation.

21. Alterations i-v would be relatively easy to find from whole genome sequencing. **For alterations that change the copy number of a region [the deletion (ii), trisomy (iii), and duplication of a region (iv)], you would look for either fewer (ii) or more (iii and iv) "reads" of the corresponding region as compared with the reads of other regions of the genome in which the normal and tumor tissue had only two copies. For the alterations that produce loss-of-heterozygosity that is copy-neutral [mitotic recombination (i) and uniparental disomy (ii)], you would compare the reads from the normal tissue and from the tumor tissue. In the regions of interest, the normal tissue should show two alleles of many of the polymorphic SNP loci, but the tumor tissue would show only one kind of allele even though the number of reads would indicate that two copies of this allele are present.**

    Point mutations that contribute to cancer would be much harder to find and may still go undetected even if you have sequenced the whole genome. You could look specifically in the regions discussed in the three scenarios in the answer to Problem 20 part (h) for alterations in known tumor-suppressor genes. You could also look for point mutations that would affect known oncogenes to produce some kind of gain-of-function mutation. But it is also possible that the tumor contains point mutations that affect cancer-related genes that are unknown or poorly characterized. Here, you would look for mutations that disrupted open reading frames of genes (like nonsense or frameshift mutations), or missense mutations that affect highly conserved amino acids within proteins. Mutations outside of the coding sequences that affected the regulation of proto-oncogenes or tumor-suppressor genes would be even harder to identify.

22. "Deep sequencing" (RNA-Seq) might reveal the identity of mRNAs or other transcripts whose levels are higher (for oncogenes) or lower (for tumor-suppressor genes) in tumorous tissue than in normal cells of the same tissue type. That is, the proportion of total mRNA "reads" for these genes in the cancerous cells would be much different from that in normal cells. As just discussed in Problem 21, if a cancer-related gene had some type of mutation in a regulatory region that affects the amount of transcript, such a change would be hard to find by just looking at the

genomic DNA sequence. RNA-Seq thus has the potential to provide medical scientists with new information relevant to a patient's cancer.

Of course, just finding an alteration in the levels of a transcript does not mean that a mutation must be in or near the corresponding gene; the problem could be in the gene encoding a *trans*-acting factor. Also, changes in mRNA levels might not be driving tumor progression, but instead might be a response to other genomic alterations that are driving the cancer. So this type of analysis can certainly provide important clues as to the genetic basis for a particular cancer, but much additional work would need to be done to understand the meaning of any change in transcript level that is seen.

23. a. Assuming that the presence of EGFRvIII (the mutational variant of the epidermal growth factor receptor protein contributes to the cancerous phenotype, then **you would classify the gene encoding EGFR as a proto-oncogene.** The reason is that the cancerous cells express both EGFRvIII and wild-type EGFR. This fact means that the EGFRvIII mutation has dominant effects. Dominant, gain-of-function mutations that change proto-oncogenes into oncogenes contribute to cancer. In contrast, loss-of-function alleles of tumor-suppressor genes are recessive in terms of their contribution to cancer.

b. **You would need to treat patients expressing EGFRvIII with a higher-than-normal dose of X-rays.** Remember that apoptosis is a cell program that allows cells with DNA damage to self-destruct. Apoptosis is necessary for the body to rid itself of cells that would be dangerous because they have accumulated a great deal of DNA damage. If the cancerous cells do not undergo apoptosis easily, then they will be relatively resistant to X-rays and not die. You would need to "zap" the cells with very high levels of X-rays to kill the cells directly. In fact, treating such patients with the usual dose of X-rays would be very dangerous, because the X rays would cause these cells to accumulate mutations including some in oncogenes and tumor-suppressor genes.

c. The wild-type allele of the *EGFR* gene is a proto-oncogene whose product promotes cell division by serving as a receptor for epidermal growth factor. The N-terminal domain is extracellular and binds the growth factor, while the C-terminal part of the protein is a kinase. Function of the kinase (to add phosphate groups to other proteins) is needed to promote cell division in the presence of the growth factor. EGFRvIII protein has the kinase domain but does not have amino acids 6 through 273 of the N-terminal domain. Thus, **the most likely explanation is that deletion of the N-terminal domain makes a form of the EGFR receptor that can signal constitutively, even in the absence of the EGF growth factor.**

d. **Iressa would be an excellent candidate drug for the treatment of glioblastomas expressing EGFRvIII.** The drug should block the kinase activity of EGFR, for example by occupying the site at which ATP binds the enzyme so that one phosphate from ATP can be transferred to target proteins. This drug should stop the constitutive signaling from the EGFRvIII protein because it would interfere with its kinase function; in this way the drug would prevent unwanted cell proliferation. One issue with this drug is of course that it would also prevent the function of the normal EGFR receptor in all cells, which would produce unwanted side effects.

chapter 19

Although Iressa theoretically makes a great deal of sense for treating glioblastomas, it has unfortunately not proven to be very effective in actual clinical studies. The reasons are unclear; perhaps not enough drug crosses the blood-brain barrier to reach the tumors in the brain. Another possibility is that many glioblastomas may have additional mutations that cause constitutive signaling from other growth factor receptors that are not inhibited by Iressa.

e. **You would need to treat patients with glioblastomas that have higher levels than normal of *ERCC1* transcription and gene product with higher-than-normal doses of cisplatin.** The reason is that these cells can easily remove cisplatin-produced DNA damage because the ERCC1 protein helps repair DNA damage. Thus, you would need use more cisplatin to achieve the amount of DNA damage that would be needed for the cells to undergo apoptosis.

You should note that cisplatin treatment may be a bad idea for patients with defects in the *p53* tumor-suppressor gene or in the genes that participate in the apoptosis pathway. In such cases, the cells could accumulate much DNA damage but they would not be removed by apoptosis; such cells might contribute to cancer progression.

Chapter 20

Variation and Selection in Populations

Synopsis

The science of **population genetics** aims to describe how the frequencies of alleles, genotypes, and phenotypes change over the course of generations in populations of organisms. This goal may seem on the surface to be purely academic, but in reality population genetics provides essential tools for understanding evolution and even the history of humanity. The insights of population genetics have important practical implications ranging from forensics (DNA fingerprinting) to the control of bacterial pathogens and insect pests.

The starting point for population genetics is the **Hardy-Weinberg law.** At the beginning of the 20th century, these two investigators realized independently that under conditions in which a new generation is formed by the union of gametes randomly obtained from a very large pool of gametes produced by the previous generation, the allele frequencies, genotype frequencies, and phenotypic frequencies would remain constant. If these conditions are met, the population will remain in an equilibrium whose genotypic frequencies are described by the equation $p^2 + 2pq + q^2 = 1$ (if the population has two alleles of an autosomal locus whose allele frequencies are p and q). The Hardy-Weinberg law approximates the behavior of many anonymous loci in human populations; these loci can thus be used in forensics because scientists can predict the probability that a random person's genotype at these loci will match that of a crime scene sample.

The genetic compositions of populations do and must change over time, indicating that the assumptions behind the Hardy-Weinberg equilibrium do not hold for any locus over the long term in actual populations. This chapter delves into two of the most important phenomena causing deviations from the idealized populations envisioned by Hardy and Weinberg. The first of these concerns the fact that real populations are not infinite in size. The smaller the population, the more important is **genetic drift** – changes in allele frequencies due to random sampling errors in choosing the gametes that happen to form the next generation. Allele frequencies can thus change rapidly in small populations. The second major reason real populations do not always obey the Hardy-Weinberg law is that some variants at some loci affect the organism's **fitness:** In other words, some mutations are not neutral, but instead confer selective advantages or disadvantages to individuals in the population. The effects of a mutation on fitness depend on the genotype and on the environment. As a result, some deleterious mutations such as that causing sickle-cell anemia can be maintained in certain populations if heterozygotes have a higher fitness than either class of homozygotes in a given environmental condition (such as tropical conditions in which the malaria parasite is frequently encountered).

The chapter concludes with a presentation of recent insights into human history gained by sequencing genomic DNAs from many individuals alive today in different populations around the world, and comparing these sequences with those obtained from skeletons of extinct human-like groups (Neanderthals and Denisovans) with whom some of our ancestors interbred.

chapter 20

Key terms

population genetics – study of the transmission of genetic variation between generations in *populations*

population – a group of individuals of the same species living in the same time and place

gene pool – the sum total of all the alleles of all genetic loci carried in all members of a population. In a population of a diploid organism, the gene pool contains $2N$ copies of a locus, where N is the number of individuals in the population.

monomorphic vs. polymorphic locus – If a locus is *monomorphic*, only one allele of the locus exists in the population; when more than one variant exists in the population, the locus is *polymorphic*.

genotype frequency – proportion of all the individuals in a population who have a particular genotype

phenotype frequency – proportion of all the individuals in a population who have a particular phenotype

allele frequency – proportion of all the copies of a locus in a population that are of a given allele type

Hardy-Weinberg equilibrium (HWE) – a condition in which allele frequencies, genotype frequencies, and phenotype frequencies remain constant between generations due to the random union at fertilization of *selectively neutral* alleles obtained from the gene pool of a very large population

Hardy-Weinberg (H-W) proportions – the genotype frequencies for a given locus observed at Hardy-Weinberg equilibrium. If two alleles of the locus exist in a population, the H-W proportions are p^2 for homozygotes of allele A, $2pq$ for Aa heterozygotes, and q^2 for aa homozygotes, where p is the allele frequency of A and q is the frequency of the a allele.

match probability – the chance that a random individual from a population will have the same genotype as a particular chosen genotype

Monte Carlo simulation – a method of modeling the effects of sampling error during a multi-event process, using a computer program that employs a random-number generator to choose an outcome for each event

genetic drift – changes in allele frequencies between generations over time as a consequence of sampling error in the choice of gametes. The effects of genetic drift become significant even in the short term when population sizes are small.

allele extinction – the loss of an allele from a population

allele fixation – When only one allele remains in a population (that is, when the locus becomes *monomorphic*), that allele is said to be *fixed*.

founder effect – a process that occurs when a few individuals leave a large population to establish a new, geographically isolated population. Due to random sampling error, the allele frequencies in the new population may differ considerably from those in the original population. If the new population is small, *genetic drift* may alter allele frequencies rapidly in the new population.

population bottleneck – a process that occurs when a catastrophe reduces a large population to a much smaller one. Just as with *founder effects,* allele frequencies before and after the catastrophe may change rapidly, and if it is small, the residual population will be subject to *genetic drift.*

migration – the movement of individuals from one population into a different population. Migrants from the first population can introduce new variants into the second population.

natural selection – the process by which nature progressively eliminates individuals with lower *fitness* and chooses individuals of higher *fitness* to survive and reproduce.

fitness – an individual's relative ability to survive (*viability*) and transmit its genes (*reproductive success*) to the next generation

neutral/deleterious/beneficial mutations – *Neutral mutations* have no advantage or disadvantage from the point of view of natural selection; *deleterious mutations* confer a selective disadvantage; *beneficial mutations* confer a selective advantage.

molecular clock – *Neutral mutations* accumulate in genomes at a relatively constant rate over evolutionary time, so the number of such differences in the genomes of two species indicates when the two species last had a common ancestor.

balancing selection/heterozygote advantage – Any process that maintains polymorphisms in a population is called a *balancing selection*. One such process is *heterozygote advantage,* where heterozygous individuals have higher fitness than individuals of either homozygous genotype.

biological vs. genetic ancestors – Your *biological ancestors* are all the people in generations prior to your own to whom you are directly related by descent (your parents, grandparents, etc.) The number of your genetic ancestors increases exponentially by a factor of two every generation going back in time. *Genetic ancestry* is a term that describes from which of your ancestors you obtained a specific region of the genome. Because you have two alleles for a given genetic locus, you have only two genetic ancestors for that locus in any generation of ancestors.

most recent common ancestor (MRCA) – the DNA sequence at a specific region of the genome (the allele of that genomic region) from which all individuals in a particular generation or population have obtained some base pair sequences. The alleles in specific individuals (the **derived alleles**) descended from the MRCA allele (the **ancestral allele**) with modifications due to mutations.

hominids versus hominins – Although the definitions of these terms have changed over time, most scientists now refer to *hominids* as the group consisting of all the Great Apes (modern humans, chimps, gorillas, and their recent extinct ancestors). The *hominins* are the group of modern humans, extinct human-like species such as Neanderthals and Denisovans, and the immediate ancestors of these species.

chapter 20

Problem Solving

Using the Hardy-Weinberg equation: $p^2 + 2pq + q^2 = 1$ is the starting point for most population genetics analyses. p and q represent the allele frequencies (for a locus with 2 alleles), while p^2, $2pq$, and q^2 are the genotype frequencies at equilibrium. For any value of either p or q, only one set of values for p^2, $2pq$, and q^2 would allow the population to be in equilibrium for the locus in question (Fig. 20.4 on p. 667). If a population has such a set of values for these genotype frequencies, then the gametes that produce the next generation can be envisioned to be infinitely large pools consisting of the two alleles present at their respective allelic frequencies, and gametes from these pools are obtained and unite at random. This idea is reflected in the fact that the Hardy-Weinberg equation can also be written as $(p + q)^2 = 1$, and that the population can be represented by the Punnett square shown in Fig. 20.3 on p. 666.

You should note several special cases in using the Hardy-Weinberg equilibrium:

Random mating leads to equilibrium in one generation: For autosomal loci, if you start with a population that is not in equilibrium, random mating within the population will cause the population to reach equilibrium in one generation, assuming that the other Hardy-Weinberg assumptions are also met.

X-linked genes: For the population to be in equilibrium, p and q must be the same in males and females. This condition occurs for X-linked genes when the genotype frequencies in females are p^2, $2pq$, and q^2 while those in males are simply p and q. Note in Fig. 20.5 on p. 668 that if a starting population has different X-linked allele frequencies in the males and females, it will take several generations of random mating to reach equilibrium.

Genes with more than two alleles: You would simply write an equation in which you would square the sum of all the allele frequencies: $(p + q + r + ...)^2$. Thus, in the case of a locus with three alleles, the equilibrium frequencies would be: p^2 (A^1A^1) + q^2 (A^2A^2) + r^2 (A^3A^3) + $2pq$ (A^1A^2) + $2pr$ (A^1A^3) + $2qr$ (A^2A^3) = 1.

Estimating allele frequencies from the frequency of a recessive homozygous genotype: In the cases of genetic diseases associated with homozygosity for a recessive allele, scientists often use the frequency of diseased individuals to estimate the frequency of the normal and mutant alleles in the population. If the population was at equilibrium, then $q = \sqrt{q^2}$, where q^2 is the proportion of individuals in the population showing the disease phenotype. Although the population cannot truly be in HWE if the fitness of one genotype is impaired, the equilibrium values nonetheless provide reasonably accurate estimates of the allelic and genotypic frequencies if the disease allele is rare.

When the population is not at HWE, then one of the Hardy-Weinberg assumptions must not hold for the particular locus in the particular population in question. The two most important exceptions are **genetic drift** (of high significance in small populations) and cases in which the **fitness** of all genotypes is not the same. Although some equations provided in the text can guide quantitative examinations of these issues, in general the questions at the end of the chapter probe your qualitative understanding of the underlying phenomena.

chapter 20

Vocabulary

1.

a.	fitness	5.	ability to survive and reproduce
b.	gene pool	10.	collection of alleles carried by all members of a population
c.	fitness cost	8.	the advantage of a particular genotype in one situation is a disadvantage in another situation
d.	allele frequencies	6.	p and q
e.	heterozygote advantage	1.	the genotype with the highest fitness is the heterozygote
f.	equilibrium frequency	9.	frequency of an allele at which $\Delta q = 0$ [* see note]
g.	genetic drift	2.	chance fluctuations in allele frequency
h.	molecular clock	3.	mutations accumulate at a relatively constant rate
i.	population bottleneck	7.	event that drastically lowers N
j.	hominins	11.	*Homo sapiens,* Neanderthals, and Denisovans
k.	fixation	4.	$p = 1.0$

[* Note: in some editions of the textbook, the Greek delta (Δ) was inadvertently left out.]

Section 20.1

2. The 3:1 ratio is seen when <u>two heterozygous individuals</u> are crossed. This ratio is <u>not</u> relevant for a <u>population</u> where the crosses are of many different sorts – some will be homozygous dominant × homozygous dominant, others homozygous dominant × heterozygous, some homozygous dominant × homozygous recessive, some heterozygous × homozygous recessive and others homozygous recessive × homozygous recessive. **The ratio of wild type : mutant progeny in the population will depend on specific allele frequencies in that population.**

   As you will see in a later section of this chapter, the idea that a dominant allele will automatically increase in frequency is also false for another reason: If a dominant allele has a deleterious effect on the fitness of an individual carrying it, the allele frequency would instead tend to decrease each generation. The term "dominance" by itself does not predict the allele's effect on fitness or its frequency in the population.

3. If a population is in Hardy-Weinberg equilibrium, then the allele frequency p is squared to give the genotype frequency p^2. Thus, **each different allele frequency of p has a different p^2 value and a different set of genotype frequencies at equilibrium** (see Fig. 20.4 on p. 667).

4. a. Calculate genotype frequencies:

   $G^G G^G$ = 120 / 200 = **0.6**; $G^G G^B$ = 60 / 200 = **0.3**; $G^B G^B$ = 20 / 200 = **0.1**.

   b. The allele frequencies can be determined in two different ways. One way is to use the genotype frequencies in part (a): Frequency of G^G = 0.6 + (1/2)(0.3) = **0.75**; frequency of G^B = 0.1 + (1/2)(0.3) = **0.25**.

   Another method is to total all alleles within each genotype:
   $G^G G^G$ = 120 individuals with 2 G^G alleles = 240 G^G alleles
   $G^G G^B$ = 60 individuals with one G^G allele = 60 G^G alleles
   $G^G G^B$ = 60 individuals with one G^B allele = 60 G^B alleles
   $G^B G^B$ = 20 individuals with two G^B alleles = 40 G^B alleles

   There are (240 + 60) = 300 G^G alleles/400 total alleles so the frequency of G^G (*p*) = **0.75**. There are (60 + 40) = 100 G^B alleles/400 total alleles so the frequency of G^B (*q*) = **0.25**.

   c. **For alleles in HWE, the expected frequency of $G^G G^G$ is p^2 = (0.75)² = 0.5625; the expected frequency of $G^G G^B$ is $2pq$ = 2(0.75)(0.25) = 0.375; the expected frequency of $G^B G^B$ = (0.25)² = 0.0625.** Note that $p^2 + 2pq = q^2$ = 0.5625 + 0.375 + 0.0625 = 1.

5. For each population, determine the allele frequencies (*p* and *q*) using the observed genotype frequencies, and then use the Hardy-Weinberg equation ($p^2 + 2pq + q^2 = 1$) to determine if the genotype frequencies predicted for the next generation remain the same as those in the original population.

   a. The frequency of *A* = 0.25 + (1/2)(0.5) = 0.5; the frequency of *a* = 0.25 + (1/2)(0.5) = 0.5. The predicted frequencies of genotypes in the next generation are *AA* [p^2 = (0.5)² = 0.25]; *Aa* [$2pq$ = 2(0.5)(0.5) = 0.5]; *aa* [q^2 = (0.5)² = 0.25]. Because these genotype frequencies predicted by the Hardy-Weinberg equilibrium are the same as those of the original population, **population a is in equilibrium**.

   b. The frequency of *A* = 0.1 + (1/2)(0.74) = 0.47; the frequency of *a* = 0.16 + (1/2)(0.74) = 0.53. The predicted frequencies of genotypes in the next generation are *AA* [p^2 = (0.47)² = 0.221]; *Aa* [$2pq$ = 2(0.47)(0.53) = 0.498]; *aa* [q^2 = (0.53)² = 0.281]. These frequencies are not the same as the original genotype frequencies, so **population b is not in equilibrium**.

   c. The frequency of *A* = 0.64 + (1/2)(0.27) = 0.775; the frequency of *a* = (1/2)(0.27) + 0.09 = 0.225. The HWE frequencies are: *AA* = (0.775)² = 0.6; *Aa* = 2(0.775)(0.225) = 0.35; *aa* = (0.225)² = 0.05. **Population c is not in equilibrium**.

   d. The frequency of *A* = 0.46 + (1/2)(0.50) = 0.71; the frequency of *a* = (1/2)(0.50) + 0.04 = 0.29. The HWE frequencies are: *AA* = (0.71)² = 0.50; *Aa* = 2(0.71)(0.29) = 0.42; *aa* = (0.29)² = 0.08. **Population d is not in equilibrium**.

   e. The frequency of *A* = 0.81 + (1/2)(0.18) = 0.9; the frequency of *a* = (1/2)(0.18) + (0.01) = 0.1. The HWE frequencies are: *AA* = (0.9)² = 0.81; *Aa* = 2(0.9)(0.1) = 0.18; *aa* = (0.1)² = 0.01. **Population e is in equilibrium;** the original genotype frequencies are the same as those predicted for the next generation by the Hardy-Weinberg equation.

6. a. There are 60 D^+D^+ flies with normal wings and 90 DD^+ flies with Delta wings. The frequencies of these two genotypes are 60/150 = 0.4 D^+D^+ and 90/150 = 0.6 DD^+. The allele frequency for D = (1/2)(0.6) = **0.3**; the allele frequency for D^+ = 0.4 + (1/2)(0.6) = **0.7**.

b. The following frequencies of F_1 zygotes will be produced: 0.49 D^+D^+ + 0.42 DD^+ + 0.09 DD = 1. The homozygous DD zygotes do not live, so the viable progeny are 0.49 + 0.42 = 0.91, or 91% of the original zygotes. If there are 160 viable adults in the F_1 generation, then there must have been 160 / 0.91 = **176 F_1 zygotes**.

c. The simplest way to answer this question is to realize that, as given the answer to part (b), 176 F_1 zygotes existed to produce 160 viable adults. Also as noted in part (b) above, the viable progeny (91% of the F_1 zygotes) are 0.49 D^+D^+ + 0.42 DD^+ = 0.91. Thus, (0.49)(176) = 86 D^+D^+ and (0.42)(176) = 74 DD^+.

Another to way to answer this question is to normalize the equation 0.49 D^+D^+ + 0.42 DD^+ = 0.91 to equal 1 before you can calculate expected numbers of each type of progeny out of 160 total. That is, 0.49 D^+D^+ / 0.91 = 0.54 D^+D^+ = proportion of the viable offspring with the D^+D^+ genotype and 0.42 D^+D / 0.91 = 0.46 D^+D = proportion of the viable offspring with the D^+D genotype. Thus, the expected numbers of viable individuals with these genotypes are: 0.54 D^+D^+ × 160 = **86 D^+D^+** and 0.46 D^+D × 160 = **74 D^+D**.

d. **No**. The lethality of the DD genotype means that **there are genotype-dependent differences in fitness, violating one of the basic tenets of the Hardy-Weinberg equilibrium**. Therefore this population will never achieve equilibrium for this locus.

7. a. The frequency of M = 0.5 + (1/2)(0.2) = 0.6; the frequency of N = 0.3 + (1/2)(0.2) = 0.4. The expected genotype frequencies in the next generation are calculated using these allele frequencies and the equation $p^2 + 2pq + q^2 = 1$: 0.36 MM + 0.48 MN + 0.16 NN = 1. These Hardy-Weinberg genotype frequencies are not the same as the genotype frequencies of the initial population, therefore **the initial population is not in equilibrium**.

b. As calculated in part (a), **the genotype frequencies in the F_1 will be: 0.36 MM + 0.48 MN + 0.16 NN = 1**. From these values, the **allele frequencies in the F_1 generation** can be calculated: M = 0.36 + (1/2)(0.48) = **0.6** and N = 0.16 + (1/2)(0.48) = **0.4**.

c. **In the next (F_2) generation, the allele and genotype frequencies will remain the same as in part (b) [M = 0.6; N = 0.4; MM = 0.36; MN = 0.48; NN = 0.16]**. HWE for this autosomal locus is reached after one generation of random mating, so the allele and genotype frequencies do not change subsequently.

8. a. First, recalculate the distribution of genotypes as fractions of the total:

$Q^F Q^F\ R^C R^C$	0.137
$Q^F Q^G\ R^C R^C$	0.068
$Q^G Q^G\ R^C R^C$	0.068
$Q^F Q^F\ R^C R^D$	0.251
$Q^F Q^G\ R^C R^D$	0.126
$Q^G Q^G\ R^C R^D$	0.126
$Q^F Q^F\ R^D R^D$	0.112
$Q^F Q^G\ R^D R^D$	0.056
$Q^G Q^G\ R^D R^D$	0.056

    Next, calculate the allele frequencies for Q and R genes separately. In this population, the frequencies of the Q gene are 0.5 $Q^F Q^F$ + 0.25 $Q^F Q^G$ + 0.25 $Q^G Q^G$ = 1. The allele frequencies are Q^F = 0.625 and Q^G = 0.375. The genotype frequencies in the next generation expected by the Hardy-Weinberg equilibrium are therefore 0.39 $Q^F Q^F$ + 0.47 $Q^F Q^G$ + 0.14 $Q^G Q^G$ = 1. **The population is not in equilibrium for the Q gene.**

    The genotype frequencies for the R gene in this population are 0.273 $R^C R^C$ + 0.503 $R^C R^D$ + 0.224 $R^D R^D$ = 1. The R^C allele frequency = 0.52 and the R^D allele frequency = 0.48. The expected genotype frequencies in the next generation (Hardy-Weinberg equilibrium) is 0.27 $R^C R^C$ + 0.5 $R^C R^D$ + 0.23 $R^D R^D$ =1. In this case, the observed genotype frequencies are very close to the Hardy-Weinberg genotype frequencies, so **the population is in equilibrium for the R gene**. Note that the same population is in HWE for one gene (R) but not for a different gene (Q).

  b. The fraction of the population that will be $Q^F Q^F$ in the next generation is the expected genotype frequency in part (a): **0.39 for the $Q^F Q^F$ genotype**.

  c. The fraction that will be $R^C R^C$ **in the next generation** will again be the expected frequency calculated in part (a) based on the allele frequency: **0.27**.

  d. This probability is not influenced by allele frequencies calculated in part (a) above. Instead, this is a standard probability question starting with parents of specific genotypes as discussed in Chapter 2 (see for example Problem 2-19). This cross is $Q^F Q^G\ R^C R^D \times Q^F Q^F\ R^C R^D$. The probability is the product of the individual probabilities for each of the genes. One parent heterozygous for the Q gene and the other is homozygous, so there is a 1/2 chance of a $Q^F Q^G$ child. Both parents are heterozygous for the R gene, so there is a 1/4 chance the child will be homozygous $R^D R^D$. There is a 1/2 chance the child will be male. The overall probability is (1/2)(1/4)(1/2) = **1/16 that the child will be a $Q^F Q^G\ R^D R^D$ male**.

9. a. If a population is in Hardy-Weinberg equilibrium, then the allele frequency is the square root of the frequency of the homozygous genotype ($\sqrt{q^2} = q$). If 1/250,000 people are affected by the autosomal recessive disorder alkaptonuria, then the genotype frequency of the recessive homozygote (q^2) = 4×10^{-6}. Therefore q = $\sqrt{4 \times 10^{-6}}$ = **0.002**.

  b. Remember that the population is in Hardy-Weinberg equilibrium, that q = 0.002, and that p (frequency of the normal allele) = 0.998. Therefore **the frequency of carriers**

chapter 20

(heterozygotes) = 2 × (0.998) × (0.002) = 0.004 = 4 × 10⁻³. **The ratio of carriers / affected individuals = (4 × 10⁻³)/ (4 × 10⁻⁶) = 1000:1.**

c. If the unaffected woman has an affected child then she must be a carrier, as was the father of the child. The woman has remarried. If her new husband is homozygous normal she can never have an affected child. However if the new husband is a carrier or is himself affected, then she can have an affected child. The total probability of an affected child is the sum of each of these individual probabilities. **The probability that she will have an affected child by this marriage** = [(0.004 probability that her new husband is a carrier) × (0.5 probability that he passes on his mutant allele) × (0.5 probability that she passes on her mutant allele)] + [(4 × 10⁻⁶ probability that her new husband is affected) × (1 probability that he passes on a mutant allele) × (0.5 probability that she passes on her mutant allele)] = [0.001] + [2 × 10⁻⁶] = 0.001002 = **0.001**. As you can see, the probability that her second husband is affected is insignificant. Therefore the probability that this woman will have an affected child with her second husband depends only on the frequency of heterozygotes in the population.

d. **No - if one of the genotypes is selected against, then the frequencies of p and q will change each generation**. The population will never reach equilibrium (see Problem 20-6 and also Fig. 20.12 on p. 675).

10. a. **The frequency of allele A^F is calculated in the following way:**

   Population 1:

   $38\ A^FA^F \times 2 = 76\ A^F$ alleles
   $44\ A^FA^S \times 1 = \underline{44\ A^F\text{ alleles}}$
   $\ 120\ A^F$ alleles in Population 1

   $120\ A^F$ alleles / 200 total A alleles = **0.6.**

   Population 2:

   $80\ A^FA^S \times 1 = 80\ A^F$ alleles

   $80\ A^F$ alleles / 200 total A alleles = **0.4.**

   b. For Population 1, the allele frequencies are $p = 0.6$ and $q = 0.4$. Genotype frequencies when the population is at equilibrium are:

   $p^2 = (0.6)^2 = 0.36$
   $2pq = 2(0.6)(0.4) = 0.48$
   $q^2 = (0.4)^2 = 0.16$

   For Population 1, which consists of 100 individuals, the equilibrium distribution of genotypes would be 36 A^FA^F, 48 A^FA^S, and 16 A^SA^S lizards. **Population 1 appears to be in equilibrium.** (Sampling error and small population size could lead to slight variation from the expected frequencies.)

   For Population 2, the allele frequency p for the A^F allele was calculated in part (a) to be 0.4, so the allele frequency q for the A^S allele is 0.6. Genotype frequencies for a population at equilibrium would be:

$p^2 = (0.4)^2 = 0.16$

$2pq = 2(0.6)(0.4) = 0.48$

$q^2 = (0.6)^2 = 0.36$

Population 2 is not at equilibrium for variants of gene A.

c. The combination of the two populations of lizards results in one population with the following allele frequencies:

A^F alleles

$A^F A^F$ (Pop 1)	38 × 2	76
$A^F A^S$ (Pop 1)	44 × 1	44
$A^F A^S$ (Pop 2)	80 × 1	80
		200

A^S alleles

$A^F A^S$ (Pop 1)	44 × 1	44
$A^F A^S$ (Pop 2)	80 × 1	80
$A^S A^S$ (Pop 1)	18 × 2	36
$A^S A^S$ (Pop 2)	20 × 2	40
		200

The allele frequencies in the new population are 200/400, or 0.5, for both p and q. **The genotype frequencies in the next generation will therefore be:**

$p^2 = (0.5)^2 = \mathbf{0.25}$

$2pq = 2(0.5)(0.5) = \mathbf{0.50}$

$q^2 = (0.5)^2 = \mathbf{0.25}$

11. a. The frequencies of the genotypes in the sailor population are: $MM = 324/400 = 0.81$, $MN = 72/400 = 0.18$ and $NN = 4/400 = 0.01$. **The frequency of the N allele = 0.01 + (1/2)(0.18) = 0.1.**

b. In order to calculate the allele and genotype frequencies in the children, you must calculate the allele frequencies of each original population and then combine them in the correct proportions. From part (a), in the sailor population the frequency of $N = 0.1$, so the frequency of $M = 0.9$. In the Polynesian population the allele frequencies are $N = 0.94$ and $M = 0.06$. When these two populations mix randomly, then 40% of the M and N alleles will come from the sailor population (400/1000) and the remaining 60% will be provided by the Polynesians. Therefore the frequency of N in the mixed population = (0.4)(0.1) + (0.6)(0.94) = 0.04 + 0.564 = 0.604. The frequency of M in the mixed population = (0.4)(0.9) + (0.6)(0.06) = 0.36 + 0.036 = 0.396. The genotype frequencies in the next generation will be: $(0.604)^2$ NN + $2(0.604)(0.396)$ MN + $(0.396)^2$ MM = 0.365 NN + 0.478 MN + 0.157 MM = 1. If there are 1000 children, then **478 children would be expected to have the MN genotype.**

chapter 20

c. The observed genotype frequencies in the children are 1000/1000 NN + 850/1000 MN + 50/1000 MM = 0.1 NN + 0.85 MN + 0.05 MM. The allele frequencies among the children are **N = 0.1 + (1/2)(0.85) = 0.525** and **M = 0.05 + (1/2)(0.85) = 0.475**.

12. a. When considering an X-linked gene, women have two alleles. Therefore the **allele and genotype frequencies in women are** calculated exactly as we have been doing for autosomal loci: $p + q = 1$ and $p^2 + 2pq + q^2 = 1$. However, men are hemizygous (they have only one allele for genes on the X chromosome), so the **men's genotype frequencies are the same as the allele frequencies: $p + q = 1$**.

   b. If 1/10,000 males is a hemophiliac, then the allele frequency of the mutant allele (q) $= 1 \times 10^{-4}$. Therefore the frequency of affected females $= q^2 = 1 \times 10^{-8}$. If there are 1×10^8 (one hundred million) women in the United States, then **only one of them should be afflicted with hemophilia**. (Note that we assume here that only females homozygous for the mutant hemophilia allele are true hemophiliacs. As discussed in Problem 4-40, rare heterozygous females show some symptoms of hemophilia.)

13. a. Remember that colorblindness is an X-linked recessive trait, so boys are hemizygous. Using the data from the boys only, the allele frequency for **C = 8324/9049 = 0.92** and for **c = 725/9049 = 0.08**.

   b. If the population is in Hardy-Weinberg equilibrium, then the girls should have the same allele frequencies as the boys. The genotype frequency of colorblindness (cc) in the girls = 40/9072 = 0.0044; if the girls are in equilibrium, then the allele frequency of $c = \sqrt{0.0044} = 0.066$. This value does not equal 0.08 [from part (a)].

   Alternately, if the population is in equilibrium, then the allele frequency of c in the boys should accurately predict the genotype frequencies in the girls. Thus c^2 should represent the frequency of colorblind girls: $(0.08)^2 = 0.0064$, but this value does not equal the cc genotype frequency of 0.0044.

   Therefore, given the information provided to this point, **this sample does not demonstrate Hardy-Weinberg equilibrium for the colorblindness gene**.

   c. Based on this new information, the frequency of the $c^p c^p$ genotype in girls = 3/9072 = 3.3×10^{-4} and **the frequency of $c^p = \sqrt{3.3 \times 10^{-4}} = 0.018$**. The frequency of the $c^d c^d$ genotype in the girls = 37/9072 = 0.0041, and **the frequency of $c^d = \sqrt{0.0041} = 0.064$**. The **frequency of the C allele** is $1 - (0.018 + 0.064) = $ **0.918**.

   d. The genotype frequencies among the boys must be the same as the allele frequencies in the girls calculated in part (c). **Thus in the boys, C = 0.918 (normal vision), c^d = 0.064 (colorblind) and c^p = 0.018 (colorblind). In the girls, the genotype frequencies are: CC = 0.843 (normal vision), Cc^d = 0.118 (normal vision), Cc^p = 0.033 (normal vision), $c^p c^p$ = 3.3×10^{-4} (colorblind), $c^d c^d$ = 0.004 (colorblind) and $c^d c^p$ = 0.002 (normal vision)**.

   e. These new results make it much more likely that **the population is in fact in equilibrium. As seen in part (c), the allele frequency of C is the same in boys and girls and the allele frequency of c in the boys is the same as the total frequencies of $c^d + c^p$ in girls**. Likewise, the frequencies of genotypes with normal vision (0.918 predicted in boys versus 0.92 observed in boys and 0.996 predicted in girls versus 0.996 observed in girls) and with colorblind vision (0.082 predicted in the boys versus 0.08 observed and 0.004 predicted in the girls versus 0.004 observed) fit the expectations of Hardy-Weinberg equilibrium.

14. a. If a gene has three alleles then $p + q + r = 1$. The genotype frequencies of a population at Hardy-Weinberg equilibrium would therefore be represented by the binomial expansion of the allele frequencies: $(p + q + r)^2 = (1)^2 = p^2 + 2pq + q^2 + 2pr + r^2 + 2qr = 1$. As discussed in the Problem Solving Tips at the beginning of this chapter, a Punnett square will give the same genotype frequencies:

	p	q	r
p	p^2	pq	pr
q	pq	q^2	qr
r	pr	qr	r^2

b. In the Armenian population, the allele frequencies are $I^A = 0.36$, $I^B = 0.104$ and $i = 0.536$. If the population is in Hardy-Weinberg equilibrium, then the binomial expansion of these allele frequencies gives: $0.130\ I^AI^A + 0.075\ I^AI^B + 0.011\ I^BI^B + 0.111\ I^Bi + 0.287\ ii + 0.386\ I^Ai = 1$. Therefore, the frequencies of the four blood types in this population are **0.516 A** ($0.130\ I^AI^A + 0.386\ I^Ai$), **0.122 B** ($0.011\ I^BI^B + 0.111\ I^Bi$), **0.075 AB** ($0.075\ I^AI^B$), **and 0.287 O** ($0.287\ ii$).

15. a. *AA* male × *AA* female: The frequency of matings within the population in which an *AA* male (genotype frequency p^2) will mate with an *AA* female (genotype frequency also p^2) with be $p^2 \times p^2 = p^4$.

b. *aa* male × *aa* female: $q^2 \times q^2 = q^4$.

c. *Aa* male × *Aa* female: $2pq \times 2pq = 4p^2q^2$.

d. *AA* homozygote × *aa* homozygote: There are two mutually exclusive ways in which an *AA* homozygote could mate with an *aa* homozygote. In the first, the *AA* homozygote is a male and the *aa* homozygote is a female; in the other, the sexes are reversed. Because these are mutually exclusive events, their individual probabilities must be summed to yield the desired probability. Thus, $(p^2 \times q^2) + (q^2 \times p^2) = 2p^2q^2$.

e. *AA* homozygote × *Aa* heterozygote: Again, two mutually exclusive events must be summed. These are: $(p^2 \times 2pq) + (2pq \times p^2) = 4p^3q$.

f. *aa* homozygote × *Aa* heterozygote: Again, two mutually exclusive events must be summed. These are: $(q^2 \times 2pq) + (2pq \times q^2) = 4pq^3$.

g. **Yes**, the six possibilities listed do account for all possible matings. You could demonstrate this in at least two ways. First, you could set up a Punnett square showing the possible matings between males and females of all possible genotypes:

chapter 20

♀↓ ♂→	AA (p^2)	Aa ($2pq$)	aa (q^2)
AA (p^2)	AA ♂ × AA ♀ $(p^2)^2 = p^4$	Aa ♂ × AA ♀ $2pq(p^2) = 2p^3q$	aa ♂ × AA ♀ $(q^2)(p^2) = p^2q^2$
Aa ($2pq$)	AA ♂ × Aa ♀ $(p^2)(2pq) = 2p^3q$	Aa ♂ × Aa ♀ $(2pq)^2 = 4p^2q^2$	aa ♂ × Aa ♀ $(q^2)(2pq) = 2pq^3$
aa (q^2)	AA ♂ × aa ♀ $(p^2)(q^2) = p^2q^2$	Aa ♂ × aa ♀ $2pq(q^2) = 2pq^3$	aa ♂ × aa ♀ $(q^2)^2 = q^4$

The Punnett square above shows that the 9 possible matings resolve into 6 combinations, all of which have been accounted for in parts (a-f).

A second method is to use numbers to verify that the 6 types of matings account for all possibilities. If we set $p = 0.2$, then $q = 0.8$. $AA = (0.2)^2 = 0.04$, $Aa = 2(0.2)(0.8) = 0.32$, and $Aa = (0.8)^2 = 0.64$. These three genotypic frequencies of course sum to 1.0 because they account for the three possible genotypes. Now we can figure out the probabilities of the 6 mating combinations using the equations determined in parts (a-f) and show that they account for all the mating possibilities because the probabilities also sum to 1.0:

$AA \times AA = p^4 = (0.2)^4 =$ 0.0016
$aa \times aa = q^4 = (0.8)^4 =$ 0.4096
$Aa \times Aa = 4p^2q^2 = 4(0.2)^2(0.8)^2 =$ 0.1024
$AA \times aa = 2p^2q^2 = 2(0.2)^2(0.8)^2 =$ 0.0512
$AA \times Aa = 4p^3q = 4(0.2)^3(0.8) =$ 0.0256
$aa \times Aa = 4pq^3 = 4(0.2)(0.8)^3 =$ <u>0.4096</u>
                                                                                                 1.0000

h. **The probability for the mating of two individuals of the same genotype is simply the product of their genotype frequencies. The probability for the mating of two individuals of different genotypes must be 2 times the product of their genotype frequencies** because there are two ways such individuals can mate (males of one genotype with females of the other genotype, and *vice versa*).

i. $AA ♂ \times AA ♀ = (p^2)(p^2) = p^4$. [Same as part (a).]
   $AA ♂ \times Aa ♀ = (p^2)(2pq) = 2p^3q$
   $AA ♂ \times aa ♀ = (p^2)(q^2) = p^2q^2$

16. a. The number of nontasters (*tt*) in the population is 1707 - 1326 = 381 individuals. If the population is at HWE, then $q^2 = 381/1707 = 0.223$ and $q = \sqrt{0.223} = 0.472$. $p = 1 - q = 1 - 0.472 = 0.528$.

   b. The genotype frequencies are: $TT = p^2 = (0.528)^2 = 0.279$. $Tt = 2pq = 2(0.528)(0.472) = 0.498$. $tt = q^2 = (0.472)^2 = 0.223$. [The genotype frequency of *tt* is also given by the fraction of nontasters in the population = 381/1707 as described in part (a).]

   c. **Non-taster × nontaster matings** = $(q^2)(q^2) = q^4 = (0.472)^4 = 0.050$

d. **Taster × nontaster matings = $2(p^2 + 2pq)(q^2) = 2(0.279 + 0.498)(0.223) = 0.347$**. Note from the answer to Problem 15 part (h) that you must multiply the product of the two genotype frequencies by 2 to account for the fact that matings between taster males and nontaster females are mutually exclusive of matings between nontaster males and taster females.

e. **Taster ♂ × nontaster ♀ matings = $(p^2 + 2pq)(q^2) = (0.279 + 0.498)(0.223) = 0.173$**.

f. Two types of taster ♂ × nontaster ♀ matings exist. These are TT ♂ × tt ♀ and Tt ♂ × tt ♀. The fraction of all possible matings that are TT ♂ × tt ♀ is $(p^2)(q^2) = (0.279)(0.223) = 0.062$. The fraction of all possible matings that are Tt ♂ × tt ♀ is $(2pq)(q^2) = (0.498)(0.223) = 0.111$. The fraction of all possible matings that are between taster males and nontaster females = $0.062 + 0.111 = 0.173$ [see also the answer to part (e)].

Of all the matings between taster males and nontaster females, $0.062/0.173 = 0.358$ are TT ♂ × tt ♀. None of the progeny of these matings will be nontasters. Of all the matings between taster males and nontaster females, $0.111/0.173 = 0.642$ are Tt ♂ × tt ♀. From these matings, half of the progeny will be tt nontasters. Thus, **among all the progeny produced by all matings between a taster male and a nontaster female, the proportion of nontasters will be $(0.5)(0.642) = 0.321$**.

g. **Taster × taster matings = $(p^2 + 2pq)(p^2 + 2pq) = (0.279 + 0.498)(0.279 + 0.498) = 0.604$**. You can check your work by showing that the sum of the answers to parts (c), (d), and (g) [nontaster × nontaster, taster × nontaster, and taster × taster] = $0.050 + 0.347 + 0.604 = 1$ (after accounting for rounding error).

17. a. Note that the gene for pattern baldness we are considering is autosomal; it is NOT sex-linked. Because the baldness mutation is dominant in men, bald men could be either AA or Aa. The nonbald men must be aa. The genotype frequency of aa (the nonbald men) = $1 - 0.51 = 0.49$. If the population is at HWE, then $q^2 = 0.49$ and $q = \sqrt{0.49} = $ **0.7** (frequency of the a allele for normal hair). $p = 1 - q = $ **0.3** (frequency of the A allele for baldness).

b. If the population is at HWE, the frequency for the baldness allele must be the same in males and females, so **$p = 0.3$**.

c. Because the baldness allele is recessive in women, **the percentage of women in the population with pattern baldness will be $p^2 = (0.3)^2 = 0.09 = 9\%$**.

d. The bald men are 51% (0.51) of the total men. The fraction of nonbald women will be $1 - 0.09 = 0.91$ of all women. **The proportion of all random matings that should be between a bald man and a nonbald woman should be $(0.51)(0.91) = 0.464 = 46.4\%$**.

e. The bald men are both AA homozygotes and Aa heterozygotes. The frequency of the AA homozygotes = $p^2 = (0.3)^2 = 0.09$. The frequency of the Aa heterozygotes is $2pq = 2(0.3)(0.7) = 0.42$. **The fraction of the total bald men who are Aa heterozygotes = $(0.42)/(0.09 + 0.42) = 0.824 = 82.4\%$**.

f. If a nonbald couple has a bald son, then the man must be aa and the woman must be Aa (because the A allele for baldness is recessive in women). From this type of

chapter 20

mating, half the progeny of either sex are Aa and half are aa. Thus, **the chance that the next son of this couple will be bald is 0.5.**

g. The woman is bald and therefore must be AA. Her daughter will be $A-$, with the unknown allele being either A or a. Because nothing is known about the father, the chance that the unknown allele will be A (and therefore **the chance that the daughter will be bald) is simply the allele frequency of the A allele in the population = p = 0.3.**

Section 20.2

18. If the deleterious effects of an allele (call it a) are completely recessive, the relative fitness values of the three genotypes will be $W_{AA} = W_{Aa} = 1$, while $1 \geq W_{aa} \geq 0$. Under such conditions, Δq, the decrease in the frequency of the deleterious allele a in one generation, will itself decrease over time. The lower the frequency of the a allele, the lower will be the frequency of aa homozygotes, and thus the less chance exists that natural selection can lower the frequency of the allele in the next generation. **In a very large population,** these facts mean that **natural selection alone can never cause extinction of the deleterious recessive allele; its frequency will decrease exponentially over successive generations, but will never reach a value of 0** (see Fig. 20.12 on p. 675). **In a small population, genetic drift will eventually eliminate deleterious alleles** because a substantial chance exists that simply through random sampling error, the gametes used to form the next generation will not include the deleterious allele once its frequency is reduced by natural selection (see Fig. 20.13 on p. 676).

19. On Tristan de Cunha in the 1960s, the frequency of the retinitis pigmentosa phenotype was $4/240 = 0.017$, whereas the frequency of this trait in Britain (from where the settlers of this island originated) is $\sim 1/6000 = 0.00017$, about 100-fold higher. **The high incidence of retinitis pigmentosa on Tristan de Cunha is likely due to three components all related to the history of the population and its small size. The first component was a founder effect:** at least one of the 15 original colonists in 1814 must have carried an allele for this condition, even though this allele was rare in Britain. **The second component was a population bottleneck** that occurred in 1885 when most of the males on the island were lost in a shipwreck. **The third component was genetic drift** due to the fact that the population on Tristan de Cunha always remained small.

The problem did not specify the nature of the allele causing retinitis pigmentosa. Some forms of this disease are known to be autosomal dominant, others are autosomal recessive, and yet others are sex-linked recessive. It turns out that the form on Tristan de Cunha is autosomal recessive. We can thus estimate the frequencies of the disease-causing allele in Britain and Tristan de Cunha during the 1960s by taking the square root of the frequencies of the affected individuals in these populations. For Britain, $q = \sqrt{0.00017} = 0.013 = 1.3\%$; for Tristan de Cunha, $q = \sqrt{0.017} = 0.13 = 13\%$.

20. a. When $p = 0.5$ and $N = 100{,}000$, the confidence interval for p in the next generation is **0.5 ± 0.0022.**

    b. When $p = 0.5$ and $N = 10$, the confidence interval for p in the next generation is **0.5 ± 0.22.**

c. **The results of parts (a) and (b) demonstrate the mathematical basis for the effects of population bottlenecks on genetic drift.** If the population is large, then in the next generation, it is highly unlikely that the allele frequency will change substantially. In the example given, in part (a), where $p = 0.5$ and $N = 100,000$, the chance is 95% that in the next generation, p will be between 0.4978 and 0.5022. But under the conditions of part (b), where the allele frequency is also 0.5 but $N = 10$, the chance is 95% that p in the next generation will be between 0.28 and 0.72. Thus, if a catastrophe occurs to lower the population size drastically, the chances are high that the allele frequency will begin to fluctuate dramatically. The equation provided in this problem allowed mathematicians to perform the Monte Carlo modeling shown in <u>Fig. 20.8</u> on p. 671.

21. a. If the population size $N = 100,000$, then the number of total alleles for an autosomal locus = $2N$ = 200,000. **If a new mutation occurs at this locus, its original allele frequency p will be 1/200,000 = 5×10^{-6}.**

    b. If the population size $N = 10$, then the number of total alleles for an autosomal locus = $2N$ = 20. **If a new mutation occurs at this locus, its original allele frequency p will be 1/20 = 0.05.**

    c. **The new mutant allele has a much higher probability of going to fixation by chance with genetic drift in the small population with $N = 10$.** In the small population, the new allele has a 0.05 probability of being fixed eventually in the population, while there is a 0.95 probability that it will be lost. In the large population, the new allele has a 0.000005 probability of being fixed eventually in the population, while the probability that it will be lost is 1 − 0.000005 = 0.999995. Note that even if a new mutation occurs in a small population, the chances are still considerably higher that it will be lost than that it will eventually go to fixation.

22. a. Convert the genotypes to frequencies: 60/150 t^+t^+ = 0.4 t^+t^+ and 90/150 t^+t = 0.6 t^+t. The **allele frequencies are: t^+ = 0.4 + (1/2)(0.6) = 0.7 and t = (1/2) (0.6) = 0.3**.

    b. First determine the frequencies of the three genotypes if all lived, then remove the inviable mice from your calculations and normalize the genotype frequencies in order to calculate the allele frequencies. The expected genotypes frequencies of the zygotes in the next generation are: 0.49 t^+t^+ + 0.42 t^+t + 0.09 tt = 1. However the tt zygotes die, so the remaining genotype frequencies are: 0.49 t^+t^+ + 0.42 t^+t = 0.91. When this equation is normalized (that is, when both sides are divided by 0.91 to make the equation equal 1), it becomes 0.538 t^+t^+ + 0.462 t^+t = 1. If 200 progeny mice are scored, the expected values are **108 normal mice and 92 tailless mice**.

    c. Pop 1 has 64 members and the genotype frequencies are 0.25 t^+t^+ and 0.75 t^+t. Thus the allele frequencies are t^+ = 0.625 and t = 0.375. Pop2 has 84 members and the genotype frequencies are 0.571 t^+t^+ and 0.429 t^+t. The allele frequencies here are t^+ = 0.785 and t = 0.215. If the populations interbreed randomly, then the Pop1 parents will provide 64/148 = 0.432 of the alleles (gametes) found in the next generation and Pop 2 will provide 0.568 or 56.8% of the alleles. Therefore the combined gamete frequencies are: t^+ = 0.432(0.625) + 0.568(0.785) = 0.716 and t = 0.432(0.375) + 0.568(0.215) = 0.284. The genotype frequencies in the next generation will be

$(0.716)^2$ t^+t^+ + $2(0.716)(0.284)$ t^+t + $(0.284)^2$ tt = 1. Of course the tt zygotes die [see part (b) above], so the normalized genotype frequencies of the two viable genotypes are **0.558 t^+t^+ and 0.442 t^+t**. {0.558 = $(0.716)^2/[(0.716)^2 + 2(0.716)(0.284)]$, while 0.442 = $[2(0.716)(0.284)]/[(0.716)^2 + 2(0.716)(0.284)]$}.

23. a. Diagram the cross:

    vg^+vg^+ × $vg\ vg$ → F_1 vg^+vg → F_2 1/4 vg^+vg^+ : 1/2 vg^+vg : 1/4 $vg\ vg$ (3 wild type : 1 vestigial). If the vestigial F_2 flies are selected against, then the remaining F_2 genotypes are 1/3 vg^+vg^+ and 2/3 vg^+vg. Therefore **the genotype frequencies in the F_2 are 0.33 vg^+vg^+ and 0.67 vg^+vg; the allele frequencies in the F_2 for vg^+ = 0.33 + (1/2)(0.67) = 0.67(p) and for vg = (1/2)(0.67) = 0.33 (q)**.

    b. The expected **genotype frequencies in the F_3 progeny are [p^2 = $(0.67)^2$ = 0.449 vg^+vg^+] + [$2pq$ = $(2)(0.67)(0.33)$ = 0.442 vg^+vg] + [q^2 = $(0.33)^2$ = 0.109 $vg\ vg$] = 1, or 0.891 wild type and 0.109 vestigial**.

    c. If the F_3 vestigial flies are selected against then the altered F_3 genotypic ratio becomes 0.449 vg^+vg^+ + 0.442 vg^+vg = 0.891. In order to calculate allele frequencies this equation must be normalized (see <u>Problem 19-22b</u>), becoming 0.504 vg^+vg^+ + 0.496 vg^+vg = 1. The F_3 allele frequencies: vg^+ = 0.504 + 1/2 (0.496) = 0.752 and vg = 1/2 (0.496) = 0.248. The genotype frequencies in the F_4 generation will be: 0.566 vg^+vg^+ + 0.373 vg^+vg + 0.062 $vg\ vg$ = 1. The genotype frequencies in the F_4 generation will be: 0.566 vg^+vg^+ + 0.373 vg^+vg + 0.062 $vg\ vg$ = 1. Therefore the **F_4 allele frequencies are vg^+** = 0.566 + 1/2 (0.373) **= 0.753 and vg** = 0.062 + 1/2 (0.373) = **0.247**.

    d. If all of the F_4 flies are allowed to mate at random, then there is no selection and **the population will be in Hardy-Weinberg equilibrium**. Therefore the F_5 genotype and allele frequencies will be the same as those in the F_4 generation in part (c) above: **0.566 vg^+vg^+ + 0.373 vg^+vg + 0.062 $vg\ vg$ = 1; vg^+ = 0.753 and vg = 0.247**.

24. a. *A1* corresponds to the *blue* line. The beneficial effects of the *A1* allele on fitness are recessive, so natural selection favors homozygotes for this allele but not heterozygotes. The effects of selection will be observed only after many generations when the allele frequency increases by drift sufficiently for homozygotes to be formed. *B1* **corresponds to the** *red* **line**. In this scenario, the beneficial effects of the *B1* allele on fitness are dominant, and the *B1* allele has an immediate strong selective advantage relative to the alternative *B2* allele. *C1* **corresponds to the** *green* **line**. In this scenario, the beneficial effects of the *C1* allele on fitness are dominant and thus immediate, but the fitness advantage of the *C1* allele is only marginally higher than that of the alternative *C2* allele.

    b. **Both the *B1* (*red*) and the *C1* allele (*green*) have dominant beneficial effects, but the *B1* allele's fitness relative to the alternative *B2* allele is much higher than the *C1* allele's fitness relative to its alternative *C1* allele. Thus, the frequency of the *B1* allele in the population will increase over the generations at a faster rate than will the frequency of the *C1* allele**.

c. **Because the beneficial effects of the *A1* allele on fitness are recessive, only *A1A1* homozygotes will have a selective advantage.** When the frequency of the *A1* allele is low, few individuals in the population will be *A1A1* homozygotes. Many generations must elapse before the frequency of the *A1* allele becomes sufficiently high (through genetic drift) so that any such *A1A1* homozygotes can be generated. In contrast, the fitness of individuals with even a single copy of *B1* or *C1* is higher than those individuals lacking *B1* or *C1*, so natural selection will immediately favor an increase in the frequency of these alleles.

d. (i) **In a small population, the operation of genetic drift means that some likelihood exists that the alleles in question would become extinct before they would become fixed.** This is particularly true for the *blue* line, because natural selection would favor increases in the frequency of *A1* only when *A1A1* homozygotes were present. Because of genetic drift, it is highly likely that a newly formed *A1* allele would be lost before homozygotes are formed. In the cases of the *B1* and *C1* alleles with dominant beneficial effects, it is still possible that newly formed alleles would be lost from the population by genetic drift despite the fitness advantages conferred by one dose of the alleles.

(ii) **Because of genetic drift in a small population, the curves would no longer be smooth and predictable.** Instead, they would be jagged and their trajectories would be unpredictable, as seen in Fig. 20.13 on p. 676.

25. a. The allele frequency of $b = \sqrt{0.25} = 0.5$, so the allele frequency of $B = 0.5$.

   b. To calculate Δq, first determine q in both generations. For all three tanks, $q = 0.5$ [see part (a)]. For tank 1, q in the next generation (q') = $\sqrt{0.16} = 0.4$. Therefore Δq for tank 1 is $q' - q = 0.4 - 0.5 = -0.1$. The same calculations are carried out for the other two tanks: q for all tanks = 0.5; q' for tank 2 = 0.5 so $\Delta q = 0$; q' for tank 3 = 0.55 and $\Delta q = 0.05$. These calculations are summarized in the table that follows:

	Tank 1	Tank 2	Tank 3
q in Generation 1	0.5	0.5	0.5
q in the next generation	$\sqrt{0.16} = 0.4$	$\sqrt{0.25} = 0.5$	$\sqrt{0.30} = 0.55$
(b) Δq	-0.1	0.0	0.05
(c) W_{Bb}	1.0	1.0	1.0
(d) W_{bb}	<1.0	1.0	>1.0

   c. If the fitness of the *BB* genotype = 1 ($W_{BB} = 1$) and the *b* allele is totally recessive, then the fitness of the *Bb* genotype (W_{Bb}) = 1 also [see row (c) in the table above]. (This conclusion assumes that the effects of the alleles of the *B/b* gene on fitness are determined exclusively through the size of the tail.)

   d. In all three tanks, the original frequency of the *b* allele (q) = 0.5. **In tank 1** the frequency of *b* in the progeny (q') = 0.4, so the frequency of the *bb* (small tail) males has decreased. The fitness of the small-tailed males decreased, so W_{bb} <1.0. **In tank 2**, $q' = 0.5$ (the *b* allele frequency remained the same), so $W_{bb} = 1.0$. **In tank 3** $q' = 0.55$ (the *b* allele frequency increased) so $W_{bb} > 1.0$. We can come to the same conclusion in a more mathematical way by using Eqn. 20.6 on p. 675 to estimate the relative fitness of the *bb* genotype given that we concluded in part (c) that $W_{BB} = W_{Bb} = 1$.

26. a. The affected genotype dies before reproductive age. Therefore the fitness value $W_{CF^- CF^-} = 0$ for the affected genotype. Carriers have no fitness advantage relative to individuals with homozygous normal genotypes, so for both of these the fitness values $W_{CF^+ CF^+}$ and $W_{CF^+ CF^-} = 1$.

    b. **The average fitness at birth of the population with respect to the cystic fibrosis trait** $\overline{W} = p^2 W_{CF^+CF^+} + 2pq W_{CF^+CF^-} + q^2 W_{CF^-CF^-} = (0.96)^2 \times 1.0 + 2(0.96)(0.04) \times 1.0 + (0.04)^2 \times 0 = \mathbf{0.9984}$. (See Eqn. 21.4a on p. 674.)

    To calculate Δq use Eqn. 20.6 on p. 675:
    $$\Delta q = \frac{pq[q(W_{aa} - W_{Aa}) - p(W_{AA} - W_{Aa})]}{\overline{W}}$$

    $p = 0.96$, $q = 0.04$, $W_{CF^+CF^+} = W_{CF^+CF^-} = 1$, $W_{CF^-CF^-} = 0$, $\overline{W} = 0.9984$. After substituting these values, $\Delta q = \mathbf{-1.54 \times 10^{-3}}$.

    c. This problem can be solved using Eqn. 20.8 on p. 678:
    $$q_e = \frac{W_{AA} - W_{Aa}}{(W_{aa} - W_{Aa}) + (W_{AA} - W_{Aa})}$$

    Here, $q_e = 0.04$, $W_{CF^+CF^-} = 1.0$ (because this genotype is presumed to be the most fit), and $W_{CF^-CF^-} = 0$. After plugging in these values, $0.4 = \{(W_{CF^+CF^+} - 1)/[(0-1) + (W_{CF^+CF^+} - 1)]\}$, so $0.4 = (W_{CF^+CF^+} - 1)/(W_{CF^+CF^+} - 2)$, and $W_{CF^+CF^+} = \mathbf{0.958}$.

    d. Interestingly, one recent study suggests the hypothesis that **CF^+/CF^- heterozygotes may be better able to survive outbreaks of cholera**. This phenomenon is conceivable because people with cholera have diarrhea that pumps water and chloride ions out of the small intestine. The CFTR protein encoded by the *CF* gene is a chloride ion channel. CF^+/CF^- heterozygotes thus lose less water than CF^+/CF^+ individuals when infected with cholera. Theoretically, the heterozygotes are therefore less likely to die of dehydration.

27. **The equilibrium frequency of the fava bean sensitivity allele (q_e) of the *G6PD* gene will be higher in the tropical regions of Africa in which malaria is endemic, compared with countries in North America in which malaria is absent.** The major factors affecting the equilibrium frequency are the relative fitness values of individuals of the three genotypes (homozygotes for the normal allele, heterozygotes, and homozygotes for the fava bean sensitivity allele). If a heterozygote advantage exists, then the fitness of the heterozygotes is higher than that of either homozygote; if the relative fitnesses of the homozygotes are known, then the equilibrium frequency (q_e) can be estimated from Eqn. 20.8 on p. 678.

    The basis of the phenomenon conferring (in malaria endemic regions) fitness advantage to individuals heterozygous for normal and fava bean sensitivity alleles of the *G6PD* gene is similar to the reason that the sickle cell allele of the *Hbβ* genes is maintained at high frequency in the same areas of the globe. The malaria parasite reproduces in red blood cells, so if these cells are "sick", then the parasite cannot grow as well. It should be said that fava beans are not the only trigger for hemolytic crises in individuals homozygous for the sensitivity allele, as other foods, chemicals and various diseases also induce hemolysis in such people.

28. **Evolutionary geneticists monitor selectively neutral polymorphisms as molecular clocks because many such polymorphisms exist and the rate at which they accumulate in genomes is relatively constant and predictable regardless of where they are located.** If a new mutation was not selectively neutral, and instead conferred a reproductive advantage or disadvantage, the rate at which it would be driven to fixation would not represent random events. Instead, the rate would largely reflect the relative fitnesses of individuals with particular genotypes, and this property is much more difficult to assess without additional knowledge.

Section 20.3

29. **African populations show much greater average DNA sequence diversity than do populations in any other part of the world.** The lineages of the most divergent current-day Africans coalesce to a branch point further back in history than the branch point for the divergence of any two populations in any regions of the world other than Africa.

    One way to think about these concepts is that humans existed in Africa for a long time (between approximately 200,000 to 60,000 years ago) before any people migrated out of Africa. During this period, many polymorphisms accumulated in various sub-populations of humans. About 60,000 years ago, people from only one or a few sub-populations of Africans left Africa to populate other parts of the world; these emigrants had only a subset of all the polymorphisms that existed at that period of time (review Fig. 20.20 on p. 683).

30. **The discovery of men with Y chromosomes that have sequences highly divergent from those of all Y chromosomes that were previously analyzed would cause scientists to calculate that the MRCA for the human Y chromosome lived on earth further back in time than estimated previously.** The MRCA for the Y chromosome, sometimes called the "Y-chromosome Adam," was not the only male living at his time, but instead he is the person from whom all current-day males have inherited their Y chromosomes in an unbroken line of patrilineal descent. Most previous estimates date the Y-chromosome Adam to between 120,000 to 200,000 years ago; this new discovery would place the date further back in history.

31. a. At k generations in the past, you would have had 2^k biological ancestors. Thus, 40 generations in the past, you would have had 2^{40} = **1.1 x 10^{13} potential ancestors.**

    b. 1.1 x 10^{13} potential ancestors, the answer for part (a), is many more than the number of people who are currently present on the earth (approximately 7 x 10^9), and is in fact several orders of magnitude larger than of people who have ever inhabited the earth. The explanation for this discrepancy is that the 2^k calculation is for the number of *potential* ancestors at any generation k, and this calculation also requires the population to be of infinite size so that the chances of matings between closely related individuals is infinitesimally small. Clearly, **many of your ancestors in** many previous generations must have been closely related to each other. This makes sense particularly during the large expanses of time when the human population was small and consisted of small tribal units with a high level of inbreeding.

32. Dion's mitochondrial genome is inherited matrilineally, so he obtained this DNA from his mother, in turn from her mother (his maternal grandmother), etc. The colors in Fig. 20.17 on p. 680 suggest that **Dion inherited his mtDNA from a great-grandmother who lived in Brazil in South America.**

His Y chromosome is inherited in a patrilineal fashion, so **he obtained his Y chromosome DNA from a great-grandfather who lived in West Africa, roughly in Nigeria.**

Dion's autosomes are pastiches of regions inherited from 8 different great-grandparents who lived in 4 different areas of the globe.

33. Figure 20.18b from p. 681 is reproduced below, with the four nodes circled and numbered in *blue*. Note that each node corresponds to positions at which one individual represented by a vertical line at the top of the node had two progeny leading to separate lineages.

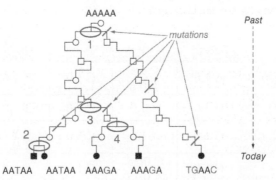

The predicted sequences at the nodes are: **Node 1 – AAAAA; Node 2 – AATAAA; Node 3 – AAAAA; and Node 4 – AAAGA.**

34. a. **A – *Pan troglodytes* (chimpanzees); B – *Homo neanderthalensis*; C – *Homo sapiens* (European Danish); D/E – *Homo sapiens* (Asian Uighurs/Native American Hopi; these two possibilities are equivalent in the cladogram); F – *Homo sapiens* (African Bantu).** The cladogram reflects the evolution of hominins and the history of human migrations on the earth (Figs. 20.19a on p. 683 and 20.22 on p. 685).

The earliest branch point led to the separation of lineages leading to chimpanzees and humans; the last time we shared a common ancestor with chimps was roughly 5 million years ago.

The branch leading to the genus *Homo* then split into separate lineages for Neanderthals (who died out about 30,000 years ago) and modern humans; according to Fig. 20.22, this split occurred between 800,000 and 500,000 years ago.

Modern humans first existed in Africa about 300,000-200,000 years ago, and some of these modern humans left Africa about 80,000-60,000 years ago. Most of these emigrants went first to the Middle East. About 40,000-30,000 years ago, some people from the Middle East and settled in Europe, while others went first to Asia and from there (about 20,000 years ago) to the Americas.

Some humans in Europe and the Middle East mated with Neanderthals they encountered at these locations, before that group of hominins died out; descendants of the progeny of those matings retain alleles of some genomic regions that they obtained from their Neanderthal ancestors.

b. **The divergence indicated with the arrow corresponds to the migration(s) leading some groups of humans out of Africa to other parts of the globe, 80,000-60,000 years ago.** After this migration, little interbreeding took place between the people remaining in Africa and those who had emigrated, so two separate lineages emerged.

c. **See the following table.** The answers are highlighted in *red* at the bottom. As an example of the reasoning involved, consider SNPα. The G allele is found in all of the chimpanzee genomic DNA samples, as well as all of the modern-day human samples. The one exception is DNA obtained from a Neanderthal skeleton. These facts suggest that G was the ancestral allele of this SNP, and that a mutation to the A allele occurred only in the Neanderthal lineage but in no other lineages (that is, this mutation is #3 in the cladogram).

	SNP α	SNP β	SNP γ	SNP δ	SNP ε	SNP ζ
H. neanderthalensis	A 1.0	C 0.4 A 0.6	C 1.0	T 1.0	T 1.0	T 1.0
P. troglodytes	G 1.0	A 1.0	C 1.0	T 1.0	G 1.0	T 1.0
African Bantu	G 1.0	A 1.0	C 1.0	A 0.7 T 0.3	T 1.0	C 1.0
European Danes	G 1.0	C 0.25 A 0.75	C 1.0	T 1.0	T 1.0	C 1.0
Native American (Hopi)	G 1.0	C 0.8 A 0.25	A 1.0	T 1.0	T 1.0	C 1.0
Asian Uighurs	G 1.0	C 0.5 A 0.5	A 0.6 C 0.4	T 1.0	T 1.0	C 1.0
Number (1-10)	3	4	8	6	1 or 2	5
Ancestral allele	G	A	C	T	Can't tell	T

35. a. **You would expect much more genetic variation in population samples of humans from Africa than in samples from Oceanic populations.** Modern humans have a much longer history in Africa than they do in Oceania, and only a subgroup of the people who lived in Africa emigrated to Oceania.

b. **The fact that Denisovan-derived variants are found among modern-day humans almost exclusively in Southeast Asia and Oceania suggests that**

chapter 20

humans encountered and mated with Denisovans in the far East during human migrations through these areas. People in Africa, Europe, and the Americas would have little or no Denisovan DNA because the ancestors of people in these areas did not encounter any Denisovans during their migrations. Presumably, Denisovans either never inhabited these regions or they died out in these regions before *Homo sapiens* arrived there.

Chapter 21

Genetics of Complex Traits

Synopsis

For many traits, multiple genes and also the environment determine phenotype. Such traits are called *quantitative* or *complex traits*, and the extent to which any individual in a population displays the trait is called that individual's *trait value* or *phenotype value*. We learned in this chapter that the fraction of a complex trait phenotype controlled by genes (as opposed to the environment and chance) is called the *heritability* of that trait. Heritability can be measured in studies where the trait values of parents and offspring are compared with the population as a whole, or in studies that compare the differences in the trait values of MZ and DZ twins. Heritability has practical implications, as a phenotype can change (evolve) as a result of genome mutation only to the extent that genes control the trait.

The genes that control complex traits – *quantitative trait loci* or *QTLs* – can be identified through association (linkage) of molecular markers with the phenotype in question. In plants, QTLs can be identified through controlled crosses. In humans, QTLs are identified in genome-wide association studies (GWASs), in which millions of SNPs are tested for association with a particular quantitative trait.

Key terms

complex traits – phenotypes that can be influenced by many factors including multiple genes and the environment

quantitative traits – complex traits where the phenotype can be measured numerically. The numbers assigned to different amounts of the trait are called **phenotypic values** or **trait values.**

normal distribution – a set of data points that scatter around a central *mean* (average) without left or right bias; an equal number of data points fall below and above the mean. Normally distributed data often form a *bell curve* graph. The phenotypic values of quantitative traits are often normally distributed.

total phenotype variance (V_P) – describes the variation in phenotypic values for a given trait in a specific population. The number V_P is calculated as the average squared difference between each individual's phenotypic value and the mean phenotypic value. $V_P = V_E + V_G$ (see below)

environmental variance (V_E) – describes the fraction of the total phenotype variance due to differences in environment as opposed to genetic differences. A numerical value for V_E can be determined experimentally for a specific population in a particular environment by measuring total phenotype variance (V_P) among a population of genetically identical individuals in a variable environment: Under these conditions, $V_P = V_E$.

genetic variance (V_G) – describes the fraction of the total phenotype variance due to differences in genes as opposed to environmental influences. A numerical value for V_G can be measured in an experiment where genetically different individuals inhabit a constant environment: Under these conditions, $V_P = V_G$.

components of genetic variance (V_A, V_D, V_I) – Genetic variance is due to the *additive effects* (V_A) of the contributions of different alleles to the trait value, the effects of *dominance interactions* (V_D) between different alleles of a single gene, and the effects of *gene interactions* (V_I; interactions between alleles of different genes): $V_G = V_A + V_D + V_I$.

heritability – the fraction of total phenotypic variability of a trait due to genetic differences, which can be approximated as V_G/V_P. Heritability is measured as the correlation between the mean trait values of genetically related individuals relative to the entire population. When those individuals do not share all of their alleles of all of their genes (for example when comparing parents and their children), what is measured is **narrow sense heritability (h^2)**, where $h^2 = V_A/V_P$. Because the effects of dominance interactions (V_D) and gene interactions (V_I) are randomized in this case, only the variation due to the additive contributions of alleles of a gene (V_A) is measured. **Broad sense heritability (H^2)** is measured when correlating the mean trait values of individuals with identical genomes (for example, identical twins). In this case, all three components of V_G contribute to the correlation observed, and so $H^2 = V_G/V_P$. When h^2 (or H^2) = 0, all of the observed variation in phenotypic values (V_P) is due to environmental influences (V_E); when h^2 (or H^2) = 1, all observed phenotypic variation for a specific trait in the population is due to genetic differences V_A (or V_G).

genetic relatedness – the average fraction of all alleles that two individuals share because the inherited them from a common ancestor. For example, the genetic relatedness of siblings is 0.5, of an uncle and a nephew is 0.25, and of first cousins 0.125.

midparent value – for a given trait, the mean phenotypic value in two parents

line of correlation – a straight line that relates the midparent values to progeny trait values. The slope of the line of correlation is the narrow sense heritability of the trait (h^2).

concordance – the extent of correlation in trait values between different individuals

cross-fostering – In heritability studies with animals, parents and progeny are sometimes mixed randomly to minimize the effects of environmental differences (parental care) on trait values.

truncation selection – selective breeding where only individuals with particular trait values become the parents of the following generation. The difference in average trait value between the selected parents and the original population is called the *selection differential* (S), and the *response to selection* (R) is the difference in the average trait value of the offspring and the original population. R/S is called the *realized heritability* (h^2). Because $R = h^2 S$, selective breeding makes sense as way to alter average trait values only if the heritability of the trait is high.

quantitative trait loci (QTLs) – genes that contribute to the phenotypic values of complex traits

chapter 21

direct QTL mapping – localizing QTLs by setting up specific experimental crosses in order to find molecular markers that correlate with specific trait values

isogenic lines – strains that are globally homozygous for alleles of all of their genes

congenic lines or **nearly isogenic lines (NILS)** – a group of isogenic strains that are genetically identical with one another except for a short region of the genome called an **introgression** that differs between any two lines

association mapping – identification of molecular markers linked to QTLs by finding a correlation between a particular allele of the marker and the quantitative trait in a population

haplotype – a particular combination of linked alleles

linkage equilibrium – when particular linked alleles associate with one another randomly

linkage disequilibrium (LD) – when particular linked alleles associate with each other at significantly higher frequency than would be expected by chance

LD blocks – stretches of genomic DNA within which clusters of linked alleles display linkage disequilibrium. Such blocks of DNA sequence are often flanked by recombination hotspots.

GWAS (Genome-Wide Association Study) – analysis of the entire genome of many individuals in a population to identify molecular markers (usually SNPs) that correlate with a specific phenotype. Such SNPs are linked to the QTLs for that trait.

Manhattan plot – display of GWAS data that resembles the Manhattan skyline. Each SNP in the genome (x-axis) is plotted against $-\log_{10} p$ on the y-axis, where p is the probability that the association of the SNP and the quantitative trait in question would be observed by chance.

allelic odds ratio – the increased risk of having a particular phenotype (usually a disease) conferred by having a particular allele of a molecular marker

genotypic odds ratio – the increased risk of having a particular phenotype (usually a disease) conferred by homozygosity or heterozygosity for particular alleles of a molecular marker associated with that trait

Problem Solving

Heritability

Heritability may be estimated experimentally.

- **By comparing parents and progeny:**

$$h^2 = |\bar{x} \text{ (progeny)} - \bar{x} \text{ (original population)}| \,/\, |\bar{x} \text{ (parents)} - \bar{x} \text{ (original population)}|$$

Note: $0 < h^2 < 1$; $\bar{x}$ = mean trait value, $\bar{x}$ (parents) = midparent value; | | is absolute value; the equation shown assumes that the denominator is not zero.

chapter 21

- **By comparing monozygotic (MZ) and dizygotic (DZ) twins:**

$H^2 = 2 \, (MZ_{concordance} - DZ_{concordance})$

Note: $0 < H^2 < 1$; concordance = fraction of twins that share the trait.

GWAS

- **χ^2 calculations in Manhattan plots.** The χ^2 test performed to test the significance of association of a SNP with a complex trait is different than the χ^2 test for "goodness of fit" that you used in Chapter 5 for analysis of testcrosses. GWAS requires a χ^2 test for "independence"; the null hypothesis is that the distribution of the SNP alleles is independent of the complex trait phenotype. The reason that the calculation is different for a GWAS is that we cannot assign Expected (E) values the same way that we did in the "goodness of fit" test for a testcross because we do not know the frequency of each molecular marker allele in the population. A *contingency table* allows the assignment of E values under these conditions, using the observed data. Problem 20(b) takes you through this procedure.

- **Significant associations in Manhattan plots.** The cut-off *p*-value for significance of associations in GWAS is $p < [0.05/(\text{number of SNPs})]$. For example, if the GWAS tested 1,000,000 SNPs, significant associations have *p*-values in the χ^2 test for independence of $< 5 \times 10^{-8}$.

- **Allelic Odds Ratio.** After you determine that a particular allele of a molecular marker is associated (significantly) with a complex trait (usually a disease), you can determine the increased risk of having the disease conferred by that particular associated allele. In the equation that follows, "Cases" are individuals with the disease, "Controls" are individuals that lack the disease, and the aSNP is the allele of the SNP associated with the trait.

$$\text{Allelic Odds Ratio} = \frac{(\text{\# Cases with aSNP}/\text{\# Cases without aSNP})}{(\text{\# Controls with aSNP}/\text{\# Controls without aSNP})}$$

- **Genotypic Odds Ratio.** You can also determine how a particular diploid genotype for a SNP associated with a trait (usually a disease) affects the chance that a person will have the disease. The genotypic odds ratio can be determined for any pair of genotypes. For example, suppose SNP^1 and SNP^2 are alternate alleles of a SNP locus and SNP^1 is associated with the disease. We can calculate the increased risk of having the disease conferred by the genotype $SNP^1 \, SNP^1$-relative to $SNP^1 \, SNP^2$ as follows (Cases and Controls are defined as above):

$$\text{Genotypic Odds Ratio} = \frac{(\text{\# Cases } SNP^1 \, SNP^1 / \text{\# Cases } SNP^1 \, SNP^2)}{(\text{\# Controls } SNP^1 \, SNP^1 / \text{\# Controls } SNP^1 \, SNP^2)}$$

chapter 21

Vocabulary

1.
   a. isogenic lines — 4. homozygous for all genomic regions
   b. QTL — 5. genes contributing to complex traits
   c. response to selection — 7. measure of evolution
   d. association mapping — 9. takes advantage of recombination over the course of a population's history
   e. MZ twins — 6. identical
   f. DZ twins — 1. fraternal
   g. congenic lines — 10. contain introgressions
   h. linkage disequilibrium — 2. blocks of association between variants at different loci
   i. heritability — 3. proportion of total phenotypic variance attributed to genetic variance
   j. genetic relatedness — 8. 0.5 for siblings

Section 21.1

2. a. Genetically identical dandelions ($V_G = 0$) grown in the controlled environment of a greenhouse (*green curve* in the following diagram) will vary in phenotype less than either genetically identical plants grown in a variable environment (Fig. 21.5a; *blue curve*) or genetically different plants grown in a constant environment (Fig. 21.5b; *red curve*):

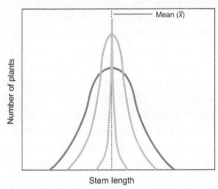

   b. The phenotypic variance is not zero ($V_P > 0$) because even in the greenhouse, the environment is not perfectly uniform; different plants will experience slightly different microenvironments.

   c. To refine the estimate of V_G obtained from the *red curve* in Fig. 21.5b [or the *red curve* in the diagram in part (a)], subtract from it the variance observed in the experiment described in part (a), depicted as the *green curve*.

chapter 21

3. a. Genetic clones have identical alleles of all of their genes, like identical (MZ) twins. **All of the phenotypic variation observed in genetic clones is due to environmental effects.** For any specific trait, $V_G = 0$ and thus $V_P = V_E$.

    b. MZ twins are genetically identical (their genetic relatedness = 1), while DZ twins have half of their alleles in common (their genetic relatedness = 0.5). **All of the phenotypic differences observed in separately adopted MZ twins are due to environmental effects.** Thus in MZ twins reared apart, $V_P = V_E$ for a particular trait. Note, however, that the value of V_E obtained in such a study may underestimate the effect of the environment on this trait in the population as whole. The reason is that the families who adopt each twin are likely more similar than two random families.

    Comparisons of MZ and DZ twins each raised in the same family can be used to estimate the heritability ($H^2 = V_G/V_P$) of specific traits. For a given phenotype, H^2 is equal to the twice the difference in concordance between MZ and DZ twins for that trait.

    c. **In studies of heritability of traits in animals, cross-fostering helps to randomize the effects of the environment among offspring of different parents.** The use of phenotypic similarity between parents and offspring as a measure of heritability assumes that all parents and offspring in the population experience the same environment.

4. a. **False.** When $H^2 =1$, all phenotypic variation is due to genes, and MZ twins are twice as similar phenotypically as are DZ twins. The reason is that MZ twins share twice as many alleles with each other (100%) as DZ twins do (50%). As H^2 decreases, MZ and DZ twins become more and more similar.

    b. **False.** It is true that MZ twins will show very little phenotypic variation if heritability is high. The phenotypic variability between DZ twins will always be higher than that of MZ twins unless $H^2 =0$. However, even when $H^2 =1$, little variation could exist between DZ twins. When heritability is high, the amount of variation in a population depends largely on the number and frequency of different alleles present for the genes that control the trait. If variant alleles are present at low frequency, the amount of phenotypic variation in the population as a whole would be low, and the variation between DZ twins would be even lower.

    c. **False.** Heritability can vary in different populations for two reasons. Remember that $H^2 = V_G/V_P = V_G / (V_G + V_E)$. Different environments in different countries could affect the total phenotypic variance (V_P) through alterations in V_E. Likewise, different gene pools in different countries could alter the total phenotypic variance through V_G.

5. a. **Comparisons of the degree of similarity between MZ and DZ twins measure the proportion of variation due to genes (heritability) only if the range of environments experienced by each group is similar.**

    b. **More accurate estimates for heritability in humans come from concordance of MZ twins raised in the same family as opposed to DZ twins raised together.** The range of environments experienced by the pairs of MZ and DZ twins would be similar, but because MZ twins are genetically identical, all differences in phenotype must be due to environmental effects. Thus, the environmental variance (V_E)

21- 6

chapter 21

observed from the concordance will be accurate. The total phenotype variance in the population (V_P) and V_E can be used to determine the variance due to genes ($V_G = V_P - V_E$) and $H^2 = V_G/V_P$ can be calculated.

6. a. **Height has the highest heritability and weight has the lowest.** MZ twins raised either together or apart differ in height about equally, and about twice as much as DZ twins or siblings. This result indicates that environment plays almost no role in height, and that genes control height ($H^2 \approx 1$). However, MZ twins reared together are much more similar in weight than MZ twins reared apart, and the weight differences of DZ twins or siblings are similar to those of MZ twins reared apart. This observation indicates that the environment largely controls weight, and genes play a minor role ($H^2 \approx 0$). The heritability of IQ is in between. MZ twins reared together are more similar to each other than MZ twins reared apart, but the latter are more similar to each other than DZ twins or siblings. Thus, both genes and environment affect IQ.

   b. **Because the phenotypic changes occurred during a period of only 42 years, the rise in height and weight must be due to environmental effects.** (Forty-two years is way too short a time period for the gene pool to change significantly through mutation, and for the purposes of this exercise, we will ignore possible changes in the gene pool in the US due to migration.) Height increased by $(68.4 - 67.5)/67.5 \approx 1.3\%$, and weight increased by $(150.3 - 135.5)/135.5 \approx 11\%$. As height has high heritability, environmental factors would not be expected to alter the average trait value very much. Weight, however, has low heritability and so it makes sense that changes in the environment could alter average weight.

7. Heritability values for a rare trait depend on the frequency of the trait in the population, on the total amount of phenotypic variation in the population, and on environmental factors that could influence the trait. In addition, different populations of the same species may have different histories, leading to the presence of different QTLs influencing the trait.

8. a. **Founder effects** could explain the relatively high incidence of particular recessive disorders and the relatively low incidence of others. If present-day Finns descend almost exclusively from a small number people, the alleles that those particular "founders" of the population introduced are the prevalent ones now.

   b. **Inbred populations like the Finns will have a smaller number of QTLs for a given trait such as schizophrenia, each of which makes a large contribution to the phenotype. The high trait values for the QTLs make them easier to identify.** On the other hand, the paucity of variation in the Finnish population means that some phenotypes will not exist in the population at all and cannot be studied.

9. **The second trait (with 12 minor loci and no major genes) would respond to selection most consistently.** The reason is that selection can take place only when new variants are introduced by mutation. The more genes that control a trait, the larger the target size for random mutation.

10. a. If n genes exist that each contribute equally to a complex trait, and each gene has two codominant (or incompletely dominant) alleles, the number of different trait values is

$2n + 1$. For example, when $n = 1$, the genotypes *AA*, *Aa*, and *aa* each have different phenotypes. And when $n = 2$, the five different trait values are explained by the following genotypes: (1) *AA BB*; (2) *Aa BB = AA Bb*; (3) *Aa Bb = aa BB = AA bb*; (4) *Aa bb = aa Bb*; (5) *aa bb*.

b. From the examples given, the formula for determining the frequency of the extreme phenotypes in the F_2 of a cross is $1/4^n$, where n = the number of gene that determine the phenotype. You can see that this makes sense if you consider the simplest case where $n = 1$. If the parental cross was between *AA* and *aa*, then the F_1 are *Aa*, and the progeny are 1/4 *AA* : 1/2 *Aa* : 1/4 *aa*. The genotypes *AA* and *aa* correspond to the extreme phenotypes. When progeny with the extreme phenotypes are each present at a frequency of 1/256, then $1/4^n = 1/256$, and $n = 4$. Thus, **4 different genes control kernel color**; the genotypes corresponding to the extreme phenotypes are *AA BB CC DD* and *aa bb cc dd*.

11. a. From Problem 10 part (a), we know that $2n + 1 = 9$ different phenotypes will exist in the F_2. As in Problem 10 part (b), the maximum and minimum leaf lengths are the parental types: *AA BB CC DD* (length = 32 cm) at a frequency of 1/256, and *A'A' B'B' C'C' D'D'* (length = 16 cm) at a frequency of 1/256.

    The remaining 7 phenotypic classes correspond to genotypes with 1 prime and 7 non-prime alleles, 2 prime and 6 non-prime alleles, 3 prime and 5 non-prime alleles, etc. The trait values for each class are easy to determine simply by adding the contributions of each of the eight alleles. For example, for F_2 with only one prime allele and seven non-prime alleles, the leaf length is $7(4) + 2 = 30$ cm.

    The frequency of each genotype is trickier to determine. First, you need to realize that in a self-cross of *AA' BB' CC' DD'* where the alleles of each gene are assorting independently, the number of different genotypes with 0, 1, 2, 3, 4, 5, 6, 7, or all 8 prime alleles can be determined using the binomial theorem. You first saw the application of the binomial theorem to genetics problems in Chapter 2, in the answer to Problem 2-49(d). Here, the binomial theorem will enable you to determine the number of different possible genotypes in the progeny self-cross, and also the number of different genotypes that correspond to each phenotypic class: 0, 1, 2, 3, 4, 5, 6, 7, or all 8 prime alleles:s

    P (**X** will occur **s** times, and **Y** will occur **t** times, in **n** trials) =

    $$\frac{n!}{s! \times t!} (p^s \times q^t)$$

    P = the probability of what is in parentheses
    p = P(**X**)
    q = P(**Y**)

    X and **Y** are the only two possibilities, so **p + q = 1**.
    Also, **s + t = n**.

    Remember that **!** means *factorial*; for example, $5! = 5 \times 4 \times 3 \times 2 \times 1$.

    As an example, let's determine the number of different genotypes that contain 2 prime alleles and 6 non-prime alleles. (These genotypes include *A'A' BB CC DD, AA' BB' CC DD, AA B'B' CC DD, AA' BB CC' DD,* and many more. You can see that

chapter 21

this problem is similar to the one in 2-49(d); in both cases, you need to determine how many different "orders" exist – here it's how many different ways a genotype can have 2 primes and 6 non-primes.) In this problem, X = prime alleles and Y = non-prime alleles. As prime and non-prime alleles are each 50% of all the alleles, the probability of X = the probability of Y = 1/2, and so p = q = 0.5. For genotypes with 2 prime and 6 non-prime alleles, s = 2, t = 6, and n = 8. The right side of the binomial equation ($p^s \times q^t$) is the probability (frequency) of each different genotype within the phenotypic class: $0.5^2 \times 0.5^6 = (1/2)^8 = 1/256$. (This means that 256 different genotypes exist among the progeny.) The left side of the binomial equation [n! / (s! × t!)] gives the number of different genotypes within each specific phenotypic class; in this case, 8! / (2! × 6!) = 28. Therefore, the frequency of progeny with 2 prime and 6 non-prime alleles is 28/256. Note that for all 9 different phenotypic classes, p = q = 0.5, n = 8, and $p^s \times q^t = (1/2)^8 = 1/256$. The only part of the calculation that differs in each case is [n! / (s! × t!)], which is the number of different genotypes in each trait value class.

The frequency and leaf length calculations for all 9 phenotypic classes of progeny are shown in the table that follows. (Note that 0! = 1.):

Trait value class	Number primes	Number non-primes	Frequency	Leaf length
1	0	8	[8! / (0! × 8!)] / 256 = 1/256	0(4) + 8(2) = 16 cm
2	1	7	[8! / (1! × 7!)] / 256 = 8/256	1(4) + 7(2) = 18 cm
3	2	6	[8! / (2! × 6!)] / 256 = 28/256	2(4) + 6(2) = 20 cm
4	3	5	[8! / (3! × 5!)] / 256 = 56/256	3(4) + 5(2) = 22 cm
5	4	4	[8! / (4! × 4!)] / 256 = 70/256	4(4) + 4(2) = 24 cm
6	5	3	[8! / (5! × 3!)] / 256 = 56/256	5(4) + 3(2) = 26 cm
7	6	2	[8! / (6! × 2!)] / 256 = 28/256	6(4) + 2(2) = 28 cm
8	7	1	[8! / (7! × 1!)] / 256 = 8/256	7(4) + 1(2) = 30 cm
9	8	0	[8! / (8! × 0!)] / 256 = 1/256	8(4) + 0(2) = 32 cm

b. The genotypic frequencies for each gene can be determined using the Hardy-Weinberg equation: $p^2 + 2pq + q^2 = 1$. For gene A, $p^2 = f(AA) = (0.9)^2 = 0.81$; $2pq = f(AA') = (2)(0.9)(0.1) = 0.18$; $q^2 = f(A'A') = (0.1)^2 = 0.01$. For gene B, $f(BB) = (0.9)^2 = 0.81$; $f(BB') = (2)(0.9)(0.1) = 0.18$; $f(B'B') = (0.1)^2 = 0.01$. For gene C, $f(CC) = (0.1)^2 = .01$; $f(CC') = (2)(0.1)(0.9) = 0.18$; $f(C'C') = (0.9)^2 = 0.81$. For gene D, $f(DD) = (0.5)^2 = 0.25$; $f(DD') = (2)(0.5)^2 = 0.50$; $f(D'D') = (0.5)^2 = 0.25$.

c. The plants with leaves that are 32 cm long have the genotype *AA BB CC DD*. In part (a), when all alleles were present at a frequency of 0.5, the frequency of this class (trait value class 9 in the preceding table) was $(0.5)^2 \times (0.5)^2 \times (0.5)^2 \times (0.5)^2 = 0.0039 = 1/256$. Here, the frequency is $(0.81) \times (0.81) \times (0.01) \times (0.25) = \mathbf{0.0016}$.

Section 21.2

12. **In all three methods, SNP genotyping is used to find SNPs linked to genes that control a trait.** In positional cloning of Mendelian disease genes, in order to identify SNP loci that are linked to the disease gene, SNP genotyping is performed for affected and unaffected family members in many different pedigrees. In direct QTL mapping, BC_1 progeny with different trait values are SNP-genotyped to determine whether they are homozygous or heterozygous for SNPs from parents with widely different phenotypic values; SNPs where homozygotes or heterozygotes have different average trait values are linked to a QTL for that trait. In GWAS, a large number of individuals that share a particular trait and a control group without that trait are SNP-genotyped in order to identify SNPs that correlate statistically with the trait.

In positional cloning, recent crossovers (during the past few generations within the pedigree) that occurred between relatively loosely-linked SNPs and the disease gene can be detected; recombination frequency between the SNP and the disease gene is used to estimate distance. In GWAS and direct QTL mapping, the association of a SNP with a complex trait depends on the history of crossing-over within the population. SNPs tightly linked to QTLs are identified because crossovers between them have occurred rarely during the population's history.

13. Coarse-scale mapping of QTLs starts with a cross of two isogenic strains with widely different trait values. The F_1 are backcrossed to one of the parents to create a BC_1 generation that is either homozygous or heterozygous for each allele of the backcross parent. BC_1 individuals are examined to determine if their trait values correlate with their genotypes at any SNPs; the SNPs that correlate are linked QTLs for that trait. Fine-scale mapping identifies the QTL. The BC_1 lines are used in further crosses to generate congenic lines – isogenic lines nearly identical to one of the original parental strains, but where each strain contains a small genomic region from the other parent called an introgression. In order to localize the QTLs, the boundaries of the introgressions in different congenic lines are determined and compared with the trait values of the lines. In both coarse and fine mapping, phenotypic values are measured in order to correlate specific regions of the genome with different trait values.

After a candidate gene is identified by fine-mapping, its identity must be validated. One method for validation is to transfer the suspect gene from a strain with one trait value to a strain with a widely different trait value and determine if the transgene changes the phenotype of the recipient in the expected manner. Another approach is to alter the candidate gene in one or both of the strains (using TALENs, for example) and to look for an effect on trait value. The success of either of these approaches depends on the contribution of the candidate gene to the trait value (a QTL with a larger contribution is easier to validate) and also on the nature of the alleles present in each strain (see Problem 16).

14. a. People with early onset of heart disease should be used for the GWAS because they are likely to have more disease-causing alleles of QTLs and/or QTLs with higher trait values than those with late onset. The control population should be individuals of the same ages as the experimental group who do not have heart disease nor a family history of heart disease, and who live in the same range of environments as the group with heart disease.

chapter 21

b. Confirming that a particular gene contributes to a complex trait can be difficult. **A first step is to determine if mutations likely to affect that gene's function are present specifically in people with the disease.** Finding such a mutation in the candidate gene can be difficult if the mutation is not in the coding region or in splice junction sequences; for example, the mutation may be in a transcriptional regulatory region far from the exons. If no obvious mutation is found, levels of transcripts or protein could be assayed using methods like RNA-Seq or Western blotting, respectively. **After a mutation is found, the mouse version of the gene could be altered to determine if the mutation affects the trait in mice.** QTLs with the highest contributions to the trait value would be the easiest to validate: Those QTLs are the most likely to be mutant in any given individual with the trait, and the mutant versions of those QTLs are most likely to affect the trait value in a mouse model.

15. a. **SNP3, SNP4, SNP5, SNP6, and SNP7 are in strong linkage disequilibrium with the disease locus.** The G allele of SNP3, the T allele of SNP4, the T allele of SNP5, the T allele of SNP6, and the C allele of SNP7 are each present in the patient genomes more than half the time (nonrandom) while they are presently randomly in the control genomes.

   b. **The Canavan disease gene is most likely within the ~400 kb genomic region bounded by SNP3 and SNP7.**

   c. **The data suggests that 2 independent mutations of the Canavan disease gene occurred.** Affected individuals 1-5 all have the haplotype "GTTTC" for SNP3-SNP7; one mutation must have occurred in a chromosome containing this haplotype. Affected individual 5 has the haplotype "GCCTG"; a different mutation in the Canavan disease gene must have occurred in a chromosome with this set of SNP alleles. Note that unaffected (Control) individual 6 has the GTTTC haplotype, but not the disease. The most likely explanation is that individual 1 is heterozygous for the disease allele. The other three Control individuals (7-9) have neither of the disease haplotypes.

   d. The Sephardic and Ashenakic Jews share the GTTTC disease gene haplotype. Therefore, **the Canavan disease gene mutation occurred on a GTTTC chromosome before the Sephardic/Ashenakic separation.** The SNP3-7 loci of additional Sephardic Jews would need to be genotyped in order to draw conclusions about the origins of GCCTG disease haplotype.

   e. When investigating association between haplotypes and a disease, focusing on a small population in which affected individuals are frequent has two **advantages: more affected individuals mean more data points, and affected individuals are likely to have the same mutation from a common ancestor and thus the same haplotype,** making the association more apparent. However, **if the subpopulation was formed recently, the regions of LD may be very large** because less time (fewer generations) has existed for recombination to occur. Larger regions of LD makes it more difficult to locate specific SNPs (specific haplotypes) that are associated with the disease gene, **and thus it would be difficult to identify the specific gene**

   f. **Particular haplotypes can be determined by examining the close relatives (parents, siblings, and children) of the individuals in the table.** If an individual is homozygous for a particular haplotype, then no comparison is necessary, but of course this will not always be the case.

Suppose that the SNP3 through SNP7 genotype of an individual is G(T/C)(T/C)T(C/G). SNP genotyping of both parents, for example, could resolve whether this individual's two haplotypes are GTTTC/GCCTG, or GCTTG/GTCTC, etc. Suppose that the parents' SNP genotypes are: mother = G(G/T)(C/T)T(C/G) and father = G(G/C)CTG. As each child must have received one SNP allele from each parent, we can infer that the configuration of the SNP alleles (the haplotypes) on each of the two homologs in the G(T/C)(T/C)T(C/G) individual are GTTTC (from mother)/GCCTG (from father). (SNP3=G from each parent; SNP4=T from mother because father doesn't have T; SNP5=T from mother because father doesn't have a T; SNP6=T from each parent; SNP7=C from mother because father doesn't have a C.) If SNP genotyping of the parents does not resolve the haplotype completely, using similar logic, analysis of the genotypes of siblings can help.

16. **The *fw2.2* gene produces a protein that is a negative regulator of growth.** The small tomato species (*S. pennellii*) has wild-type *fw2.2* alleles, and the large tomato species (*S. lycospersicum*) has a loss-of-function *fw2.2* allele. As the *fw2.2* gene has a high trait value, transfer of *S. pennellii*'s wild-type *fw2.2* allele into *S. lycospersicum* decreased the *S. lycospersicum* average tomato size significantly. **Because of the nature of the *fw2.2* alleles in each species, the converse experiment, transfer of the *S. lycospersicum* *fw2.2* allele into *S. pennellii* would little or no effect; the *fw2.2* allele of *S. lycospersicum* has little or no function, and *S. pennellii* is already homozygous for wild-type *fw2-2* alleles.**

17. a. **Some chromosomes in the disease group in Fig. 21.13 (p.207) do not carry the *red* disease-causing variant because this disease is a complex trait and the *red* variant is a single QTL whose trait value is far less than <100%;** many other genes (QTLs) contribute to this disease, and a person without this particular variant can have the disease provided that they have other QTLs with a sufficiently high combined trait value.

  b. **Some chromosomes in the control group carry the *red* variant because that one QTL does not have a trait value that is sufficiently high to cause the disease phenotype on its own.** As explain in part (a), this disease is a complex trait, and so by definition it is controlled by multiple genes. A person can have the *red* variant and yet not have the disease as long the combined trait value of all of their disease QTLs is beneath a cut-off value for the disease phenotype.

  c. **Scientists would look for alleles whose frequencies in the affected group are greater than their frequencies in the control group. In the example in Fig. 21.13, a mutation that caused the disease occurred originally on the *blue* chromosome.**

  d. **The disease-causing mutation in Region 1 likely occurred earlier than the mutation in Region 2. The shorter region of LD in Region 1 means that more recombination has occurred** between the chromosome on which the disease mutation was introduced originally and other chromosomes in the population that have different variants of linked loci. Thus, more time must have elapsed between the original mutation event and the present day.

  e. As **recombination is not uniform in the human genome (hotspots exist),** the length of a region of a chromosome containing SNPs in LD with a disease allele is not a perfect indicator of how long ago the disease mutation occurred. For example, suppose two different disease alleles are introduced on two different chromosomes during the same year in history; one of the genes is flanked by recombination hotspots and the other is not. The length of the region of associated SNPs will be much shorter for the disease gene flanked by recombination hotspots than the disease gene that is not, even though both mutations were introduced at the same historical time.

18. a. **Recombination hotspots** are located between LD blocks.

    b. **Limited genotyping has predictive value due to the existence of haplotypes**. The allele of SNP3 predicts SNP1-SNP5 and SNP7, and the alleles of either SNP8 or SNP9 predict SNP7 through SNP9.

    c. **Common SNPs** are typically used for genotyping because by very definition they are most likely to be found in any population.

    d. **Lower frequency SNPs** omitted from typical genotyping studies but identified by genome sequencing **could provide additional clues to the ancestry of populations.**

19. Not all diseased individuals have the same SNP genotype **because the disease is a complex trait.** As explained in Problem 17, parts (a) and (b) above, a particular SNP is linked to a QTL for the disease that makes only a small contribution to the disease phenotype. In addition, **the SNP itself is unlikely to be the QTL (the disease-causing mutation); the SNP is closely linked to it.** Thus, in some individuals, the particular SNP variant most often associated with the mutation may have recombined away from it.

20. a. The C allele of the SNP appears to correlate with high blood pressure: The C allele appears more frequently in people with the condition (1025/1927 Cases) than in people without it (725/1647 Controls). To determine if the correlation is statistically significant, we need to perform a "χ^2 test for independence". This version of the χ^2 test is somewhat different than the "χ^2 test for goodness of fit" that we performed in Chapter 5 to assess linkage of two genes.

    In Chapter 5, we used the χ^2 statistic to ask, for example, how well the phenotypic distribution of testcross progeny ($AB/ab \times ab/ab$) fit a 1 AB/ab : 1 ab/ab : 1 Ab/ab : 1 aB/ab ratio – the expected proportions for the null hypothesis of independent assortment of the alleles of gene A with those of gene B (no linkage). The testcrosses were set up such that the allele frequencies of A and a, and B and b in gametes were each 50%, and our expected ratio (given the null hypothesis) was based on the assumption that all of the alleles and allele combinations would give rise to equally viable progeny. Thus, deviations from 1:1:1:1 were assumed to be due either to linkage, sampling error, or a combination of both.

    In the present problem, the situation is different in a way that prevents us from using the χ^2 test for goodness of fit: Because we do not know the frequencies of the SNP alleles C and T in the population (as we did for the alleles A and a, and B and b in the testcrosses), a null hypothesis that no correlation (linkage) exists between the SNP allele and the high blood pressure phenotype does not allow us to assign expected values for the numbers of cases and controls with either the C or T allele in the way we did it in testcrosses.

Instead of testing the fit of the observed data to a specific expected ratio, we can test the data against the null hypothesis that the distribution of C and T in the Cases and Controls is the same; in other words, the null hypothesis is that C and T, and Cases and Controls, associate with one another randomly – the occurrence of a particular SNP allele and high blood pressure are *independent*. Given the null hypothesis of independence, the E (Expected) values for each of the four observed classes (Cases with SNP C, Cases with SNP T, Controls with SNP C, and Controls with SNP T) can be calculated using the data in the table. [Note that the total number of individuals observed is (1025 + 725 + 902 +922) = 3574.] For example, if the SNP C allele is distributed the same way in the Cases and Controls, the E value for SNP C in the Cases is simply the frequency of the occurrence of SNP C in all the individuals observed [(1025 + 725)/3574] multiplied by the total number of Cases (1025 + 902): E = [(1025 + 725)/3574] × (1025 + 902) = 943.6. Similarly, the E value for SNP C in the Controls is the fraction of all individuals with SNP C multiplied by the total number of Controls: E = [(1025 + 725)/3574] × (725 + 922) = 806.4. The same logic is used to calculate the E values for the Cases and Controls with SNP T.

The χ^2 value for each of the four classes is calculated as $\chi^2 = (O-E)^2/E$, where O = Observed data, and the four individual χ^2 values are summed to obtain the χ^2 statistic for the entire data set. The χ^2 calculations are shown in the so-called "contingency table" that follows:

	Cases	Controls	
C	O = 1025 E = [(1750/3574) × 1927] = 943.6 χ^2 = [(1025 - 943.6)² / 943.6] = **7.02**	O = 725 E = [(1750/3574) × 1647] = 806.4 χ^2 = [(725 - 806.4)² / 806.4] = **8.22**	Total C = 1750
T	O = 902 E = [(1824/3574) × 1927] = 983.4 χ^2 = [(902 - 983.4)² / 983.4] = **6.74**	O = 922 E = [(1824/3574) × 1647] = 840.6 χ^2 = [(922 - 840.6)² / 840.6] = **7.88**	Total T = 1824
	Total Cases = 1927	Total Controls = 1647	Total Individuals = 3574

Note that in the contingency table, there is one degree of freedom (df = 1). The reason is that because of the way in which we calculated the E values, if you set one value for E, the other three are determined. (You can prove this to yourself with a little bit of algebra; OK maybe a lot of algebra!) The χ^2 value is (7.02 + 8.22 + 6.74 + 7.88) = 29.86. A χ^2 **value of 29.86 with df = 1 means that** $p = 5 \times 10^{-8}$**; the probability that the observed association of SNP C and high blood pressure is really random (chance) is only 0.00000005 or 1/20,000,000. As for a single SNP,** $p < 0.05$ **(1/20) is considered significant, the association observed is highly significant.**

b. **In the context of a GWAS where the SNP in part (a) was 1 of 1,000,000 SNPs**

chapter 21

assayed, the p value of 5×10^{-8} puts the SNP on the borderline for **significance**; $p < 5 \times 10^{-8}$ is the cut-off for significance in when 1,000,000 SNPs are assayed in a GWAS. The reason that $p < 0.05$ cannot be used when assaying 1,000,000 SNPs is that approximately 1/20 of the SNPs assayed (50,000 SNPs) would be associated with the trait simply by chance. For a GWAS that includes 1,000,000 SNPs, the probability value is typically corrected to $< 0.05/1,000,000 = 5 \times 10^{-8}$ so that only 1/20,000,000 SNPs would be associate randomly with the trait. In other words, the chance that even one SNP association is random is 1/20.

c. **The C allele is associated with high blood pressure.** As stated in part (a), C appears more frequently in people with the condition (1025/1927 = 53% of the Cases) than in people without it (725/1647 = 44% of the Controls).

d. The allelic odds ratio is the increased risk of having high blood pressure conferred by having the C allele of the SNP. The ratio is calculated as [(Cases with C/Cases with T) /(Controls with C/Controls with T)], which is [(1025/725)/(902/922)] = 1.41/0.978 = **1.44.** The allelic odds ratio of 1.44 means that if you have a C allele of the SNP, you are 1.44× more likely to have high blood pressure than if you don't have a C allele.

21. Identifying a gene associated with a particular complex trait can be difficult, and so there is no simple answer to this question. **The LD block could be examined for genes whose inferred functions suggest connection to the phenotype. The sequences of candidate genes in individuals who display the phenotype could be compared with those of individuals who do not.** Variations in DNA sequences predicted to affect gene function could be investigated, as well as the function of the candidate gene.

22. **GWASs in general have looked for associations between common (frequent) SNPs and height, and so the "missing heritability" could be explained by more rare variants caused by more recent mutations. This hypothesis could be tested by performing a GWAS using moderately rare or very rare SNPs.**

23. a. Your SNP genotypes can tell you if you have a Mendelian genetic disease allele if the gene and mutation(s) are known, as well as the inheritance patterns of the mutant allele (dominant or recessive), and the disease gene is 100% penetrant.

   b. For a disease that is a complex trait, the answer depends on the number of **QTLs known to be associated with the disease and their trait values.** Only if you have several QTLs with high trait values would you be likely to have a disease. For most diseases that are complex traits, the sum total trait values of identified QTLs are too low to be of much predictive value.

   c. **The different companies may have examined different SNPs and also the models regarding the genetic basis for the trait used by all the companies may have been different and all are based on incomplete knowledge.**

   d. **The accuracy of risk estimates for complex traits will undoubtedly improve with time, but will remain far from perfect for the foreseeable future.** The problem of "missing heritability" might be solved by GWASs based on higher frequency SNPs; these variants might have higher trait values than the common SNPs found to be associated with complex traits to date.